Relativity and Cosmology

Relativity and Cosmology

Volume 5 of *Modern Classical Physics*

KIP S. THORNE *and* **ROGER D. BLANDFORD**

PRINCETON UNIVERSITY PRESS

Princeton and Oxford

Published by Princeton University Press
41 William Street, Princeton, New Jersey 08540
6 Oxford Street, Woodstock, Oxfordshire OX20 1TR

press.princeton.edu

ISBN (pbk.) 978-0-691-20739-1
ISBN (e-book) 978-0-691-21554-9

British Library Cataloging-in-Publication Data is available

Editorial: Ingrid Gnerlich and Arthur Werneck
Production Editorial: Mark Bellis
Text and Cover Design: Wanda España
Production: Jacqueline Poirier
Publicity: Matthew Taylor and Amy Stewart
Copyeditor: Cyd Westmoreland

This book has been composed in MinionPro, Whitney, and Ratio Modern by Windfall Software, Carlisle, Massachusetts, using ZzTEX

Printed on acid-free paper.

Printed in China

10 9 8 7 6 5 4 3 2 1

A NOTE TO READERS

This book is the fifth in a series of volumes that together comprise a unified work titled *Modern Classical Physics*. Each volume is designed to be read independently of the others and can be used as a textbook in an advanced undergraduate- to graduate-level course on the subject of the title or for self-study. However, as the five volumes are highly complementary to one another, we hope that reading one volume may inspire the reader to investigate others in the series—or the full, unified work—and thereby explore the rich scope of modern classical physics.

To Carolee and Liz

CONTENTS

T2 Track Two; see page xvii

BOXES

PREFACE

We, the authors of this book—Roger Blandford and Kip Thorne—have devoted a large portion of our half-century careers to teaching, mentoring, and research in relativity and cosmology. During that time, we and our students have had the joy of participating in remarkable transformations of human knowledge about our cosmological universe and the curved-spacetime arena that underpins its wonders.

Those transformations have included, among many others:

- deepened insights into the physical interpretation of the mathematical formulas of general relativity—Einstein's laws that govern gravity and the curvature of spacetime;
- orders-of-magnitude improvements in experimental tests of general relativity, many involving neutron stars;
- major new insights into black holes—the quintessential examples of objects made from curved spacetime;
- the spectacularly successful half-century quest to create gravitational-wave astronomy, and with it to begin exploring aspects of the universe that could not be discerned in any other way; and
- amazing revelations about the birth of our universe, its fundamental constituents (matter, radiation, dark matter, dark energy), their evolution from the Planck era of the universe's birth to the present era, and details of how they gave rise to what we now see around us: stars, galaxies, clusters of galaxies, and the remarkably speckled cosmic microwave radiation.

This textbook is a pedagogical introduction to relativity and cosmology, including these remarkable transformations of human knowledge. It differs from most other relativity textbooks in ways that we think important, including (among many others):

- We highlight the transformations described above.
- We pay close attention to the physical interpretation of general relativity in terms of measurements made by observers who often accelerate and rotate relative to local inertial frames in manners that elucidate the physics of a situation (e.g., an observer at rest in a laboratory on our rotating Earth, or one hovering just above the horizon of a spinning black hole).
- We elucidate the physics of the Riemann curvature tensor in terms of such observers' measurements, including the tidal gravitational forces that Riemann produces and the frame-dragging precession of gyroscopes that it induces. We show how the Riemann tensor's pattern of tidal forces can be embodied in tendex lines (analogs of electric field lines), and its frame-dragging pattern can be embodied in frame-drag vortex lines (analogs of magnetic field lines). And we use these field lines, for example, to elucidate the dynamics of spacetime curvature around colliding black holes and how that dynamics gives rise to gravitational waves.
- We give a detailed analysis of an incoming gravitational wave's interaction with a laser interferometer gravitational wave detector such as LIGO, to produce, ultimately, an output laser light intensity entering a photodetector. We analyze this twice, in two different gauges (the proper reference frame of the interferometer's accelerating beam splitter and the transverse-traceless gauge in which gravitational waves are usually described). And we use these two different analyses to elucidate the physical interpretation of general relativity.
- We present a detailed analysis of the evolution of our inhomogeneous universe from its earliest moments until today—the evolution that has given rise to the galaxies and mottled cosmic microwave radiation that we now observe, as well as the detailed structure of our own galaxy whose fossil record we can study in exquisite detail. We show that much of this evolution can be understood analytically without resorting to detailed numerical simulations, thereby elucidating the underlying physics.

GUIDANCE FOR READERS

The amount and variety of material covered in this book may seem overwhelming. If so, keep in mind that

- *the primary goals of this book* are to teach the fundamental concepts and principles of relativity and cosmology (which are not so extensive that they should overwhelm); to illustrate those concepts and principles in action; and through our illustrations, to give the reader some physical understanding of curved spacetime and our universe.

We have aimed this book at advanced undergraduates and first- and second-year graduate students, and also working scientists, of whom we expect only (1) a typical physics or engineering student's facility with applied mathematics (and no differential geometry), and (2) a typical undergraduate-level understanding of classical mechanics, electromagnetism, and thermodynamics. Familiarity with quantum mechanics at the undergraduate level will also occasionally be helpful, particularly in Chapters 27 and 28, but it is not essential. Nearly everything in this book can be understood without quantum mechanics.

Although this book is brief enough for a one-quarter or one-semester course, it goes sufficiently deeply into its topics to form a foundation for understanding portions of cutting-edge research in relativity and cosmology and participating in that research.

For those readers who would like an even briefer introduction to relativity and cosmology, we have labeled as "Track Two" sections that can be skipped on a first reading or skipped entirely—but are sufficiently interesting that most readers may choose to browse or study them. Track-Two sections are identified by the symbol **T2**.

This book is the fifth and last volume of a series—five volumes that together constitute a single treatise, *Modern Classical Physics* (or "MCP," as we shall call it). The full treatise was published in 2017 as an embarrassingly thick single book. (The electronic edition is a good deal lighter.) For readers' convenience, we have placed, at the end of this volume, the Table of Contents, Preface, and Acknowledgments of MCP. The five separate textbooks of this decomposition are

- Volume 1: *Statistical Physics*,
- Volume 2: *Optics*,
- Volume 3: *Elasticity and Fluid Dynamics*,
- Volume 4: *Plasma Physics*, and
- Volume 5: *Relativity and Cosmology*.

These individual volumes are much more suitable for human transport and for use in individual courses than their one-volume parent treatise, MCP.

The present volume is enriched by extensive cross-references to the other four volumes—cross-references that elucidate the rich interconnections of various areas of physics.

In this and the other four volumes, we have retained the chapter numbers from MCP and, for the body of each volume, MCP's pagination.. In fact, the body of this volume and its appendix are identical to the corresponding MCP chapters, aside from corrections of errata (which are tabulated at the MCP website http://press.princeton.edu/titles/MCP.html) and a minor amount of updating that has not changed pagination. For readers' cross-referencing convenience, a list of the chapters in each of the five volumes appears immediately after this Preface.

EXERCISES

Exercises are a major component of this volume, as well as of the other four volumes of MCP. The exercises come in five types:

1. *Practice*. Exercises that provide practice at mathematical manipulations (e.g., of tensors).
2. *Derivation*. Exercises that fill in details of arguments skipped over in the text.
3. *Example*. Exercises that lead the reader step by step through the details of some important extension or application of the material in the text.
4. *Problem*. Exercises with few, if any, hints, in which the task of figuring out how to set up the calculation and get started on it often is as difficult as doing the calculation itself.
5. *Challenge*. Especially difficult exercises whose solution may require reading other books or articles as a foundation for getting started.

We urge readers to try working many of the exercises—especially the examples, which should be regarded as continuations of the text and which contain many of the most illuminating applications. Exercises that we regard as especially important are designated by **.

UNITS

In this volume, we normally use SI units with their vacuum permitivities ϵ_0 and μ_0. However, sometimes, when setting the speed of light c to unity (geometrized units), we lapse into Gaussian units, as then ϵ_0, μ_0, and c disappear from electromagnetic equations—a convenience typically embraced by relativity researchers.

BRIEF OUTLINE OF THIS BOOK

The primary focus of this volume is general relativity. However, we have included, as an appendix, a pedagogical introduction to special relativity (Chap. 2 of MCP). Readers who are already comfortable with special relativity and with its mathematical language (differential geometry in flat spacetime, i.e., tensor analysis) can begin their reading and study at the front of this book, with MCP Chap. 24. Readers without that comfort level might best begin with the special-relativity appendix (MCP Chap. 2).

Our approach to general relativity (and, in fact, to all of classical physics) is very geometric. In MCP, we espouse the power of geometry for almost all of physics. The essence of this approach is that all the laws of physics must be expressible in coordinate-independent, geometric language. That geometric language usually elucidates the physics more clearly than coordinate-based language, and it often circumvents lengthy, coordinate-based analytical calculations. Readers familiar with special relativity but unaccustomed to our geometric language may find it useful to browse or read our special-relativity appendix (consisting of MCP Chap. 2) before launching into Chap. 24.

In Chap. 24, we review and elaborate those aspects of special relativity and flat-spacetime differential geometry that are crucial for the transition to general relativity, most importantly:

- The generalization of differential geometry from flat-spacetime Lorentz coordinates to curvilinear coordinate systems and general sets of basis vectors, both in flat spacetime and curved. (Here we meet the possibility of curved spacetime without yet introducing tools to quantify the curvature.)
- An exploration, in depth, of the stress-energy tensor—the geometric object that (as we will see in Chap. 25) generates the curvature of spacetime.
- An exploration of the reference frame of an accelerated, rotating observer in flat spacetime. This is the foundation for understanding physical measurements by such observers in curved spacetime, in later chapters.

In Chap. 25, we develop the basic concepts of general relativity, including

- spacetime curvature (the Riemann tensor) and its embodiment of tidal gravity,
- the Einstein field equation that quantifies how the stress-energy tensor generates spacetime curvature, and
- the laws of nongravitational physics in curved spacetime (e.g., Maxwell's equations).

Here we see the power of thinking about measurements made by accelerated observers, and the power and ambiguities of Einstein's Equivalence Principle in lifting the laws of physics from flat spacetime into curved spacetime. In the last part of this chapter, we explore relativity's predictions about weak gravitational fields by linearizing around flat spacetime. As our most important application (among many), we explore conservation laws for mass, momentum, and angular momentum, not only when gravity is weak everywhere but also for the total mass, momentum, and angular momentum of a strongly gravitating object, such as a black hole.

In Chap. 26, we use the tools from Chap. 25 to develop exact relativistic descriptions of black holes, nonrotating relativistic stars (e.g., neutron stars), and the spherical implosion of a nonrotating star to form a black hole. We begin by exploring the Schwarzschild solution to Einstein's field equation in vacuum. Here we learn how to read out from the mathematics both the nature of the coordinate system being used and the coordinate-independent physical properties of the curved spacetime. We get experience solving Einstein's field equation by using it to deduce the detailed structure of a static neutron star, and we discover that outside the star, its spacetime geometry is that of Schwarzschild. We explore in detail the implosion of a spherical, nonrotating star to form a black hole, and discover that the hole's spacetime geometry is also that of Schwarzschild. Finally, we write down the Kerr solution to the Einstein field equation, which describes a spinning, quiescent black hole, and explore its physical properties in depth—thereby deducing almost all the now-famous properties of such a black hole and some remarkable properties that are not so famous. This discussion suffices to understand and appreciate, for example, the remarkable image of the plasma orbiting the black hole in the nearby galaxy M87, made by the Event Horizon Telescope.

In Chap. 27, we describe the best experimental tests of general relativity that have been performed on Earth, in the solar system, and in binary pulsars as of the writing of this textbook, and we analyze these experiments in ways that elucidate the physical interpretation of relativity's mathematics. Then we turn to gravitational waves. We first develop their mathematical and physical descriptions when the waves are weak and propagate through a spacetime that, aside from the waves, is flat, and we deduce the properties of the quantum mechanical gravitons that carry the waves. For the more realistic situation of gravitational waves propagating through curved spacetime, we use an expansion in powers of the waves' wavelength λ divided by the radius of curvature of the spacetime R (λ/R; a geometric optics approximation) to deduce the detailed physical properties of the waves' propagation, including the energy and momentum that they carry. We then analyze the generation of gravitational waves, most importantly by two stars or black holes orbiting each other and spiraling together due to radiation reaction. Finally, we discuss the detection of gravitational waves, with special focus on the types of gravitational-wave interferometers that today (2020) are detecting about six black hole collisions per month, and we analyze these interferometers' interaction with the incoming waves in two different ways (as described above). Elsewhere, in Volumes 1, 2, and 3 of MCP (cross-referenced in this chapter), we use these interferometers' designs and their noise to illustrate fundamental principles of statistical physics, optics, and elasticity.

Chapter 28 is devoted to cosmology. By contrast with most relativity textbooks, we specialize to the actual universe we inhabit and show how an impressive range of accurate measurements and observations are fully consistent with an extremely simple description. In this description, the dynamics is dominated by a cosmological constant and an unidentified dark matter. This description also accommodates the evolution of photons and neutrinos and the growth of the perturbations in the matter that we see around us today, perturbations that ultimately became galaxies, stars, and planets. This deduced history of the universe allows us to infer the conditions at the very earliest times and seems to be fully consistent with the theory of inflation, which invokes a phase of exponential expansion that homogenized the universe at very early times and set down the scale-free, initial perturbations. It is remarkable that the physical laws that go into this analysis are not only those of general relativity, but also of fluid mechanics, plasma physics, statistical physics, and optics, the subfields of physics treated in the other four volumes of MCP. Cosmology is surely the grandest application of Modern Classical Physics!

CONTENTS OF *MODERN CLASSICAL PHYSICS*, VOLUMES 1–5

Volume 1: Statistical Physics

Volume 2: Optics

Volume 3: Elasticity and Fluid Dynamics

Volume 4: Plasma Physics

Volume 5: Relativity and Cosmology

PART VII

GENERAL RELATIVITY

We have reached the final part of this book, in which we present an introduction to the basic concepts of general relativity and its most important applications. This subject, although a little more challenging than the material that we have covered so far, is nowhere near as formidable as its reputation. Indeed, if you have mastered the techniques developed in the first five parts, the path to the Einstein field equations should be short and direct.

The general theory of relativity is the crowning achievement of classical physics, the last great fundamental theory created prior to the discovery of quantum mechanics. Its formulation by Albert Einstein in 1915 marks the culmination of the great intellectual adventure undertaken by Newton 250 years earlier. Einstein created it after many wrong turns and with little experimental guidance, almost by pure thought. Unlike the special theory, whose physical foundations and logical consequences were clearly appreciated by physicists soon after Einstein's 1905 formulation, the unique and distinctive character of the general theory only came to be widely appreciated long after its creation. Ultimately, in hindsight, rival classical theories of gravitation came to seem unnatural, inelegant, and arbitrary by comparison [see Will (1993b) for a popular account and Pais (1982) for a more scholarly treatment].

Experimental tests of Einstein's theory also were slow to come. Only since 1970 have there been striking tests of high enough precision to convince most empiricists that—in all probability and in its domain of applicability—general relativity is essentially correct. Despite these tests, it is still very poorly tested compared to, for example, quantum electrodynamics.

We begin our discussion of general relativity in Chap. 24 with a review and an elaboration of special relativity as developed in Chap. 2, focusing on those concepts that are crucial for the transition to general relativity. Our elaboration includes (i) an extension of differential geometry to curvilinear coordinate systems and general bases both in the flat spacetime of special relativity and in the curved spacetime that is the venue for general relativity; (ii) an in-depth exploration of the stress-energy tensor, which in general relativity generates the curvature of spacetime; and (iii) construction and exploration of the reference frames of accelerated observers (e.g., physicists who reside on Earth's surface).

In Chap. 25, we turn to the basic concepts of general relativity, including spacetime curvature, the Einstein field equation that governs the generation of spacetime curvature, the laws of physics in curved spacetime, and weak-gravity limits of general relativity.

In the remaining chapters, we explore applications of general relativity to stars, black holes, gravitational waves, experimental tests of the theory, and cosmology. We begin in Chap. 26 by studying the spacetime curvature around and inside highly compact stars (such as neutron stars). We then discuss the implosion of massive stars and describe the circumstances under which the implosion inevitably produces a black hole. We explore the surprising and, initially, counterintuitive properties of black holes (both nonspinning and spinning holes), and we learn about the many-fingered nature of time in general relativity. In Chap. 27, we study experimental tests of general relativity and then turn to gravitational waves (i.e., ripples in the curvature of spacetime that propagate with the speed of light). We explore the properties of these waves, and their close analogy with electromagnetic waves, and their production by binary stars and merging black holes. We also describe projects to detect them (both on Earth and in space) and the prospects and success for using them to explore observationally the "warped side of the universe" and the nature of ultrastrong spacetime curvature. Finally, in Chap. 28,[1] we draw on all the previous parts of this book, combining them with general relativity to describe the universe on the largest of scales and longest of times: cosmology. It is here, more than anywhere else in classical physics, that we are conscious of reaching a frontier where the still-promised land of quantum gravity beckons.

1. Chapter 28 is very different in style from the rest of the book. It presents a minimalist treatment of the now standard description of the universe at large. This is a huge subject from which we have ruthlessly excised history, observational justification, and didacticism. Our goal is limited to showing that much of what is now widely accepted about the origin and evolution of the cosmos can be explained directly and quantitatively using the ideas developed in this book.

From Special to General Relativity

The Theory of Relativity confers an absolute meaning on a magnitude which in classical theory has only a relative significance: the velocity of light. The velocity of light is to the Theory of Relativity as the elementary quantum of action is to the Quantum Theory: it is its absolute core.

MAX PLANCK (1949)

24.1 Overview

24.1

We begin our discussion of general relativity in this chapter with a review, and elaboration of relevant material already covered in earlier chapters. In Sec. 24.2, we give a brief encapsulation of special relativity drawn largely from Chap. 2, emphasizing those aspects that underpin the transition to general relativity. Then in Sec. 24.3 we collect, review, and extend the fundamental ideas of differential geometry that have been scattered throughout the book and that we shall need as foundations for the mathematics of *spacetime curvature* (Chap. 25). Most importantly, we generalize differential geometry to encompass coordinate systems whose coordinate lines are not orthogonal and bases that are not orthonormal.

Einstein's field equation (to be studied in Chap. 25) is a relationship between the curvature of spacetime and the matter that generates it, akin to the Maxwell equations' relationship between the electromagnetic field and the electric currents and charges that generate it. The matter in Einstein's equation is described by the stress-energy tensor that we introduced in Sec. 2.13. We revisit the stress-energy tensor in Sec. 24.4 and develop a deeper understanding of its properties.

In general relativity one often wishes to describe the outcome of measurements made by observers who refuse to fall freely—for example, an observer who hovers in a spaceship just above the horizon of a black hole, or a gravitational-wave experimenter in an Earthbound laboratory. As a foundation for treating such observers, in Sec. 24.5 we examine measurements made by accelerated observers in the flat spacetime of special relativity.

24.2 Special Relativity Once Again

24.2

Our viewpoint on general relativity is unapologetically geometrical. (Other viewpoints, e.g., those of particle theorists such as Feynman and Weinberg, are quite different.) Therefore, a prerequisite for our treatment of general relativity is understanding special relativity in geometric language. In Chap. 2, we discussed the foundations of

BOX 24.1. READERS' GUIDE

- This chapter relies significantly on:
  - Chap. 2 on special relativity, which now should be regarded as Track One.
  - The discussion of connection coefficients in Sec. 11.8.
- This chapter is a foundation for the presentation of general relativity theory and cosmology in Chaps. 25–28.

special relativity with this in mind. In this section we briefly review the most important points.

We suggest that any reader who has not studied Chap. 2 read Sec. 24.2 first, to get an overview and flavor of what will be important for our development of general relativity, and then (or in parallel with reading Sec. 24.2) read those relevant sections of Chap. 2 that the reader does not already understand.

24.2.1

24.2.1 Geometric, Frame-Independent Formulation

review of the geometric, frame-independent formulation of special relativity

In Secs. 1.1.1 and 2.2.2, we learned that *every law of physics must be expressible as a geometric, frame-independent relationship among geometric, frame-independent objects.* This is equally true in Newtonian physics, in special relativity, and in general relativity. The key difference between the three is the geometric arena: in Newtonian physics, the arena is 3-dimensional Euclidean space; in special relativity, it is 4-dimensional Minkowski spacetime; in general relativity (Chap. 25), it is 4-dimensional curved spacetime (see Fig. 1 in the Introduction to Part I and the associated discussion).

Principle of Relativity—laws as geometric relations between geometric objects

In special relativity, the demand that the laws be geometric relationships among geometric objects that live in Minkowski spacetime is the *Principle of Relativity*; see Sec. 2.2.2. Examples of the geometric objects are:

examples of geometric objects: points, curves, proper time ticked by an ideal clock, vectors, tensors, scalar product

1. A point $\mathcal{P}$ in spacetime (which represents an *event*); Sec. 2.2.1.
2. A parameterized curve in spacetime, such as the world line $\mathcal{P}(\tau)$ of a particle, for which the parameter τ is the particle's *proper time* (i.e., the time measured by an ideal clock[1] that the particle carries; Fig. 24.1); Sec. 2.4.1.

1. Recall that an ideal clock is one that ticks uniformly when compared, e.g., to the period of the light emitted by some standard type of atom or molecule, and that has been made impervious to accelerations. Thus two ideal clocks momentarily at rest with respect to each other tick at the same rate independent of their relative acceleration; see Secs. 2.2.1 and 2.4.1. For greater detail, see Misner, Thorne, and Wheeler (1973, pp. 23–29, 395–399).

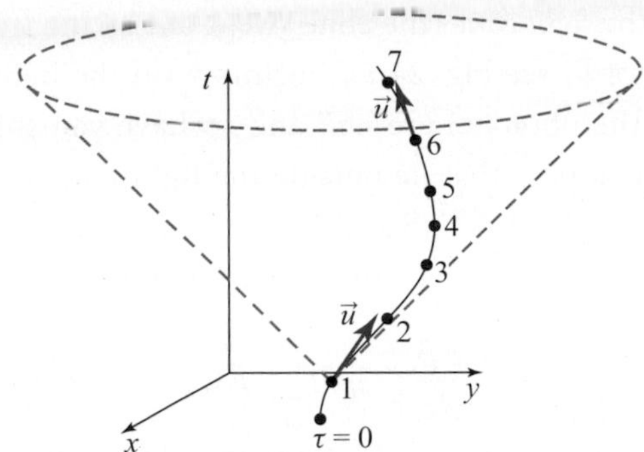

FIGURE 24.1 The world line $\mathcal{P}(\tau)$ of a particle in Minkowski spacetime and the tangent vector $\vec{u} = d\mathcal{P}/d\tau$ to this world line; $\vec{u}$ is the particle's 4-velocity. The bending of the world line is produced by some force that acts on the particle, such as the Lorentz force embodied in Eq. (24.3). Also shown is the light cone emitted from the event $\mathcal{P}(\tau = 1)$. Although the axes of an (arbitrary) inertial reference frame are shown, no reference frame is needed for the definition of the world line, its tangent vector $\vec{u}$, or the light cone. Nor is one needed for the formulation of the Lorentz force law.

3. Vectors, such as the particle's 4-velocity $\vec{u} = d\mathcal{P}/d\tau$ [the tangent vector to the curve $\mathcal{P}(\tau)$] and the particle's 4-momentum $\vec{p} = m\vec{u}$ (with m the particle's rest mass); Secs. 2.2.1 and 2.4.1.
4. Tensors, such as the electromagnetic field tensor $\boldsymbol{F}(\_, \_)$; Secs. 1.3 and 2.3.

Recall that a tensor is a linear real-valued function of vectors; when one puts vectors $\vec{A}$ and $\vec{B}$ into the two slots of $\boldsymbol{F}$, one obtains a real number (a scalar) $\boldsymbol{F}(\vec{A}, \vec{B})$ that is linear in $\vec{A}$ and in $\vec{B}$ so, for example: $\boldsymbol{F}(\vec{A}, b\vec{B} + c\vec{C}) = b\boldsymbol{F}(\vec{A}, \vec{B}) + c\boldsymbol{F}(\vec{A}, \vec{C})$. When one puts a vector $\vec{B}$ into just one of the slots of $\boldsymbol{F}$ and leaves the other empty, one obtains a tensor with one empty slot, $\boldsymbol{F}(\_, \vec{B})$, that is, a vector. The result of putting a vector into the slot of a vector is the scalar product: $\vec{D}(\vec{B}) = \vec{D} \cdot \vec{B} = \boldsymbol{g}(\vec{D}, \vec{B})$, where $\boldsymbol{g}(\_, \_)$ is the metric.

spacetime metric

In Secs. 2.3 and 2.4.1, we tied our definitions of the inner product and the spacetime metric to the ticking of ideal clocks: If $\Delta\vec{x}$ is the vector separation of two neighboring events $\mathcal{P}(\tau)$ and $\mathcal{P}(\tau + \Delta\tau)$ along a particle's world line, then

$$\boxed{\boldsymbol{g}(\Delta\vec{x}, \Delta\vec{x}) \equiv \Delta\vec{x} \cdot \Delta\vec{x} \equiv -(\Delta\tau)^2.} \tag{24.1}$$

This relation for any particle with any timelike world line, together with the linearity of $\boldsymbol{g}(\_, \_)$ in its two slots, is enough to determine $\boldsymbol{g}$ completely and to guarantee that it is symmetric: $\boldsymbol{g}(\vec{A}, \vec{B}) = \boldsymbol{g}(\vec{B}, \vec{A})$ for all $\vec{A}$ and $\vec{B}$. Since the particle's 4-velocity $\vec{u}$ is

$$\vec{u} = \frac{d\mathcal{P}}{d\tau} = \lim_{\Delta\tau\to 0} \frac{\mathcal{P}(\tau + \Delta\tau) - \mathcal{P}(\tau)}{\Delta\tau} \equiv \lim_{\Delta\tau\to 0} \frac{\Delta\vec{x}}{\Delta\tau}, \tag{24.2}$$

Eq. (24.1) implies that $\vec{u} \cdot \vec{u} = \boldsymbol{g}(\vec{u}, \vec{u}) = -1$ (Sec. 2.4.1).

light cone; timelike, null, and spacelike vectors

The 4-velocity $\vec{u}$ is an example of a *timelike* vector (Sec. 2.2.3); it has a negative inner product with itself (negative "squared length"). This shows up pictorially in the fact that $\vec{u}$ lies inside the *light cone* (the cone swept out by the trajectories of photons emitted from the tail of $\vec{u}$; see Fig. 24.1). Vectors $\vec{k}$ on the light cone (the tangents to the world lines of the photons) are *null* and so have vanishing squared lengths: $\vec{k} \cdot \vec{k} = \boldsymbol{g}(\vec{k}, \vec{k}) = 0$; vectors $\vec{A}$ that lie outside the light cone are *spacelike* and have positive squared lengths: $\vec{A} \cdot \vec{A} > 0$ (Sec. 2.2.3).

An example of a physical law in 4-dimensional geometric language is the Lorentz force law (Sec. 2.4.2):

Lorentz force law

$$\frac{d\vec{p}}{d\tau} = q\boldsymbol{F}(\_\_, \vec{u}). \tag{24.3}$$

Here q is the particle's charge (a scalar), and both sides of this equation are vectors, or equivalently, first-rank tensors (i.e., tensors with just one slot). As we learned in Secs. 1.5.1 and 2.5.3, it is convenient to give names to slots. When we do so, we can rewrite the Lorentz force law as

$$\frac{dp^\alpha}{d\tau} = qF^{\alpha\beta}u_\beta. \tag{24.4}$$

slot-naming index notation

Here α is the name of the slot of the vector $d\vec{p}/d\tau$, α and β are the names of the slots of $\boldsymbol{F}$, β is the name of the slot of $\mathbf{u}$. The double use of β with one up and one down on the right-hand side of the equation represents the insertion of $\vec{u}$ into the β slot of $\boldsymbol{F}$, whereby the two β slots disappear, and we wind up with a vector whose slot is named α. As we learned in Sec. 1.5, this slot-naming index notation is isomorphic to the notation for components of vectors, tensors, and physical laws in some reference frame. However, no reference frames are needed or involved when one formulates the laws of physics in geometric, frame-independent language as above.

Those readers who do not feel completely comfortable with these concepts, statements, and notation should reread the relevant portions of Chaps. 1 and 2.

EXERCISES

Exercise 24.1 *Practice: Frame-Independent Tensors*

Let $\boldsymbol{A}$, $\boldsymbol{B}$ be second-rank tensors.

(a) Show that $\boldsymbol{A} + \boldsymbol{B}$ is also a second-rank tensor.

(b) Show that $\boldsymbol{A} \otimes \boldsymbol{B}$ is a fourth-rank tensor.

(c) Show that the contraction of $\boldsymbol{A} \otimes \boldsymbol{B}$ on its first and fourth slots is a second-rank tensor. (If necessary, consult Secs. 1.5 and 2.5 for discussions of contraction.)

(d) Write the following quantities in slot-naming index notation: the tensor $\boldsymbol{A} \otimes \boldsymbol{B}$, and the simultaneous contraction of this tensor on its first and fourth slots and on its second and third slots.

24.2.2

24.2.2 Inertial Frames and Components of Vectors, Tensors, and Physical Laws

inertial reference frame

In special relativity, a key role is played by *inertial reference frames*, Sec. 2.2.1. An inertial frame is an (imaginary) latticework of rods and clocks that moves through spacetime freely (inertially, without any force acting on it). The rods are orthogonal to one another and attached to inertial-guidance gyroscopes, so they do not rotate. These rods are used to identify the spatial, Cartesian coordinates $(x^1, x^2, x^3) = (x, y, z)$ of an event $\mathcal{P}$ [which we also denote by lowercased Latin indices $x^j(\mathcal{P})$, with j running over 1, 2, 3]. The latticework's clocks are ideal and are synchronized with one another by the Einstein light-pulse process. They are used to identify the temporal coordinate $x^0 = t$ of an event $\mathcal{P}$: $x^0(\mathcal{P})$ is the time measured by that latticework clock whose world line passes through $\mathcal{P}$, at the moment of passage. The spacetime coordinates of $\mathcal{P}$ are denoted by lowercased Greek indices x^α, with α running over 0, 1, 2, 3. An inertial frame's spacetime coordinates $x^\alpha(\mathcal{P})$ are called *Lorentz coordinates* or *inertial coordinates.*

Lorentz (inertial) coordinates

In the real universe, spacetime curvature is small in regions well removed from concentrations of matter (e.g., in intergalactic space), so special relativity is highly accurate there. In such a region, frames of reference (rod-clock latticeworks) that are nonaccelerating and nonrotating with respect to cosmologically distant galaxies (and hence with respect to a local frame in which the cosmic microwave radiation looks isotropic) constitute good approximations to inertial reference frames.

orthonormal basis vectors of an inertial frame

Associated with an inertial frame's Lorentz coordinates are basis vectors $\vec{e}_\alpha$ that point along the frame's coordinate axes (and thus are orthogonal to one another) and have unit length (making them orthonormal); see Sec. 2.5. This orthonormality is embodied in the inner products

$$\boxed{\vec{e}_\alpha \cdot \vec{e}_\beta = \eta_{\alpha\beta},} \tag{24.5}$$

where by definition:

$$\boxed{\eta_{00} = -1, \quad \eta_{11} = \eta_{22} = \eta_{33} = +1, \quad \eta_{\alpha\beta} = 0 \quad \text{if } \alpha \neq \beta.} \tag{24.6}$$

geometrized units

Here and throughout Part VII (as in Chap. 2), we set the speed of light to unity (i.e., we use the geometrized units introduced in Sec. 1.10), so spatial lengths (e.g., along the x-axis) and time intervals (e.g., along the t-axis) are measured in the same units, seconds or meters, with $1\text{ s} = 2.99792458 \times 10^8$ m.

In Sec. 2.5 (see also Sec. 1.5), we used the basis vectors of an inertial frame to build a component representation of tensor analysis. The fact that the inner products of timelike vectors with each other are negative (e.g., $\vec{e}_0 \cdot \vec{e}_0 = -1$), while those of spacelike vectors are positive (e.g., $\vec{e}_1 \cdot \vec{e}_1 = +1$), forced us to introduce two types of components: *covariant* (indices down) and *contravariant* (indices up). The covariant components of a tensor are computable by inserting the basis vectors into the tensor's slots: $u_\alpha = \vec{u}(\vec{e}_\alpha) \equiv \vec{u} \cdot \vec{e}_\alpha$; $F_{\alpha\beta} = \boldsymbol{F}(\vec{e}_\alpha, \vec{e}_\beta)$. For example, in our Lorentz basis the covariant components of the metric are $g_{\alpha\beta} = \boldsymbol{g}(\vec{e}_\alpha, \vec{e}_\beta) = \vec{e}_\alpha \cdot \vec{e}_\beta = \eta_{\alpha\beta}$. The contravariant components of a tensor were related to the covariant components via "index lowering" with the aid of the metric, $F_{\alpha\beta} = g_{\alpha\mu} g_{\beta\nu} F^{\mu\nu}$, which simply said that one reverses the sign when lowering a time index and makes no change of sign when lowering a space index. This lowering rule implied that the contravariant components of the metric in a Lorentz basis are the same numerically as the covariant

covariant and contravariant components of vectors and tensors

components, $g^{\alpha\beta} = \eta_{\alpha\beta}$, and that they can be used to raise indices (i.e., to perform the trivial sign flip for temporal indices): $F^{\mu\nu} = g^{\mu\alpha} g^{\nu\beta} F_{\alpha\beta}$. As we saw in Sec. 2.5, tensors can be expressed in terms of their contravariant components as $\vec{p} = p^\alpha \vec{e}_\alpha$, and $\boldsymbol{F} = F^{\alpha\beta} \vec{e}_\alpha \otimes \vec{e}_\beta$, where $\otimes$ represents the tensor product [Eqs. (1.5)].

component equations are same as slot-naming-index-notation equations

We also learned in Chap. 2 that any frame-independent geometric relation among tensors can be rewritten as a relation among those tensors' components in any chosen Lorentz frame. When one does so, the resulting component equation takes precisely the same form as the slot-naming-index-notation version of the geometric relation (Sec. 1.5.1). For example, the component version of the Lorentz force law says $dp^\alpha/d\tau = qF^{\alpha\beta}u_\beta$, which is identical to Eq. (24.4). The only difference is the interpretation of the symbols. In the component equation $F^{\alpha\beta}$ are the components of $\boldsymbol{F}$ and the repeated β in $F^{\alpha\beta}u_\beta$ is to be summed from 0 to 3. In the geometric relation $F^{\alpha\beta}$ means $\boldsymbol{F}(\_\_, \_\_)$, with the first slot named α and the second β, and the repeated β in $F^{\alpha\beta}u_\beta$ implies the insertion of $\vec{u}$ into the second slot of $\boldsymbol{F}$ to produce a single-slotted tensor (i.e., a vector) whose slot is named α.

components of 4-velocity in an inertial frame

3 + 1 split

As we saw in Sec. 2.6, a particle's 4-velocity $\vec{u}$ (defined originally without the aid of any reference frame; Fig. 24.1) has components, in any inertial frame, given by $u^0 = \gamma$, $u^j = \gamma v^j$, where $v^j = dx^j/dt$ is the particle's ordinary velocity and $\gamma \equiv 1\Big/\sqrt{1 - \delta_{ij}v^i v^j}$. Similarly, the particle's energy $E \equiv p^0$ is $m\gamma$, and its spatial momentum is $p^j = m\gamma v^j$ (i.e., in 3-dimensional geometric notation: $\mathbf{p} = m\gamma\mathbf{v}$). This is an example of the manner in which a choice of Lorentz frame produces a "3+1" split of the physics: a split of 4-dimensional spacetime into 3-dimensional space (with Cartesian coordinates x^j) plus 1-dimensional time $t = x^0$; a split of the particle's 4-momentum $\vec{p}$ into its 3-dimensional spatial momentum $\mathbf{p}$ and its 1-dimensional energy $\mathcal{E} = p^0$; and similarly a split of the electromagnetic field tensor $\boldsymbol{F}$ into the 3-dimensional electric field $\mathbf{E}$ and 3-dimensional magnetic field $\mathbf{B}$ (cf. Secs. 2.6 and 2.11).

Principle of Relativity restated: laws take same form in every inertial frame

The Principle of Relativity (all laws expressible as geometric relations between geometric objects in Minkowski spacetime), when translated into 3+1 language, says that, when the laws of physics are expressed in terms of components in a specific Lorentz frame, the form of those laws must be independent of one's choice of frame. When translated into operational terms, it says that, if two observers in two different Lorentz frames are given identical written instructions for a self-contained physics experiment, then their two experiments must yield the same results to within their experimental accuracies (Sec. 2.2.2).

Lorentz transformations

Principle of Relativity restated: laws are Lorentz invariant

The components of tensors in one Lorentz frame are related to those in another by a Lorentz transformation (Sec. 2.7), so the Principle of Relativity can be restated as saying that, when expressed in terms of Lorentz-frame components, *the laws of physics must be Lorentz-invariant* (unchanged by Lorentz transformations). This is the version of the Principle of Relativity that one meets in most elementary treatments of special relativity. However, as the above discussion shows, it is a mere shadow of the true Principle of Relativity—the shadow cast into Lorentz frames when one performs a 3+1 split. The ultimate, fundamental version of the Principle of Relativity is the one that needs no frames at all for its expression: *all the laws of physics are expressible as geometric relations among geometric objects that reside in Minkowski spacetime.*

ultimate version of Principle of Relativity

24.2.3 Light Speed, the Interval, and Spacetime Diagrams

24.2.3

One set of physical laws that must be the same in all inertial frames is Maxwell's equations. Let us discuss the implications of Maxwell's equations and the Principle of Relativity for the speed of light c. (For a more detailed discussion, see Sec. 2.2.2.) According to Maxwell, c can be determined by performing nonradiative laboratory experiments; it is not necessary to measure the time it takes light to travel along some path; see Box 2.2. The Principle of Relativity requires that such experiments must give the same result for c, independent of the reference frame in which the measurement apparatus resides, so the speed of light must be independent of reference frame. It is this frame independence that enables us to introduce geometrized units with $c = 1$.

light speed is the same in all inertial frames

Another example of frame independence (Lorentz invariance) is provided by the *interval between two events* (Sec. 2.2.3). The components $g_{\alpha\beta} = \eta_{\alpha\beta}$ of the metric imply that, if $\Delta\vec{x}$ is the vector separating the two events and Δx^α are its components in some Lorentz coordinate system, then the squared length of $\Delta\vec{x}$ [also called the *interval* and denoted $(\Delta s)^2$] is given by

$$\boxed{\begin{aligned}(\Delta s)^2 \equiv \Delta\vec{x}\cdot\Delta\vec{x} = \boldsymbol{g}(\Delta\vec{x}, \Delta\vec{x}) = g_{\alpha\beta}\Delta x^\alpha \Delta x^\beta \\ = -(\Delta t)^2 + (\Delta x)^2 + (\Delta y)^2 + (\Delta z)^2.\end{aligned}} \tag{24.7}$$

interval between two events

Since $\Delta\vec{x}$ is a geometric, frame-independent object, so must be the interval. This implies that the equation $(\Delta s)^2 = -(\Delta t)^2 + (\Delta x)^2 + (\Delta y)^2 + (\Delta z)^2$ by which one computes the interval between the two chosen events in one Lorentz frame must give the same numerical result when used in any other frame (i.e., this expression must be Lorentz invariant). This *invariance of the interval* is the starting point for most introductions to special relativity—and, indeed, we used it as a starting point in Sec. 2.2.

invariance of the interval

Spacetime diagrams play a major role in our development of general relativity. Accordingly, it is important that the reader feel very comfortable with them. We recommend reviewing Fig. 2.7 and Ex. 2.14.

spacetime diagrams

EXERCISES

Exercise 24.2 *Example: Invariance of a Null Interval*
You have measured the intervals between a number of adjacent events in spacetime and thereby have deduced the metric $\boldsymbol{g}$. Your friend claims that the metric is some other frame-independent tensor $\tilde{\boldsymbol{g}}$ that differs from $\boldsymbol{g}$. Suppose that your correct metric $\boldsymbol{g}$ and his wrong one $\tilde{\boldsymbol{g}}$ agree on the forms of the light cones in spacetime (i.e., they agree as to which intervals are null, which are spacelike, and which are timelike), but they give different answers for the value of the interval in the spacelike and timelike cases: $\boldsymbol{g}(\Delta\vec{x}, \Delta\vec{x}) \neq \tilde{\boldsymbol{g}}(\Delta\vec{x}, \Delta\vec{x})$. Prove that $\tilde{\boldsymbol{g}}$ and $\boldsymbol{g}$ differ solely by

a scalar multiplicative factor, $\tilde{\boldsymbol{g}} = a\boldsymbol{g}$ for some scalar a. We say that $\tilde{\boldsymbol{g}}$ and $\boldsymbol{g}$ are *conformal to each other.* [Hint: Pick some Lorentz frame and perform computations there, then lift yourself back up to a frame-independent viewpoint.]

Exercise 24.3 *Problem: Causality*
If two events occur at the same spatial point but not simultaneously in one inertial frame, prove that the temporal order of these events is the same in all inertial frames. Prove also that in all other frames the temporal interval Δt between the two events is larger than in the first frame, and that there are no limits on the events' spatial or temporal separation in the other frames. Give *two* proofs of these results, one algebraic and the other via spacetime diagrams.

24.3

24.3 Differential Geometry in General Bases and in Curved Manifolds

The differential geometry (tensor-analysis) formalism reviewed in the last section is inadequate for general relativity in several ways.

First, in general relativity we need to use bases $\vec{e}_\alpha$ that are not orthonormal (i.e., for which $\vec{e}_\alpha \cdot \vec{e}_\beta \neq \eta_{\alpha\beta}$). For example, near a spinning black hole there is much power in using a time basis vector $\vec{e}_t$ that is tied in a simple way to the metric's time-translation symmetry and a spatial basis vector $\vec{e}_\phi$ that is tied to its rotational symmetry. This time basis vector has an inner product with itself $\vec{e}_t \cdot \vec{e}_t = g_{tt}$ that is influenced by the slowing of time near the hole (so $g_{tt} \neq -1$); and $\vec{e}_\phi$ is not orthogonal to $\vec{e}_t$ ($\vec{e}_t \cdot \vec{e}_\phi = g_{t\phi} \neq 0$), as a result of the dragging of inertial frames by the hole's spin. In this section, we generalize our formalism to treat such nonorthonormal bases.

Second, in the curved spacetime of general relativity (and in any other curved space, e.g., the 2-dimensional surface of Earth), the definition of a vector as an arrow connecting two points (Secs. 1.2 and 2.2.1) is suspect, as it is not obvious on what route the arrow should travel nor that the linear algebra of tensor analysis should be valid for such arrows. In this section, we refine the concept of a vector to deal with this problem. In the process we introduce the concept of a *tangent space* in which the linear algebra of tensors takes place—a different tangent space for tensors that live at different points in the space.

Third, once we have been forced to think of a tensor as residing in a specific tangent space at a specific point in the space, the question arises: how can one transport tensors from the tangent space at one point to the tangent space at an adjacent point? Since the notion of a gradient of a vector depends on comparing the vector at two different points and thus depends on the details of transport, we have to rework the notion of a gradient and the gradient's connection coefficients.

Fourth, when doing an integral, one must add contributions that live at different points in the space, so we must also rework the notion of integration.

We tackle each of these four issues in turn in the following four subsections.

24.3.1 Nonorthonormal Bases

24.3.1

Consider an n-dimensional *manifold*, that is, a space that, in the neighborhood of any point, has the same topological and smoothness properties as n-dimensional Euclidean space, though it might not have a locally Euclidean or locally Lorentz metric and perhaps has no metric at all. If the manifold has a metric (e.g., 4-dimensional spacetime, 3-dimensional Euclidean space, and the 2-dimensional surface of a sphere) it is called "Riemannian." In this chapter, all manifolds we consider will be Riemannian.

manifold

At some point $\mathcal{P}$ in our chosen n-dimensional manifold with metric, introduce a set of basis vectors $\{\vec{e}_1, \vec{e}_2, \ldots, \vec{e}_n\}$ and denote them generally as $\vec{e}_\alpha$. We seek to generalize the formalism of Sec. 24.2 in such a way that the index-manipulation rules for components of tensors are unchanged. For example, we still want it to be true that covariant components of any tensor are computable by inserting the basis vectors into the tensor's slots, $F_{\alpha\beta} = \mathbf{F}(\vec{e}_\alpha, \vec{e}_\beta)$, and that the tensor itself can be reconstructed from its contravariant components: $\mathbf{F} = F^{\mu\nu}\vec{e}_\mu \otimes \vec{e}_\nu$. We also require that the two sets of components are computable from each other via raising and lowering with the metric components: $F_{\alpha\beta} = g_{\alpha\mu} g_{\beta\nu} F^{\mu\nu}$. The only thing we do not want to preserve is the orthonormal values of the metric components: we must allow the basis to be nonorthonormal and thus $\vec{e}_\alpha \cdot \vec{e}_\beta = g_{\alpha\beta}$ to have arbitrary values (except that the metric should be nondegenerate, so no linear combination of the $\vec{e}_\alpha$s vanishes, which means that the matrix $||g_{\alpha\beta}||$ should have nonzero determinant).

tensors in a nonorthonormal basis

We can easily achieve our goal by introducing a second set of basis vectors, denoted $\{\vec{e}^1, \vec{e}^2, \ldots, \vec{e}^n\}$, which is *dual* to our first set in the sense that

dual sets of basis vectors

$$\boxed{\vec{e}^\mu \cdot \vec{e}_\beta \equiv \mathbf{g}(\vec{e}^\mu, \vec{e}_\beta) = \delta^\mu{}_\beta.} \tag{24.8}$$

Here $\delta^\alpha{}_\beta$ is the Kronecker delta. This duality relation actually constitutes a *definition* of the e^μ once the $\vec{e}_\alpha$ have been chosen. To see this, regard $\vec{e}^\mu$ as a tensor of rank one. This tensor is defined as soon as its value on each and every vector has been determined. Expression (24.8) gives the value $\vec{e}^\mu(\vec{e}_\beta) = \vec{e}^\mu \cdot \vec{e}_\beta$ of $\vec{e}^\mu$ on each of the four basis vectors $\vec{e}_\beta$; and since every other vector can be expanded in terms of the $\vec{e}_\beta$s and $\vec{e}^\mu(\_)$ is a linear function, Eq. (24.8) thereby determines the value of $\vec{e}^\mu$ on every other vector.

The duality relation (24.8) says that $\vec{e}^1$ is always perpendicular to all the $\vec{e}_\alpha$s except $\vec{e}_1$, and its scalar product with $\vec{e}_1$ is unity—and similarly for the other basis vectors. This interpretation is illustrated for 3-dimensional Euclidean space in Fig. 24.2. In Minkowski spacetime, if the $\vec{e}_\alpha$ are an orthonormal Lorentz basis, then duality dictates that $\vec{e}^0 = -\vec{e}_0$, and $\vec{e}^j = +\vec{e}_j$.

The duality relation (24.8) leads immediately to the same index-manipulation formalism as we have been using, if one defines the contravariant, covariant, and mixed components of tensors in the obvious manner:

covariant, contravariant, and mixed components of a tensor

$$\boxed{F^{\mu\nu} = \mathbf{F}(\vec{e}^\mu, \vec{e}^\nu), \quad F_{\alpha\beta} = \mathbf{F}(\vec{e}_\alpha, \vec{e}_\beta), \quad F^\mu{}_\beta = \mathbf{F}(\vec{e}^\mu, \vec{e}_\beta);} \tag{24.9}$$

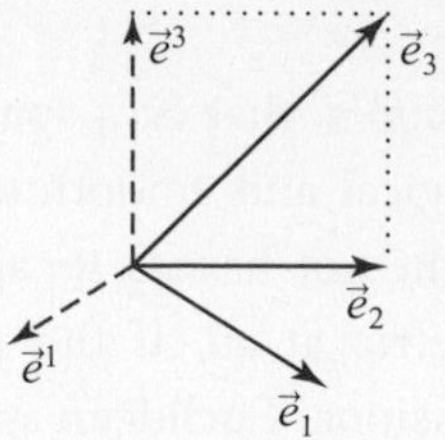

FIGURE 24.2 Nonorthonormal basis vectors $\vec{e}_j$ in Euclidean 3-space and two members $\vec{e}^1$ and $\vec{e}^3$ of the dual basis. The vectors $\vec{e}_1$ and $\vec{e}_2$ lie in the horizontal plane, so $\vec{e}^3$ is orthogonal to that plane (i.e., it points vertically upward), and its inner product with $\vec{e}_3$ is unity. Similarly, the vectors $\vec{e}_2$ and $\vec{e}_3$ span a vertical plane, so $\vec{e}^1$ is orthogonal to that plane (i.e., it points horizontally), and its inner product with $\vec{e}_1$ is unity.

see Ex. 24.4. Among the consequences of this duality are the following:

covariant and contravariant components of the metric

1. The matrix of contravariant components of the metric is inverse to that of the covariant components, $||g^{\mu\nu}|| = ||g_{\alpha\beta}||^{-1}$, so that

$$\boxed{g^{\mu\beta} g_{\beta\nu} = \delta^\mu{}_\nu.} \tag{24.10}$$

   This relation guarantees that when one raises an index on a tensor $F_{\alpha\beta}$ with $g^{\mu\beta}$ and then lowers it back down with $g_{\beta\mu}$, one recovers one's original covariant components $F_{\alpha\beta}$ unaltered.

reconstructing a tensor from its components

2. One can reconstruct a tensor from its components by lining up the indices in a manner that accords with the rules of index manipulation:

$$\boxed{\mathbf{F} = F^{\mu\nu}\vec{e}_\mu \otimes \vec{e}_\nu = F_{\alpha\beta}\vec{e}^\alpha \otimes \vec{e}^\beta = F^\mu{}_\beta \vec{e}_\mu \otimes \vec{e}^\beta.} \tag{24.11}$$

component equations are same as slot-naming-index-notation equations

3. The component versions of tensorial equations are identical in mathematical symbology to the slot-naming-index-notation versions:

$$\boxed{\mathbf{F}(\vec{p}, \vec{q}) = F^{\alpha\beta} p_\alpha p_\beta.} \tag{24.12}$$

Associated with any coordinate system $x^\alpha(\mathcal{P})$ there is a *coordinate basis* whose basis vectors are defined by

coordinate basis

$$\boxed{\vec{e}_\alpha \equiv \frac{\partial \mathcal{P}}{\partial x^\alpha}.} \tag{24.13}$$

Since the derivative is taken holding the other coordinates fixed, the basis vector $\vec{e}_\alpha$ points along the α coordinate axis (the axis on which x^α changes and all the other coordinates are held fixed).

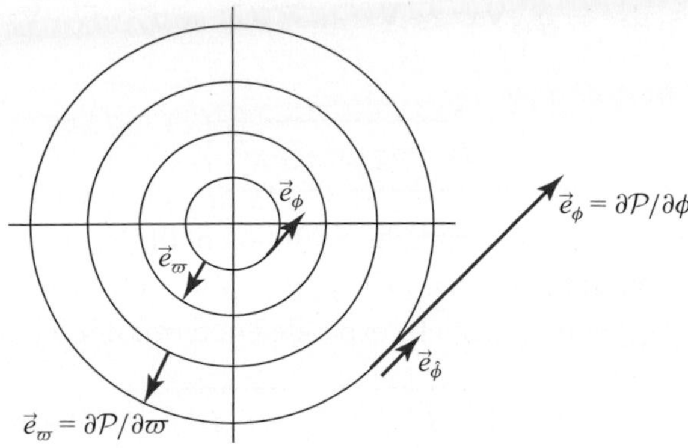

FIGURE 24.3 A circular coordinate system $\{\varpi, \phi\}$ and its coordinate basis vectors $\vec{e}_\varpi = \partial\mathcal{P}/\partial\varpi$, $\vec{e}_\phi = \partial\mathcal{P}/\partial\phi$ at several locations in the coordinate system. Also shown is the orthonormal basis vector $\vec{e}_{\hat\phi}$.

orthogonal curvilinear coordinates

In an orthogonal curvilinear coordinate system [e.g., circular polar coordinates (ϖ, ϕ) in Euclidean 2-space; Fig. 24.3], this coordinate basis is quite different from the coordinate system's orthonormal basis. For example, $\vec{e}_\phi = (\partial\mathcal{P}/\partial\phi)_\varpi$ is a very long vector at large radii and a very short one at small radii; the corresponding unit-length vector is $\vec{e}_{\hat\phi} = (1/\varpi)\vec{e}_\phi = (1/\varpi)\partial/\partial\phi$ (i.e., the derivative with respect to physical distance along the ϕ direction). By contrast, $\vec{e}_\varpi = (\partial\mathcal{P}/\partial\varpi)_\phi$ already has unit length, so the corresponding orthonormal basis vector is simply $\vec{e}_{\hat\varpi} = \vec{e}_\varpi$. The metric components in the coordinate basis are readily seen to be $g_{\phi\phi} = \varpi^2$, $g_{\varpi\varpi} = 1$, and $g_{\varpi\phi} = g_{\phi\varpi} = 0$, which are in accord with the equation for the squared distance (interval) between adjacent points: $ds^2 = g_{ij}dx^i dx^j = d\varpi^2 + \varpi^2 d\phi^2$. Of course, the metric components in the orthonormal basis are $g_{\hat i \hat j} = \delta_{ij}$.

Henceforth, we use hats to identify orthonormal bases; bases whose indices do not have hats will typically (though not always) be coordinate bases.

We can construct the basis $\{\vec{e}^\mu\}$ that is dual to the coordinate basis $\{\vec{e}_\alpha\} = \{\partial\mathcal{P}/\partial x^\alpha\}$ by taking the gradients of the coordinates, viewed as scalar fields $x^\alpha(\mathcal{P})$:

the basis dual to a coordinate basis

$$\boxed{\vec{e}^\mu = \vec\nabla x^\mu.} \tag{24.14}$$

It is straightforward to verify the duality relation (24.8) for these two bases:

$$\vec{e}^\mu \cdot \vec{e}_\alpha = \vec{e}_\alpha \cdot \vec\nabla x^\mu = \nabla_{\vec{e}_\alpha} x^\mu = \nabla_{\partial\mathcal{P}/\partial x^\alpha} x^\mu = \frac{\partial x^\mu}{\partial x^\alpha} = \delta^\mu{}_\alpha. \tag{24.15}$$

In any coordinate system, the expansion of the metric in terms of the dual basis, $\boldsymbol{g} = g_{\alpha\beta}\vec{e}^\alpha \otimes \vec{e}^\beta = g_{\alpha\beta}\vec\nabla x^\alpha \otimes \vec\nabla x^\beta$, is intimately related to the line element $ds^2 = g_{\alpha\beta}dx^\alpha dx^\beta$. Consider an infinitesimal vectorial displacement $d\vec{x} = dx^\alpha(\partial/\partial x^\alpha)$. Insert this displacement into the metric's two slots to obtain the interval ds^2 along

$d\vec{x}$. The result is $ds^2 = g_{\alpha\beta}\nabla x^\alpha \otimes \nabla x^\beta(d\vec{x}, d\vec{x}) = g_{\alpha\beta}(d\vec{x}\cdot\nabla x^\alpha)(d\vec{x}\cdot\nabla x^\beta) = g_{\alpha\beta}dx^\alpha dx^\beta$:

the line element for the invariant interval along a displacement vector

$$\boxed{ds^2 = g_{\alpha\beta}dx^\alpha dx^\beta.} \tag{24.16}$$

Here the second equality follows from the definition of the tensor product $\otimes$, and the third from the fact that for any scalar field ψ, $d\vec{x}\cdot\nabla\psi$ is the change $d\psi$ along $d\vec{x}$.

Any two bases $\{\vec{e}_\alpha\}$ and $\{\vec{e}_{\bar{\mu}}\}$ can be expanded in terms of each other:

transformation matrices linking two bases

$$\boxed{\vec{e}_\alpha = \vec{e}_{\bar{\mu}} L^{\bar{\mu}}{}_\alpha, \quad \vec{e}_{\bar{\mu}} = \vec{e}_\alpha L^\alpha{}_{\bar{\mu}}.} \tag{24.17}$$

(By convention the first index on L is always placed up, and the second is always placed down.) The quantities $||L^{\bar{\mu}}{}_\alpha||$ and $||L^\alpha{}_{\bar{\mu}}||$ are transformation matrices, and since they operate in opposite directions, they must be the inverse of each other:

$$\boxed{L^{\bar{\mu}}{}_\alpha L^\alpha{}_{\bar{\nu}} = \delta^{\bar{\mu}}{}_{\bar{\nu}}, \quad L^\alpha{}_{\bar{\mu}} L^{\bar{\mu}}{}_\beta = \delta^\alpha{}_\beta.} \tag{24.18}$$

These $||L^{\bar{\mu}}{}_\alpha||$ are the generalizations of Lorentz transformations to arbitrary bases [cf. Eqs. (2.34) and (2.35a)]. As in the Lorentz-transformation case, the transformation laws (24.17) for the basis vectors imply corresponding transformation laws for components of vectors and tensors—laws that entail lining up indices in the obvious manner:

transformation of tensor components between bases

$$\boxed{\begin{gathered} A_{\bar{\mu}} = L^\alpha{}_{\bar{\mu}} A_\alpha, \quad T^{\bar{\mu}\bar{\nu}}{}_{\bar{\rho}} = L^{\bar{\mu}}{}_\alpha L^{\bar{\nu}}{}_\beta L^\gamma{}_{\bar{\rho}} T^{\alpha\beta}{}_\gamma, \\ \text{and similarly in the opposite direction.} \end{gathered}} \tag{24.19}$$

For coordinate bases, these $L^{\bar{\mu}}{}_\alpha$ are simply the partial derivatives of one set of coordinates with respect to the other:

transformation matrices between coordinate bases

$$\boxed{L^{\bar{\mu}}{}_\alpha = \frac{\partial x^{\bar{\mu}}}{\partial x^\alpha}, \quad L^\alpha{}_{\bar{\mu}} = \frac{\partial x^\alpha}{\partial x^{\bar{\mu}}},} \tag{24.20}$$

as one can easily deduce via

$$\vec{e}_\alpha = \frac{\partial \mathcal{P}}{\partial x^\alpha} = \frac{\partial x^\mu}{\partial x^\alpha}\frac{\partial \mathcal{P}}{\partial x^\mu} = \vec{e}_\mu \frac{\partial x^\mu}{\partial x^\alpha}. \tag{24.21}$$

In many physics textbooks a tensor is *defined* as a set of components $F_{\alpha\beta}$ that obey the transformation laws

$$F_{\alpha\beta} = F_{\mu\nu}\frac{\partial x^\mu}{\partial x^\alpha}\frac{\partial x^\nu}{\partial x^\beta}. \tag{24.22}$$

This definition (valid only in a coordinate basis) is in accord with Eqs. (24.19) and (24.20), though it hides the true and very simple nature of a tensor as a linear function of frame-independent vectors.

EXERCISES

Exercise 24.4 *Derivation: Index-Manipulation Rules from Duality*

For an arbitrary basis $\{\vec{e}_\alpha\}$ and its dual basis $\{\vec{e}^\mu\}$, use (i) the duality relation (24.8), (ii) the definition (24.9) of components of a tensor, and (iii) the relation $\vec{A}\cdot\vec{B} = \boldsymbol{g}(\vec{A}, \vec{B})$ between the metric and the inner product to deduce the following results.

(a) The relations

$$\vec{e}^\mu = g^{\mu\alpha}\vec{e}_\alpha, \quad \vec{e}_\alpha = g_{\alpha\mu}\vec{e}^\mu. \tag{24.23}$$

(b) The fact that indices on the components of tensors can be raised and lowered using the components of the metric:

$$F^{\mu\nu} = g^{\mu\alpha}F_\alpha{}^\nu, \quad p_\alpha = g_{\alpha\beta}p^\beta. \tag{24.24}$$

(c) The fact that a tensor can be reconstructed from its components in the manner of Eq. (24.11).

Exercise 24.5 *Practice: Transformation Matrices for Circular Polar Bases*

Consider the circular polar coordinate system $\{\varpi, \phi\}$ and its coordinate bases and orthonormal bases as shown in Fig. 24.3 and discussed in the associated text. These coordinates are related to Cartesian coordinates $\{x, y\}$ by the usual relations: $x = \varpi\cos\phi$, $y = \varpi\sin\phi$.

(a) Evaluate the components ($L^x{}_\varpi$, etc.) of the transformation matrix that links the two coordinate bases $\{\vec{e}_x, \vec{e}_y\}$ and $\{\vec{e}_\varpi, \vec{e}_\phi\}$. Also evaluate the components ($L^\varpi{}_x$, etc.) of the inverse transformation matrix.

(b) Similarly, evaluate the components of the transformation matrix and its inverse linking the bases $\{\vec{e}_x, \vec{e}_y\}$ and $\{\vec{e}_{\hat\varpi}, \vec{e}_{\hat\phi}\}$.

(c) Consider the vector $\vec{A} \equiv \vec{e}_x + 2\vec{e}_y$. What are its components in the other two bases?

24.3.2 Vectors as Directional Derivatives; Tangent Space; Commutators

24.3.2

As discussed in the introduction to Sec. 24.3, the notion of a vector as an arrow connecting two points is problematic in a curved manifold and must be refined. As a first step in the refinement, let us consider the tangent vector $\vec{A}$ to a curve $\mathcal{P}(\zeta)$ at some point $\mathcal{P}_o \equiv \mathcal{P}(\zeta = 0)$. We have defined that tangent vector by the limiting process:

tangent vector to a curve

$$\boxed{\vec{A} \equiv \frac{d\mathcal{P}}{d\zeta} \equiv \lim_{\Delta\zeta\to 0}\frac{\mathcal{P}(\Delta\zeta) - \mathcal{P}(0)}{\Delta\zeta}} \tag{24.25}$$

[Eq. (24.2)]. In this definition the difference $\mathcal{P}(\zeta) - \mathcal{P}(0)$ means the tiny arrow reaching from $\mathcal{P}(0) \equiv \mathcal{P}_o$ to $\mathcal{P}(\Delta\zeta)$. In the limit as $\Delta\zeta$ becomes vanishingly small, these two points get arbitrarily close together. In such an arbitrarily small region of the manifold, the effects of the manifold's curvature become arbitrarily small and

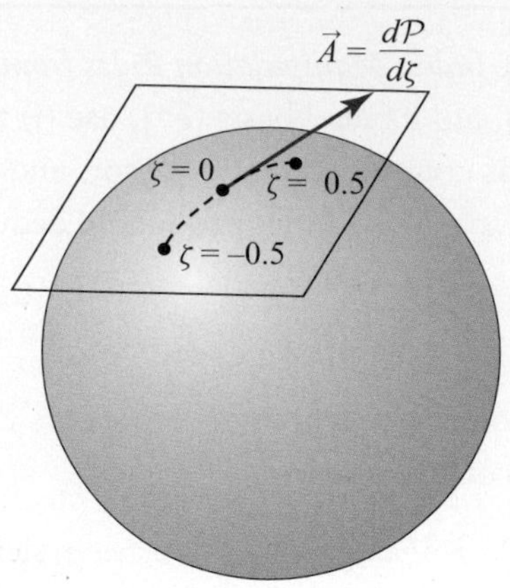

FIGURE 24.4 A curve $\mathcal{P}(\zeta)$ on the surface of a sphere and the curve's tangent vector $\vec{A} = d\mathcal{P}/d\zeta$ at $\mathcal{P}(\zeta = 0) \equiv \mathcal{P}_o$. The tangent vector lives in the tangent space at $\mathcal{P}_o$ (i.e., in the flat plane that is tangent to the sphere there, as seen in the flat Euclidean 3-space in which the sphere's surface is embedded).

negligible (just think of an arbitrarily tiny region on the surface of a sphere), so the notion of the arrow should become sensible. However, before the limit is completed, we are required to divide by $\Delta\zeta$, which makes our arbitrarily tiny arrow big again. What meaning can we give to this?

One way to think about it is to imagine embedding the curved manifold in a higher-dimensional flat space (e.g., embed the surface of a sphere in a flat 3-dimensional Euclidean space, as shown in Fig. 24.4). Then the tiny arrow $\mathcal{P}(\Delta\zeta) - \mathcal{P}(0)$ can be thought of equally well as lying on the sphere, or as lying in a surface that is tangent to the sphere and is flat, as measured in the flat embedding space. We can give meaning to $[\mathcal{P}(\Delta\zeta) - \mathcal{P}(0)]/\Delta\zeta$ if we regard this expression as a formula for lengthening an arrow-type vector in the flat tangent surface; correspondingly, we must regard the resulting tangent vector $\vec{A}$ as an arrow living in the tangent surface.

tangent space at a point

The (conceptual) flat tangent surface at the point $\mathcal{P}_o$ is called the *tangent space* to the curved manifold at that point. It has the same number of dimensions n as the manifold itself (two in the case of the surface of the sphere in Fig. 24.4). Vectors at $\mathcal{P}_o$ are arrows residing in that point's tangent space, tensors at $\mathcal{P}_o$ are linear functions of these vectors, and all the linear algebra of vectors and tensors that reside at $\mathcal{P}_o$ occurs in this tangent space. For example, the inner product of two vectors $\vec{A}$ and $\vec{B}$ at $\mathcal{P}_o$ (two arrows living in the tangent space there) is computed via the standard relation $\vec{A} \cdot \vec{B} = \boldsymbol{g}(\vec{A}, \vec{B})$ using the metric $\boldsymbol{g}$ that also resides in the tangent space. (Scalars reside in both the manifold and the tangent space.)

This pictorial way of thinking about the tangent space and vectors and tensors that reside in it is far too heuristic to satisfy most mathematicians. Therefore, mathematicians have insisted on making it much more precise at the price of greater abstraction. Mathematicians define the tangent vector to the curve $\mathcal{P}(\zeta)$ to be the derivative $d/d\zeta$

that differentiates scalar fields along the curve. This derivative operator is well defined by the rules of ordinary differentiation: if $\psi(\mathcal{P})$ is a scalar field in the manifold, then $\psi[\mathcal{P}(\zeta)]$ is a function of the real variable ζ, and its derivative $(d/d\zeta)\psi[\mathcal{P}(\zeta)]$ evaluated at $\zeta = 0$ is the ordinary derivative of elementary calculus. Since the derivative operator $d/d\zeta$ differentiates in the manifold along the direction in which the curve is moving, it is often called the *directional derivative* along $\mathcal{P}(\zeta)$. Mathematicians notice that all the directional derivatives at a point $\mathcal{P}_o$ of the manifold form a vector space (they can be multiplied by scalars and added and subtracted to get new vectors), and so the mathematicians define this vector space to be the tangent space at $\mathcal{P}_o$.

directional derivative

This mathematical procedure turns out to be isomorphic to the physicists' more heuristic way of thinking about the tangent space. In physicists' language, if one introduces a coordinate system in a region of the manifold containing $\mathcal{P}_o$ and constructs the corresponding coordinate basis $\vec{e}_\alpha = \partial\mathcal{P}/\partial x^\alpha$, then one can expand any vector in the tangent space as $\vec{A} = A^\alpha \partial\mathcal{P}/\partial x^\alpha$. One can also construct, in physicists' language, the directional derivative along $\vec{A}$; it is $\partial_{\vec{A}} \equiv A^\alpha \partial/\partial x^\alpha$. Evidently, the components A^α of the physicist's vector $\vec{A}$ (an arrow) are identical to the coefficients A^α in the coordinate-expansion of the directional derivative $\partial_{\vec{A}}$. Therefore a one-to-one correspondence exists between the directional derivatives $\partial_{\vec{A}}$ at $\mathcal{P}_o$ and the vectors $\vec{A}$ there, and a complete isomorphism holds between the tangent-space manipulations that a mathematician performs treating the directional derivatives as vectors, and those that a physicist performs treating the arrows as vectors.

tangent vector as directional derivative along a curve

"Why not abandon the fuzzy concept of a vector as an arrow, and *redefine the vector $\vec{A}$ to be the same as the directional derivative $\partial_{\vec{A}}$?*" mathematicians have demanded of physicists. Slowly, over the past century, physicists have come to see the merit in this approach. (i) It does, indeed, make the concept of a vector more rigorous than before. (ii) It simplifies a number of other concepts in mathematical physics (e.g., the commutator of two vector fields; see below). (iii) It facilitates communication with mathematicians. (iv) It provides a formalism that is useful for calculation. With these motivations in mind, and because one always gains conceptual and computational power by having multiple viewpoints at one's fingertips (see Feynman, 1966, p. 160), we henceforth shall regard vectors both as arrows living in a tangent space and as directional derivatives. Correspondingly, we assert the equalities:

$$\boxed{\frac{\partial\mathcal{P}}{\partial x^\alpha} = \frac{\partial}{\partial x^\alpha}, \quad \vec{A} = \partial_{\vec{A}},} \tag{24.26}$$

and often expand vectors in a coordinate basis using the notation

$$\boxed{\vec{A} = A^\alpha \frac{\partial}{\partial x^\alpha}.} \tag{24.27}$$

This directional-derivative viewpoint on vectors makes natural the concept of the *commutator* of two vector fields $\vec{A}$ and $\vec{B}$: $[\vec{A}, \vec{B}]$ is the vector that, when viewed

commutator of two vector fields

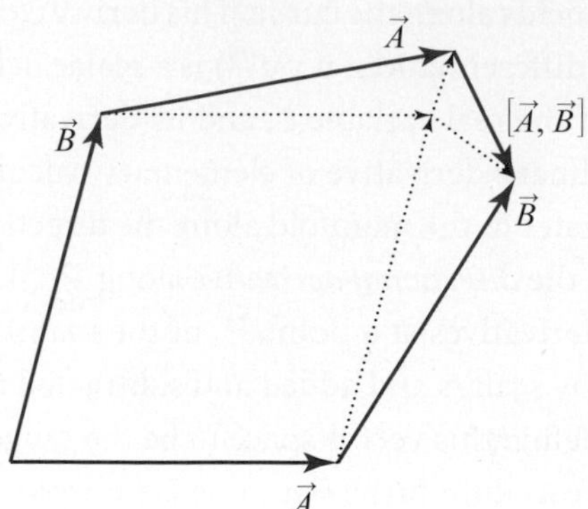

FIGURE 24.5 The commutator $[\vec{A}, \vec{B}]$ of two vector fields. The vectors are assumed to be so small that the curvature of the manifold is negligible in the region of the diagram, so all the vectors can be drawn lying in the manifold itself rather than in their respective tangent spaces. In evaluating the two terms in the commutator (24.28), a locally orthonormal coordinate basis is used, so $A^\alpha \partial B^\beta/\partial x^\alpha$ is the amount by which the vector $\vec{B}$ changes when one travels along $\vec{A}$ (i.e., it is the rightward-and-downward pointing dashed arrow in the upper right), and $B^\alpha \partial A^\beta/\partial x^\alpha$ is the amount by which $\vec{A}$ changes when one travels along $\vec{B}$ (i.e., it is the rightward-and-upward pointing dashed arrow). According to Eq. (24.28), the difference of these two dashed arrows is the commutator $[\vec{A}, \vec{B}]$. As the diagram shows, this commutator closes the quadrilateral whose legs are $\vec{A}$ and $\vec{B}$. If the commutator vanishes, then there is no gap in the quadrilateral, which means that in the region covered by this diagram, one can construct a coordinate system in which $\vec{A}$ and $\vec{B}$ are coordinate basis vectors.

as a differential operator, is given by $[\partial_{\vec{A}}, \partial_{\vec{B}}]$—where the latter quantity is the same commutator as one meets elsewhere in physics (e.g., in quantum mechanics). Using this definition, we can compute the components of the commutator in a coordinate basis:

$$\boxed{[\vec{A}, \vec{B}] \equiv \left[A^\alpha \frac{\partial}{\partial x^\alpha}, B^\beta \frac{\partial}{\partial x^\beta}\right] = \left(A^\alpha \frac{\partial B^\beta}{\partial x^\alpha} - B^\alpha \frac{\partial A^\beta}{\partial x^\alpha}\right) \frac{\partial}{\partial x^\beta}.} \tag{24.28}$$

This is an operator equation where the final derivative is presumed to operate on a scalar field, just as in quantum mechanics. From this equation we can read off the components of the commutator in any coordinate basis; they are $A^\alpha B^\beta{}_{,\alpha} - B^\alpha A^\beta{}_{,\alpha}$, where the comma denotes partial differentiation. Figure 24.5 uses this equation to deduce the geometric meaning of the commutator: it is the fifth leg needed to close a quadrilateral whose other four legs are constructed from the vector fields $\vec{A}$ and $\vec{B}$. In other words, it is "the change in $\vec{B}$ relative to $\vec{A}$," and as such it is a type of derivative of $\vec{B}$ along $\vec{A}$, called the *Lie derivative*: $\mathcal{L}_{\vec{A}}\vec{B} \equiv [\vec{A}, \vec{B}]$ (cf. footnote 2 in Chap. 14).

coordinate bases have vanishing commutators

The commutator is useful as a tool for distinguishing between coordinate bases and noncoordinate bases (also called nonholonomic bases). In a coordinate basis, the basis vectors are just the coordinate system's partial derivatives, $\vec{e}_\alpha = \partial/\partial x^\alpha$, and since partial derivatives commute, it must be that $[\vec{e}_\alpha, \vec{e}_\beta] = 0$. Conversely (as Fig. 24.5 shows), if one has a basis with vanishing commutators $[\vec{e}_\alpha, \vec{e}_\beta] = 0$, then it is possible to construct a coordinate system for which this is the coordinate basis. In a noncoordinate basis, at least one of the commutators $[\vec{e}_\alpha, \vec{e}_\beta]$ will be nonzero.

24.3.3 Differentiation of Vectors and Tensors; Connection Coefficients

24.3.3

In a curved manifold, the differentiation of vectors and tensors is rather subtle. To elucidate the problem, let us recall how we defined such differentiation in Minkowski spacetime or Euclidean space (Sec. 1.7). Converting to the notation used in Eq. (24.25), we began by defining the directional derivative of a tensor field $\boldsymbol{F}(\mathcal{P})$ along the tangent vector $\vec{A} = d/d\zeta$ to a curve $\mathcal{P}(\zeta)$:

$$\boxed{\nabla_{\vec{A}} \boldsymbol{F} \equiv \lim_{\Delta\zeta \to 0} \frac{\boldsymbol{F}[\mathcal{P}(\Delta\zeta)] - \boldsymbol{F}[\mathcal{P}(0)]}{\Delta\zeta}.} \tag{24.29}$$

directional derivative of a tensor field

This definition is problematic, because $\boldsymbol{F}[\mathcal{P}(\Delta\zeta)]$ lives in a different tangent space than does $\boldsymbol{F}[\mathcal{P}(0)]$. To make the definition meaningful, we must identify some connection between the two tangent spaces, when their points $\mathcal{P}(\Delta\zeta)$ and $\mathcal{P}(0)$ are arbitrarily close together. That connection is equivalent to identifying a rule for transporting $\boldsymbol{F}$ from one tangent space to the other.

In flat space or flat spacetime, and when $\boldsymbol{F}$ is a vector $\vec{F}$, that transport rule is obvious: keep $\vec{F}$ parallel to itself and keep its length fixed during the transport. In other words, keep constant its components in an orthonormal coordinate system (Cartesian coordinates in Euclidean space, Lorentz coordinates in Minkowski spacetime). This is called the *law of parallel transport*. For a tensor $\boldsymbol{F}$, the parallel transport law is the same: keep its components fixed in an orthonormal coordinate basis.

Now, just as the curvature of Earth's surface prevents one from placing a Cartesian coordinate system on it, so nonzero curvature of any other manifold prevents one from introducing orthonormal coordinates; see Sec. 25.3. However, in an arbitrarily small region on Earth's surface, one can introduce coordinates that are arbitrarily close to Cartesian (as surveyors well know); the fractional deviations from Cartesian need be no larger than $O(L^2/R^2)$, where L is the size of the region and R is Earth's radius (see Sec. 25.3). Similarly, in curved spacetime, in an arbitrarily small region, one can introduce coordinates that are arbitrarily close to Lorentz, differing only by amounts quadratic in the size of the region—and similarly for a *local* orthonormal coordinate basis in any curved manifold.

When defining $\nabla_{\vec{A}} \boldsymbol{F}$, one is sensitive only to first-order changes of quantities, not second, so the parallel transport used in defining it in a flat manifold, based on constancy of components in an orthonormal coordinate basis, must also work in a *local* orthonormal coordinate basis of any curved manifold: In Eq. (24.29), one must transport $\boldsymbol{F}$ from $\mathcal{P}(\Delta\zeta)$ to $\mathcal{P}(0)$, holding its components fixed in a locally orthonormal coordinate basis (parallel transport), and then take the difference in the tangent space at $\mathcal{P}_o = \mathcal{P}(0)$, divide by $\Delta\zeta$, and let $\Delta\zeta \to 0$. The result is a tensor at $\mathcal{P}_o$: the directional derivative $\nabla_{\vec{A}} \boldsymbol{F}$ of $\boldsymbol{F}$.

gradient of a tensor field

Having made the directional derivative meaningful, one can proceed as in Secs. 1.7 and 2.10: define the gradient of $\boldsymbol{F}$ by $\nabla_{\vec{A}}\boldsymbol{F} = \vec{\nabla}\boldsymbol{F}(\_\_, \_\_, \vec{A})$ [i.e., put $\vec{A}$ in the last—differentiation—slot of $\vec{\nabla}\boldsymbol{F}$; Eq. (1.15b)].

As in Chap. 2, in any basis we denote the components of $\vec{\nabla}\boldsymbol{F}$ by $F_{\alpha\beta;\gamma}$. And as in Sec. 11.8 (elasticity theory), we can compute these components in any basis with the aid of that basis's *connection coefficients*.

In Sec. 11.8, we restricted ourselves to an orthonormal basis in Euclidean space and thus had no need to distinguish between covariant and contravariant indices; all indices were written as subscripts. Now, dealing with nonorthonormal bases in spacetime, we must distinguish covariant and contravariant indices. Accordingly, by analogy with Eq. (11.68), we define the connection coefficients $\Gamma^{\mu}{}_{\alpha\beta}$ as

connection coefficients for a basis and its dual

$$\boxed{\nabla_\beta \vec{e}_\alpha \equiv \nabla_{\vec{e}_\beta}\vec{e}_\alpha = \Gamma^{\mu}{}_{\alpha\beta}\vec{e}_\mu.} \tag{24.30}$$

The duality between bases $\vec{e}^{\nu} \cdot \vec{e}_\alpha = \delta^{\nu}{}_{\alpha}$ then implies

$$\boxed{\nabla_\beta \vec{e}^{\mu} \equiv \nabla_{\vec{e}_\beta}\vec{e}^{\mu} = -\Gamma^{\mu}{}_{\alpha\beta}\vec{e}^{\alpha}.} \tag{24.31}$$

Note the sign flip, which is required to keep $\nabla_\beta(\vec{e}^{\mu} \cdot \vec{e}_\alpha) = 0$, and note that the differentiation index always goes last on Γ. Duality also implies that Eqs. (24.30) and (24.31) can be rewritten as

$$\Gamma^{\mu}{}_{\alpha\beta} = \vec{e}^{\mu} \cdot \nabla_\beta \vec{e}_\alpha = -\vec{e}_\alpha \cdot \nabla_\beta \vec{e}^{\mu}. \tag{24.32}$$

With the aid of these connection coefficients, we can evaluate the components $A_{\alpha;\beta}$ of the gradient of a vector field in any basis. We just compute

$$\begin{aligned} A^{\mu}{}_{;\beta}\vec{e}_\mu &= \nabla_\beta \vec{A} = \nabla_\beta(A^{\mu}\vec{e}_\mu) = (\nabla_\beta A^{\mu})\vec{e}_\mu + A^{\mu}\nabla_\beta\vec{e}_\mu \\ &= A^{\mu}{}_{,\beta}\vec{e}_\mu + A^{\mu}\Gamma^{\alpha}{}_{\mu\beta}\vec{e}_\alpha \\ &= (A^{\mu}{}_{,\beta} + A^{\alpha}\Gamma^{\mu}{}_{\alpha\beta})\vec{e}_\mu. \end{aligned} \tag{24.33}$$

In going from the first line to the second, we have used the notation

$$A^{\mu}{}_{,\beta} \equiv \partial_{\vec{e}_\beta} A^{\mu}; \tag{24.34}$$

that is, *the comma denotes the result of letting a basis vector act as a differential operator on the component of the vector*. In going from the second line of (24.33) to the third, we have renamed some summed-over indices. By comparing the first and last expressions in Eq. (24.33), we conclude that

components of the gradient of a vector field

$$\boxed{A^{\mu}{}_{;\beta} = A^{\mu}{}_{,\beta} + A^{\alpha}\Gamma^{\mu}{}_{\alpha\beta}.} \tag{24.35}$$

The first term in this equation describes the changes in $\vec{A}$ associated with changes of its component A^{μ}; the second term *corrects for* artificial changes of A^{μ} that are induced by turning and length changes of the basis vector $\vec{e}_\mu$. We shall use the short-hand terminology that the second term "corrects the index μ."

By a similar computation, we conclude that in any basis the covariant components of the gradient are

$$\boxed{A_{\alpha;\beta} = A_{\alpha,\beta} - \Gamma^{\mu}{}_{\alpha\beta} A_{\mu},} \tag{24.36}$$

where again $A_{\alpha,\beta} \equiv \partial_{\vec{e}_\beta} A_\alpha$. Notice that, when the index being corrected is down [α in Eq. (24.36)], the connection coefficient has a minus sign; when it is up [μ in Eq. (24.35)], the connection coefficient has a plus sign. This is in accord with the signs in Eqs. (24.30) and (24.31).

These considerations should make obvious the following equations for the components of the gradient of a second rank tensor field:

components of the gradient of a tensor field

$$\begin{aligned} F^{\alpha\beta}{}_{;\gamma} &= F^{\alpha\beta}{}_{,\gamma} + \Gamma^{\alpha}{}_{\mu\gamma} F^{\mu\beta} + \Gamma^{\beta}{}_{\mu\gamma} F^{\alpha\mu}, \\ F_{\alpha\beta;\gamma} &= F_{\alpha\beta,\gamma} - \Gamma^{\mu}{}_{\alpha\gamma} F_{\mu\beta} - \Gamma^{\mu}{}_{\beta\gamma} F_{\alpha\mu}, \\ F^{\alpha}{}_{\beta;\gamma} &= F^{\alpha}{}_{\beta,\gamma} + \Gamma^{\alpha}{}_{\mu\gamma} F^{\mu}{}_{\beta} - \Gamma^{\mu}{}_{\beta\gamma} F^{\alpha}{}_{\mu}. \end{aligned} \tag{24.37}$$

Notice that each index of $\boldsymbol{F}$ must be corrected, the correction has a sign dictated by whether the index is up or down, the differentiation index always goes last on the Γ, and all other indices can be deduced by requiring that the free indices in each term be the same and all other indices be summed.

If we have been given a basis, then how can we compute the connection coefficients? We can try to do so by drawing pictures and examining how the basis vectors change from point to point—a method that is fruitful in spherical and cylindrical coordinates in Euclidean space (Sec. 11.8). However, in other situations this method is fraught with peril, so we need a firm mathematical prescription. It turns out that the following prescription works (see Ex. 24.7 for a proof).

1. Evaluate the commutation coefficients $c_{\alpha\beta}{}^{\rho}$ of the basis, which are defined by the two equivalent relations:

commutation coefficients for a basis

$$\boxed{[\vec{e}_\alpha, \vec{e}_\beta] \equiv c_{\alpha\beta}{}^{\rho} \vec{e}_\rho, \quad c_{\alpha\beta}{}^{\rho} \equiv \vec{e}^{\rho} \cdot [\vec{e}_\alpha, \vec{e}_\beta].} \tag{24.38a}$$

   (Note that in a coordinate basis the commutation coefficients will vanish. Warning: Commutation coefficients also appear in the theory of Lie groups; there it is conventional to use a different ordering of indices than here: $c_{\alpha\beta}{}^{\rho}{}_{\text{here}} = c^{\rho}{}_{\alpha\beta\,\text{Lie groups}}$.)
2. Lower the last index on the commutation coefficients using the metric components in the basis:

$$\boxed{c_{\alpha\beta\gamma} \equiv c_{\alpha\beta}{}^{\rho} g_{\rho\gamma}.} \tag{24.38b}$$

3. Compute the quantities

formulas for computing connection coefficients

$$\boxed{\Gamma_{\alpha\beta\gamma} \equiv \frac{1}{2}(g_{\alpha\beta,\gamma} + g_{\alpha\gamma,\beta} - g_{\beta\gamma,\alpha} + c_{\alpha\beta\gamma} + c_{\alpha\gamma\beta} - c_{\beta\gamma\alpha}).} \tag{24.38c}$$

Here the commas denote differentiation with respect to the basis vectors as though the metric components were scalar fields [as in Eq. (24.34)]. Notice that the pattern of indices is the same on the gs and on the cs. It is a peculiar pattern—one of the few aspects of index gymnastics that cannot be reconstructed by merely lining up indices. In a coordinate basis the c terms will vanish, so $\Gamma_{\alpha\beta\gamma}$ will be symmetric in its last two indices. In an orthonormal basis $g_{\mu\nu}$ are constant, so the g terms will vanish, and $\Gamma_{\alpha\beta\gamma}$ will be antisymmetric in its first two indices. And in a Cartesian or Lorentz coordinate basis, which is both coordinate and orthonormal, both the c terms and the g terms will vanish, so $\Gamma_{\alpha\beta\gamma}$ will vanish.

4. Raise the first index on $\Gamma_{\alpha\beta\gamma}$ to obtain the connection coefficients

$$\boxed{\Gamma^{\mu}{}_{\beta\gamma} = g^{\mu\alpha}\Gamma_{\alpha\beta\gamma}.} \tag{24.38d}$$

In a coordinate basis, the $\Gamma^{\mu}{}_{\beta\gamma}$ are sometimes called *Christoffel symbols*, though we will use the name connection coefficients independent of the nature of the basis.

The first three steps in the above prescription for computing the connection coefficients follow from two key properties of the gradient $\vec{\nabla}$. First, the gradient of the metric tensor vanishes:

vanishing gradient of the metric tensor

$$\boxed{\vec{\nabla}\boldsymbol{g} = 0.} \tag{24.39}$$

Second, for any two vector fields $\vec{A}$ and $\vec{B}$, the gradient is related to the commutator by

relation of gradient to commutator

$$\boxed{\nabla_{\vec{A}}\vec{B} - \nabla_{\vec{B}}\vec{A} = [\vec{A}, \vec{B}].} \tag{24.40}$$

For a derivation of these relations and then a derivation of the prescription 1–4, see Exs. 24.6 and 24.7.

The gradient operator $\vec{\nabla}$ is an example of a geometric object that is not a tensor. The connection coefficients $\Gamma^{\mu}{}_{\beta\gamma} = \vec{e}^{\mu} \cdot \left(\nabla_{\vec{e}_\gamma}\vec{e}_\beta\right)$ can be regarded as the components of $\vec{\nabla}$; because it is not a tensor, these components do not obey the tensorial transformation law (24.19) when switching from one basis to another. Their transformation law is far more complicated and is rarely used. Normally one computes them from scratch in the new basis, using the above prescription or some other, equivalent prescription (cf. Misner, Thorne, and Wheeler, 1973, Chap. 14). For most curved spacetimes that one meets in general relativity, these computations are long and tedious and therefore are normally carried out on computers using symbolic manipulation software, such as Maple, Matlab, or Mathematica, or such programs as GR-Tensor and MathTensor that run under Maple or Mathematica. Such software is easily found on the Internet using a search engine. A particularly simple Mathematica program for use with coordinate bases is presented and discussed in Appendix C of Hartle (2003) and is available on that book's website: http://web.physics.ucsb.edu/~gravitybook/.

EXERCISES

Exercise 24.6 *Derivation: Properties of the Gradient* $\vec{\nabla}$

(a) Derive Eq. (24.39). [Hint: At a point $\mathcal{P}$ where $\vec{\nabla} \boldsymbol{g}$ is to be evaluated, introduce a locally orthonormal coordinate basis (i.e., locally Cartesian or locally Lorentz). When computing in this basis, the effects of curvature show up only to second order in distance from $\mathcal{P}$. Show that in this basis, the components of $\vec{\nabla} \boldsymbol{g}$ vanish, and from this infer that $\vec{\nabla} \boldsymbol{g}$, viewed as a frame-independent third-rank tensor, vanishes.]

(b) Derive Eq. (24.40). [Hint: Again work in a locally orthonormal coordinate basis.]

Exercise 24.7 *Derivation and Example: Prescription for Computing Connection Coefficients*

Derive the prescription 1–4 [Eqs. (24.38)] for computing the connection coefficients in any basis. [Hints: (i) In the chosen basis, from $\vec{\nabla} \boldsymbol{g} = 0$ infer that $\Gamma_{\alpha\beta\gamma} + \Gamma_{\beta\alpha\gamma} = g_{\alpha\beta,\gamma}$. Notice that this determines the part of $\Gamma_{\alpha\beta\gamma}$ that is symmetric in its first two indices. Show that the number of independent components of $\Gamma_{\alpha\beta\gamma}$ thereby determined is $\frac{1}{2}n^2(n+1)$, where n is the manifold's dimension. (ii) From Eq. (24.40) infer that $\Gamma_{\gamma\beta\alpha} - \Gamma_{\gamma\alpha\beta} = c_{\alpha\beta\gamma}$, which fixes the part of Γ antisymmetric in the last two indices. Show that the number of independent components thereby determined is $\frac{1}{2}n^2(n-1)$. (iii) Infer that the number of independent components determined by (i) and (ii) together is n^3, which is the entirety of $\Gamma_{\alpha\beta\gamma}$. By somewhat complicated algebra, deduce Eq. (24.38c) for $\Gamma_{\alpha\beta\gamma}$. (The algebra is sketched in Misner, Thorne, and Wheeler, 1973, Ex. 8.15.) (iv) Then infer the final answer, Eq. (24.38d), for $\Gamma^{\mu}{}_{\beta\gamma}$.]

Exercise 24.8 *Practice: Commutation and Connection Coefficients for Circular Polar Bases*

Consider the circular polar coordinates $\{\varpi, \phi\}$ of Fig. 24.3 and their associated bases.

(a) Evaluate the commutation coefficients $c_{\alpha\beta}{}^{\rho}$ for the coordinate basis $\{\vec{e}_{\varpi}, \vec{e}_{\phi}\}$, and also for the orthonormal basis $\{\vec{e}_{\hat{\varpi}}, \vec{e}_{\hat{\phi}}\}$.

(b) Compute by hand the connection coefficients for the coordinate basis and also for the orthonormal basis, using Eqs. (24.38). [Note: The answer for the orthonormal basis was worked out pictorially in our study of elasticity theory; Fig. 11.15 and Eq. (11.70).]

(c) Repeat this computation using symbolic manipulation software on a computer.

Exercise 24.9 *Practice: Connection Coefficients for Spherical Polar Coordinates*

(a) Consider spherical polar coordinates in 3-dimensional space, and verify that the nonzero connection coefficients, assuming an orthonormal basis, are given by Eq. (11.71).

(b) Repeat the exercise in part (a) assuming a coordinate basis with

$$\mathbf{e}_r \equiv \frac{\partial}{\partial r}, \quad \mathbf{e}_\theta \equiv \frac{\partial}{\partial \theta}, \quad \mathbf{e}_\phi \equiv \frac{\partial}{\partial \phi}. \tag{24.41}$$

(c) Repeat both computations in parts (a) and (b) using symbolic manipulation software on a computer.

Exercise 24.10 *Practice: Index Gymnastics—Geometric Optics*
This exercise gives the reader practice in formal manipulations that involve the gradient operator. In the geometric-optics (eikonal) approximation of Sec. 7.3, for electromagnetic waves in Lorenz gauge, one can write the 4-vector potential in the form $\vec{A} = \vec{\mathcal{A}} e^{i\varphi}$, where $\vec{\mathcal{A}}$ is a slowly varying amplitude and φ is a rapidly varying phase. By the techniques of Sec. 7.3, one can deduce from the vacuum Maxwell equations that the wave vector, defined by $\vec{k} \equiv \vec{\nabla}\varphi$, is null: $\vec{k} \cdot \vec{k} = 0$.

(a) Rewrite all the equations in the above paragraph in slot-naming index notation.

(b) Using index manipulations, show that the wave vector $\vec{k}$ (which is a vector field, because the wave's phase φ is a scalar field) satisfies the geodesic equation $\nabla_{\vec{k}}\vec{k} = 0$ (cf. Sec. 24.5.2). The geodesics, to which $\vec{k}$ is the tangent vector, are the rays discussed in Sec. 7.3, along which the waves propagate.

24.3.4

24.3.4 Integration

Our desire to use general bases and work in curved manifolds gives rise to two new issues in the definition of integrals.

The first issue is that the volume elements used in integration involve the Levi-Civita tensor [Eqs. (2.43), (2.52), and (2.55)], so we need to know the components of the Levi-Civita tensor in a general basis. It turns out (see, e.g., Misner, Thorne, and Wheeler, 1973, Ex. 8.3) that the covariant components differ from those in an orthonormal basis by a factor $\sqrt{|g|}$ and the contravariant by $1/\sqrt{|g|}$, where

$$\boxed{g \equiv \det ||g_{\alpha\beta}||} \tag{24.42}$$

is the determinant of the matrix whose entries are the covariant components of the metric. More specifically, let us denote by $[\alpha\beta \ldots \nu]$ the value of $\epsilon_{\alpha\beta\ldots\nu}$ in an orthonormal basis of our n-dimensional space [Eq. (2.43)]:

$$[12\ldots n] = +1,$$

$$[\alpha\beta\ldots\nu] = \begin{cases} +1 & \text{if } \alpha, \beta, \ldots, \nu \text{ is an even permutation of } 1, 2, \ldots, n \\ -1 & \text{if } \alpha, \beta, \ldots, \nu \text{ is an odd permutation of } 1, 2, \ldots, n \\ 0 & \text{if } \alpha, \beta, \ldots, \nu \text{ are not all different.} \end{cases} \tag{24.43}$$

(In spacetime the indices must run from 0 to 3 rather than 1 to $n = 4$.) Then in a general right-handed basis the components of the Levi-Civita tensor are

$$\epsilon_{\alpha\beta\ldots\nu} = \sqrt{|g|}\,[\alpha\beta\ldots\nu], \quad \epsilon^{\alpha\beta\ldots\nu} = \pm\frac{1}{\sqrt{|g|}}\,[\alpha\beta\ldots\nu], \tag{24.44}$$

components of Levi-Civita tensor in an arbitrary basis

where the $\pm$ is plus in Euclidean space and minus in spacetime. In a left-handed basis the sign is reversed.

As an example of these formulas, consider a spherical polar coordinate system (r, θ, ϕ) in 3-dimensional Euclidean space, and use the three infinitesimal vectors $dx^j(\partial/\partial x^j)$ to construct the volume element $d\Sigma$ [cf. Eq. (1.26)]:

$$dV = \epsilon\left(dr\frac{\partial}{\partial r}, d\theta\frac{\partial}{\partial\theta}, d\phi\frac{\partial}{\partial\phi}\right) = \epsilon_{r\theta\phi}drd\theta d\phi = \sqrt{g}\,drd\theta d\phi = r^2\sin\theta drd\theta d\phi. \tag{24.45}$$

Here the second equality follows from linearity of ϵ and the formula for computing its components by inserting basis vectors into its slots; the third equality follows from our formula (24.44) for the components. The fourth equality entails the determinant of the metric coefficients, which in spherical coordinates are $g_{rr} = 1$, $g_{\theta\theta} = r^2$, and $g_{\phi\phi} = r^2\sin^2\theta$; all other g_{jk} vanish, so $g = r^4\sin^2\theta$. The resulting volume element $r^2\sin\theta drd\theta d\phi$ should be familiar and obvious.

The second new integration issue we must face is that such integrals as

$$\int_{\partial\mathcal{V}} T^{\alpha\beta}d\Sigma_\beta \tag{24.46}$$

[cf. Eqs. (2.55), (2.56)] involve constructing a vector $T^{\alpha\beta}d\Sigma_\beta$ in each infinitesimal region $d\Sigma_\beta$ of the surface of integration $\partial\mathcal{V}$ and then adding up the contributions from all the infinitesimal regions. A major difficulty arises because each contribution lives in a different tangent space. To add them together, we must first transport them all to the same tangent space at some single location in the manifold. How is that transport to be performed? The obvious answer is "by the same parallel transport technique that we used in defining the gradient." However, when defining the gradient, we only needed to perform the parallel transport over an infinitesimal distance, and now we must perform it over long distances. When the manifold is curved, long-distance parallel transport gives a result that depends on the route of the transport, and in general there is no way to identify any preferred route (see, e.g., Misner, Thorne, and Wheeler, 1973, Sec. 11.4).

As a result, *integrals such as Eq. (24.46) are ill-defined in a curved manifold. The only integrals that are well defined in a curved manifold are those such as* $\int_{\partial\mathcal{V}} S^\alpha d\Sigma_\alpha$, *whose infinitesimal contributions* $S^\alpha d\Sigma_\alpha$ *are scalars* (i.e., integrals whose value is a scalar). This fact will have profound consequences in curved spacetime for the laws of conservation of energy, momentum, and angular momentum (Secs. 25.7 and 25.9.4).

integrals in a curved manifold are well defined only if infinitesimal contributions are scalars

EXERCISES

Exercise 24.11 *Practice: Integration—Gauss's Theorem*

In 3-dimensional Euclidean space Maxwell's equation $\nabla \cdot \mathbf{E} = \rho_e/\epsilon_0$ can be combined with Gauss's theorem to show that the electric flux through the surface $\partial\mathcal{V}$ of a sphere is equal to the charge in the sphere's interior $\mathcal{V}$ divided by ϵ_0:

$$\int_{\partial\mathcal{V}} \mathbf{E}\cdot d\boldsymbol{\Sigma} = \int_{\mathcal{V}} (\rho_e/\epsilon_0)\, dV. \tag{24.47}$$

Introduce spherical polar coordinates so the sphere's surface is at some radius $r = R$. Consider a surface element on the sphere's surface with vectorial legs $d\phi\partial/\partial\phi$ and $d\theta\partial/\partial\theta$. Evaluate the components $d\Sigma_j$ of the surface integration element $d\boldsymbol{\Sigma} = \boldsymbol{\epsilon}(\ldots, d\theta\partial/\partial\theta, d\phi\partial/\partial\phi)$. (Here $\boldsymbol{\epsilon}$ is the Levi-Civita tensor.) Similarly, evaluate dV in terms of vectorial legs in the sphere's interior. Then use these results for $d\Sigma_j$ and dV to convert Eq. (24.47) into an explicit form in terms of integrals over r, θ, and ϕ. The final answer should be obvious, but the above steps in deriving it are informative.

24.4

24.4 The Stress-Energy Tensor Revisited

In Sec. 2.13.1, we defined the stress-energy tensor $\boldsymbol{T}$ of any matter or field as a symmetric, second-rank tensor that describes the flow of 4-momentum through spacetime. More specifically, the total 4-momentum $\vec{P}$ that flows through some small 3-volume $\vec{\Sigma}$ (defined in Sec. 2.12.1), going from the negative side of $\vec{\Sigma}$ to its positive side, is

stress-energy tensor

$$\boxed{\boldsymbol{T}(\_, \vec{\Sigma}) = (\text{total 4-momentum } \vec{P} \text{ that flows through } \vec{\Sigma}); \quad T^{\alpha\beta}\Sigma_\beta = P^\alpha} \tag{24.48}$$

[Eq. (2.66)]. Of course, this stress-energy tensor depends on the location $\mathcal{P}$ of the 3-volume in spacetime [i.e., it is a tensor field $\boldsymbol{T}(\mathcal{P})$].

From this geometric, frame-independent definition of the stress-energy tensor, we were able to read off the physical meaning of its components in any inertial reference frame [Eqs. (2.67)]: T^{00} is the total energy density, including rest mass-energy; $T^{j0} = T^{0j}$ is the j-component of momentum density, or equivalently, the j-component of energy flux; and T^{jk} are the components of the stress tensor, or equivalently, of the momentum flux.

In Sec. 2.13.2, we formulated the law of conservation of 4-momentum in a local form and a global form. The local form,

local form of 4-momentum conservation

$$\boxed{\vec{\nabla}\cdot\boldsymbol{T} = 0,} \tag{24.49}$$

says that, in any chosen Lorentz frame, the time derivative of the energy density plus the divergence of the energy flux vanishes, $\partial T^{00}/\partial t + \partial T^{0j}/\partial x^j = 0$, and similarly for the momentum, $\partial T^{j0}/\partial t + \partial T^{jk}/\partial x^k = 0$. The global form, $\int_{\partial\mathcal{V}} T^{\alpha\beta} d\Sigma_\beta = 0$ [Eq. (2.71)], says that all the 4-momentum that enters a closed 4-volume $\mathcal{V}$ in spacetime through its boundary $\partial\mathcal{V}$ in the past must ultimately exit through $\partial\mathcal{V}$ in the future (Fig. 2.11). Unfortunately, this global form requires transporting vectorial contributions $T^{\alpha\beta} d\Sigma_\beta$ to a common location and adding them, which cannot be done in a route-independent way in curved spacetime (see the end of Sec. 24.3.4). Therefore (as we shall discuss in greater detail in Secs. 25.7 and 25.9.4), the global conservation law becomes problematic in curved spacetime.

The stress-energy tensor and local 4-momentum conservation play major roles in our development of general relativity. Almost all of our examples will entail perfect fluids.

Recall [Eq. (2.74a)] that in the local rest frame of a perfect fluid, there is no energy flux or momentum density, $T^{j0} = T^{0j} = 0$, but there is a total energy density (including rest mass) ρ and an isotropic pressure P:

$$T^{00} = \rho, \quad T^{jk} = P\delta^{jk}. \tag{24.50}$$

From this special form of $T^{\alpha\beta}$ in the fluid's local rest frame, one can derive a geometric, frame-independent expression for the fluid's stress-energy tensor $\boldsymbol{T}$ in terms of its 4-velocity $\vec{u}$, the metric tensor $\boldsymbol{g}$, and the rest-frame energy density ρ and pressure P:

stress-energy tensor for a perfect fluid

$$\boxed{\boldsymbol{T} = (\rho + P)\vec{u} \otimes \vec{u} + P\boldsymbol{g}; \quad T^{\alpha\beta} = (\rho + P)u^\alpha u^\beta + P g^{\alpha\beta}} \tag{24.51}$$

[Eq. (2.74b)]; see Ex. 2.26. This expression for the stress-energy tensor of a perfect fluid is an example of a geometric, frame-independent description of physics.

The equations of relativistic fluid dynamics for a perfect fluid are obtained by inserting the stress-energy tensor (24.51) into the law of 4-momentum conservation $\vec{\nabla} \cdot \boldsymbol{T} = 0$, and augmenting with the law of rest-mass conservation. We explored this in brief in Ex. 2.26, and in much greater detail in Sec. 13.8. Applications that we have explored are the relativistic Bernoulli equation and ultrarelativistic jets (Sec. 13.8.2) and relativistic shocks (Ex. 17.9). In Sec. 13.8.3, we explored in detail the slightly subtle way in which a fluid's nonrelativistic energy density, energy flux, and stress tensor arise from the relativistic perfect-fluid stress-energy tensor (24.51).

These issues for a perfect fluid are so important that readers are encouraged to review them (except possibly the applications) in preparation for our foray into general relativity.

Four other examples of the stress-energy tensor are those for the electromagnetic field (Ex. 2.28), for a kinetic-theory swarm of relativistic particles (Secs. 3.4.2 and 3.5.3), for a point particle (Box 24.2), and for a relativistic fluid with viscosity and diffusive heat conduction (Ex. 24.13). However, we shall not do much with any of these during our study of general relativity, except viscosity and heat conduction in Sec. 28.5.

T2

BOX 24.2. STRESS-ENERGY TENSOR FOR A POINT PARTICLE T2

For a point particle that moves through spacetime along a world line $\mathcal{P}(\zeta)$ [where ζ is the affine parameter such that the particle's 4-momentum is $\vec{p} = d/d\zeta$, Eq. (2.14)], the stress-energy tensor vanishes everywhere except on the world line itself. Correspondingly, $\boldsymbol{T}$ must be expressed in terms of a Dirac delta function. The relevant delta function is a scalar function of two points in spacetime, $\delta(\mathcal{Q}, \mathcal{P})$, with the property that when one integrates over the point $\mathcal{P}$, using the 4-dimensional volume element $d\Sigma$ (which in any inertial frame just reduces to $d\Sigma = dt dx dy dz$), one obtains

$$\boxed{\int_{\mathcal{V}} f(\mathcal{P})\delta(\mathcal{Q}, \mathcal{P}) d\Sigma = f(\mathcal{Q}).} \tag{1}$$

Here $f(\mathcal{P})$ is an arbitrary scalar field, and the region $\mathcal{V}$ of 4-dimensional integration must include the point $\mathcal{Q}$. One can easily verify that in terms of Lorentz coordinates this delta function can be expressed as

$$\boxed{\delta(\mathcal{Q}, \mathcal{P}) = \delta(t_{\mathcal{Q}} - t_{\mathcal{P}})\delta(x_{\mathcal{Q}} - x_{\mathcal{P}})\delta(y_{\mathcal{Q}} - y_{\mathcal{P}})\delta(z_{\mathcal{Q}} - z_{\mathcal{P}}),} \tag{2}$$

where the deltas on the right-hand side are ordinary 1-dimensional Dirac delta functions. [Proof: Simply insert Eq. (2) into Eq. (1), replace $d\Sigma$ by $dt_{\mathcal{Q}} dx_{\mathcal{Q}} dy_{\mathcal{Q}} dz_{\mathcal{Q}}$, and perform the four integrations.]

The general definition (24.48) of the stress-energy tensor $\boldsymbol{T}$ implies that the integral of a point particle's stress-energy tensor over any 3-surface $\mathcal{S}$ that slices through the particle's world line just once, at an event $\mathcal{P}(\zeta_o)$, must be equal to the particle's 4-momentum at the intersection point:

$$\int_{\mathcal{S}} T^{\alpha\beta} d\Sigma_\beta = p^\alpha(\zeta_o). \tag{3}$$

It is a straightforward but sophisticated exercise (Ex. 24.12) to verify that the following frame-independent expression has this property:

$$\boxed{\boldsymbol{T}(\mathcal{Q}) = \int_{-\infty}^{+\infty} \vec{p}(\zeta) \otimes \vec{p}(\zeta)\, \delta[\mathcal{Q}, \mathcal{P}(\zeta)]\, d\zeta.} \tag{4}$$

Here the integral is along the world line $\mathcal{P}(\zeta)$ of the particle, and $\mathcal{Q}$ is the point at which $\boldsymbol{T}$ is being evaluated. Therefore, Eq. (4) is the point-particle stress-energy tensor.

EXERCISES T2

Exercise 24.12 *Derivation: Stress-Energy Tensor for a Point Particle* T2

Show that the point-particle stress-energy tensor (4) of Box 24.2 satisfies that box's Eq. (3), as claimed.

Exercise 24.13 *Example: Stress-Energy Tensor for a Viscous Fluid with Diffusive Heat Conduction*

This exercise serves two roles: It develops the relativistic stress-energy tensor for a viscous fluid with diffusive heat conduction, and in the process it allows the reader to gain practice in index gymnastics.

In our study of elasticity theory, we introduced the concept of the irreducible tensorial parts of a second-rank tensor in Euclidean space (Box 11.2). Consider a relativistic fluid flowing through spacetime with a 4-velocity $\vec{u}(\mathcal{P})$. The fluid's gradient $\vec{\nabla}\vec{u}$ ($u_{\alpha;\beta}$ in slot-naming index notation) is a second-rank tensor in spacetime. With the aid of the 4-velocity itself, we can break it down into irreducible tensorial parts as follows:

$$u_{\alpha;\beta} = -a_\alpha u_\beta + \frac{1}{3}\theta P_{\alpha\beta} + \sigma_{\alpha\beta} + \omega_{\alpha\beta}. \tag{24.52}$$

Here: (i)

$$P_{\alpha\beta} \equiv g_{\alpha\beta} + u_\alpha u_\beta \tag{24.53}$$

is a tensor that projects vectors into the 3-space orthogonal to $\vec{u}$ (it can also be regarded as that 3-space's metric; see Ex. 2.10); (ii) $\sigma_{\alpha\beta}$ is symmetric, trace-free, and orthogonal to the 4-velocity; and (iii) $\omega_{\alpha\beta}$ is antisymmetric and orthogonal to the 4-velocity.

(a) Show that the rate of change of $\vec{u}$ along itself, $\nabla_{\vec{u}}\vec{u}$ (i.e., the fluid 4-acceleration) is equal to the vector $\vec{a}$ that appears in the decomposition (24.52). Show, further, that $\vec{a} \cdot \vec{u} = 0$.

(b) Show that the divergence of the 4-velocity, $\boldsymbol{\nabla} \cdot \vec{u}$, is equal to the scalar field θ that appears in the decomposition (24.52). As we shall see in part (d), this is the fluid's rate of expansion.

(c) The quantities $\sigma_{\alpha\beta}$ and $\omega_{\alpha\beta}$ are the relativistic versions of a Newtonian fluid's shear and rotation tensors, which we introduced in Sec. 13.7.1. Derive equations for these tensors in terms of $u_{\alpha;\beta}$ and $P_{\mu\nu}$.

(d) Show that, as viewed in a Lorentz reference frame where the fluid is moving with speed small compared to the speed of light, to first order in the fluid's ordinary velocity $v^j = dx^j/dt$, the following statements are true: (i) $u^0 = 1$, $u^j = v^j$; (ii) θ is the nonrelativistic rate of expansion of the fluid, $\theta = \boldsymbol{\nabla} \cdot \mathbf{v} \equiv v^j{}_{,j}$ [Eq. (13.67a)]; (iii) σ_{jk} is the fluid's nonrelativistic shear [Eq. (13.67b)]; and (iv) ω_{jk} is the fluid's nonrelativistic rotation tensor [denoted r_{ij} in Eq. (13.67c)].

(e) At some event $\mathcal{P}$ where we want to know the influence of viscosity on the fluid's stress-energy tensor, introduce the fluid's local rest frame. Explain why, in that frame, the only contributions of viscosity to the components of the stress-energy tensor are $T^{jk}_{\rm visc} = -\zeta\theta g^{jk} - 2\mu\sigma^{jk}$, where ζ and μ are the coefficients of bulk and shear viscosity, respectively; the contributions to T^{00} and $T^{j0} = T^{0j}$ vanish. [Hint: See Eq. (13.73) and associated discussions.]

(f) From nonrelativistic fluid mechanics, infer that, in the fluid's rest frame at $\mathcal{P}$, the only contributions of diffusive heat conductivity to the stress-energy tensor are $T^{0j}_{\rm cond} = T^{j0}_{\rm cond} = -\kappa\,\partial T/\partial x^j$, where κ is the fluid's thermal conductivity and T is its temperature. [Hint: See Eq. (13.74) and associated discussion.] Actually, this expression is not fully correct. If the fluid is accelerating, there is a correction term: $\partial T/\partial x^j$ gets replaced by $\partial T/\partial x^j + a^j T$, where a^j is the acceleration. After reading Sec. 24.5 and especially Ex. 24.16, explain this correction.

(g) Using the results of parts (e) and (f), deduce the following geometric, frame-invariant form of the fluid's stress-energy tensor:

$$T_{\alpha\beta} = (\rho + P)u_\alpha u_\beta + P g_{\alpha\beta} - \zeta\theta g_{\alpha\beta} - 2\mu\sigma_{\alpha\beta} - 2\kappa u_{(\alpha} P_{\beta)}{}^{\mu}(T_{;\mu} + a_\mu T). \quad (24.54)$$

Here the subscript parentheses in the last term mean to symmetrize in the α and β slots.

From the divergence of this stress-energy tensor, plus the first law of thermodynamics and the law of rest-mass conservation, one can derive the full theory of relativistic fluid mechanics for a fluid with viscosity and heat flow (see, e.g., Misner, Thorne, and Wheeler, 1973, Ex. 22.7). This particular formulation of the theory, including Eq. (24.54), is due to Carl Eckart (1940). Landau and Lifshitz (1959) have given a slightly different formulation. For discussion of the differences, and of causal difficulties with both formulations and the difficulties' repair, see, for example, the reviews by Israel and Stewart (1980), Andersson and Comer (2007, Sec. 14), and López-Monsalvo (2011, Sec. 4).

24.5

24.5 The Proper Reference Frame of an Accelerated Observer

Physics experiments and astronomical measurements almost always use an apparatus that accelerates and rotates. For example, if the apparatus is in an Earthbound laboratory and is attached to the laboratory floor and walls, then it accelerates upward (relative to freely falling particles) with the negative of the "acceleration of gravity," and it rotates (relative to inertial gyroscopes) because of the rotation of Earth. It is useful, in studying such an apparatus, to regard it as attached to an accelerating, rotating reference frame. As preparation for studying such reference frames in the presence of gravity, we study them in flat spacetime. For a somewhat more sophisticated treatment, see Misner, Thorne, and Wheeler (1973, pp. 163–176, 327–332).

Consider an observer with 4-velocity $\vec{U}$, who moves along an accelerated world line through flat spacetime (Fig. 24.6) so she has a nonzero 4-acceleration:

$$\vec{a} = \vec{\nabla}_{\vec{U}}\vec{U}. \quad (24.55)$$

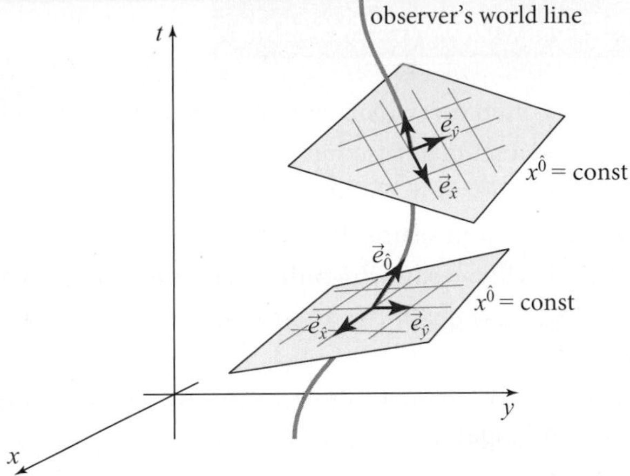

FIGURE 24.6 The proper reference frame of an accelerated observer. The spatial basis vectors $\vec{e}_{\hat{x}}$, $\vec{e}_{\hat{y}}$, and $\vec{e}_{\hat{z}}$ are orthogonal to the observer's world line and rotate, relative to local gyroscopes, as they move along the world line. The flat 3-planes spanned by these basis vectors are surfaces of constant coordinate time $x^{\hat{0}} \equiv$ (proper time as measured by the observer's clock at the event where the 3-plane intersects the observer's world line); in other words, they are the observer's slices of simultaneity and "3-space." In each of these flat 3-planes the spatial coordinates $\{\hat{x}, \hat{y}, \hat{z}\}$ are Cartesian, with $\partial/\partial\hat{x} = \vec{e}_{\hat{x}}$, $\partial/\partial\hat{y} = \vec{e}_{\hat{y}}$, and $\partial/\partial\hat{z} = \vec{e}_{\hat{z}}$.

proper reference frame of an accelerated observer

Have that observer construct, in the vicinity of her world line, a coordinate system $\{x^{\hat{\alpha}}\}$ (called her *proper reference frame*) with these properties: (i) The spatial origin is centered on her world line at all times (i.e., her world line is given by $x^{\hat{j}} = 0$). (ii) Along her world line, the time coordinate $x^{\hat{0}}$ is the same as the proper time ticked by an ideal clock that she carries. (iii) In the immediate vicinity of her world line, the spatial coordinates $x^{\hat{j}}$ measure physical distance along the axes of a little Cartesian latticework that she carries (and that she regards as purely spatial, which means it lies in the 3-plane orthogonal to her world line). These properties dictate that, in the immediate vicinity of her world line, the metric has the form $ds^2 = \eta_{\hat{\alpha}\hat{\beta}} dx^{\hat{\alpha}} dx^{\hat{\beta}}$, where $\eta_{\hat{\alpha}\hat{\beta}}$ are the Lorentz-basis metric coefficients, Eq. (24.6); in other words, all along her world line the coordinate basis vectors are orthonormal:

$$g_{\hat{\alpha}\hat{\beta}} = \frac{\partial}{\partial x^{\hat{\alpha}}} \cdot \frac{\partial}{\partial x^{\hat{\beta}}} = \eta_{\hat{\alpha}\hat{\beta}} \quad \text{at } x^{\hat{j}} = 0. \tag{24.56}$$

Moreover, properties (i) and (ii) dictate that along the observer's world line, the basis vector $\vec{e}_{\hat{0}} \equiv \partial/\partial x^{\hat{0}}$ differentiates with respect to her proper time, and thus is identically equal to her 4-velocity $\vec{U}$:

$$\vec{e}_{\hat{0}} = \frac{\partial}{\partial x^{\hat{0}}} = \vec{U}. \tag{24.57}$$

rotating and nonrotating proper reference frames

There remains freedom as to how the observer's latticework is oriented spatially. The observer can lock it to the gyroscopes of an *inertial-guidance system* that she carries (Box 24.3), in which case we say that it is "nonrotating"; or she can rotate it relative to such gyroscopes. For generality, we assume that the latticework rotates.

BOX 24.3. INERTIAL GUIDANCE SYSTEMS

Aircraft and rockets often carry inertial guidance systems, which consist of an accelerometer and a set of gyroscopes.

The accelerometer measures the system's 4-acceleration $\vec{a}$ (in relativistic language). Equivalently, it measures the system's Newtonian 3-acceleration **a** relative to inertial coordinates in which the system is momentarily at rest. As we see in Eq. (24.58), these quantities are two different ways of thinking about the same thing.

Each gyroscope is constrained to remain at rest in the aircraft or rocket by a force that is applied at its center of mass. Such a force exerts no torque around the center of mass, so the gyroscope maintains its direction (does not precess) relative to an inertial frame in which it is momentarily at rest.

As the accelerating aircraft or rocket turns, its walls rotate with some angular velocity $\vec{\Omega}$ relative to these inertial-guidance gyroscopes. This is the angular velocity discussed in the text between Eqs. (24.57) and (24.58).

From the time-evolving 4-acceleration $\vec{a}(\tau)$ and angular velocity $\vec{\Omega}(\tau)$, a computer can calculate the aircraft's (or rocket's) world line and its changing orientation.

Its angular velocity, as measured by the observer (by comparing the latticework's orientation with inertial-guidance gyroscopes), is a 3-dimensional spatial vector $\mathbf{\Omega}$ in the 3-plane orthogonal to her world line; and as viewed in 4-dimensional spacetime, it is a 4-vector $\vec{\Omega}$ whose components in the observer's reference frame are $\Omega^{\hat{j}} \neq 0$ and $\Omega^{\hat{0}} = 0$. Similarly, the latticework's acceleration, as measured by an inertial-guidance accelerometer attached to it (Box 24.3), is a 3-dimensional spatial vector **a** that can be thought of as a 4-vector with components in the observer's frame:

$$a^{\hat{0}} = 0, \quad a^{\hat{j}} = (\hat{j}\text{-component of the measured } \mathbf{a}). \tag{24.58}$$

This 4-vector is the observer's 4-acceleration, as one can verify by computing the 4-acceleration in an inertial frame in which the observer is momentarily at rest.

constructing coordinates of proper reference frame

Geometrically, the coordinates of the proper reference frame are constructed as follows. Begin with the basis vectors $\vec{e}_{\hat{\alpha}}$ along the observer's world line (Fig. 24.6)—basis vectors that satisfy Eqs. (24.56) and (24.57), and that rotate with angular velocity $\vec{\Omega}$ relative to gyroscopes. Through the observer's world line at time $x^{\hat{0}}$ construct the flat 3-plane spanned by the spatial basis vectors $\vec{e}_{\hat{j}}$. Because $\vec{e}_{\hat{j}} \cdot \vec{e}_{\hat{0}} = 0$, this 3-plane is orthogonal to the world line. All events in this 3-plane are given the same value of coordinate time $x^{\hat{0}}$ as the event where it intersects the world line; thus the 3-plane is a surface of constant coordinate time $x^{\hat{0}}$. The spatial coordinates in this flat 3-plane are ordinary, Cartesian coordinates $x^{\hat{j}}$ with $\vec{e}_{\hat{j}} = \partial/\partial x^{\hat{j}}$.

24.5.1 Relation to Inertial Coordinates; Metric in Proper Reference Frame; Transport Law for Rotating Vectors

24.5.1

It is instructive to examine the coordinate transformation between these proper-reference-frame coordinates $x^{\hat{\alpha}}$ and the coordinates x^{μ} of an inertial reference frame. We pick a very special inertial frame for this purpose. Choose an event on the observer's world line, near which the coordinate transformation is to be constructed; adjust the origin of the observer's proper time, so this event is $x^{\hat{0}} = 0$ (and of course $x^{\hat{j}} = 0$); and choose the inertial frame to be one that, arbitrarily near this event, coincides with the observer's proper reference frame. If we were doing Newtonian physics, then the coordinate transformation from the proper reference frame to the inertial frame would have the form (accurate through terms quadratic in $x^{\hat{\alpha}}$):

$$x^i = x^{\hat{i}} + \frac{1}{2}a^{\hat{i}}(x^{\hat{0}})^2 + \epsilon^{\hat{i}}{}_{\hat{j}\hat{k}}\Omega^{\hat{j}}x^{\hat{k}}x^{\hat{0}}, \quad x^0 = x^{\hat{0}}. \tag{24.59}$$

Here the term $\frac{1}{2}a^{\hat{i}}(x^{\hat{0}})^2$ is the standard expression for the vectorial displacement produced after time $x^{\hat{0}}$ by the acceleration $a^{\hat{i}}$; and the term $\epsilon^{\hat{i}}{}_{\hat{j}\hat{k}}\Omega^{\hat{j}}x^{\hat{k}}x^{\hat{0}}$ is the standard expression for the displacement produced by the rotation rate (rotational angular velocity) $\Omega^{\hat{j}}$ during a short time $x^{\hat{0}}$. In relativity theory there is only one departure from these familiar expressions (up through quadratic order): after time $x^{\hat{0}}$ the acceleration has produced a velocity $v^{\hat{j}} = a^{\hat{j}}x^{\hat{0}}$ of the proper reference frame relative to the inertial frame; correspondingly, there is a Lorentz-boost correction to the transformation of time: $x^0 = x^{\hat{0}} + v^{\hat{j}}x^{\hat{j}} = x^{\hat{0}}(1 + a_{\hat{j}}x^{\hat{j}})$ [cf. Eq. (2.37c)], accurate only to quadratic order. Thus, the full transformation to quadratic order is

inertial coordinates related to those of the proper reference frame of an accelerated, rotating observer

$$x^i = x^{\hat{i}} + \frac{1}{2}a^{\hat{i}}(x^{\hat{0}})^2 + \epsilon^{\hat{i}}{}_{\hat{j}\hat{k}}\Omega^{\hat{j}}x^{\hat{k}}x^{\hat{0}},$$
$$x^0 = x^{\hat{0}}(1 + a_{\hat{j}}x^{\hat{j}}). \tag{24.60a}$$

From this transformation and the form of the metric, $ds^2 = -(dx^0)^2 + \delta_{ij}dx^i dx^j$ in the inertial frame, we easily can evaluate the form of the metric, accurate to linear order in $\mathbf{x}$, in the proper reference frame:

metric in proper reference frame of an accelerated, rotating observer

$$\boxed{ds^2 = -(1 + 2\mathbf{a}\cdot\mathbf{x})(dx^{\hat{0}})^2 + 2(\boldsymbol{\Omega}\times\mathbf{x})\cdot d\mathbf{x}\, dx^{\hat{0}} + \delta_{jk}dx^{\hat{j}}dx^{\hat{k}}} \tag{24.60b}$$

(Ex. 24.14a). Here the notation is that of 3-dimensional vector analysis, with $\mathbf{x}$ the 3-vector whose components are $x^{\hat{j}}$, $d\mathbf{x}$ that with components $dx^{\hat{j}}$, $\mathbf{a}$ that with components $a^{\hat{j}}$, and $\boldsymbol{\Omega}$ that with components $\Omega^{\hat{j}}$.

Because the transformation (24.60a) was constructed near an arbitrary event on the observer's world line, the metric (24.60b) is valid near any and every event on the world line (i.e., it is valid all along the world line). In fact, it is the leading order in an expansion in powers of the spatial separation $x^{\hat{j}}$ from the world line. For higher-order terms in this expansion see, for example, Ni and Zimmermann (1978).

Notice that precisely on the observer's world line, the metric coefficients $g_{\hat{\alpha}\hat{\beta}}$ [the coefficients of $dx^{\hat{\alpha}}dx^{\hat{\beta}}$ in Eq. (24.60b)] are $g_{\hat{\alpha}\hat{\beta}} = \eta_{\hat{\alpha}\hat{\beta}}$, in accord with Eq. (24.56). However, as one moves farther away from the observer's world line, the effects of the acceleration $a^{\hat{j}}$ and rotation $\Omega^{\hat{j}}$ cause the metric coefficients to deviate more and more strongly from $\eta_{\hat{\alpha}\hat{\beta}}$.

From the metric coefficients of Eq. (24.60b), one can compute the connection coefficients $\Gamma^{\hat{\alpha}}{}_{\hat{\beta}\hat{\gamma}}$ on the observer's world line, and from these connection coefficients, one can infer the rates of change of the basis vectors along the world line: $\nabla_{\vec{U}}\vec{e}_{\hat{\alpha}} = \nabla_{\hat{0}}\vec{e}_{\hat{\alpha}} = \Gamma^{\hat{\mu}}{}_{\hat{\alpha}\hat{0}}\vec{e}_{\hat{\mu}}$. The result is (Ex. 24.14b):

equations for transport of proper reference frame's basis vectors along observer's world line

$$\nabla_{\vec{U}}\vec{e}_{\hat{0}} \equiv \nabla_{\vec{U}}\vec{U} = \vec{a}, \tag{24.61a}$$

$$\nabla_{\vec{U}}\vec{e}_{\hat{j}} = (\vec{a}\cdot\vec{e}_{\hat{j}})\vec{U} + \boldsymbol{\epsilon}(\vec{U}, \vec{\Omega}, \vec{e}_{\hat{j}}, \_). \tag{24.61b}$$

Equation (24.61b) is the general "law of transport" for constant-length vectors that are orthogonal to the observer's world line and that the observer thus sees as purely spatial. For the spin vector $\vec{S}$ of an inertial-guidance gyroscope (Box 24.3), the transport law is Eq. (24.61b) with $\vec{e}_{\hat{j}}$ replaced by $\vec{S}$ and with $\vec{\Omega} = 0$:

Fermi-Walker transport for the spin of an inertial-guidance gyroscope

$$\boxed{\nabla_{\vec{U}}\vec{S} = \vec{U}(\vec{a}\cdot\vec{S}).} \tag{24.62}$$

This is called *Fermi-Walker transport.* The term on the right-hand side of this transport law is required to keep the spin vector always orthogonal to the observer's 4-velocity: $\nabla_{\vec{U}}(\vec{S}\cdot\vec{U}) = 0$. For any other vector $\vec{A}$ that rotates relative to inertial-guidance gyroscopes, the transport law has, in addition to this "keep-it-orthogonal-to $\vec{U}$" term, a second term, which is the 4-vector form of $d\mathbf{A}/dt = \boldsymbol{\Omega}\times\mathbf{A}$:

transport law for a vector that is orthogonal to observer's 4-velocity and rotates relative to gyroscopes

$$\nabla_{\vec{U}}\vec{A} = \vec{U}(\vec{a}\cdot\vec{A}) + \boldsymbol{\epsilon}(\vec{U},\ \vec{\Omega},\ \vec{A},\ \_). \tag{24.63}$$

Equation (24.61b) is this general transport law with $\vec{A}$ replaced by $\vec{e}_{\hat{j}}$.

24.5.2 Geodesic Equation for a Freely Falling Particle

Consider a particle with 4-velocity $\vec{u}$ that moves freely through the neighborhood of an accelerated observer. As seen in an inertial reference frame, the particle travels through spacetime on a straight line, also called a *geodesic* of flat spacetime. Correspondingly, a geometric, frame-independent version of its *geodesic law of motion* is

geodesic law of motion for freely falling particle

$$\boxed{\nabla_{\vec{u}}\vec{u} = 0} \tag{24.64}$$

(i.e., the particle parallel transports its 4-velocity $\vec{u}$ along $\vec{u}$). It is instructive to examine the component form of this geodesic equation in the proper reference frame of the observer. Since the components of $\vec{u}$ in this frame are $u^{\alpha} = dx^{\alpha}/d\tau$, where τ is the particle's proper time (not the observer's proper time), the components $u^{\hat{\alpha}}{}_{;\hat{\mu}}u^{\hat{\mu}} = 0$ of the geodesic equation (24.64) are

$$u^{\hat{\alpha}}{}_{,\hat{\mu}}u^{\hat{\mu}} + \Gamma^{\hat{\alpha}}{}_{\hat{\mu}\hat{\nu}}u^{\hat{\mu}}u^{\hat{\nu}} = \left(\frac{\partial}{\partial x^{\hat{\mu}}}\frac{dx^{\hat{\alpha}}}{d\tau}\right)\frac{dx^{\hat{\mu}}}{d\tau} + \Gamma^{\hat{\alpha}}{}_{\hat{\mu}\hat{\nu}}u^{\hat{\mu}}u^{\hat{\nu}} = 0; \tag{24.65}$$

or equivalently,

$$\boxed{\frac{d^2x^{\hat{\alpha}}}{d\tau^2} + \Gamma^{\hat{\alpha}}{}_{\hat{\mu}\hat{\nu}}\frac{dx^{\hat{\mu}}}{d\tau}\frac{dx^{\hat{\nu}}}{d\tau} = 0.} \tag{24.66}$$

Suppose, for simplicity, that the particle is moving slowly relative to the observer, so its ordinary velocity $v^{\hat{j}} = dx^{\hat{j}}/dx^{\hat{0}}$ is nearly equal to $u^{\hat{j}} = dx^{\hat{j}}/d\tau$ and is small compared to unity (the speed of light), and $u^{\hat{0}} = dx^{\hat{0}}/d\tau$ is nearly unity. Then to first order in the ordinary velocity $v^{\hat{j}}$, the spatial part of the geodesic equation (24.66) becomes

$$\frac{d^2x^{\hat{i}}}{(dx^{\hat{0}})^2} = -\Gamma^{\hat{i}}{}_{\hat{0}\hat{0}} - (\Gamma^{\hat{i}}{}_{\hat{j}\hat{0}} + \Gamma^{\hat{i}}{}_{\hat{0}\hat{j}})v^{\hat{j}}. \tag{24.67}$$

By computing the connection coefficients from the metric coefficients of Eq. (24.60b) (Ex. 24.14), we bring this low-velocity geodesic law of motion into the form

geodesic equation for slowly moving particle in proper reference frame of accelerated, rotating observer

$$\frac{d^2x^{\hat{i}}}{(dx^{\hat{0}})^2} = -a^{\hat{i}} - 2\epsilon^{\hat{i}}{}_{\hat{j}\hat{k}}\Omega^{\hat{j}}v^{\hat{k}}, \quad \text{that is,} \quad \frac{d^2\mathbf{x}}{(dx^{\hat{0}})^2} = -\mathbf{a} - 2\boldsymbol{\Omega}\times\mathbf{v}. \tag{24.68}$$

This is the standard nonrelativistic form of the law of motion for a free particle as seen in a rotating, accelerating reference frame. The first term on the right-hand side is the inertial acceleration due to the failure of the frame to fall freely, and the second term is the Coriolis acceleration due to the frame's rotation. There would also be a centrifugal acceleration if we had kept terms of higher order in distance away from the observer's world line, but this acceleration has been lost due to our linearizing the metric (24.60b) in that distance.

This analysis shows how the elegant formalism of tensor analysis gives rise to familiar physics. In the next few chapters we will see it give rise to less familiar, general relativistic phenomena.

EXERCISES

Exercise 24.14 *Derivation: Proper Reference Frame*

(a) Show that the coordinate transformation (24.60a) brings the metric $ds^2 = \eta_{\alpha\beta}dx^\alpha dx^\beta$ into the form of Eq. (24.60b), accurate to linear order in separation $x^{\hat{j}}$ from the origin of coordinates.

(b) Compute the connection coefficients for the coordinate basis of Eq. (24.60b) at an arbitrary event on the observer's world line. Do so first by hand calculations, and then verify your results using symbolic-manipulation software on a computer.

(c) Using the connection coefficients from part (b), show that the rate of change of the basis vectors $\mathbf{e}_{\hat{\alpha}}$ along the observer's world line is given by Eq. (24.61).

(d) Using the connection coefficients from part (b), show that the low-velocity limit of the geodesic equation [Eq. (24.67)] is given by Eq. (24.68).

24.5.3

24.5.3 Uniformly Accelerated Observer

transformation between inertial coordinates and uniformly accelerated coordinates

As an important example (cf. Ex. 2.16), consider an observer whose accelerated world line, written in some inertial (Lorentz) coordinate system $\{t, x, y, z\}$, is

$$t = (1/\kappa)\sinh(\kappa\tau), \quad x = (1/\kappa)\cosh(\kappa\tau), \quad y = z = 0. \tag{24.69}$$

Here τ is proper time along the world line, and κ is the magnitude of the observer's 4-acceleration: $\kappa = |\vec{a}|$ (which is constant along the world line; see Ex. 24.15, where the reader can derive the various claims made in this subsection and the next).

The world line (24.69) is depicted in Fig. 24.7 as a thick, solid hyperbola that asymptotes to the past light cone at early times and to the future light cone at late times. The dots along the world line mark events that have proper times $\tau = -1.2, -0.9, -0.6, -0.3, 0.0, +0.3, +0.6, +0.9, +1.2$ (in units of $1/\kappa$). At each of these dots, the 3-plane orthogonal to the world line is represented by a dashed line (with the 2 dimensions out of the plane of the paper suppressed from the diagram). This 3-plane is labeled by its coordinate time $x^{\hat{0}}$, which is equal to the proper time of the dot. The basis vector $\vec{e}_{\hat{1}}$ is chosen to point along the observer's 4-acceleration, so $\vec{a} = \kappa\vec{e}_{\hat{1}}$. The coordinate $x^{\hat{1}}$ measures proper distance along the straight line that starts out tangent to $\vec{e}_{\hat{1}}$. The other two basis vectors $\vec{e}_{\hat{2}}$ and $\vec{e}_{\hat{3}}$ point out of the plane of the figure and are parallel transported along the world line: $\nabla_{\vec{U}}\vec{e}_{\hat{2}} = \nabla_{\vec{U}}\vec{e}_{\hat{3}} = 0$. In addition, $x^{\hat{2}}$ and $x^{\hat{3}}$ are measured along straight lines, in the orthogonal 3-plane, that start out tangent to these vectors. This construction implies that the resulting proper reference frame has vanishing rotation, $\vec{\Omega} = 0$ (Ex. 24.15), and that $x^{\hat{2}} = y$ and $x^{\hat{3}} = z$, where y and z are coordinates in the $\{t, x, y, z\}$ Lorentz frame that we used to define the world line [Eqs. (24.69)].

Usually, when constructing an observer's proper reference frame, one confines attention to the immediate vicinity of her world line. However, in this special case it is instructive to extend the construction (the orthogonal 3-planes and their resulting spacetime coordinates) outward arbitrarily far. By doing so, we discover that the 3-planes all cross at location $x^{\hat{1}} = -1/\kappa$, which means the coordinate system $\{x^{\hat{\alpha}}\}$ becomes singular there. This singularity shows up in a vanishing $g_{\hat{0}\hat{0}}(x^{\hat{1}} = -1/\kappa)$ for the spacetime metric, written in that coordinate system:

singularity of uniformly accelerated coordinates

spacetime metric in uniformly accelerated coordinates

$$ds^2 = -(1 + \kappa x^{\hat{1}})^2 (dx^{\hat{0}})^2 + (dx^{\hat{1}})^2 + (dx^{\hat{2}})^2 + (dx^{\hat{3}})^2. \tag{24.70}$$

[Note that for $|x^{\hat{1}}| \ll 1/\kappa$ this metric agrees with the general proper-reference-frame metric (24.60b).] From Fig. 24.7, it should be clear that this coordinate system can only cover smoothly one quadrant of Minkowski spacetime: the quadrant $x > |t|$.

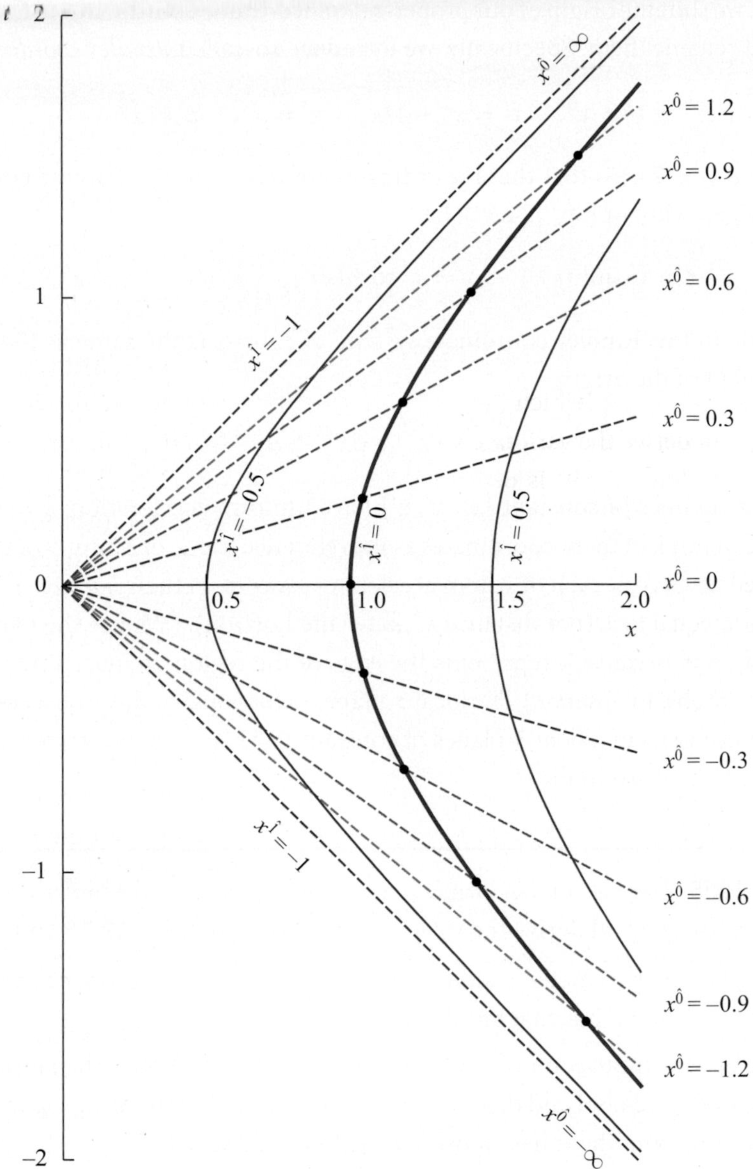

FIGURE 24.7 The proper reference frame of a uniformly accelerated observer. All lengths and times are measured in units of $1/\kappa$. We show only 2 dimensions of the reference frame—those in the 2-plane of the observer's curved world line.

24.5.4 Rindler Coordinates for Minkowski Spacetime

24.5.4

The spacetime metric (24.70) in our observer's proper reference frame resembles the metric in the vicinity of a black hole, as expressed in coordinates of observers who accelerate so as to avoid falling into the hole. In preparation for discussing this in

Rindler coordinates

Chap. 26, we shift the origin of our proper-reference-frame coordinates to the singular point and rename them. Specifically, we introduce so-called *Rindler coordinates*:

$$t' = x^{\hat{0}}, \quad x' = x^{\hat{1}} + 1/\kappa, \quad y' = x^{\hat{2}}, \quad z' = x^{\hat{3}}. \tag{24.71}$$

It turns out (Ex. 24.15) that these coordinates are related to the Lorentz coordinates that we began with, in Eqs. (24.69), by

$$t = x' \sinh(\kappa t'), \quad x = x' \cosh(\kappa t'), \quad y = y', \quad z = z'. \tag{24.72}$$

The metric in this Rindler coordinate system, of course, is the same as (24.70) with displacement of the origin:

spacetime metric in Rindler coordinates

$$ds^2 = -(\kappa x')^2 dt'^2 + dx'^2 + dy'^2 + dz'^2. \tag{24.73}$$

The world lines of constant $\{x', y', z'\}$ have uniform acceleration: $\vec{a} = (1/x')\vec{e}_{x'}$. Thus we can think of these coordinates as the reference frame of a family of uniformly accelerated observers, each of whom accelerates away from their *horizon* $x' = 0$ with acceleration equal to 1/(her distance x' above the horizon). (We use the name "horizon" for $x' = 0$, because it represents the edge of the region of spacetime that these observers are able to observe.) The local 3-planes orthogonal to these observers' world lines all mesh to form global 3-planes of constant t'. This is a major factor in making the metric (24.73) so simple.

horizon of Rindler coordinates

EXERCISES

Exercise 24.15 *Derivation: Uniformly Accelerated Observer and Rindler Coordinates*
In this exercise you will derive the various claims made in Secs. 24.5.3 and 24.5.4.

(a) Show that the parameter τ along the world line (24.69) is proper time and that the 4-acceleration has magnitude $|\vec{a}| = 1/\kappa$.

(b) Show that the unit vectors $\vec{e}_{\hat{j}}$ introduced in Sec. 24.5.3 all obey the Fermi-Walker transport law (24.62) and therefore, by virtue of Eq. (24.61b), the proper reference frame built from them has vanishing rotation rate: $\vec{\Omega} = 0$.

(c) Show that the coordinates $x^{\hat{2}}$ and $x^{\hat{3}}$ introduced in Sec. 24.5.3 are equal to the y and z coordinates of the inertial frame used to define the observer's world line [Eqs. (24.69)].

(d) Show that the proper-reference-frame coordinates constructed in Sec. 24.5.3 are related to the original $\{t, x, y, z\}$ coordinates by

$$t = (x^{\hat{1}} + 1/\kappa)\sinh(\kappa x^{\hat{0}}), \quad x = (x^{\hat{1}} + 1/\kappa)\cosh(\kappa x^{\hat{0}}), \quad y = x^{\hat{2}}, \quad z = x^{\hat{3}}; \tag{24.74}$$

and from this, deduce the form (24.70) of the Minkowski spacetime metric in the observer's proper reference frame.

(e) Show that, when converted to Rindler coordinates by moving the spatial origin, the coordinate transformation (24.74) becomes (24.72), and the metric (24.70) becomes (24.73).

(f) Show that observers at rest in the Rindler coordinate system (i.e., who move along world lines of constant $\{x', y', z'\}$) have 4-acceleration $\vec{a} = (1/x')\vec{e}_{x'}$.

Exercise 24.16 *Example: Gravitational Redshift*

Inside a laboratory on Earth's surface the effects of spacetime curvature are so small that current technology cannot measure them. Therefore, experiments performed in the laboratory can be analyzed using special relativity. (This fact is embodied in Einstein's equivalence principle; end of Sec. 25.2.)

(a) Explain why the spacetime metric in the proper reference frame of the laboratory's floor has the form

$$ds^2 = (1 + 2gz)(dx^{\hat{0}})^2 + dx^2 + dy^2 + dz^2, \tag{24.75}$$

plus terms due to the slow rotation of the laboratory walls, which we neglect in this exercise. Here g is the acceleration of gravity measured on the floor.

(b) An electromagnetic wave is emitted from the floor, where it is measured to have wavelength λ_o, and is received at the ceiling. Using the metric (24.75), show that, as measured in the proper reference frame of an observer on the ceiling, the received wave has wavelength $\lambda_r = \lambda_o(1 + gh)$, where h is the height of the ceiling above the floor (i.e., the light is *gravitationally redshifted* by $\Delta\lambda/\lambda_o = gh$). [Hint: Show that all crests of the wave must travel along world lines that have the same shape, $z = F(x^{\hat{0}} - x_e^{\hat{0}})$, where F is some function, and $x_e^{\hat{0}}$ is the coordinate time at which the crest is emitted from the floor. You can compute the shape function F if you wish, but it is not needed to derive the gravitational redshift; only its universality is needed.]

The first high-precision experiments to test this prediction were by Robert Pound and his student Glen Rebka and postdoc Joseph Snider, in a tower at Harvard University in the 1950s and 1960s. They achieved 1% accuracy. We discuss this gravitational redshift in Sec. 27.2.1.

Exercise 24.17 *Example: Rigidly Rotating Disk*

Consider a thin disk with radius R at $z = 0$ in a Lorentz reference frame. The disk rotates rigidly with angular velocity Ω. In the early years of special relativity there was much confusion over the geometry of the disk: In the inertial frame it has physical radius (proper distance from center to edge) R and physical circumference $C = 2\pi R$. But Lorentz contraction dictates that, as measured on the disk, the circumference should be $\sqrt{1 - v^2}\, C$ (with $v = \Omega R$), and the physical radius, R, should be unchanged. This seemed weird. How could an obviously flat disk in flat spacetime have a curved, non-Euclidean geometry, with physical circumference divided by physical radius smaller than 2π? In this exercise you will explore this issue.

(a) Consider a family of observers who ride on the edge of the disk. Construct a circular curve, orthogonal to their world lines, that travels around the disk (at $\sqrt{x^2+y^2}=R$). This curve can be thought of as lying in a 3-surface of constant time $x^{\hat{0}}$ of the observers' proper reference frames. Show that it spirals upward in a Lorentz-frame spacetime diagram, so it cannot close on itself after traveling around the disk. Thus the 3-planes, orthogonal to the observers' world lines at the edge of the disk, cannot mesh globally to form global 3-planes (by contrast with the case of the uniformly accelerated observers in Sec. 24.5.4 and Ex. 24.15).

(b) Next, consider a 2-dimensional family of observers who ride on the surface of the rotating disk. Show that at each radius $\sqrt{x^2+y^2}=\text{const}$, the constant-radius curve that is orthogonal to their world lines spirals upward in spacetime with a different slope. Show this means that even locally, the 3-planes orthogonal to each of their world lines cannot mesh to form larger 3-planes—thus there does not reside in spacetime any 3-surface orthogonal to these observers' world lines. There is no 3-surface that has the claimed non-Euclidean geometry.

Bibliographic Note

For a very readable presentation of most of this chapter's material, from much the same point of view, see Hartle (2003, Chap. 20). For an equally elementary introduction from a somewhat different viewpoint, see Schutz (2009, Chaps. 1–4). A far more detailed and somewhat more sophisticated introduction, largely but not entirely from our viewpoint, will be found in Misner, Thorne, and Wheeler (1973, Chaps. 1–6). More sophisticated treatments from rather different viewpoints than ours are given in Wald (1984, Chaps. 1, 2, and Sec. 3.1), and Carroll (2004, Chaps. 1, 2). A treasure trove of exercises on this material, with solutions, is in Lightman et al. (1975, Chaps. 6–8). See also the bibliography for Chap. 2.

For a detailed and sophisticated discussion of accelerated observers and the measurements they make, see Gourgoulhon (2013).

Fundamental Concepts of General Relativity

The physical world is represented as a four dimensional continuum. If in this I adopt a Riemannian metric, and look for the simplest laws which such a metric can satisfy, I arrive at the relativistic gravitation theory of empty space.

ALBERT EINSTEIN (1934)

25.1 History and Overview

25.1

Newton's theory of gravity is logically incompatible with the special theory of relativity. Newtonian gravity presumes the existence of a universal, frame-independent 3-dimensional space in which lives the Newtonian potential Φ and a universal, frame-independent time t with respect to which the propagation of Φ is instantaneous. By contrast, special relativity insists that the concepts of time and of 3-dimensional space are frame dependent, so that instantaneous propagation of Φ in one frame would mean noninstantaneous propagation in another.

The most straightforward way to remedy this incompatibility is to retain the assumption that gravity is described by a scalar field Φ but modify Newton's instantaneous, action-at-a-distance field equation

$$\left(\frac{\partial^2}{\partial x^2} + \frac{\partial^2}{\partial y^2} + \frac{\partial^2}{\partial z^2}\right)\Phi = 4\pi G\rho \tag{25.1}$$

(where G is Newton's gravitation constant and ρ is the mass density) to read

$$\Box\Phi \equiv g^{\alpha\beta}\Phi_{;\alpha\beta} = -4\pi G T^{\mu}{}_{\mu}, \tag{25.2a}$$

where $\Box \equiv \vec{\nabla}\cdot\vec{\nabla}$ is the squared gradient (i.e., d'alembertian or wave operator) in Minkowski spacetime, and $T^{\mu}{}_{\mu}$ is the trace (i.e., contraction on its two slots) of the stress-energy tensor. This modified field equation at first sight is attractive and satisfactory (but see Ex. 25.1): (i) it satisfies Einstein's Principle of Relativity in that it is expressed as a geometric, frame-independent relationship among geometric objects; and (ii) in any Lorentz frame it takes the form [with factors of $c =$ (speed of light) restored]:

$$\left(-\frac{1}{c^2}\frac{\partial^2}{\partial t^2} + \frac{\partial^2}{\partial x^2} + \frac{\partial^2}{\partial y^2} + \frac{\partial^2}{\partial z^2}\right)\Phi = \frac{4\pi G}{c^2}(T^{00} - T^{xx} - T^{yy} - T^{zz}), \tag{25.2b}$$

which reduces to the Newtonian field equation (25.1) in the kinds of situation contemplated by Newton [energy density predominantly due to rest-mass density, $T^{00} \cong \rho c^2$;

BOX 25.1. READERS' GUIDE

- This chapter relies significantly on:
  - Chap. 2 on special relativity; and
  - Chap. 24 on the transition from special relativity to general relativity.
- This chapter is a foundation for the applications of general relativity theory in Chaps. 26–28.

stress negligible compared to rest mass-energy density, $|T^{jk}| \ll \rho c^2$; and $1/c \times$ (time rate of change of Φ) negligible compared to spatial gradient of Φ].

Not surprisingly, most theoretical physicists in the decade following Einstein's formulation of special relativity (1905–1915) presumed that gravity would be correctly describable, in the framework of special relativity, by this type of modification of Newton's theory, or something resembling it. For a brief historical account, see Pais (1982, Chap. 13). To Einstein, by contrast, it seemed clear that the correct description of gravity should involve a generalization of special relativity rather than an incorporation into special relativity: since an observer in a local, freely falling reference frame near Earth should not feel any gravitational acceleration at all, local freely falling frames (local inertial frames) should in some sense be the domain of special relativity, and gravity should somehow be described by the relative acceleration of such frames.

key idea of general relativity: gravity described by relative acceleration of local inertial frames

Although the seeds of this idea were in Einstein's mind as early as 1907 [see the discussion of the equivalence principle in Einstein (1907)], it required 8 years for him to bring them to fruition. A first crucial step, which took half the 8 years, was for Einstein to conquer his initial aversion to Minkowski's geometric formulation of special relativity and to realize that a curvature of Minkowski's 4-dimensional spacetime is the key to understanding the relative acceleration of freely falling frames. The second crucial step was to master the mathematics of differential geometry, which describes spacetime curvature, and using that mathematics, to formulate a logically self-consistent theory of gravity. This second step took an additional 4 years and culminated in Einstein's (1915, 1916a) general theory of relativity. For a historical account of Einstein's 8-year struggle toward general relativity, see, for example, Pais (1982, Part IV). For selected quotations from Einstein's technical papers during this 8-year period, which tell the story of his struggle, see Misner, Thorne, and Wheeler (1973, Sec. 17.7). For his papers themselves with scholarly annotations, see Einstein (1989, vols. 2–4, 6).

the mathematics of general relativity: differential geometry in curved spacetime

It is remarkable that Einstein was led, not by experiment, but by philosophical and aesthetic arguments, to reject the incorporation of gravity into special relativity [Eqs. (25.2) and Ex. 25.1], and to insist instead on describing gravity by curved spacetime. Only after the full formulation of his general relativity did experiments begin to confirm that he was right and that the advocates of special-relativistic gravity were wrong, and only a half century after general relativity was formulated did the experimental evidence become extensive and strong. For detailed discussions see, for example, Will (1993a,b, 2014).

The mathematical tools, the diagrams, and the phrases by which we describe general relativity have changed somewhat in the century since Einstein formulated his theory. Indeed, we can even assert that we understand the theory more deeply than did Einstein. However, the basic ideas are unchanged, and general relativity's claim to be the most elegant and aesthetic of physical theories has been reinforced and strengthened by our growing insights.

general relativity as a framework for all the laws of physics

General relativity is not merely a theory of gravity. Like special relativity before it, the general theory is a framework in which to formulate all the laws of physics, classical and quantum—but now with gravity included. However, there is one remaining, crucial, gaping hole in this framework. It is incapable of functioning—indeed, it fails completely—when conditions become so extreme that space and time themselves must be quantized. In those extreme conditions general relativity must be married in some deep, as-yet-ill-understood way, with quantum theory, to produce an all-inclusive quantum theory of gravity—a theory that, one may hope, will be a "theory of everything." To this we shall return, briefly, in Chaps. 26 and 28.

In this chapter, we present, in modern language, the foundations of general relativity. Our presentation is proudly geometrical, as this seems to us the most powerful approach to general relativity for most situations. By contrast, some outstanding physicists, particularly Weinberg (1972, especially his preface), prefer a field-theoretic approach. Our presentation relies heavily on the geometric concepts, viewpoint, and formalism developed in Chaps. 2 and 24.

We begin in Sec. 25.2 with a discussion of three concepts that are crucial to Einstein's viewpoint on gravity: a local Lorentz frame (the closest thing there is, in the presence of gravity, to special relativity's "global" Lorentz frame), the extension of the Principle of Relativity to deal with gravitational situations, and Einstein's equivalence principle by which one can "lift" laws of physics out of the flat spacetime of special relativity and into the curved spacetime of general relativity. In Sec. 25.3, we see how gravity prevents the meshing of local Lorentz frames to form global Lorentz frames and infer from this that spacetime must be curved. In Sec. 25.4, we lift into curved spacetime the law of motion for free test particles, and in Sec. 25.5, we see how spacetime curvature pushes two freely moving test particles together or apart, and we use this phenomenon to make contact between spacetime curvature and the Newtonian "tidal gravitational field" (gradient of the Newtonian gravitational acceleration). In Sec. 25.6, we study some mathematical and geometric properties of the tensor field that embodies spacetime curvature: the Riemann tensor. In Sec. 25.7, we examine "curvature coupling delicacies" that plague the lifting of laws of physics from flat spacetime to curved spacetime. In Sec. 25.8, we meet the Einstein field equation, which describes the manner in which spacetime curvature is produced by the total stress-energy tensor of all matter and nongravitational fields. In Sec. 25.9.1, we examine in some detail how Newton's laws of gravity arise as a weak-gravity, slow-motion, low-stress limit of general relativity. In Sec. 25.9.2, we develop an approximation to general relativity called "linearized theory" that is valid when gravity is weak but speeds and stresses may be high, and in Sec. 25.9.3, we use this approximation to deduce the weak relativistic gravitational field outside a stationary (unchanging) source. Finally, in Secs. 25.9.4 and 25.9.5, we examine the conservation laws for energy, momentum, and angular momentum of gravitating bodies that live in "asymptotically flat" regions of spacetime.

EXERCISES

Exercise 25.1 *Example: A Special Relativistic, Scalar-Field Theory of Gravity*

Equation (25.2a) is the field equation for a special relativistic theory of gravity with gravitational potential Φ. To complete the theory, one must describe the forces that the field Φ produces on matter.

(a) One conceivable choice for the force on a test particle of rest mass m is the following generalization of the familiar Newtonian expression:

$$\nabla_{\vec{u}}\vec{p} = -m\vec{\nabla}\Phi; \quad \text{that is,} \quad \frac{dp_\alpha}{d\tau} = -m\Phi_{,\alpha} \quad \text{in a Lorentz frame,} \tag{25.3}$$

where τ is proper time along the particle's world line, $\vec{p}$ is the particle's 4-momentum, $\vec{u}$ is its 4-velocity, and $\vec{\nabla}\Phi$ is the spacetime gradient of the gravitational potential. Show that this equation of motion reduces, in a Lorentz frame and for low particle velocities, to the standard Newtonian equation of motion. Show, however, that this equation of motion is flawed in that the gravitational field will alter the particle's rest mass—in violation of extensive experimental evidence that the rest mass of an elementary particle is unique and conserved.

(b) Show that the equation of motion (25.3), when modified to read:

$$\nabla_{\vec{u}}\vec{p} = -(\boldsymbol{g} + \vec{u}\otimes\vec{u})\cdot m\vec{\nabla}\Phi;$$

$$\text{that is,} \quad \frac{dp^\alpha}{d\tau} = -(g^{\alpha\beta} + u^\alpha u^\beta)m\Phi_{,\beta} \quad \text{in a Lorentz frame,} \tag{25.4}$$

preserves the particle's rest mass. In this equation of motion $\vec{u}$ is the particle's 4-velocity, $\boldsymbol{g}$ is the metric, and $\boldsymbol{g} + \vec{u}\otimes\vec{u}$ projects $\vec{\nabla}\Phi$ into the 3-space orthogonal to the particle's world line (cf. Fig. 24.6 and Ex. 2.10).

(c) Show, by treating a zero-rest-mass particle as the limit of a particle of finite rest mass ($\vec{p} = m\vec{u}$ and $\zeta = \tau/m$ finite as τ and m go to zero), that the theory sketched in parts (a) and (b) predicts that in any Lorentz reference frame, $p^\alpha e^\Phi$ (with $\alpha = 0, 1, 2, 3$) are constant along the zero-rest-mass particle's world line. Explain why this prediction implies that there will be no gravitational deflection of light by the Sun, which conflicts severely with experiments that were done after Einstein formulated his general theory of relativity (see Sec. 27.2.3). (There was no way, experimentally, to rule out this theory in the epoch, ca. 1914, when Einstein was doing battle with his colleagues over whether gravity should be treated by adding a gravitational force to special relativity or should be treated as a geometric extension of special relativity.)

25.2 Local Lorentz Frames, the Principle of Relativity, and Einstein's Equivalence Principle

25.2

One of Einstein's greatest insights was to recognize that special relativity is valid not globally, but only locally, inside local, freely falling (inertial) reference frames. Figure 25.1 shows an example of a *local inertial frame:* the interior of a Space Shuttle in Earth orbit, where an astronaut has set up a freely falling (from his viewpoint "freely floating") latticework of rods and clocks. This latticework is constructed by all the rules appropriate to a special relativistic, inertial (Lorentz) reference frame (Secs. 2.2.1 and 24.2.2): (i) the latticework moves freely through spacetime, so no forces act on it, and its rods are attached to gyroscopes so they do not rotate; (ii) the measuring rods are orthogonal to one another, with their intervals of length uniform compared, for example, to the wavelength of light (orthonormal lattice); (iii) the clocks are densely packed in the lattice, they tick uniformly relative to ideal atomic standards (they are ideal clocks), and they are synchronized by the Einstein light-pulse process. However, there is one crucial change from special relativity: The latticework must be *small enough* that one can neglect the effects of inhomogeneities of gravity (which general relativity will associate with spacetime curvature; and which, e.g., would cause two freely floating particles, one nearer Earth than the other, to gradually move apart, even though initially they are at rest with respect to each other). The necessity for smallness is embodied in the word "local" of "local inertial frame," and we shall quantify it with ever greater precision as we move through this chapter.

local inertial (Lorentz) frame

We use the phrases *local Lorentz frame* and *local inertial frame* interchangeably to describe the above type of synchronized, orthonormal latticework. The spacetime coordinates $\{t, x, y, z\}$ that the latticework provides (in the manner of Sec. 2.2.1) we call, interchangeably, *local Lorentz coordinates* and *local inertial coordinates.*

local inertial (Lorentz) coordinates

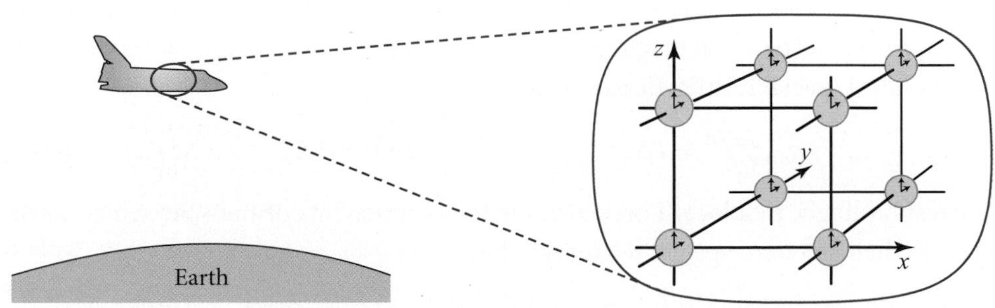

FIGURE 25.1 A local inertial frame (local Lorentz frame) inside a Space Shuttle that is orbiting Earth.

Since in the presence of gravity, inertial reference frames must be restricted to be local, the inertial-frame version of the Principle of Relativity (Sec. 2.2.2) must similarly be restricted: *all the* local, *nongravitational laws of physics are the same in every* local *inertial frame, everywhere and everywhen in the universe*. Here, by "local" laws we mean those laws, classical or quantum, that can be expressed entirely in terms of quantities confined to (measurable in) a local inertial frame. The exclusion of gravitational laws from this version of the Principle of Relativity is necessary, because gravity is to be described by a curvature of spacetime, which (by definition; see below) cannot show up in a local inertial frame. This version of the Principle of Relativity can be described in operational terms using the same language as for the special relativistic version (Secs. 2.2.2 and 24.2.2): If two different observers, in two different local Lorentz frames, in different (or the same) regions of the universe, are given identical written instructions for a physics experiment that can be performed within the confines of their local Lorentz frames, then their two experiments must yield the same results to within their experimental accuracies.

Principle of Relativity in presence of gravity

It is worth emphasizing that the Principle of Relativity is asserted to hold everywhere and everywhen in the universe: the local laws of physics must have the same form in the early universe, a fraction of a second after the big bang, as they have on Earth today, and as they have at the center of the Sun or inside a black hole.

Einstein's equivalence principle

It is reasonable to expect that *the specific forms that the local, nongravitational laws of physics take in general relativistic local Lorentz frames are the same as they take in the (global) Lorentz frames of special relativity*. This assertion is a modern version of Einstein's equivalence principle. (Einstein's original version states that local physical measurements in a uniformly accelerated reference frame cannot be distinguished from those in a uniform gravitational field. How is this related to the modern version? See Ex. 26.11.) In the next section, we use this principle to deduce some properties of the general relativistic spacetime metric. In Sec. 25.7 we use it to deduce the forms of some nongravitational laws of physics in curved spacetime, and we discover delicacies (ambiguities) in this principle of equivalence triggered by spacetime curvature.

25.3

25.3 The Spacetime Metric, and Gravity as a Curvature of Spacetime

The Einstein equivalence principle guarantees that nongravitational physics in a local Lorentz frame can be described using a spacetime metric $\boldsymbol{g}$, which gives for the invariant interval between neighboring events with separation vector $\vec{\xi} = \Delta x^\alpha \partial/\partial x^\alpha$, the standard special relativistic expression

$$\vec{\xi}^2 = g_{\alpha\beta}\xi^\alpha\xi^\beta = (\Delta s)^2 = -(\Delta t)^2 + (\Delta x)^2 + (\Delta y)^2 + (\Delta z)^2. \tag{25.5}$$

Correspondingly, in a local Lorentz frame the components of the spacetime metric take on their standard special relativity values:

$$g_{\alpha\beta} = \eta_{\alpha\beta} \equiv \{-1 \text{ if } \alpha=\beta=0, \quad +1 \text{ if } \alpha=\beta=(x,\ y,\ \text{or } z), \quad 0 \text{ otherwise}\}. \tag{25.6}$$

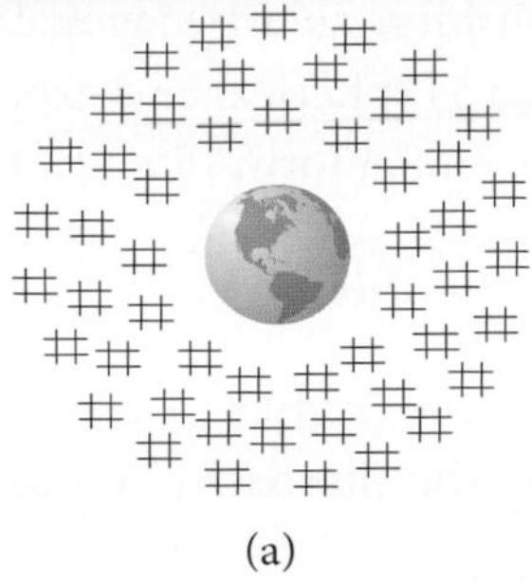

(a)

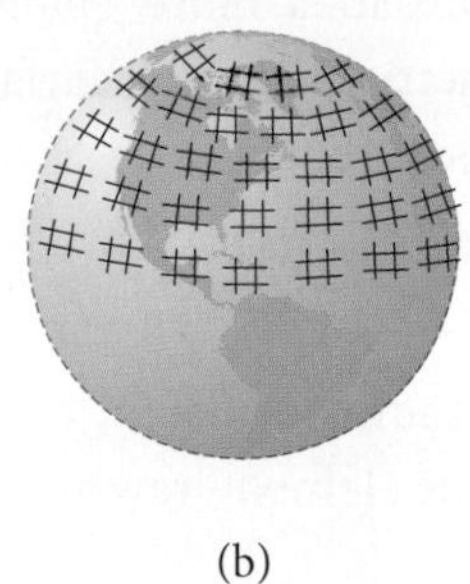

(b)

FIGURE 25.2 (a) A family of local Lorentz frames, all momentarily at rest above Earth's surface. (b) A family of local, 2-dimensional Euclidean coordinate systems on Earth's surface. The nonmeshing of Lorentz frames in (a) is analogous to the nonmeshing of Euclidean coordinates in (b) and motivates attributing gravity to a curvature of spacetime.

Turn, now, to a first look at gravity-induced constraints on the size of a local Lorentz frame. Above Earth, set up a family of local Lorentz frames scattered over the entire region from two Earth radii out to four Earth radii, with all the frames initially at rest with respect to Earth (Fig. 25.2a). From experience—or, if you prefer, from Newton's theory of gravity which after all is quite accurate near Earth—we know that, as time passes, these frames will all fall toward Earth. If (as a pedagogical aid) we drill holes through Earth to let the frames continue falling after reaching its surface, the frames will all pass through Earth's center and fly out the opposite side.

Obviously, two adjacent frames, which initially were at rest with respect to each other, acquire a relative velocity during their fall, which causes them to interpenetrate and pass through each other. Gravity is the cause of their relative velocity.

If these two adjacent frames could be meshed to form a larger Lorentz frame, then as time passes they would always remain at rest relative to each other. Thus, a meshing to form a larger Lorentz frame is impossible. The gravity-induced relative velocity prevents it. In brief: gravity prevents the meshing of local Lorentz frames to form global Lorentz frames.

nonmeshing of local Lorentz frames

This situation is closely analogous to the nonmeshing of local, 2-dimensional, Euclidean coordinate systems on the surface of Earth (Figure 25.2b): the curvature of Earth prevents a Euclidean mesh—thereby giving grief to mapmakers and surveyors. This analogy suggested to Einstein in 1912 a powerful new viewpoint on gravity. Just as the curvature of Earth prevents the meshing of local Euclidean coordinates on Earth's surface, so it must be that a curvature of spacetime prevents the meshing of local Lorentz frames in the spacetime above Earth—or anywhere else, for that matter. And since it is already known that gravity is the cause of the nonmeshing of Lorentz frames, it must be that *gravity is a manifestation of spacetime curvature.*

gravity is a manifestation of spacetime curvature

To make this idea more quantitative, consider, as a pedagogical tool, the 2-dimensional metric of Earth's surface, idealized as spherical and expressed in terms of a spherical polar coordinate system in line-element form [Eq. (2.24)]:

$$ds^2 = R^2 d\theta^2 + R^2 \sin^2\theta d\phi^2. \tag{25.7a}$$

Here R is the radius of Earth, or equivalently, the "radius of curvature" of Earth's surface. This line element, rewritten in terms of the alternative coordinates

$$x \equiv R\phi, \quad y \equiv R\left(\frac{\pi}{2} - \theta\right), \tag{25.7b}$$

has the form

$$ds^2 = \cos^2(y/R)dx^2 + dy^2 = dx^2 + dy^2 + O(y^2/R^2)dx^2, \tag{25.7c}$$

where as usual, $O(y^2/R^2)$ means "terms of order y^2/R^2 or smaller." Notice that the metric coefficients have the standard Euclidean form $g_{jk} = \delta_{jk}$ all along the equator ($y = 0$); but as one moves away from the equator, they begin to differ from Euclidean by fractional amounts of $O(y^2/R^2) = O[y^2/(\text{radius of curvature of Earth})^2]$. Thus, local Euclidean coordinates can be meshed and remain Euclidean all along the equator—or along any other great circle—but Earth's curvature forces the coordinates to cease being Euclidean when one moves off the chosen great circle, thereby causing the metric coefficients to differ from δ_{jk} by amounts $\Delta g_{jk} = O[(\text{distance from great circle})^2/(\text{radius of curvature})^2]$.

nonmeshing of local Euclidean coordinates on Earth's curved surface

Turn next to a specific example of curved spacetime: that of a $k = 0$ Robertson-Walker model for our expanding universe (to be studied in depth in Chap. 28). In spherical coordinates $\{\eta, \chi, \theta, \phi\}$, the 4-dimensional metric of this curved spacetime, described as a line element, can take the form

$$ds^2 = a^2(\eta)[-d\eta^2 + d\chi^2 + \chi^2(d\theta^2 + \sin^2\theta d\phi^2)]. \tag{25.8a}$$

cosmological example of nonmeshing

Here a, the "expansion factor of the universe," is a monotonic increasing function of the "time" coordinate η (not to be confused with the flat metric $\eta_{\alpha\beta}$). This line element, rewritten near $\chi = 0$ in terms of the alternative coordinates

$$t = \int_0^\eta a d\eta + \frac{1}{2}\chi^2\frac{da}{d\eta}, \quad x = a\chi \sin\theta \cos\phi, \quad y = a\chi \sin\theta \sin\phi, \quad z = a\chi \cos\theta, \tag{25.8b}$$

takes the form (Ex. 25.2)

$$ds^2 = \eta_{\alpha\beta}dx^\alpha dx^\beta + O\left(\frac{x^2 + y^2 + z^2}{\mathcal{R}^2}\right) dx^\alpha dx^\beta, \tag{25.8c}$$

where $\mathcal{R}$ is a quantity that, by analogy with the radius of curvature R of Earth's surface, can be identified as a radius of curvature of spacetime:

$$\frac{1}{\mathcal{R}^2} = O\left(\frac{\dot{a}^2}{a^2}\right) + O\left(\frac{\ddot{a}}{a}\right), \quad \text{where } \dot{a} \equiv \left(\frac{da}{dt}\right)_{x=y=z=0}, \quad \ddot{a} \equiv \left(\frac{d^2a}{dt^2}\right)_{x=y=z=0}. \tag{25.8d}$$

From the form of the metric coefficients in Eq. (25.8d), we see that, all along the world line $x = y = z = 0$, the coordinates are precisely Lorentz, but as one moves away from that world line they cease to be Lorentz, and the metric coefficients begin to differ from $\eta_{\alpha\beta}$ by amounts $\Delta g_{\alpha\beta} = \text{O}[(\text{distance from the chosen world line})^2/(\text{radius of curvature of spacetime})^2]$. This result is completely analogous to our equatorial Euclidean coordinates on Earth's surface. The curvature of Earth's surface prevented our local Euclidean coordinates from remaining Euclidean as we moved away from the equator; here the curvature of spacetime prevents our local Lorentz coordinates from remaining Lorentz as we move away from the chosen world line.

spacetime curvature forces Lorentz coordinates to be only locally Lorentz

Notice that the chosen world line is the spatial origin of our local Lorentz coordinates. Thus we can think of those coordinates as provided by a tiny, spatial latticework of rods and clocks, like that of Figure 25.1. The latticework remains locally Lorentz for all time (as measured by its own clocks), but it ceases to be locally Lorentz when one moves a finite spatial distance away from the spatial origin of the latticework.

This behavior is generic. One can show [see, e.g., Misner, Thorne, and Wheeler (1973, Sec. 13.6, esp. item (5) on p. 331)] specialized to vanishing acceleration and rotation] that, if any freely falling observer, anywhere in spacetime, sets up a little latticework of rods and clocks in accord with our standard rules and keeps the latticework's spatial origin on her free-fall world line, then the coordinates provided by the latticework will be locally Lorentz, with metric coefficients

metric coefficients in a local Lorentz frame

$$g_{\alpha\beta} = \left\{ \begin{array}{ll} \eta_{\alpha\beta} + \text{O}\left(\frac{\delta_{jk}x^jx^k}{\mathcal{R}^2}\right), \\ \eta_{\alpha\beta} \quad \text{at spatial origin} \end{array} \right\} \text{ in a local Lorentz frame,} \tag{25.9a}$$

where $\mathcal{R}$ is the radius of curvature of spacetime. Notice that, because the deviations of the metric from $\eta_{\alpha\beta}$ are of second order in the distance from the spatial origin, the first derivatives of the metric coefficients are of first order: $g_{\alpha\beta,k} = \text{O}(x^j/\mathcal{R}^2)$. This, plus the vanishing of the commutation coefficients in our coordinate basis, implies that the connection coefficients of the local Lorentz frame's coordinate basis are [Eqs. (24.38c) and (24.38d)][1]

connection coefficients in a local Lorentz frame

$$\Gamma^\alpha{}_{\beta\gamma} = \left\{ \begin{array}{l} \text{O}\left(\frac{\sqrt{\delta_{jk}x^jx^k}}{\mathcal{R}^2}\right), \\ 0 \quad \text{at spatial origin} \end{array} \right\} \text{ in a local Lorentz frame.} \tag{25.9b}$$

1. In any manifold, coordinates for which the metric and connection have the form of Eqs. (25.9) in the vicinity of some chosen geodesic (the "spatial origin") are called *Fermi coordinates* or sometimes *Fermi normal coordinates.*

It is instructive to compare Eq. (25.9a) for the metric in the local Lorentz frame of a freely falling observer in curved spacetime with Eq. (24.60b) for the metric in the proper reference frame of an accelerated observer in flat spacetime. Whereas the spacetime curvature in Eq. (25.9a) produces corrections to $g_{\alpha\beta} = \eta_{\alpha\beta}$ of second order in distance from the world line, the acceleration and spatial rotation of the reference frame in Eq. (24.60b) produce corrections of first order. This remains true when one studies accelerated observers in curved spacetime (e.g., Sec. 26.3.2). In their proper reference frames, the metric coefficients $g_{\alpha\beta}$ contain both the first-order terms of Eq. (24.60b) due to acceleration and rotation [e.g., Eq. (26.26)], and the second-order terms of Eq. (25.9a) due to spacetime curvature.

EXERCISES

Exercise 25.2 *Derivation: Local Lorentz Frame in Robertson-Walker Universe*
By inserting the coordinate transformation (25.8b) into the Robertson-Walker metric (25.8a), derive the metric (25.8c), (25.8d) for a local Lorentz frame.

25.4

25.4 Free-Fall Motion and Geodesics of Spacetime

To make more precise the concept of spacetime curvature, we need to study quantitatively the relative acceleration of neighboring, freely falling particles. Before we can carry out such a study, however, we must understand quantitatively the motion of a single freely falling particle in curved spacetime. That is the objective of this section.

In a global Lorentz frame of flat, special relativistic spacetime, a free particle moves along a straight world line—one with the form

$$(t, x, y, z) = (t_o, x_o, y_o, z_o) + (p^0, p^x, p^y, p^z)\zeta ; \quad \text{that is,} \quad x^\alpha = x^\alpha_o + p^\alpha \zeta. \tag{25.10a}$$

Here the p^α are the Lorentz-frame components of the particle's 4-momentum; ζ is the affine parameter such that $\vec{p} = d/d\zeta$, so $p^\alpha = dx^\alpha/d\zeta$ [Eq. (2.10) and subsequent material]; and x^α_o are the coordinates of the particle when its affine parameter is $\zeta = 0$. The straight-line motion (25.10a) can be described equally well by the statement that the Lorentz-frame components p^α of the particle's 4-momentum are constant (i.e., are independent of ζ):

$$\frac{dp^\alpha}{d\zeta} = 0. \tag{25.10b}$$

Even nicer is the frame-independent description, which says that, as the particle moves, it parallel-transports its tangent vector $\vec{p}$ along its world line:

$$\nabla_{\vec{p}}\vec{p} = 0, \quad \text{or equivalently,} \quad p^\alpha{}_{;\beta} p^\beta = 0. \tag{25.10c}$$

For a particle with nonzero rest mass m, which has $\vec{p} = m\vec{u}$ and $\zeta = \tau/m$ (with $\vec{u} = d/d\tau$ its 4-velocity and τ its proper time), Eq. (25.10c) is equivalent to $\nabla_{\vec{u}}\vec{u} = 0$.

This is the *geodesic* form of the particle's law of motion [Eq. (24.64)]; Eq. (25.10c) is the extension of that geodesic law to a particle that may have vanishing rest mass. Recall that the word *geodesic* refers to the particle's straight world line.

This geodesic description of the motion is readily carried over into curved spacetime using the equivalence principle. Let $\mathcal{P}(\zeta)$ be the world line of a freely moving particle in curved spacetime. At a specific event $\mathcal{P}_o = \mathcal{P}(\zeta_o)$ on that world line, introduce a local Lorentz frame (so the frame's spatial origin, carried by the particle, passes through $\mathcal{P}_o$ as time progresses). Then the equivalence principle tells us that the particle's law of motion must be the same in this local Lorentz frame as it is in the global Lorentz frame of special relativity [Eq. (25.10b)]:

$$\left(\frac{dp^\alpha}{d\zeta}\right)_{\zeta=\zeta_o} = 0. \tag{25.11a}$$

More powerful than this local-Lorentz-frame description of the motion is a description that is frame independent. We can easily deduce such a description from Eq. (25.11a). Since the connection coefficients vanish at the spatial origin of the local Lorentz frame where Eq. (25.11a) is being evaluated [cf. Eq. (25.9b)], Eq. (25.11a) can be written equally well, in our local Lorentz frame, as

$$0 = \left(\frac{dp^\alpha}{d\zeta} + \Gamma^\alpha{}_{\beta\gamma} p^\beta \frac{dx^\gamma}{d\zeta}\right)_{\zeta=\zeta_o} = \left((p^\alpha{}_{,\gamma} + \Gamma^\alpha{}_{\beta\gamma} p^\beta)\frac{dx^\gamma}{d\zeta}\right)_{\zeta=\zeta_o} = (p^\alpha{}_{;\gamma} p^\gamma)_{\zeta=\zeta_o}. \tag{25.11b}$$

Thus, as the particle passes through the spatial origin of our local Lorentz coordinate system, the components of the directional derivative of its 4-momentum along itself vanish. Now, if two 4-vectors have components that are equal in one basis, their components are guaranteed [by the tensorial transformation law (24.19)] to be equal in all bases; correspondingly, the two vectors, viewed as frame-independent, geometric objects, must be equal. Thus, since Eq. (25.11b) says that the components of the 4-vector $\nabla_{\vec p}\vec p$ and the zero vector are equal in our chosen local Lorentz frame, it must be true that

geodesic equation of motion for a freely falling particle in curved spacetime

$$\boxed{\nabla_{\vec p}\, \vec p = 0} \tag{25.11c}$$

at the moment when the particle passes through the point $\mathcal{P}_o = \mathcal{P}(\zeta_o)$. Moreover, since $\mathcal{P}_o$ is an arbitrary point (event) along the particle's world line, it must be that Eq. (25.11c) is a geometric, frame-independent equation of motion for the particle, valid everywhere along its world line. Notice that this geometric, frame-independent equation of motion $\nabla_{\vec p}\vec p = 0$ in curved spacetime is precisely the same as that [Eq. (25.10c)] for flat spacetime. We generalize this conclusion to other laws of physics in Sec. 25.7.

Our equation of motion (25.11c) for a freely moving point particle says, in words, that the particle *parallel transports* its 4-momentum along its world line. As in flat spacetime, so also in curved spacetime, if the particle has finite rest mass, we can rewrite the equation of motion $\nabla_{\vec{p}}\vec{p} = 0$ as

geodesic equation for particle with finite rest mass

$$\boxed{\nabla_{\vec{u}}\vec{u} = 0,} \tag{25.11d}$$

where $\vec{u} = \vec{p}/m = d/d\tau$ is the particle's 4-velocity, and $\tau = m\zeta$ is proper time along the particle's world line.

In any curved manifold, not just in spacetime, the relation $\nabla_{\vec{u}}\vec{u} = 0$ (or $\nabla_{\vec{p}}\vec{p} = 0$) is called the *geodesic equation*, and the curve to which $\vec{u}$ is the tangent vector is called a *geodesic*. If the geodesic is spacelike, its tangent vector can be normalized such that $\vec{u} = d/ds$, with s the proper distance along the geodesic—the obvious analog of $\vec{u} = d/d\tau$ for a timelike geodesic.

On the surface of a sphere, such as Earth, the geodesics are the great circles; they are the unique curves along which local Euclidean coordinates can be meshed, keeping one of the two Euclidean coordinates constant along the curve [cf. Eq. (25.7c)]. They are also the trajectories generated by an airplane's inertial guidance system, which guides the plane along the straightest trajectory it can. Similarly, in spacetime the trajectories of freely falling particles are geodesics. They are the unique curves along which local Lorentz coordinates can be meshed, keeping the three spatial coordinates constant along the curve and letting the time vary, thereby producing a local Lorentz reference frame [Eqs. (25.9)]. They are also the spacetime trajectories along which inertial guidance systems guide a spacecraft.

The geodesic equation $\nabla_{\vec{p}}\vec{p} = 0$ for a particle in spacetime guarantees that the square of the 4-momentum will be conserved along the particle's world line; in slot-naming index notation, we have:

$$(g_{\alpha\beta}p^\alpha p^\beta)_{;\gamma}p^\gamma = 2g_{\alpha\beta}p^\alpha p^\beta{}_{;\gamma}p^\gamma = 0. \tag{25.12}$$

Here the standard Leibniz rule for differentiating products has been used; this rule follows from the definition (24.29) of the frame-independent directional derivative of a tensor; it also can be deduced in a local Lorentz frame, where $\Gamma^\alpha{}_{\mu\nu} = 0$, so each gradient with a ";" reduces to a partial derivative with a ",". In Eq. (25.12) the term involving the gradient of the metric has been discarded, since it vanishes [Eq. (24.39)], and the two terms involving derivatives of p^α and p^β, being equal, have been combined. In index-free notation the frame-independent relation (25.12) says

$$\nabla_{\vec{p}}(\vec{p}\cdot\vec{p}) = 2\vec{p}\cdot\nabla_{\vec{p}}\vec{p} = 0. \tag{25.13}$$

conservation of rest mass for freely falling particle

This is a pleasing result, since the square of the 4-momentum is the negative of the particle's squared rest mass, $\vec{p}\cdot\vec{p} = -m^2$, which surely should be conserved along the particle's free-fall world line! Note that, as in flat spacetime, so also in curved, for a particle of finite rest mass the free-fall trajectory (the geodesic world line) is timelike, $\vec{p}\cdot\vec{p} = -m^2 < 0$, while for a zero-rest-mass particle it is null, $\vec{p}\cdot\vec{p} = 0$. Spacetime also supports spacelike geodesics [i.e., curves with tangent vectors $\vec{p}$ that satisfy the geodesic equation (25.11c) and are spacelike, $\vec{p} \cdot \vec{p} > 0$]. Such curves can be thought of as the world lines of freely falling "tachyons" (i.e., faster-than-light particles)—though it seems unlikely that such particles exist in Nature. Note that the constancy of $\vec{p} \cdot \vec{p}$ along a geodesic implies that a geodesic can never change its character: if initially timelike, it always remains timelike; if initially null, it remains null; if initially spacelike, it remains spacelike.

The geodesic world line of a freely moving particle has three very important properties:

1. When written in a coordinate basis, the geodesic equation, $\nabla_{\vec{p}}\vec{p} = 0$, becomes the following differential equation for the particle's world line $x^\alpha(\zeta)$ in the coordinate system (Ex. 25.3):

$$\boxed{\frac{d^2x^\alpha}{d\zeta^2} + \Gamma^\alpha{}_{\mu\nu}\frac{dx^\mu}{d\zeta}\frac{dx^\nu}{d\zeta} = 0.} \tag{25.14}$$

coordinate representation of geodesic equation

Here $\Gamma^\alpha{}_{\mu\nu}$ are the connection coefficients of the coordinate system's coordinate basis. [Equation (24.66) was a special case of this.] Note that these are four coupled equations ($\alpha = 0, 1, 2, 3$) for the four coordinates x^α as functions of affine parameter ζ along the geodesic. If the initial position, x^α at $\zeta = 0$, and initial tangent vector (particle momentum), $p^\alpha = dx^\alpha/d\zeta$ at $\zeta = 0$, are specified, then these four equations will determine uniquely the coordinates $x^\alpha(\zeta)$ as a function of ζ along the geodesic.

2. Consider a spacetime that possesses a symmetry, which is embodied in the fact that the metric coefficients in some coordinate system are independent of one of the coordinates x^A. Associated with that symmetry there will be a so-called *Killing vector field* $\vec{\xi} = \partial/\partial x^A$ and a conserved quantity $p_A \equiv \vec{p} \cdot \partial/\partial x^A$ for free-particle motion. Exercises 25.4 and 25.5 discuss Killing vector fields, derive this conservation law, and develop a familiar example.

Killing vectors and conserved quantities for geodesic motion

3. Among all timelike curves linking two events $\mathcal{P}_0$ and $\mathcal{P}_1$ in spacetime, those whose proper time lapse (timelike length) is stationary under small variations of the curve are timelike geodesics; see Ex. 25.6. In other words, timelike geodesics are the curves that satisfy the action principle (25.19). Now, one can always send a photon from $\mathcal{P}_0$ to $\mathcal{P}_1$ by bouncing it off a set of strategically located mirrors, and that photon path is the limit of a timelike curve as the curve becomes null. Therefore, there exist timelike curves from $\mathcal{P}_0$ to $\mathcal{P}_1$ with vanishingly small length, so no timelike geodesics can be an absolute minimum of the proper time lapse, but one can be an absolute maximum.

action principle of stationary proper time lapse for geodesic motion

EXERCISES

Exercise 25.3 *Derivation and Problem: Geodesic Equation in an Arbitrary Coordinate System*

Show that in an arbitrary coordinate system $x^\alpha(\mathcal{P})$ the geodesic equation (25.11c) takes the form of Eq. (25.14).

Exercise 25.4 ***Derivation and Example: Constant of Geodesic Motion in a Spacetime with Symmetry*

(a) Suppose that in some coordinate system the metric coefficients are independent of some specific coordinate x^A: $g_{\alpha\beta,A} = 0$ (e.g., in spherical polar coordinates $\{t, r, \theta, \phi\}$ in flat spacetime $g_{\alpha\beta,\phi} = 0$, so we could set $x^A = \phi$). Show that

$$p_A \equiv \vec{p} \cdot \frac{\partial}{\partial x^A} \tag{25.15}$$

is a constant of the motion for a freely moving particle [p_ϕ = (conserved z-component of angular momentum) in the above, spherically symmetric example]. [Hint: Show that the geodesic equation can be written in the form

$$\frac{dp_\alpha}{d\zeta} - \Gamma_{\mu\alpha\nu} p^\mu p^\nu = 0, \tag{25.16}$$

where $\Gamma_{\mu\alpha\nu}$ is the covariant connection coefficient of Eqs. (24.38c), (24.38d) with $c_{\alpha\beta\gamma} = 0$, because we are using a coordinate basis.] Note the analogy of the constant of the motion p_A with Hamiltonian mechanics: there, if the hamiltonian is independent of x^A, then the generalized momentum p_A is conserved; here, if the metric coefficients are independent of x^A, then the covariant component p_A of the momentum is conserved. For an elucidation of the connection between these two conservation laws, see Ex. 25.7c.

(b) As an example, consider a particle moving freely through a time-independent, Newtonian gravitational field. In Ex. 25.18, we learn that such a gravitational field can be described in the language of general relativity by the spacetime metric

$$ds^2 = -(1 + 2\Phi)dt^2 + (\delta_{jk} + h_{jk})dx^j dx^k, \tag{25.17}$$

where $\Phi(x, y, z)$ is the time-independent Newtonian potential, and h_{jk} are contributions to the metric that are independent of the time coordinate t and have magnitude of order $|\Phi|$. That the gravitational field is weak means $|\Phi| \ll 1$ (or, in conventional—SI or cgs—units, $|\Phi/c^2| \ll 1$). The coordinates being used are Lorentz, aside from tiny corrections of order $|\Phi|$, and as this exercise and Ex. 25.18 show, they coincide with the coordinates of the Newtonian theory of gravity. Suppose that the particle has a velocity $v^j \equiv dx^j/dt$ through this coordinate system that is $\lesssim |\Phi|^{\frac{1}{2}}$ and thus is small compared to the speed of light. Because the metric is independent of the time coordinate t, the component p_t of the particle's 4-momentum must be conserved along its world line. Since throughout physics, the conserved quantity associated with time-translation invariance is always the energy, we expect that p_t, when evaluated accurate to first order in $|\Phi|$, must be equal to the particle's conserved Newtonian energy, $E = m\Phi + \frac{1}{2}mv^j v^k \delta_{jk}$, aside from some multiplicative and additive constants. Show that this, indeed, is true, and evaluate the constants.

Exercise 25.5 *Example: Killing Vector Field*
A *Killing vector field*[2] is a coordinate-independent tool for exhibiting symmetries of the metric. It is any vector field $\vec{\xi}$ that satisfies

$$\xi_{\alpha;\beta} + \xi_{\beta;\alpha} = 0 \tag{25.18}$$

(i.e., any vector field whose symmetrized gradient vanishes).

(a) Let $\vec{\xi}$ be a vector field that might or might not be Killing. Show, by construction, that it is possible to introduce a coordinate system in which $\vec{\xi} = \partial/\partial x^A$ for some coordinate x^A.

(b) Show that in the coordinate system of part (a) the symmetrized gradient of $\vec{\xi}$ is $\xi_{\alpha;\beta} + \xi_{\beta;\alpha} = \partial g_{\alpha\beta}/\partial x^A$. From this infer that a vector field $\vec{\xi}$ is Killing if and only if there exists a coordinate system in which (i) $\vec{\xi} = \partial/\partial x^A$ and (ii) the metric is independent of x^A.

(c) Use Killing's equation (25.18) to show, without introducing a coordinate system, that, if $\vec{\xi}$ is a Killing vector field and $\vec{p}$ is the 4-momentum of a freely falling particle, then $\vec{\xi} \cdot \vec{p}$ is conserved along the particle's geodesic world line. This is the same conservation law as we proved in Ex. 25.4a using a coordinate-dependent calculation.

Exercise 25.6 *Problem: Timelike Geodesic as Path of Extremal Proper Time*
By introducing a specific but arbitrary coordinate system, show that among all timelike world lines that a particle could take to get from event $\mathcal{P}_0$ to event $\mathcal{P}_1$, the one or ones whose proper time lapse is stationary under small variations of path are the free-fall geodesics. In other words, an action principle for a timelike geodesic $\mathcal{P}(\lambda)$ [i.e., $x^\alpha(\lambda)$ in any coordinate system x^α] is

$$\boxed{\delta \int_{\mathcal{P}_0}^{\mathcal{P}_1} d\tau = \delta \int_0^1 \left(-g_{\alpha\beta} \frac{dx^\alpha}{d\lambda} \frac{dx^\beta}{d\lambda} \right)^{\frac{1}{2}} d\lambda = 0,} \tag{25.19}$$

2. Named after Wilhelm Killing, the mathematician who introduced it.

where λ is an arbitrary parameter, which by construction ranges from 0 at $\mathcal{P}_0$ to 1 at $\mathcal{P}_1$. [Note: Unless, after the variation, you choose the arbitrary parameter λ to be "affine" ($\lambda = a\zeta + b$, where a and b are constants and ζ is such that $\vec{p} = d/d\zeta$), your equation for $d^2x^\alpha/d\lambda^2$ will not look quite like Eq. (25.14).]

Exercise 25.7 *Problem: Super-Hamiltonian for Free Particle Motion*

(a) Show that, among all curves $\mathcal{P}(\zeta)$ that could take a particle from event $\mathcal{P}_0 = \mathcal{P}(0)$ to event $\mathcal{P}_1 = \mathcal{P}(\zeta_1)$ (for some ζ_1), those that satisfy the action principle

$$\delta \frac{1}{2}\int_0^{\zeta_1} g_{\alpha\beta}\frac{dx^\alpha}{d\zeta}\frac{dx^\beta}{d\zeta}\, d\zeta = 0 \tag{25.20}$$

are geodesics, and ζ is the affine parameter along the geodesic related to the particle's 4-momentum by $\vec{p} = d/d\zeta$. [Note: In this action principle, by contrast with Eq. (25.19), the integration parameter is necessarily affine.]

(b) The lagrangian $\mathcal{L}(x^\mu, dx^\mu/d\zeta)$ associated with this action principle is $\frac{1}{2}g_{\alpha\beta}(dx^\alpha/d\zeta)(dx^\beta/d\zeta)$, where the coordinates $\{x^0, x^1, x^2, x^3\}$ appear in the metric coefficients and their derivatives appear explicitly. Using standard principles of Hamiltonian mechanics, show that the momentum canonically conjugate to x^μ is $p_\mu = g_{\mu\nu}dx^\nu/d\zeta$—which in fact is the covariant component of the particle's 4-momentum. Show, further, that the hamiltonian associated with the particle's lagrangian is

super-hamiltonian for geodesic motion

$$\mathcal{H} = \frac{1}{2}g^{\mu\nu}p_\mu p_\nu. \tag{25.21}$$

Explain why this guarantees that Hamilton's equations $dx^\alpha/d\zeta = \partial\mathcal{H}/\partial p_\alpha$ and $dp_\alpha/d\zeta = -\partial\mathcal{H}/\partial x^\alpha$ are satisfied if and only if $x^\alpha(\zeta)$ is a geodesic with affine parameter ζ and p_α is the tangent vector to the geodesic.

(c) Show that Hamilton's equations guarantee that, if the metric coefficients are independent of some coordinate x^A, then p_A is a conserved quantity. This is the same conservation law as we derived by other methods in Exs. 25.4 and 25.5.

$\mathcal{H} = \frac{1}{2}g^{\mu\nu}p_\mu p_\nu$ is often called the geodesic's *super-hamiltonian*. It turns out that, often, the easiest way to compute geodesics numerically (e.g., for particle motion around a black hole) is to solve the super-hamiltonian's Hamilton equations (see, e.g., Levin and Perez-Giz, 2008).

25.5

25.5 Relative Acceleration, Tidal Gravity, and Spacetime Curvature

Now that we understand the motion of an individual freely falling particle in curved spacetime, we are ready to study the effects of gravity on the relative motions of such particles. Before doing so in general relativity, let us recall the Newtonian description.

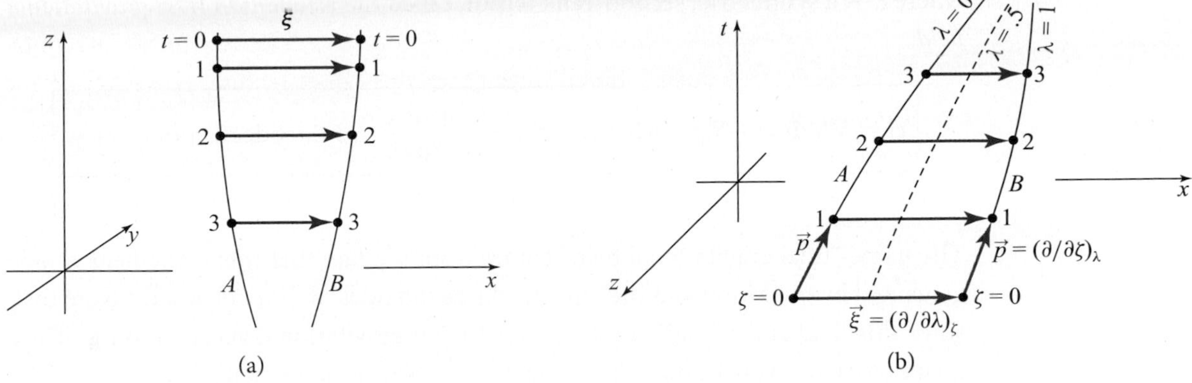

FIGURE 25.3 The effects of tidal gravity on the relative motions of two freely falling particles. (a) In Euclidean 3-space using Newton's theory of gravity. (b) In spacetime using Einstein's theory of gravity, general relativity.

25.5.1 Newtonian Description of Tidal Gravity 25.5.1

Consider, as shown in Fig. 25.3a, two point particles, A and B, falling freely through 3-dimensional Euclidean space under the action of an external Newtonian potential Φ (i.e., a potential generated by other masses, not by the particles themselves). At Newtonian time $t = 0$ the particles are separated by only a small distance and are moving with the same velocity: $\mathbf{v}_A = \mathbf{v}_B$. As time passes, however, the two particles, being at slightly different locations in space, experience slightly different gravitational potentials Φ and gravitational accelerations $\mathbf{g} = -\boldsymbol{\nabla}\Phi$ and thence develop slightly different velocities: $\mathbf{v}_A \neq \mathbf{v}_B$. To quantify this, denote by $\boldsymbol{\xi}$ the vector separation of the two particles in Euclidean 3-space. The components of $\boldsymbol{\xi}$ on any Euclidean basis (e.g., that of Fig. 25.3a) are $\xi^j = x^j_B - x^j_A$, where x^j_I is the coordinate location of particle I. Correspondingly, the rate of change of ξ^j with respect to Newtonian time is $d\xi^j/dt = v^j_B - v^j_A$ (i.e., the relative velocity of the two particles is the difference of their velocities). The second time derivative of the relative separation (i.e., the relative acceleration of the two particles) is thus given by

$$\frac{d^2\xi^j}{dt^2} = \frac{d^2x^j_B}{dt^2} - \frac{d^2x^j_A}{dt^2} = -\left(\frac{\partial\Phi}{\partial x^j}\right)_B + \left(\frac{\partial\Phi}{\partial x^j}\right)_A = -\frac{\partial^2\Phi}{\partial x^j\partial x^k}\xi^k, \tag{25.22}$$

accurate to first order in the separation ξ^k. This equation gives the components of the relative acceleration in an arbitrary Euclidean basis. Rewritten in geometric, basis-independent language, this equation is

relative acceleration of freely falling particles

$$\boxed{\frac{d^2\boldsymbol{\xi}}{dt^2} = -\boldsymbol{\mathcal{E}}(\_, \boldsymbol{\xi}); \quad \text{or} \quad \frac{d^2\xi^j}{dt^2} = -\mathcal{E}^j{}_k\xi^k,} \tag{25.23}$$

where $\mathcal{E}$ is a symmetric, second-rank tensor, called the *Newtonian tidal gravitational field:*

Newtonian tidal gravitational field

$$\mathcal{E} \equiv \boldsymbol{\nabla}\boldsymbol{\nabla}\Phi = -\boldsymbol{\nabla}\mathbf{g}; \quad \text{that is,} \quad \mathcal{E}_{jk} = \frac{\partial^2 \Phi}{\partial x^j \partial x^k} \quad \text{in Euclidean coordinates.} \tag{25.24}$$

The name "tidal gravitational field" comes from the fact that this is the field which, generated by the Moon and the Sun, produces the tides on Earth's oceans. Note that, since this field is the gradient of the Newtonian gravitational acceleration $\mathbf{g}$, it is a quantitative measure of the inhomogeneities of Newtonian gravity.

Equation (25.23) shows quantitatively how the tidal gravitational field produces the relative acceleration of our two particles. As a specific application, one can use it to compute, in Newtonian theory, the relative accelerations and thence relative motions of two neighboring local Lorentz frames as they fall toward and through the center of Earth (Fig. 25.2a and associated discussion).

25.5.2

25.5.2 Relativistic Description of Tidal Gravity

Now turn to the general relativistic description of the relative motions of two free particles. As shown in Fig. 25.3b, the particles, labeled A and B, move along geodesic world lines with affine parameters ζ and 4-momentum tangent vectors $\vec{p} = d/d\zeta$. The origins of ζ along the two world lines can be chosen however we wish, so long as events with the same ζ on the two world lines, $\mathcal{P}_A(\zeta)$ and $\mathcal{P}_B(\zeta)$, are close enough to each other that we can perform power-series expansions in their separation, $\vec{\xi}(\zeta) = \mathcal{P}_B(\zeta) - \mathcal{P}_A(\zeta)$, and keep only the leading terms. As in our Newtonian analysis, we require that the two particles initially have vanishing relative velocity, $\nabla_{\vec{p}}\vec{\xi} = 0$, and we compute the tidal-gravity-induced relative acceleration $\nabla_{\vec{p}}\nabla_{\vec{p}}\vec{\xi}$.

As a tool in our calculation, we introduce into spacetime a 2-dimensional surface that contains our two geodesics A and B, and also contains an infinity of other geodesics in between and alongside them. On that surface, we introduce two coordinates, $\zeta =$ (affine parameter along each geodesic) and $\lambda =$ (a parameter that labels the geodesics); see Fig. 25.3b. Geodesic A carries the label $\lambda = 0$; geodesic B is $\lambda = 1$; $\vec{p} = (\partial/\partial\zeta)_{\lambda=\text{const}}$ is a vector field that, evaluated on any geodesic (A, B, or other curve of constant λ), is equal to the 4-momentum of the particle that moves along that geodesic; and $\vec{\xi} \equiv (\partial/\partial\lambda)_{\zeta=\text{const}}$ is a vector field that, when evaluated on geodesic A (i.e., at $\lambda = 0$), we identify as a rigorous version of the separation vector $\mathcal{P}_B(\zeta) - \mathcal{P}_A(\zeta)$ that we wish to study. This identification requires, for good accuracy, that the geodesics be close together and be so parameterized that $\mathcal{P}_A(\zeta)$ is close to $\mathcal{P}_B(\zeta)$.

Our objective is to compute the relative acceleration of particles B and A, $\nabla_{\vec{p}}\nabla_{\vec{p}}\vec{\xi}$, evaluated at $\lambda = 0$. The quantity $\nabla_{\vec{p}}\vec{\xi}$, which we wish to differentiate a second time in that computation, is one of the terms in the following expression for the commutator of the vector fields $\vec{p}$ and $\vec{\xi}$ [Eq. (24.40)]:

$$\boxed{[\vec{p}, \vec{\xi}] = \nabla_{\vec{p}}\vec{\xi} - \nabla_{\vec{\xi}}\vec{p}.} \tag{25.25}$$

Because $\vec{p} = (\partial/\partial\zeta)_\lambda$ and $\vec{\xi} = (\partial/\partial\lambda)_\zeta$, these two vector fields commute, and Eq. (25.25) tells us that $\nabla_{\vec{p}}\vec{\xi} = \nabla_{\vec{\xi}}\vec{p}$. Correspondingly, the relative acceleration of our two particles can be expressed as

$$\nabla_{\vec{p}}\nabla_{\vec{p}}\vec{\xi} = \nabla_{\vec{p}}\nabla_{\vec{\xi}}\vec{p} = (\nabla_{\vec{p}}\nabla_{\vec{\xi}} - \nabla_{\vec{\xi}}\nabla_{\vec{p}})\vec{p}. \tag{25.26}$$

Here the second equality results from adding on, for use below, a term that vanishes because $\nabla_{\vec{p}}\vec{p} = 0$ (geodesic equation).

This first part of our calculation was performed efficiently using index-free notation. The next step will be easier if we introduce indices as names for slots. Then expression (25.26) takes the form

$$(\xi^\alpha{}_{;\beta}p^\beta)_{;\gamma}p^\gamma = (p^\alpha{}_{;\gamma}\xi^\gamma)_{;\delta}p^\delta - (p^\alpha{}_{;\gamma}p^\gamma)_{;\delta}\xi^\delta, \tag{25.27}$$

which can be evaluated by using the rule for differentiating products and then renaming indices and collecting terms. The result is

$$(\xi^\alpha{}_{;\beta}p^\beta)_{;\gamma}p^\gamma = (p^\alpha{}_{;\gamma\delta} - p^\alpha{}_{;\delta\gamma})\xi^\gamma p^\delta + p^\alpha{}_{;\gamma}(\xi^\gamma{}_{;\delta}p^\delta - p^\gamma{}_{;\delta}\xi^\delta). \tag{25.28}$$

The second term in this expression vanishes, since it is just the commutator of $\vec{\xi}$ and $\vec{p}$ [Eq. (25.25)] written in slot-naming index notation, and as we noted above, $\vec{\xi}$ and $\vec{p}$ commute. The resulting equation,

$$(\xi^\alpha{}_{;\beta}p^\beta)_{;\gamma}p^\gamma = (p^\alpha{}_{;\gamma\delta} - p^\alpha{}_{;\delta\gamma})\xi^\gamma p^\delta, \tag{25.29}$$

reveals that the relative acceleration of the two particles is caused by noncommutation of the two slots of a double gradient (slots here named γ and δ). In the flat spacetime of special relativity, the two slots would commute[3] and there would be no relative acceleration. Spacetime curvature prevents them from commuting and thereby causes the relative acceleration.

Now, one can show that for any vector field $\vec{p}(\mathcal{P})$, $p^\alpha{}_{;\gamma\delta} - p^\alpha{}_{;\delta\gamma}$ is linear in p^α; see Ex. 25.8. Thus there must exist a fourth-rank tensor field $\boldsymbol{R}(\_, \_, \_, \_)$ such that

$$\boxed{p^\alpha{}_{;\gamma\delta} - p^\alpha{}_{;\delta\gamma} = -R^\alpha{}_{\beta\gamma\delta}p^\beta} \tag{25.30}$$

Riemann curvature tensor

3. In flat spacetime, in global Lorentz coordinates, $\Gamma^\alpha{}_{\beta\gamma} = 0$ everywhere, so $p^\alpha{}_{;\gamma\delta} = \partial^2 p^\alpha/\partial x^\gamma \partial x^\delta$. Because partial derivatives commute, expression (25.29) vanishes.

for any $\vec{p}$. The tensor $\boldsymbol{R}$ can be regarded as responsible for the failure of gradients to commute, so it must be some aspect of spacetime curvature. It is called the *Riemann curvature tensor.*

Inserting Eq. (25.30) into Eq. (25.29) and writing the result in both slot-naming index notation and abstract notation, we obtain

equation of geodesic deviation

$$(\xi^{\alpha}{}_{;\beta} p^{\beta})_{;\gamma} p^{\gamma} = -R^{\alpha}{}_{\beta\gamma\delta} p^{\beta} \xi^{\gamma} p^{\delta}, \quad \nabla_{\vec{p}} \nabla_{\vec{p}} \vec{\xi} = -\boldsymbol{R}(\_\_, \vec{p}, \vec{\xi}, \vec{p}). \tag{25.31}$$

This is the equation of relative acceleration for freely moving test particles. It is also called the *equation of geodesic deviation,* because it describes the manner in which spacetime curvature $\boldsymbol{R}$ forces geodesics that are initially parallel (the world lines of freely moving particles with zero initial relative velocity) to deviate from one another (Fig. 25.3b).

EXERCISES

Exercise 25.8 *Derivation: Linearity of the Commutator of the Double Gradient*

(a) Let a and b be scalar fields with arbitrary but smooth dependence on location in curved spacetime, and let $\vec{A}$ and $\vec{B}$ be vector fields. Show that

$$(aA^{\alpha} + bB^{\alpha})_{;\gamma\delta} - (aA^{\alpha} + bB^{\alpha})_{;\delta\gamma} = a(A^{\alpha}{}_{;\gamma\delta} - A^{\alpha}{}_{;\delta\gamma}) + b(B^{\alpha}{}_{;\gamma\delta} - B^{\alpha}{}_{;\delta\gamma}). \tag{25.32}$$

[Hint: The double gradient of a scalar field commutes, as one can easily see in a local Lorentz frame.]

(b) Use Eq. (25.32) to show that (i) the commutator of the double gradient is independent of how the differentiated vector field varies from point to point and depends only on the value of the field at the location where the commutator is evaluated, and (ii) the commutator is linear in that value. Thereby conclude that there must exist a fourth-rank tensor field $\boldsymbol{R}$ such that Eq. (25.30) is true for any vector field $\vec{p}$.

25.5.3

25.5.3 Comparison of Newtonian and Relativistic Descriptions

It is instructive to compare this relativistic description of the relative acceleration of freely moving particles with the Newtonian description. For this purpose we consider a region of spacetime, such as our solar system, in which the Newtonian description of gravity is highly accurate; and there we study the relative acceleration of two free particles from the viewpoint of a local Lorentz frame in which the particles are both initially at rest.

In the Newtonian description, the transformation from a Newtonian universal reference frame (e.g., that of the center of mass of the solar system) to the chosen local Lorentz frame is achieved by introducing new Euclidean coordinates that are uniformly accelerated relative to the old ones, with just the right uniform acceleration to annul the gravitational acceleration at the center of the local Lorentz frame. This transformation adds a spatially homogeneous constant to the Newtonian acceleration, $\mathbf{g} = -\nabla\Phi$, but leaves unchanged the tidal field, $\boldsymbol{\mathcal{E}} = \nabla\nabla\Phi$. Correspondingly, the Newtonian equation of relative acceleration in the local Lorentz frame retains its standard Newtonian form, $d^2\xi^j/dt^2 = -\mathcal{E}^j{}_k\xi^k$ [Eq. (25.23)], with the components of the tidal field computable equally well in the original universal reference frame or in the local Lorentz frame, using the standard relation $\mathcal{E}^j{}_k = \mathcal{E}_{jk} = \partial^2\Phi/\partial x^j\partial x^k$.

As an aid in making contact between the relativistic and the Newtonian descriptions, we convert from using the 4-momentum $\vec{p}$ as the relativistic tangent vector and ζ as the relativistic parameter along the particles' world lines to using the 4-velocity $\vec{u} = \vec{p}/m$ and the proper time $\tau = m\zeta$. This conversion brings the relativistic equation of relative acceleration (25.31) into the form

geodesic deviation for particles with finite rest mass

$$\nabla_{\vec{u}}\nabla_{\vec{u}}\vec{\xi} = -\boldsymbol{R}(\ldots, \vec{u}, \vec{\xi}, \vec{u}). \tag{25.33}$$

Because the particles are (momentarily) at rest near the origin of the local Lorentz frame, their 4-velocities are $\vec{u} \equiv d/d\tau = \partial/\partial t$, which implies that the components of their 4-velocities are $u^0 = 1$, $u^j = 0$, and their proper times τ are equal to coordinate time t, which in turn coincides with the time t of the Newtonian analysis: $\tau = t$. In the relativistic analysis, as in the Newtonian, the separation vector $\vec{\xi}$ will have only spatial components, $\xi^0 = 0$ and $\xi^j \neq 0$. (If this were not so, we could make it so by readjusting the origin of proper time for particle B, so $\vec{\xi} \cdot \vec{p} = m\vec{\xi} \cdot \vec{u} = 0$, whence $\xi^0 = 0$; Fig. 25.3b.) These facts, together with the vanishing of all the connection coefficients and derivatives of them ($\Gamma^j{}_{k0,0} = 0$) that appear in $(\xi^j{}_{;\beta}u^\beta)_{;\gamma}u^\gamma$ at the origin of the local Lorentz frame [cf. Eqs. (25.9)], imply that the local Lorentz components of the equation of relative acceleration (25.33) take the form

$$\frac{d^2\xi^j}{dt^2} = -R^j{}_{0k0}\xi^k. \tag{25.34}$$

By comparing this with the Newtonian equation of relative acceleration (25.23), we infer that, in the Newtonian limit, in the local rest frame of the two particles, we have

space-time-space-time components of Riemann tensor become tidal field in Newtonian limit of general relativity

$$\boxed{R^j{}_{0k0} = \mathcal{E}_{jk} = \frac{\partial^2\Phi}{\partial x^j\partial x^k}}. \tag{25.35}$$

Thus, the Riemann curvature tensor is the relativistic generalization of the Newtonian tidal field. This conclusion and the above equations make quantitative the statement that gravity is a manifestation of spacetime curvature.

components of Riemann tensor outside a Newtonian, gravitating body

Outside a spherical body with weak (Newtonian) gravity, such as Earth, the Newtonian potential is $\Phi = -GM/r$, where G is Newton's gravitation constant, M is the body's mass, and r is the distance from its center. If we introduce Cartesian coordinates with origin at the body's center and with the point at which the Riemann tensor is to be measured lying on the z-axis at $\{x, y, z\} = \{0, 0, r\}$, then Φ near that point is

$\Phi = -GM/(z^2 + x^2 + y^2)^{\frac{1}{2}}$, and on the z-axis the only nonzero $R^j{}_{0k0}$ components, as computed from Eq. (25.35), are

$$\boxed{R^z{}_{0z0} = \frac{-2GM}{r^3}, \quad R^x{}_{0x0} = R^y{}_{0y0} = \frac{+GM}{r^3}.} \tag{25.36}$$

Correspondingly, for two particles separated from each other in the radial (z) direction, the relative acceleration (25.34) is $d^2\xi^j/dt^2 = (2GM/r^3)\xi^j$ (i.e., the particles are pulled apart by the body's tidal gravitational field). Similarly, for two particles separated from each other in a transverse direction (in the x-y plane), we have $d^2\xi^j/dt^2 = -(GM/r^3)\xi^j$ (i.e., the particles are pushed together by the body's tidal gravitational field). There thus is a radial tidal stretch and a lateral tidal squeeze; the lateral squeeze has half the strength of the radial stretch but occurs in two lateral dimensions compared to the one radial dimension. This stretch and squeeze, produced by the Sun and the Moon, are responsible for the tides on Earth's oceans; Ex. 25.9.

EXERCISES

Exercise 25.9 ***Example: Ocean Tides*

(a) Place a local Lorentz frame at the center of Earth, and let $\mathcal{E}_{jk}$ be the tidal field there, produced by the Newtonian gravitational fields of the Sun and the Moon. For simplicity, treat Earth as precisely spherical. Show that the gravitational acceleration (relative to Earth's center) at some location on or near Earth's surface (radius r) is

$$g_j = -\frac{GM}{r^2}n^j - \mathcal{E}^j{}_k r n^k, \tag{25.37}$$

where M is Earth's mass, and n^j is a unit vector pointing from Earth's center to the location at which g_j is evaluated.

(b) Show that this gravitational acceleration is minus the gradient of the Newtonian potential

$$\Phi = -\frac{GM}{r} + \frac{1}{2}\mathcal{E}_{jk}\, r^2 n^j n^k. \tag{25.38}$$

(c) Consider regions of Earth's oceans that are far from any coast and have ocean depth large compared to the heights of ocean tides. If Earth were nonrotating, then explain why the analysis of Sec. 13.3 predicts that the ocean surface in these regions would be a surface of constant Φ. Explain why this remains true to good accuracy also for the rotating Earth.

(d) Show that in these ocean regions, the Moon creates high tides pointing toward and away from itself and low tides in the transverse directions on Earth; and similarly for the Sun. Compute the difference between high and low tides produced by the Moon and by the Sun, and the difference of the total tide when the Moon and the Sun are in approximately the same direction in the sky. Your answers are reasonably accurate for deep ocean regions far from any coast, but near a coast, the tides are typically larger and sometimes far larger, and they are shifted in phase relative to the positions of the Moon and Sun. Why?

25.6 Properties of the Riemann Curvature Tensor

25.6

We now pause in our study of the foundations of general relativity to examine a few properties of the Riemann curvature tensor $\boldsymbol{R}$.

As a tool for deriving other things, we begin by evaluating the components of the Riemann tensor at the spatial origin of a local Lorentz frame (i.e., at a point where $g_{\alpha\beta} = \eta_{\alpha\beta}$ and $\Gamma^{\alpha}{}_{\beta\gamma}$ vanishes, but its derivatives do not). For any vector field $\vec{p}$, a straightforward computation reveals

$$p^{\alpha}{}_{;\gamma\delta} - p^{\alpha}{}_{;\delta\gamma} = (\Gamma^{\alpha}{}_{\beta\gamma,\delta} - \Gamma^{\alpha}{}_{\beta\delta,\gamma})p^{\beta}. \tag{25.39}$$

By comparing with Eq. (25.30), we can read off the local-Lorentz components of the Riemann tensor:

$$R^{\alpha}{}_{\beta\gamma\delta} = \Gamma^{\alpha}{}_{\beta\delta,\gamma} - \Gamma^{\alpha}{}_{\beta\gamma,\delta} \quad \text{at the spatial origin of a local Lorentz frame.} \tag{25.40}$$

From this expression we infer that, at a spatial distance $\sqrt{\delta_{ij}x^i x^j}$ from the origin of a local Lorentz frame, the connection coefficients and the metric have magnitudes

influence of spacetime curvature on connection and metric in a local Lorentz frame

$$\Gamma^{\alpha}{}_{\beta\gamma} = \mathrm{O}(R^{\mu}{}_{\nu\lambda\rho}\sqrt{\delta_{ij}x^i x^j}), \quad g_{\alpha\beta} - \eta_{\alpha\beta} = \mathrm{O}(R^{\mu}{}_{\nu\lambda\rho}\,\delta_{ij}x^i x^j) \quad \text{in a local Lorentz frame.} \tag{25.41}$$

Comparison with Eqs. (25.9) shows that the radius of curvature of spacetime (a concept defined only semiquantitatively) is of order the inverse square root of the components of the Riemann tensor in a local Lorentz frame:

radius of curvature of spacetime

$$\boxed{\mathcal{R} = \mathrm{O}\left(\frac{1}{|R^{\alpha}{}_{\beta\gamma\delta}|^{\frac{1}{2}}}\right) \quad \text{in a local Lorentz frame.}} \tag{25.42}$$

By comparison with Eq. (25.36), we see that at radius r outside a weakly gravitating body of mass M, the radius of curvature of spacetime is

$$\boxed{\mathcal{R} \sim \left(\frac{r^3}{GM}\right)^{\frac{1}{2}} = \left(\frac{c^2 r^3}{GM}\right)^{\frac{1}{2}},} \tag{25.43}$$

where the factor c (speed of light) in the second expression makes the formula valid in conventional units. For further discussion, see Ex. 25.10.

components of Riemann tensor in a local Lorentz frame

Using the components (25.40) of the Riemann tensor in a local Lorentz frame in terms of the connection coefficients, and using expressions (24.38) for the connection coefficients in terms of the metric components and commutation coefficients together with the vanishing of the commutation coefficients (because a local Lorentz frame is a coordinate basis), one easily can show that

$$R_{\alpha\beta\gamma\delta} = \frac{1}{2}(g_{\alpha\delta,\beta\gamma} + g_{\beta\gamma,\alpha\delta} - g_{\alpha\gamma,\beta\delta} - g_{\beta\delta,\alpha\gamma}) \quad \text{in a local Lorentz frame.} \tag{25.44}$$

From these expressions, plus the commutation of partial derivatives $g_{\alpha\gamma,\beta\delta} = g_{\alpha\gamma,\delta\beta}$ and the symmetry of the metric, one readily can show that in a local Lorentz frame the components of the Riemann tensor have the following symmetries:

three symmetries of Riemann tensor

$$R_{\alpha\beta\gamma\delta} = -R_{\beta\alpha\gamma\delta}, \quad R_{\alpha\beta\gamma\delta} = -R_{\alpha\beta\delta\gamma}, \quad R_{\alpha\beta\gamma\delta} = +R_{\gamma\delta\alpha\beta} \tag{25.45a}$$

(antisymmetry in first pair of indices, antisymmetry in second pair of indices, and symmetry under interchange of the pairs). When one computes the value of the tensor on four vectors, $\mathbf{R}(\vec{A}, \vec{B}, \vec{C}, \vec{D})$ using component calculations in this frame, one trivially sees that these symmetries produce corresponding symmetries under interchange of the vectors inserted into the slots, and thence under interchange of the slots themselves. (This is always the case: any symmetry that the components of a tensor exhibit in any special basis will induce the same symmetry on the slots of the geometric, frame-independent tensor.) The resulting symmetries for $\mathbf{R}$ are given by Eq. (25.45a) with the "Escher mind-flip" (Sec. 1.5.1) in which the indices switch from naming components in a special frame to naming slots. The Riemann tensor is antisymmetric under interchange of its first two slots, antisymmetric under interchange of the last two, and symmetric under interchange of the two pairs.

One additional symmetry can be verified by calculation in the local Lorentz frame [i.e., from Eq. (25.44)]:[4]

a fourth symmetry of Riemann tensor

$$R_{\alpha\beta\gamma\delta} + R_{\alpha\gamma\delta\beta} + R_{\alpha\delta\beta\gamma} = 0. \tag{25.45b}$$

Riemann tensor has 20 independent components generically

One can show that the full set of symmetries (25.45) reduces the number of independent components of the Riemann tensor, in 4-dimensional spacetime, from $4^4 = 256$ to "just" 20.

Of these 20 independent components, 10 are contained in the *Ricci curvature tensor*, which is the contraction of the Riemann tensor on its first and third slots:

Ricci tensor and its symmetry

$$R_{\alpha\beta} \equiv R^{\mu}{}_{\alpha\mu\beta}, \tag{25.46}$$

and which, by the symmetries (25.45) of Riemann, is itself symmetric:

4. Note that this cyclic symmetry is the same as occurs in the second of Maxwell's equations (2.48), $\epsilon^{\alpha\beta\gamma\delta}F_{\gamma\delta;\beta} = 0$; it is also the same as occurs in the Jacobi identity for commutators $\left[\vec{B}, [\vec{C}, \vec{D}]\right] + \left[\vec{C}, [\vec{D}, \vec{B}]\right] + \left[\vec{D}, [\vec{B}, \vec{C}]\right] = 0$.

$$\boxed{R_{\alpha\beta} = R_{\beta\alpha}.} \tag{25.47}$$

The other 10 independent components of Riemann are contained in the Weyl curvature tensor:

Weyl tensor

$$\boxed{C^{\mu\nu}{}_{\rho\sigma} = R^{\mu\nu}{}_{\rho\sigma} - 2g^{[\mu}{}_{[\rho}\, R^{\nu]}{}_{\sigma]} + \frac{1}{3} g^{[\mu}{}_{[\rho}\, g^{\nu]}{}_{\sigma]}\, R.} \tag{25.48}$$

scalar curvature

Here the square brackets denote antisymmetrization, $A_{[\alpha\beta]} \equiv \frac{1}{2}(A_{\alpha\beta} - A_{\beta\alpha})$, and R is the contraction of the Ricci tensor on its two slots,

$$\boxed{R \equiv R^{\alpha}{}_{\alpha},} \tag{25.49}$$

and is called the *curvature scalar* or *scalar curvature.* The Weyl curvature tensor $C^{\mu\nu}{}_{\rho\sigma}$ has vanishing contraction on every pair of slots and has the same symmetries as the Riemann tensor; Ex. 25.12.

components of Riemann tensor in an arbitrary basis

One often needs to know the components of the Riemann curvature tensor in some non-local-Lorentz basis. Exercise 25.11 derives the following equation for them in an arbitrary basis:

$$\boxed{R^{\alpha}{}_{\beta\gamma\delta} = \Gamma^{\alpha}{}_{\beta\delta,\gamma} - \Gamma^{\alpha}{}_{\beta\gamma,\delta} + \Gamma^{\alpha}{}_{\mu\gamma}\Gamma^{\mu}{}_{\beta\delta} - \Gamma^{\alpha}{}_{\mu\delta}\Gamma^{\mu}{}_{\beta\gamma} - \Gamma^{\alpha}{}_{\beta\mu}c_{\gamma\delta}{}^{\mu}.} \tag{25.50}$$

Here $\Gamma^{\alpha}{}_{\beta\gamma}$ are the connection coefficients in the chosen basis; $\Gamma^{\alpha}{}_{\beta\gamma,\delta}$ is the result of letting the basis vector $\vec{e}_\delta$ act as a differential operator on $\Gamma^{\alpha}{}_{\beta\gamma}$, as though $\Gamma^{\alpha}{}_{\beta\gamma}$ were a scalar; and $c_{\gamma\delta}{}^{\mu}$ are the basis vectors' commutation coefficients. Calculations with this equation are usually long and tedious, and so are carried out using symbolic-manipulation software on a computer. See, for example, the simple Mathematica program (specialized to a coordinate basis) in Hartle (2003, Appendix C), also available on that textbook's website: http://web.physics.ucsb.edu/~gravitybook/mathematica.html.

EXERCISES

Exercise 25.10 *Example: Orders of Magnitude for the Radius of Curvature of Spacetime*
With the help of the Newtonian limit (25.35) of the Riemann curvature tensor, show that near Earth's surface the radius of curvature of spacetime has a magnitude $\mathcal{R} \sim$ (1 astronomical unit) $\equiv$ (distance from the Sun to Earth). What is the radius of curvature of spacetime near the Sun's surface? Near the surface of a white-dwarf star? Near the surface of a neutron star? Near the surface of a one-solar-mass black hole? In intergalactic space?

Exercise 25.11 *Derivation: Components of the Riemann Tensor in an Arbitrary Basis*
By evaluating expression (25.30) in an arbitrary basis (which might not even be a coordinate basis), derive Eq. (25.50) for the components of the Riemann tensor. In your derivation keep in mind that commas denote partial derivations *only* in a

coordinate basis; in an arbitrary basis they denote the result of letting a basis vector act as a differential operator [cf. Eq. (24.34)].

Exercise 25.12 *Derivation: Weyl Curvature Tensor*
Show that the Weyl curvature tensor (25.48) has vanishing contraction on all its slots and has the same symmetries as Riemann: Eqs. (25.45). From these properties, show that Weyl has just 10 independent components. Write the Riemann tensor in terms of the Weyl tensor, the Ricci tensor, and the scalar curvature.

Exercise 25.13 *Problem: Curvature of the Surface of a Sphere*
On the surface of a sphere, such as Earth, introduce spherical polar coordinates in which the metric, written as a line element, takes the form

$$ds^2 = a^2(d\theta^2 + \sin^2\theta d\phi^2), \tag{25.51}$$

where a is the sphere's radius.

(a) Show (first by hand and then by computer) that the connection coefficients for the coordinate basis $\{\partial/\partial\theta, \partial/\partial\phi\}$ are

$$\Gamma^{\theta}{}_{\phi\phi} = -\sin\theta\cos\theta, \quad \Gamma^{\phi}{}_{\theta\phi} = \Gamma^{\phi}{}_{\phi\theta} = \cot\theta, \quad \text{all others vanish.} \tag{25.52a}$$

(b) Show that the symmetries (25.45) of the Riemann tensor guarantee that its only nonzero components in the above coordinate basis are

$$R_{\theta\phi\theta\phi} = R_{\phi\theta\phi\theta} = -R_{\theta\phi\phi\theta} = -R_{\phi\theta\theta\phi}. \tag{25.52b}$$

(c) Show, first by hand and then by computer, that

$$R_{\theta\phi\theta\phi} = a^2\sin^2\theta. \tag{25.52c}$$

(d) Show that in the basis

$$\{\vec{e}_{\hat{\theta}}, \vec{e}_{\hat{\phi}}\} = \left\{\frac{1}{a}\frac{\partial}{\partial\theta}, \frac{1}{a\sin\theta}\frac{\partial}{\partial\phi}\right\}, \tag{25.52d}$$

the components of the metric, the Riemann tensor, the Ricci tensor, the curvature scalar, and the Weyl tensor are

$$g_{\hat{j}\hat{k}} = \delta_{jk}, \quad R_{\hat{\theta}\hat{\phi}\hat{\theta}\hat{\phi}} = \frac{1}{a^2}, \quad R_{\hat{j}\hat{k}} = \frac{1}{a^2}g_{\hat{j}\hat{k}}, \quad R = \frac{2}{a^2}, \quad C_{\hat{\theta}\hat{\phi}\hat{\theta}\hat{\phi}} = 0, \tag{25.52e}$$

respectively. The first of these implies that the basis is orthonormal; the rest imply that the curvature is independent of location on the sphere, as it should be by spherical symmetry. [The θ dependence in the coordinate components of Riemann, Eq. (25.52c), like the θ dependence in the metric component $g_{\phi\phi}$, is a result of the θ dependence in the length of the coordinate basis vector $\vec{e}_\phi$: $|\vec{e}_\phi| = a\sin\theta$.]

Exercise 25.14 *Problem: Geodesic Deviation on a Sphere*

Consider two neighboring geodesics (great circles) on a sphere of radius a, one the equator and the other a geodesic slightly displaced from the equator (by $\Delta\theta = b$) and parallel to it at $\phi = 0$. Let $\vec{\xi}$ be the separation vector between the two geodesics, and note that at $\phi = 0$, $\vec{\xi} = b\partial/\partial\theta$. Let l be proper distance along the equatorial geodesic, so $d/dl = \vec{u}$ is its tangent vector.

(a) Show that $l = a\phi$ along the equatorial geodesic.

(b) Show that the equation of geodesic deviation (25.31) reduces to

$$\frac{d^2\xi^\theta}{d\phi^2} = -\xi^\theta, \quad \frac{d^2\xi^\phi}{d\phi^2} = 0. \tag{25.53}$$

(c) Solve Eq. (25.53), subject to the above initial conditions, to obtain

$$\xi^\theta = b\cos\phi, \quad \xi^\phi = 0. \tag{25.54}$$

Verify, by drawing a picture, that this is precisely what one would expect for the separation vector between two great circles.

25.7 Delicacies in the Equivalence Principle, and Some Nongravitational Laws of Physics in Curved Spacetime 25.7

Suppose that one knows a local, special relativistic, nongravitational law of physics in geometric, frame-independent form—for example, the expression for the stress-energy tensor of a perfect fluid in terms of its 4-velocity $\vec{u}$ and its rest-frame mass-energy density ρ and pressure P:

$$\boldsymbol{T} = (\rho + P)\vec{u} \otimes \vec{u} + P\boldsymbol{g} \tag{25.55}$$

[Eq. (24.51)]. Then the equivalence principle guarantees that in general relativity this law will assume the same geometric, frame-independent form. One can see that this is so by the same method as we used to derive the general relativistic equation of motion $\nabla_{\vec{p}}\,\vec{p} = 0$ for free particles [Eq. (25.11c) and associated discussion]:

1. Rewrite the special relativistic law in terms of components in a global Lorentz frame $[T^{\alpha\beta} = (\rho + P)u^\alpha u^\beta + Pg^{\alpha\beta}]$.
2. Infer from the equivalence principle that this same component form of the law will hold, unchanged, in a local Lorentz frame in general relativity.
3. Deduce that this component law is the local-Lorentz-frame version of the original geometric law $[\boldsymbol{T} = (\rho + P)\vec{u} \otimes \vec{u} + P\boldsymbol{g}]$, now lifted into general relativity.

equivalence principle implies that local nongravitational laws are the same in general relativity as in special relativity

Thus, when the local, nongravitational laws of physics are known in frame-independent form, one need not distinguish between whether they are special relativistic or general relativistic.

In this conclusion the word *local* is crucial. The equivalence principle is strictly valid only at the spatial origin of a local Lorentz frame; correspondingly, it is in danger of failure for any law of physics that cannot be formulated solely in terms of quantities that reside at the spatial origin (i.e., along a timelike geodesic). For the above example, $\boldsymbol{T} = (\rho + P)\vec{u} \otimes \vec{u} + P\boldsymbol{g}$, there is no problem; and for the local law of 4-momentum conservation, $\vec{\nabla} \cdot \boldsymbol{T} = 0$, there is no problem. However, for the global law of 4-momentum conservation

examples of the influence of spacetime curvature on nonlocal laws:

$$\int_{\partial\mathcal{V}} T^{\alpha\beta} d\Sigma_\beta = 0 \tag{25.56}$$

[Eq. (2.71) and Fig. 2.11], there is serious trouble. This law is severely nonlocal, since it involves integration over a finite, closed 3-surface $\partial\mathcal{V}$ in spacetime. Thus the equivalence principle fails for it. The failure shows up especially clearly when one notices (as we discussed in Sec. 24.3.4) that the quantity $T^{\alpha\beta} d\Sigma_\beta$, which the integral is trying to add up over $\partial\mathcal{V}$, has one empty slot, named α (i.e., it is a vector). This means that, to compute the integral (25.56), we must transport the contributions $T^{\alpha\beta} d\Sigma_\beta$ from the various tangent spaces in which they normally live to the tangent space of some single, agreed-on location, where they are to be added. The result of that transport depends on the route used, and in general no preferred route is available. As a result, the integral (25.56) is ill defined, and in general relativity we lose the global conservation law for 4-momentum!—except in special situations, one of which is discussed in Sec. 25.9.5.

no global conservation of 4-momentum in general relativity, generically

25.7.1

25.7.1 Curvature Coupling in the Nongravitational Laws T2

Another instructive example is the law by which a freely moving particle transports its spin angular momentum. The spin angular momentum is readily defined in the instantaneous local Lorentz rest frame of the particle's center of mass; there it is a 4-vector $\vec{S}$ with vanishing time component (so $\vec{S}$ is orthogonal to the particle's 4-velocity), with space components given by the familiar integral

$$S_i = \int_{\text{interior of body}} \epsilon_{ijk} x^j T^{k0}\, dV, \tag{25.57}$$

where the T^{k0} are components of the momentum density. In special relativity, the law of angular momentum conservation (e.g., Misner, Thorne, and Wheeler, 1973, Sec. 5.11) guarantees that the Lorentz-frame components S^α of this spin angular momentum remain constant, so long as no external torques act on the particle. This conservation law can be written in special relativistic, frame-independent notation as Eq. (24.62), specialized to a nonaccelerated particle:

$$\nabla_{\vec{u}}\, \vec{S} = 0; \tag{25.58}$$

T2

that is, the spin vector $\vec{S}$ is parallel-transported along the world line of the freely falling particle (which has 4-velocity $\vec{u}$). If this were a local law of physics, it would take this same form, unchanged, in general relativity (i.e., in curved spacetime). Whether the law is local or not clearly depends on the size of the particle. If the particle is vanishingly small in its own rest frame, then the law is local, and Eq. (25.58) will be valid in general relativity. However, if the particle has finite size, the law (25.58) is in danger of failing—and, indeed, it does fail if the particle's finite size is accompanied by a finite quadrupole moment. In that case, the coupling of the quadrupole moment $\mathcal{I}_{\alpha\beta}$ to the curvature of spacetime $R^{\alpha}{}_{\beta\gamma\delta}$ produces a torque on the "particle," so Eq. (25.58) acquires a driving term on the right-hand side:

$$\boxed{S^{\alpha}{}_{;\mu}u^{\mu} = \epsilon^{\alpha\beta\gamma\delta}\mathcal{I}_{\beta\mu}R^{\mu}{}_{\nu\gamma\zeta}u_{\delta}u^{\nu}u^{\zeta};} \tag{25.59}$$

see Ex. 25.16. Earth is a good example: the Riemann tensor $R^{\alpha}{}_{\beta\gamma\delta}$ produced at Earth by the Moon and the Sun couples to Earth's centrifugal-flattening-induced quadrupole moment $\mathcal{I}_{\mu\nu}$. The resulting torque (25.59) causes Earth's spin axis to precess relative to the distant stars, with a precession period of 26,000 years—sufficiently fast to show up clearly in historical records[5] as well as in modern astronomical measurements.

tidally induced precession of Earth's spin axis

This example illustrates the fact that, if a small amount of nonlocality is present in a physical law, then, when lifted from special relativity into general relativity, the law may acquire a small *curvature-coupling* modification.

What is the minimum amount of nonlocality that can produce curvature-coupling modifications in physical laws? As a rough rule of thumb, the minimum amount is double gradients. Because the connection coefficients vanish at the origin of a local Lorentz frame, the local Lorentz components of a single gradient are the same as the components in a global Lorentz frame (e.g., $A^{\alpha}{}_{;\beta} = \partial A^{\alpha}/\partial x^{\beta}$). However, because spacetime curvature prevents the spatial derivatives of the connection coefficients from vanishing at the origin of a local Lorentz frame, any law that involves double gradients is in danger of acquiring curvature-coupling corrections when lifted into general relativity. As an example, it turns out that the vacuum wave equation for the electromagnetic vector 4-potential, which in Lorenz gauge (Jackson, 1999, Sec. 6.3) takes the form $A^{\alpha;\mu}{}_{\mu} = 0$ in flat spacetime, becomes in curved spacetime:

double gradients as a source of curvature coupling in physical laws

example: curvature coupling in wave equation for electromagnetic 4-potential

$$\boxed{A^{\alpha;\mu}{}_{\mu} = R^{\alpha\mu}A_{\mu},} \tag{25.60}$$

where $R^{\alpha\mu}$ is the Ricci curvature tensor; see Ex. 25.15. [In Eq. (25.60)—and always—all indices that follow the semicolon represent differentiation slots: $A^{\alpha;\mu}{}_{\mu} \equiv A^{\alpha;\mu}{}_{;\mu}$.]

The curvature-coupling ambiguities that occur when one lifts slightly nonlocal laws from special relativity into general relativity using the equivalence principle are very similar to "factor-ordering ambiguities" that occur when one lifts a hamiltonian from classical mechanics into quantum mechanics using the correspondence

curvature-coupling ambiguities in general relativity are analogous to factor-ordering ambiguities in the quantum mechanical correspondence principle

5. For example, Earth's north pole did not point toward the star Polaris in the era of the ancient Egyptian civilization. Hipparchus of Nicaea discovered this precession in 127 BC by comparing his own observations of the stars with those of earlier astronomers.

T2

principle. In the case of the equivalence principle, the curvature coupling can be regarded as stemming from double gradients that commute in special relativity but do not commute in general relativity. In the case of the correspondence principle, the factor-ordering difficulties result because quantities that commute classically (e.g., position x and momentum p) do not commute quantum mechanically ($\hat{x}\hat{p} \neq \hat{p}\hat{x}$), so when the products of such quantities appear in a classical hamiltonian one does not know, a priori, their correct order in the quantum hamiltonian [does xp become $\hat{x}\hat{p}$, or $\hat{p}\hat{x}$, or $\frac{1}{2}(\hat{x}\hat{p} + \hat{p}\hat{x})$?]. (However, in each case, general relativity or quantum mechanics, the true curvature coupling or true factor ordering is unambiguous. The ambiguity is solely in the prescription for deducing it via the equivalence principle or the correspondence principle.)

EXERCISES

Exercise 25.15 *Example and Derivation: Curvature Coupling in the Electromagnetic Wave Equation* T2

Since Maxwell's equations, written in terms of the classically measurable electromagnetic field tensor $\boldsymbol{F}$ [Eqs. (2.48)] involve only single gradients, it is reasonable to expect them to be lifted into curved spacetime without curvature-coupling additions. Assume this is true. It can be shown that: (i) if one writes the electromagnetic field tensor $\boldsymbol{F}$ in terms of a 4-vector potential $\vec{A}$ as

$$F_{\alpha\beta} = A_{\beta;\alpha} - A_{\alpha;\beta}, \tag{25.61}$$

then half of the curved-spacetime Maxwell equations, $F_{\alpha\beta;\gamma} + F_{\beta\gamma;\alpha} + F_{\gamma\alpha;\beta} = 0$ [the second of Eqs. (2.48)] are automatically satisfied; (ii) $\boldsymbol{F}$ is unchanged by gauge transformations in which a gradient is added to the vector potential, $\vec{A} \to \vec{A} + \vec{\nabla}\psi$; and (iii) by such a gauge transformation one can impose the Lorenz-gauge condition $\vec{\nabla} \cdot \vec{A} = 0$ on the vector potential.

Show that, when the charge-current 4-vector vanishes, $\vec{J} = 0$, the other half of the Maxwell equations, $F^{\alpha\beta}{}_{;\beta} = 0$ [the first of Eqs. (2.48)] become, in Lorenz gauge and in curved spacetime, the wave equation with curvature coupling [Eq. (25.60)].

Exercise 25.16 *Example and Derivation: Curvature-Coupling Torque* T2

(a) In the Newtonian theory of gravity, consider an axisymmetric, spinning body (e.g., Earth) with spin angular momentum S_j and time-independent mass distribution $\rho(\mathbf{x})$, interacting with an externally produced tidal gravitational field $\mathcal{E}_{jk}$ (e.g., that of the Sun and the Moon). Show that the torque around the body's center of mass, exerted by the tidal field, and the resulting evolution of the body's spin are

$$\frac{dS_i}{dt} = -\epsilon_{ijk}\mathcal{I}_{jl}\mathcal{E}_{kl}. \tag{25.62}$$

Here

$$\mathcal{I}_{kl} = \int \rho \left(x_k x_l - \frac{1}{3} r^2 \delta_{kl} \right) dV \tag{25.63}$$

is the body's mass quadrupole moment, with $r = \sqrt{\delta_{ij}x_i x_j}$ the distance from the center of mass. T2

(b) For the centrifugally flattened Earth interacting with the tidal fields of the Moon and the Sun, estimate in order of magnitude the spin-precession period produced by this torque. [The observed precession period is 26,000 years.]

(c) Show that when rewritten in the language of general relativity, and in frame-independent, geometric language, Eq. (25.62) takes the form (25.59) discussed in the text. As part of showing this, explain the meaning of $\mathcal{I}_{\beta\mu}$ in that equation.

For a derivation and discussion of this relativistic curvature-coupling torque when the spinning body is a black hole rather than a Newtonian body, see, for example, Thorne and Hartle (1985).

25.8 The Einstein Field Equation

25.8

One crucial issue remains to be studied in this overview of the foundations of general relativity: What is the physical law that determines the curvature of spacetime? Einstein's search for that law, his *Einstein field equation,* occupied a large fraction of his efforts during the years 1913, 1914, and 1915. Several times he thought he had found it, but each time his proposed law turned out to be fatally flawed; for some flavor of his struggle, see the excerpts from his writings in Misner, Thorne, and Wheeler (1973, Sec. 17.7).

In this section, we briefly examine one segment of Einstein's route toward his field equation: the segment motivated by contact with Newtonian gravity.

Newtonian gravity

The Newtonian potential Φ is a close analog of the general relativistic spacetime metric $\boldsymbol{g}$. From Φ we can deduce everything about Newtonian gravity, and from $\boldsymbol{g}$ we can deduce everything about spacetime curvature. In particular, by differentiating Φ twice we can obtain the Newtonian tidal field $\boldsymbol{\mathcal{E}}$ [Eq. (25.24)], and by differentiating the components of $\boldsymbol{g}$ twice we can obtain the components of the relativistic generalization of $\boldsymbol{\mathcal{E}}$: the Riemann curvature tensor [Eq. (25.44) in a local Lorentz frame; Eq. (25.50) in an arbitrary basis].

In Newtonian gravity, Φ is determined by Newton's field equation

$$\boxed{\nabla^2\Phi = 4\pi G\rho,} \qquad (25.64)$$

which can be rewritten in terms of the tidal field, $\mathcal{E}_{jk} = \partial^2\Phi/\partial x^j \partial x^k$, as

$$\mathcal{E}^j{}_j = 4\pi G\rho. \qquad (25.65)$$

deducing the Einstein field equation from Newtonian gravity

Note that this equates a piece of the tidal field—its contraction or *trace*—to the density of mass. By analogy we can expect the Einstein field equation to equate a piece of the Riemann curvature tensor (the analog of the Newtonian tidal field) to some tensor analog of the Newtonian mass density. Further guidance comes from the demand that in nearly Newtonian situations (e.g., in the solar system), the Einstein field equation should reduce to Newton's field equation. To exploit that guidance, we can (i) write the Newtonian tidal field for nearly Newtonian situations in terms of general relativity's Riemann tensor, $\mathcal{E}_{jk} = R_{j0k0}$ [Eq. (25.35); valid in a local Lorentz frame], (ii) then take the trace and note that by its symmetries $R^0{}_{000} = 0$ so that $\mathcal{E}^j{}_j = R^\alpha{}_{0\alpha 0} = R_{00}$, and (iii) thereby infer that the Newtonian limit of the Einstein equation should read, in a local Lorentz frame:

$$R_{00} = 4\pi G\rho. \tag{25.66}$$

Here R_{00} is the time-time component of the Ricci curvature tensor, which can be regarded as a piece of the Riemann tensor. An attractive proposal for the Einstein field equation should now be obvious. Since the equation should be geometric and frame independent, and since it must have the Newtonian limit (25.66), it presumably should say $R_{\alpha\beta} = 4\pi G \times$(a second-rank symmetric tensor that generalizes the Newtonian mass density ρ). The obvious generalization of ρ is the stress-energy tensor $T_{\alpha\beta}$, so a candidate is

a failed field equation

$$R_{\alpha\beta} = 4\pi G T_{\alpha\beta}. \tag{25.67}$$

Einstein flirted extensively with this proposal for the field equation during 1913–1915. However, it, like several others he studied, was fatally flawed. When expressed in a coordinate system in terms of derivatives of the metric components $g_{\mu\nu}$, it becomes (because $R_{\alpha\beta}$ and $T_{\alpha\beta}$ both have 10 independent components) 10 independent differential equations for the 10 $g_{\mu\nu}$. This is too many equations. By an arbitrary change of coordinates, $x^\alpha_{\text{new}} = F^\alpha(x^0_{\text{old}}, x^1_{\text{old}}, x^2_{\text{old}}, x^3_{\text{old}})$, involving four arbitrary functions F^0, F^1, F^2, and F^3, one should be able to impose on the metric components four arbitrary conditions, analogous to gauge conditions in electromagnetism (e.g., one should be able to set $g_{00} = -1$ and $g_{0j} = 0$ everywhere). Correspondingly, the field equations should constrain only 6, not 10 of the components of the metric (the 6 g_{ij} in our example).

the need for a divergence-free curvature tensor

In November 1915, Einstein (1915), and independently Hilbert (1915) [who was familiar with Einstein's struggle as a result of private conversations and correspondence] discovered the resolution of this dilemma. Because the local law of 4-momentum conservation guarantees $T^{\alpha\beta}{}_{;\beta} = 0$ independent of the field equation, if we replace the Ricci tensor in Eq. (25.67) by a constant (to be determined) times some new curvature tensor $G^{\alpha\beta}$ that is also automatically divergence free independently of the field equation ($G^{\alpha\beta}{}_{;\beta} \equiv 0$), then the new field equation $G^{\alpha\beta} = \kappa T^{\alpha\beta}$ (with $\kappa =$ constant) will not constrain all 10 components of the metric. Rather, the four equations, $(G^{\alpha\beta} - \kappa T^{\alpha\beta})_{;\beta} = 0$ with $\alpha = 0, 1, 2, 3$, will automatically be satisfied; they will not constrain the metric components in any way, and only six independent constraints on the metric components will remain in the field equation, precisely the desired number.

It turns out, in fact, that from the Ricci tensor and the scalar curvature one can construct a curvature tensor $G^{\alpha\beta}$ with the desired property:

$$\boxed{G^{\alpha\beta} \equiv R^{\alpha\beta} - \frac{1}{2}Rg^{\alpha\beta}.} \tag{25.68}$$

Einstein curvature tensor

Today we call this the *Einstein curvature tensor.* That it has vanishing divergence, independently of how one chooses the metric,

$$\boxed{\vec{\nabla} \cdot \boldsymbol{G} \equiv 0,} \tag{25.69}$$

Bianchi identity and contracted Bianchi identity

is called the *contracted Bianchi identity,* since it can be obtained by contracting the following *Bianchi identity* on the tensor $\epsilon_\alpha{}^{\beta\mu\nu}\epsilon_\nu{}^{\gamma\delta\epsilon}$:

$$\boxed{R^\alpha{}_{\beta\gamma\delta;\epsilon} + R^\alpha{}_{\beta\delta\epsilon;\gamma} + R^\alpha{}_{\beta\epsilon\gamma;\delta} = 0.} \tag{25.70}$$

[This Bianchi identity holds true for the Riemann curvature tensor of any and every manifold (i.e., of any and every smooth space; Ex. 25.17). For an extensive discussion of the Bianchi identities (25.70) and (25.69) and their geometric interpretation, see Misner, Thorne, and Wheeler (1973, Chap. 15).]

The Einstein field equation should then equate a multiple of $T^{\alpha\beta}$ to the Einstein tensor $G^{\alpha\beta}$:

$$G^{\alpha\beta} = \kappa T^{\alpha\beta}. \tag{25.71a}$$

The proportionality constant κ is determined from the Newtonian limit as follows. By rewriting the field equation (25.71a) in terms of the Ricci tensor

$$R^{\alpha\beta} - \frac{1}{2}g^{\alpha\beta}R = \kappa T^{\alpha\beta}, \tag{25.71b}$$

then taking the trace to obtain $R = -\kappa g_{\mu\nu}T^{\mu\nu}$ and inserting this back into (25.71a), we obtain

$$R^{\alpha\beta} = \kappa\left(T^{\alpha\beta} - \frac{1}{2}g^{\alpha\beta}g_{\mu\nu}T^{\mu\nu}\right). \tag{25.71c}$$

In nearly Newtonian situations and in a local Lorentz frame, the mass-energy density $T^{00} \cong \rho$ is far greater than the momentum density T^{j0} and is also far greater than the stress T^{jk}. Correspondingly, the time-time component of the field equation (25.71c) becomes

$$R^{00} = \kappa\left(T^{00} - \frac{1}{2}\eta^{00}\eta_{00}T^{00}\right) = \frac{1}{2}\kappa T^{00} = \frac{1}{2}\kappa\rho. \tag{25.71d}$$

By comparing with the correct Newtonian limit (25.66) and noting that in a local Lorentz frame $R_{00} = R^{00}$, we see that $\kappa = 8\pi G$, whence the Einstein field equation is

$$\boxed{G^{\alpha\beta} = 8\pi G T^{\alpha\beta}.} \tag{25.72}$$

Einstein field equation

EXERCISES

Exercise 25.17 *Derivation and Example: Bianchi Identities*

(a) Derive the Bianchi identity (25.70) in 4-dimensional spacetime. [Hint: (i) Introduce a local Lorentz frame at some arbitrary event. (ii) In that frame show, from Eq. (25.50), that the components $R_{\alpha\beta\gamma\delta}$ of Riemann have the form of Eq. (25.44) plus corrections that are quadratic in the distance from the origin. (iii) Compute the left-hand side of Eq. (25.70), with index α down, at the origin of that frame, and show that it is zero. (iv) Then argue that because the origin of the frame was an arbitrary event in spacetime, and because the left-hand side of Eq. (25.70) is an arbitrary component of a tensor, the left-hand side viewed as a frame-independent geometric object must vanish at all events in spacetime.]

(b) By contracting the Bianchi identity (25.70) on $\epsilon_\alpha{}^{\beta\mu\nu}\epsilon_\nu{}^{\gamma\delta\epsilon}$, derive the contracted Bianchi identity (25.69).

These derivations are easily generalized to an arbitrary manifold with any dimension by replacing the 4-dimensional local Lorentz frame by a locally orthonormal coordinate system (cf. the third and fourth paragraphs of Sec. 24.3.3).

25.8.1

25.8.1 Geometrized Units

By now the reader should be accustomed to our use of geometrized units in which the speed of light is unity. Just as setting $c = 1$ has simplified greatly the mathematical notation in Chaps. 2 and 24, so also subsequent notation is greatly simplified if we set Newton's gravitation constant G to unity. This further geometrization of our units corresponds to equating mass units to length units via the relation

geometrized units: setting Newton's gravitational constant to one

$$1 = \frac{G}{c^2} = 7.426 \times 10^{-28}\ \text{m kg}^{-1}; \quad \text{so} \quad 1\,\text{kg} = 7.426 \times 10^{-28}\ \text{m}. \tag{25.73}$$

Any equation can readily be converted from conventional units to geometrized units by removing all factors of c and G; it can readily be converted back by inserting whatever factors of c and G one needs to make both sides of the equation dimensionally the same. Table 25.1 lists a few important numerical quantities in both conventional units and geometrized units.

In geometrized units, the Einstein field equation (25.72) assumes the following standard form, to which we appeal extensively in the three coming chapters:

the Einstein field equation in geometrized units

$$G^{\alpha\beta} = 8\pi T^{\alpha\beta}; \quad \text{or} \quad \mathbf{G} = 8\pi \mathbf{T}. \tag{25.74}$$

25.9

25.9 Weak Gravitational Fields

All the foundations of general relativity are now in our hands. In this concluding section of the chapter, we explore their predictions for the properties of weak gravitational

TABLE 25.1: Some useful quantities in conventional and geometrized units

Quantity	Conventional units	Geometrized units
Speed of light, c	2.998×10^8 m sec^{-1}	one
Newton's gravitation constant, G	6.674×10^{-11} m^3 kg^{-1} s^{-2}	one
G/c^2	7.426×10^{-28} m kg^{-1}	one
c^5/G	3.628×10^{52} W	one
Planck's reduced constant $\hbar$	1.055×10^{-34} kg m^2 s^{-1}	$(1.616 \times 10^{-35}$ m$)^2$
Sun's mass, $M_\odot$	1.989×10^{30} kg	1.477 km
Sun's radius, $R_\odot$	6.957×10^8 m	6.957×10^8 m
Earth's mass, $M_\oplus$	5.972×10^{24} kg	4.435 mm
Earth's mean radius, $R_\oplus$	6.371×10^6 m	6.371×10^6 m

Note: 1 Mpc = 10^6 parsecs (pc), 1 pc = 3.262 light-years (lt-yr), 1 lt-yr = 0.9461×10^{16} m, 1 AU = 1.496×10^{11} m. For other useful astronomical constants, see Cox (2000).

fields, beginning with the Newtonian limit of general relativity and then moving on to other situations.

25.9.1 Newtonian Limit of General Relativity

25.9.1

A general relativistic gravitational field (spacetime curvature) is said to be *weak* if there exist "nearly globally Lorentz" coordinate systems in which the metric coefficients differ only slightly from unity:

$$\boxed{g_{\alpha\beta} = \eta_{\alpha\beta} + h_{\alpha\beta}, \quad \text{with } |h_{\alpha\beta}| \ll 1.} \tag{25.75a}$$

conditions required for general relativity to become Newton's theory of gravity (the Newtonian limit)

The Newtonian limit requires that gravity be weak in this sense throughout the system being studied. It further requires a slow-motion constraint, which has three aspects:

1. The sources of gravity must have slow enough motions that the following holds for some specific choice of the nearly globally Lorentz coordinates:

$$|h_{\alpha\beta,t}| \ll |h_{\alpha\beta,j}|; \tag{25.75b}$$

2. the sources' motions must be slow enough that in this frame the momentum density is very small compared to the energy density:

$$|T^{j0}| \ll T^{00} \equiv \rho; \tag{25.75c}$$

3. and any particles on which the action of gravity is to be studied must move with low velocities, so they must have 4-velocities satisfying

$$|u^j| \ll u^0. \tag{25.75d}$$

Finally, the Newtonian limit requires that the stresses in the gravitating bodies be small compared to their mass densities:

$$|T^{jk}| \ll T^{00} \equiv \rho. \tag{25.75e}$$

When conditions (25.75) are all satisfied, then to leading nontrivial order in the small dimensionless quantities $|h_{\alpha\beta}|$, $|h_{\alpha\beta,t}|/|h_{\alpha\beta,j}|$, $|T^{j0}|/T^{00}$, $|u^j|/u^0$, and $|T^{jk}|/T^{00}$ the laws of general relativity reduce to those of Newtonian theory.

The details of this reduction are an exercise for the reader (Ex. 25.18); here we give an outline.

details of the Newtonian limit:

The low-velocity constraint $|u^j|/u^0 \ll 1$ on the 4-velocity of a particle, together with its normalization $u^\alpha u^\beta g_{\alpha\beta} = -1$ and the near flatness of the metric (25.75a), implies that

$$u^0 \cong 1, \quad u^j \cong v^j \equiv \frac{dx^j}{dt}. \tag{25.76}$$

Since $u^0 = dt/d\tau$, the first of these relations implies that in our nearly globally Lorentz coordinate system the coordinate time is very nearly equal to the proper time of our slow-speed particle. In this way, we recover the "universal time" of Newtonian theory. The universal, Euclidean space is that of our nearly Lorentz frame, with $h_{\mu\nu}$ completely ignored because of its smallness. The universal time and universal Euclidean space become the arena in which Newtonian physics is formulated.

Equation (25.76) for the components of a particle's 4-velocity, together with $|v^j| \ll 1$ and $|h_{\mu\nu}| \ll 1$, imply that the geodesic equation for a freely moving particle to leading nontrivial order is

$$\frac{dv^j}{dt} \cong \frac{1}{2}\frac{\partial h_{00}}{\partial x^j}, \quad \text{where } \frac{d}{dt} \equiv \frac{\partial}{\partial t} + \mathbf{v}\cdot\boldsymbol{\nabla}. \tag{25.77}$$

(Because our spatial coordinates are Cartesian, we can put the spatial index j up on one side of the equation and down on the other without creating any danger of error.)

By comparing Eq. (25.77) with Newton's equation of motion for the particle, we deduce that h_{00} must be related to the Newtonian gravitational potential by

$$h_{00} = -2\Phi, \tag{25.78}$$

so the spacetime metric in our nearly globally Lorentz coordinate system must be

spacetime metric in the Newtonian limit

$$\boxed{ds^2 = -(1+2\Phi)dt^2 + (\delta_{jk} + h_{jk})dx^j dx^k + 2h_{0j}dt\, dx^j.} \tag{25.79}$$

Because gravity is weak, only those parts of the Einstein tensor that are linear in $h_{\alpha\beta}$ are significant; quadratic and higher-order contributions can be ignored. Now, by the same mathematical steps as led us to Eq. (25.44) for the components of the Riemann tensor in a local Lorentz frame, one can show that the components of the linearized Riemann tensor in our nearly global Lorentz frame have that same form (i.e., setting $g_{\alpha\beta} = \eta_{\alpha\beta} + h_{\alpha\beta}$):

$$R_{\alpha\beta\gamma\delta} = \frac{1}{2}(h_{\alpha\delta,\beta\gamma} + h_{\beta\gamma,\alpha\delta} - h_{\alpha\gamma,\beta\delta} - h_{\beta\delta,\alpha\gamma}). \tag{25.80}$$

From this equation and the slow-motion constraint $|h_{\alpha\beta,t}| \ll |h_{\alpha\beta,j}|$, we infer that the space-time-space-time components of Riemann are

tidal gravity in the Newtonian limit

$$\boxed{R_{j0k0} = -\frac{1}{2}h_{00,jk} = \Phi_{,jk} = \mathcal{E}_{jk}.} \tag{25.81}$$

In the last step we have used Eq. (25.78). We have thereby recovered the relation between the Newtonian tidal field $\mathcal{E}_{jk} \equiv \Phi_{,jk}$ and the relativistic tidal field R_{j0k0}. That relation can now be used, via the train of arguments in the preceding section, to show that the Einstein field equation, $G^{\mu\nu} = 8\pi T^{\mu\nu}$, reduces to the Newtonian field equation, $\nabla^2\Phi = 4\pi T^{00} \equiv 4\pi\rho$.

This analysis leaves the details of h_{0j} and h_{jk} unknown, because the Newtonian limit is insensitive to them.

EXERCISES

Exercise 25.18 *Derivation: Newtonian Limit of General Relativity*

Consider a system that can be covered by nearly globally Lorentz coordinates in which the Newtonian-limit constraints (25.75) are satisfied. For such a system, flesh out the details of the text's derivation of the Newtonian limit. More specifically, do the following.

(a) Derive Eq. (25.76) for the components of the 4-velocity of a particle.

(b) Show that the geodesic equation reduces to Eq. (25.77).

(c) Show that to linear order in the metric perturbation $h_{\alpha\beta}$, the components of the Riemann tensor take the form of Eq. (25.80).

(d) Show that in the slow-motion limit the space-time-space-time components of Riemann take the form of Eq. (25.81).

25.9.2 Linearized Theory

25.9.2

There are many systems in the universe that have weak gravity [Eq. (25.75a)], but for which the slow-motion approximations (25.75b)–(25.75d) and/or weak-stress approximation (25.75e) fail. Examples are high-speed particles and electromagnetic fields. For such systems, we need a generalization of Newtonian theory that drops the slow-motion and weak-stress constraints but keeps the weak-gravity constraint:

details of linearized theory (weak-gravity limit of general relativity):

spacetime metric and metric perturbation

$$\boxed{g_{\alpha\beta} = \eta_{\alpha\beta} + h_{\alpha\beta}, \quad \text{with } |h_{\alpha\beta}| \ll 1.} \tag{25.82}$$

The obvious generalization is a linearization of general relativity in $h_{\alpha\beta}$, with no other approximations being made—the so-called *linearized theory of gravity.* In this subsection we develop it.

In formulating linearized theory we can regard the metric perturbation $h_{\mu\nu}$ as a gravitational field that lives in flat spacetime, and correspondingly, we can carry out our mathematics as though we were in special relativity. In other words, linearized theory can be regarded as a field theory of gravity in flat spacetime—a variant of the type of theory that Einstein toyed with and then rejected (Sec. 25.1).

In linearized theory, the Riemann tensor takes the form of Eq. (25.80), but we have no right to simplify it further to the form of Eq. (25.81), so we must follow a different route to the linearized Einstein field equation.

Contracting the first and third indices in Eq. (25.80), we obtain an expression for the linearized Ricci tensor $R_{\mu\nu}$ in terms of $h_{\alpha\beta}$. Contracting once again, we obtain the scalar curvature R, and then from Eq. (25.68) we obtain for the Einstein tensor and the Einstein field equation:

linearized Einstein field equation

$$\begin{aligned} 2G_{\mu\nu} &= h_{\mu\alpha,\nu}{}^{\alpha} + h_{\nu\alpha,\mu}{}^{\alpha} - h_{\mu\nu,\alpha}{}^{\alpha} - h_{,\mu\nu} - \eta_{\mu\nu}(h_{\alpha\beta}{}^{,\alpha\beta} - h_{,\alpha}{}^{\alpha}) \\ &= 16\pi T_{\mu\nu}. \end{aligned} \tag{25.83}$$

Here all indices, subscript or superscript, that follow the comma are partial-derivative indices [e.g., $h_{,\beta}{}^{\beta} = (\partial^2 h/\partial x^\beta \partial x^\alpha)\eta^{\alpha\beta}$], and

$$h \equiv \eta^{\alpha\beta} h_{\alpha\beta} \tag{25.84}$$

is the "trace" of the metric perturbation. We can simplify the field equation (25.83) by reexpressing it in terms of the quantity

trace-reversed metric perturbation

$$\boxed{\bar{h}_{\mu\nu} \equiv h_{\mu\nu} - \frac{1}{2} h \eta_{\mu\nu}.} \tag{25.85}$$

One can easily check that this quantity has the opposite trace to that of $h_{\mu\nu}$ ($\bar{h} \equiv \bar{h}_{\alpha\beta}\eta^{\alpha\beta} = -h$), so it is called the *trace-reversed metric perturbation*. In terms of it, the field equation (25.83) becomes

$$-\bar{h}_{\mu\nu,\alpha}{}^{\alpha} - \eta_{\mu\nu}\bar{h}_{\alpha\beta,}{}^{\alpha\beta} + \bar{h}_{\mu\alpha,\nu}{}^{\alpha} + \bar{h}_{\nu\alpha,\mu}{}^{\alpha} = 16\pi T_{\mu\nu}. \tag{25.86}$$

We can simplify this field equation further by specializing our coordinates. We introduce a new nearly globally Lorentz coordinate system that is related to the old one by

infinitesimal coordinate transformation (gauge change)

$$\boxed{x^\alpha_{\text{new}}(\mathcal{P}) = x^\alpha_{\text{old}}(\mathcal{P}) + \xi^\alpha(\mathcal{P}),} \tag{25.87}$$

where ξ^α is a small vectorial displacement of the coordinate grid. This change of coordinates via four arbitrary functions ($\alpha = 0, 1, 2, 3$) produces a change of the functional form of the metric perturbation $h_{\alpha\beta}$ to

influence of gauge change on metric perturbation

$$\boxed{h^{\text{new}}_{\mu\nu} = h^{\text{old}}_{\mu\nu} - \xi_{\mu,\nu} - \xi_{\nu,\mu}} \tag{25.88}$$

(Ex. 25.19), and a corresponding change of the trace-reversed metric perturbation. This is linearized theory's analog of a *gauge transformation* in electromagnetic theory. Just as an electromagnetic gauge change generated by a scalar field ψ alters the vector potential, $A_\mu^{\text{new}} = A_\mu^{\text{old}} - \psi_{,\mu}$, so the linearized-theory gauge change generated by ξ^α alters $h_{\mu\nu}$ and $\bar{h}_{\mu\nu}$; and just as the force-producing electromagnetic field tensor $F_{\mu\nu}$ is unaffected by an electromagnetic gauge change, so the tidal-force-producing linearized Riemann tensor is left unaffected by the gravitational gauge change (Ex. 25.19).

By a special choice of the four functions ξ^α (Ex. 25.19), we can impose the following four gauge conditions on $\bar{h}_{\mu\nu}$:

$$\boxed{\bar{h}_{\mu\nu,}{}^{\nu} = 0.} \tag{25.89}$$

gravitational Lorenz gauge condition

These are linearized theory's analog of the electromagnetic Lorenz gauge condition $A_{\mu,}{}^{\mu} = 0$, so they are called the *gravitational Lorenz gauge*. Just as the flat-spacetime Maxwell equations take the remarkably simple wave-equation form $A_{\mu,\alpha}{}^{\alpha} = 4\pi J_\mu$ in Lorenz gauge, so also the linearized Einstein equation (25.86) takes the corresponding simple wave-equation form in gravitational Lorenz gauge:

$$\boxed{-\bar{h}_{\mu\nu,\alpha}{}^{\alpha} = 16\pi T_{\mu\nu}.} \tag{25.90}$$

linearized Einstein field equation in Lorenz gauge

By the same method as one uses in electromagnetic theory (e.g., Jackson, 1999, Sec. 6.4), one can solve this gravitational field equation for the field $\bar{h}_{\mu\nu}$ produced by an arbitrary stress-energy-tensor source:

$$\boxed{\bar{h}_{\mu\nu}(t, \mathbf{x}) = \int \frac{4T_{\mu\nu}(t - |\mathbf{x} - \mathbf{x}'|, \mathbf{x}')}{|\mathbf{x} - \mathbf{x}'|} dV_{x'}.} \tag{25.91}$$

Lorenz-gauge trace-reversed metric perturbation as retarded integral over stress-energy tensor

The quantity in the numerator is the stress-energy source evaluated at the "retarded time" $t' = t - |\mathbf{x} - \mathbf{x}'|$. This equation for the field, and the wave equation (25.90) that underlies it, show explicitly that dynamically changing distributions of stress-energy must generate *gravitational waves*, which propagate outward from their source at the speed of light (Einstein, 1916b; Einstein, 1918). We study these gravitational waves in Chap. 27.

EXERCISES

Exercise 25.19 *Derivation: Gauge Transformations in the Linearized Theory*

(a) Show that the "infinitesimal" coordinate transformation (25.87) produces the change (25.88) of the linearized metric perturbation and that it leaves the Riemann tensor (25.80) unchanged.

(b) Exhibit a differential equation for the ξ^α that brings the metric perturbation into gravitational Lorenz gauge [i.e., that makes $h_{\mu\nu}^{\text{new}}$ obey the Lorenz gauge condition (25.89)].

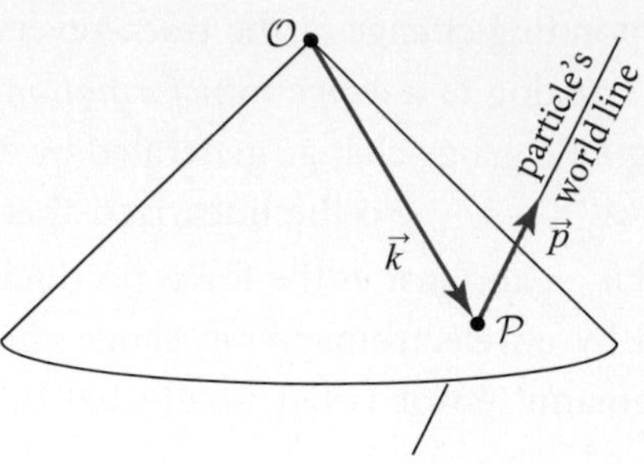

FIGURE 25.4 The past light cone of an observation event $\mathcal{O}$, the world line of a particle, and two 4-vectors: the particle's 4-momentum $\vec{p}$ at the point $\mathcal{P}$ where it passes through $\mathcal{O}$'s past light cone, and the past-directed null vector $\vec{k}$ that reaches from $\mathcal{O}$ to $\mathcal{P}$.

(c) Show that in gravitational Lorenz gauge, the Einstein field equation (25.86) reduces to Eq. (25.90).

Exercise 25.20 *Example: Gravitational Field of a Rapidly Moving Particle* T2
In this exercise we illustrate linearized theory by computing the gravitational field of a moving particle with finite rest mass and then that of a zero-rest-mass particle that moves with the speed of light.

(a) From Eq. (25.91), deduce that, for a particle with mass M at rest at the origin, the only nonvanishing component of $\bar{h}^{\mu\nu}$ is $\bar{h}^{00} = 4M/r$.

(b) Regarding $\bar{h}^{\mu\nu}$ as a field that lives in flat spacetime, show that it can be written in frame-independent, special relativistic form as

$$\bar{h}^{\mu\nu} = \frac{4p^{\mu}p^{\nu}}{\vec{k}\cdot\vec{p}}, \tag{25.92}$$

where $\vec{p}$ is the particle's 4-momentum, and $\vec{k}$ is the past-directed null vector that reaches from the observation event $\mathcal{O}$ to the event $\mathcal{P}$ at which the particle's world line passes through the observer's past light cone; see Fig. 25.4. Equation (25.92) is a very powerful formula. It is an explicit form of the solution (25.91) to the wave equation (25.90) not only for a particle that moves inertially (and thus could be at rest in our original reference frame) but also for an arbitrarily accelerated particle. Explain why.

(c) In the Lorentz rest frame of the observer, let the particle move along the x-axis with speed v, so its world line is $\{x = vt,\ y = z = 0\}$, and its 4-momentum has components $p^0 = M\gamma$ and $p^x = Mv\gamma$, with $\gamma = 1/\sqrt{1-v^2}$. Show that, expressed in terms of the observation event's coordinates (t, x, y, z),

$$\vec{k}\cdot\vec{p} = \gamma M R, \quad \text{where } R = \sqrt{(1-v^2)(y^2+z^2) + (x-vt)^2}. \tag{25.93}$$

Show, further, that the linearized spacetime metric is T2

$$ds^2 = \left(1 + \frac{2M}{\gamma R}\right)(-dt^2 + dx^2 + dy^2 + dz^2) + \frac{4\gamma M}{R}(dt - v dx)^2. \quad (25.94)$$

(d) Take the limit of a zero-rest-mass particle moving at the speed of light by sending $m \to 0$, $v \to 1$, and $M\gamma \to \mathcal{E}$ (the particle's energy). [The limit of $1/R = 1/\sqrt{(1-v^2)(y^2+z^2) + (x-vt)^2}$ is quite tricky. It turns out to be $1/|x-t| - \delta(x-t)\ln(y^2+z^2)$, where δ is the Dirac delta function (Aichelberg and Sexl, 1971).] Show that the resulting metric is

$$\begin{aligned} ds^2 = & -dt^2 + dx^2 + dy^2 + dz^2 \\ & + 4\mathcal{E}\left(\frac{1}{|x-t|} - \ln(y^2+z^2)\delta(x-t)\right)(dx - dt)^2. \end{aligned}$$

By a change of coordinates, get rid of the $1/|x-t|$ term, thereby obtaining our final form for the metric of a zero-rest-mass particle:

$$ds^2 = -dt^2 + dx^2 + dy^2 + dz^2 - 4\mathcal{E}\ln(x^2+y^2)\,\delta(x-t)(dx-dt)^2. \quad (25.95)$$

Equation (25.95) turns out to be an exact solution of the fully nonlinear Einstein field equation for a zero-rest-mass particle; it is called the Aichelberg-Sexl *ultraboost* solution. Just as, when a charged particle is accelerated to near light speed, its electric field lines are compressed into its transverse plane, so the metric (25.95) has all its deviations from flat spacetime concentrated in the particle's transverse plane.

25.9.3 Gravitational Field outside a Stationary, Linearized Source of Gravity T2

25.9.3

stationary source of gravity

Let us specialize to a time-independent source of weak gravity (so $T_{\mu\nu,t} = 0$ in our chosen nearly globally Lorentz frame) and compute its external gravitational field as a power series in 1/(distance to source). We place our origin of coordinates at the source's center of mass, so

mass-centered coordinates

$$\int x^j T^{00} dV = 0, \quad (25.96)$$

and in the same manner as in electromagnetic theory, we expand

$$\frac{1}{|\mathbf{x} - \mathbf{x}'|} = \frac{1}{r} + \frac{x^j x^{j'}}{r^3} + \ldots, \quad (25.97)$$

where $r \equiv |\mathbf{x}|$ is the distance of the field point from the source's center of mass. Inserting Eq. (25.97) into the general solution (25.91) of the Einstein equation and

T2

taking note of the conservation laws $T^{\alpha j}{}_{,j} = 0$, we obtain for the source's external field:

trace-reversed metric perturbation

$$\bar{h}_{00} = \frac{4M}{r} + \mathrm{O}\left(\frac{1}{r^3}\right), \quad \bar{h}_{0j} = -\frac{2\epsilon_{jkm} J^k x^m}{r^3} + \mathrm{O}\left(\frac{1}{r^3}\right), \quad \bar{h}_{ij} = \mathrm{O}\left(\frac{1}{r^3}\right). \tag{25.98a}$$

Here M and J^k are the source's mass and angular momentum:

source's mass and angular momentum

$$M \equiv \int T^{00} dV, \quad J_k \equiv \int \epsilon_{kab} x^a T^{0b} dV \tag{25.98b}$$

(see Ex. 25.21). This expansion in $1/r$, as in the electromagnetic case, is a multipolar expansion. At order $1/r$ the field is spherically symmetric and the monopole moment is the source's mass M. At order $1/r^2$ there is a "magnetic-type dipole moment," the source's spin angular momentum J_k. These are the leading-order moments in two infinite sets: the "mass multipole" moments (analog of electric moments), and the "mass-current multipole" moments (analog of magnetic moments). For details on all the higher-order moments, see, for example, Thorne (1980).

The metric perturbation can be computed by reversing the trace reversal: $h_{\alpha\beta} = \bar{h}_{\alpha\beta} - \frac{1}{2}\eta_{\alpha\beta}\bar{h}$. Thereby we obtain for the spacetime metric, $g_{\alpha\beta} = \eta_{\alpha\beta} + h_{\alpha\beta}$, at linear order, outside the source:

spacetime metric

$$\begin{aligned} ds^2 = &-\left(1 - \frac{2M}{r}\right) dt^2 - \frac{4\epsilon_{jkm} J^k x^m}{r^3} dt dx^j \\ &+ \left(1 + \frac{2M}{r}\right) \delta_{jk} dx^j dx^k + \mathrm{O}\left(\frac{1}{r^3}\right) dx^\alpha dx^\beta \end{aligned}. \tag{25.98c}$$

In spherical polar coordinates, with the polar axis along the direction of the source's angular momentum, the leading-order terms take the form

$$\begin{aligned} ds^2 = &-\left(1 - \frac{2M}{r}\right) dt^2 - \frac{4J}{r} \sin^2\theta dt d\phi \\ &+ \left(1 + \frac{2M}{r}\right) (dr^2 + r^2 d\theta^2 + r^2 \sin^2\theta d\phi^2) \end{aligned}, \tag{25.98d}$$

where $J \equiv |\mathbf{J}|$ is the magnitude of the source's angular momentum.

reading off the source's mass and angular momentum from its metric

This is a very important result. It tells us that we can "read off" the mass M and angular momentum J^k from the asymptotic form of the source's metric. More specifically: (i) The mass M shows up in g_{00} in just the way we expect from the Newtonian limit—by comparing Eqs. (25.98c) and (25.79), we see that $\Phi = -M/r$, and from our experience with Newtonian gravity, we conclude that M is the mass that governs the Keplerian orbits of planets around our gravitational source. (ii) The angular momentum J^k shows up in $g_{0j} = -(2/r^3)\epsilon_{jkm} J^k x^m$. The physical manifestation of this g_{0j} is a gravitational torque on gyroscopes.

reading off mass from Keplerian orbits of planets

Consider an inertial-guidance gyroscope whose center of mass is at rest in the coordinate system of Eq. (25.98c) (i.e., at rest relative to the gravitating source). The transport law for the gyroscope's spin is $\nabla_{\vec{u}}\vec{S} = \vec{u}(\vec{a}\cdot\vec{S})$ [Eq. (24.62) boosted from special relativity to general relativity via the equivalence principle]. Here $\vec{u}$ is the gyroscope's 4-velocity (so $u^j = 0$, $u^0 = 1/\sqrt{1-2M/r} \simeq 1 + M/r \simeq 1$), and $\vec{a}$ is its 4-acceleration. The spatial components of this transport law are

$$S^j{}_{,t}u^0 \simeq S^j{}_{,t} = -\Gamma^j{}_{k0}S^k u^0 \simeq -\Gamma^j{}_{k0}S^k \simeq -\Gamma_{jk0}S^k \simeq \frac{1}{2}(g_{0k,j} - g_{0j,k})S^k. \quad (25.99)$$

Here each "$\simeq$" means "is equal, up to fractional corrections of order M/r." By inserting g_{0j} from the line element (25.98c) and performing some manipulations with Levi-Civita tensors, we can bring Eq. (25.99) into the form (cf. Ex. 26.19 and Sec. 27.2.5)

reading off angular momentum from gyroscopic precession (frame dragging)

$$\boxed{\frac{d\mathbf{S}}{dt} = \boldsymbol{\Omega}_{\rm prec} \times \mathbf{S}, \quad \text{where } \boldsymbol{\Omega}_{\rm prec} = \frac{1}{r^3}[-\mathbf{J} + 3(\mathbf{J}\cdot\mathbf{n})\mathbf{n}].} \quad (25.100)$$

Here $\mathbf{n} = \mathbf{e}_{\hat{r}}$ is the unit radial vector pointing away from the gravitating source.

Equation (25.100) says that the gyroscope's spin angular momentum rotates (precesses) with angular velocity $\boldsymbol{\Omega}_{\rm prec}$ in the coordinate system (which is attached to distant inertial frames, i.e., to distant galaxies and quasars). This is sometimes called a *gravitomagnetic precession,* because the off-diagonal term g_{j0} in the metric, when thought of as a 3-vector, is $-2\mathbf{J}\times\mathbf{n}/r^2$, which has the same form as the vector potential of a magnetic dipole; and the gyroscopic precession is similar to that of a magnetized spinning body interacting with that magnetic dipole. It is also called a *frame-dragging precession,* because one can regard the source's angular momentum as dragging inertial frames into precession and regard those inertial frames as being locked to inertial-guidance gyroscopes, such as $\mathbf{S}$. And it is sometimes called Lense-Thirring precession after the physicists who first discovered it mathematically, with Einstein's help (Pfister, 2007).

Figure 25.5 shows this frame-dragging precessional angular velocity $\boldsymbol{\Omega}_{\rm prec}$ as a vector field attached to the source. Notice that it has precisely the same form as a dipolar magnetic field in electromagnetic theory. In Sec. 27.2.5, we discuss the magnitude of this frame dragging in the solar system and the experiments that have measured it.

For a time-independent body with *strong* internal gravity (e.g., a black hole), the distant gravitational field will have the same general form [Eqs. (25.98)] as for a weakly gravitating body, but the constants M and J^k that appear in the metric will not be expressible as the integrals (25.98b) over the body's interior. Nevertheless, they will be measurable by the same techniques as for a weakly gravitating body (Kepler's laws and frame dragging), and they can be interpreted as the body's total mass and angular momentum. We explore this in the next chapter [Eq. (26.71)], where the body will be a spinning black hole.

T2

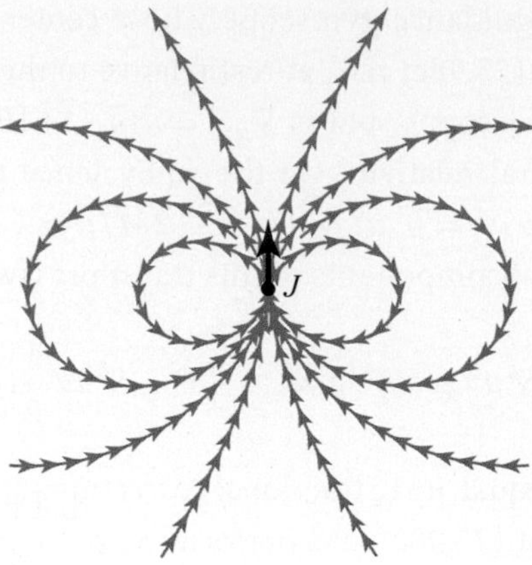

FIGURE 25.5 The precessional angular velocity $\mathbf{\Omega}_{\text{prec}}$ [Eq. (25.100)] of an inertial-guidance gyroscope at rest outside a stationary, linearized source of gravity that has angular momentum **J**. The arrows are all drawn with the same length rather than proportional to the magnitude of $\mathbf{\Omega}_{\text{prec}}$.

EXERCISES

Exercise 25.21 *Derivation and Example: External Field of a Stationary, Linearized Source* T2

Derive Eqs. (25.98a) for the trace-reversed metric perturbation outside a stationary (time-independent), linearized source of gravity. More specifically, do the following.

(a) First derive $\bar{h}_{00}$. In your derivation identify a dipolar term of the form $4D_j x^j/r^3$, and show that by placing the origin of coordinates at the center of mass, Eq. (25.96), one causes the dipole moment D_j to vanish.

(b) Next derive $\bar{h}_{0j}$. The two terms in Eq. (25.97) should give rise to two terms. The first of these is $4P_j/r$, where P_j is the source's linear momentum. Show, using the gauge condition $\bar{h}^{0\mu}{}_{,\mu} = 0$ [Eq. (25.89)], that if the momentum is nonzero, then the mass dipole term of part (a) must have a nonzero time derivative, which violates our assumption of stationarity. Therefore, the linear momentum must vanish for this source. Show that the second term gives rise to the $\bar{h}_{0j}$ of Eq. (25.98a). [Hint: You will have to add a perfect divergence, $-\frac{1}{2}(T^{0a}x^j x^m)_{,a}$ to the integrand $T^{0j}x^m$.]

(c) Finally derive $\bar{h}_{ij}$. [Hint: Show that $T^{ij} = (T^{ia}x^i)_{,a}$ and thence that the volume integral of T^{ij} vanishes; similarly for $T^{ij}x^k$.]

Exercise 25.22 *Derivation and Problem: Differential Precession and Frame-Drag Field* T2

(a) Derive the equation $\Delta\Omega_i = \mathcal{B}_{ij}\xi^j$ for the precession angular velocity of a gyroscope at the tip of $\boldsymbol{\xi}$ as measured in an inertial frame at its tail. Here $\mathcal{B}_{ij}$ is the frame-drag field introduced in Box 25.2. [For a solution, see Nichols et al. (2011, Sec. III.C).]

T2

BOX 25.2. DECOMPOSITION OF RIEMANN: TIDAL AND FRAME-DRAG FIELDS T2

In any local Lorentz frame, and also in a Lorentz frame of the linearized theory, the electromagnetic field tensor $F_{\mu\nu}$ can be decomposed into two spatial vector fields: the electric field $E_i = F_{i0}$ and magnetic field $B_i = \frac{1}{2}\epsilon_{ipq}F^{pq}$ (Sec. 2.11). Similarly, in vacuum (for simplicity) the Riemann curvature tensor can be decomposed into two spatial tensor fields: the *tidal field* $\mathcal{E}_{ij} = R_{i0j0}$ and the *frame-drag field* $\mathcal{B}_{ij} = \frac{1}{2}\epsilon_{ipq}R^{pq}{}_{j0}$. The symmetries (25.45) of Riemann, and the fact that in vacuum it is trace-free, imply that both $\mathcal{E}_{ij}$ and $\mathcal{B}_{jk}$ are symmetric and trace-free (STF). In the 3-space of the chosen frame, they are the irreducible tensorial parts of the vacuum Riemann tensor (cf. Box 11.2).

In a local Lorentz frame for strong gravity, and also in the linearized theory for weak gravity, the Bianchi identities (25.70) take on the following Maxwell-like form [Nichols et al., 2011, Eqs. (2.4), (2.15)], in which the superscript S means to symmetrize:

$$\nabla\cdot\boldsymbol{\mathcal{E}} = 0, \quad \nabla\cdot\boldsymbol{\mathcal{B}} = 0, \quad \frac{\partial\boldsymbol{\mathcal{E}}}{\partial t} - (\nabla\times\boldsymbol{\mathcal{B}})^{\mathrm{S}} = 0, \quad \frac{\partial\boldsymbol{\mathcal{B}}}{\partial t} + (\nabla\times\boldsymbol{\mathcal{E}})^{\mathrm{S}} = 0. \tag{1}$$

This has motivated some physicists to call the tidal field $\boldsymbol{\mathcal{E}}$ and the frame-drag field $\boldsymbol{\mathcal{B}}$ the "electric" and "magnetic" parts of the vacuum Riemann tensor. We avoid this language because of the possibility of confusing these second-rank tensorial gravitational fields with their truly electromagnetic vector-field counterparts **E** and **B**.

The tidal and frame-drag fields get their names from the forces they produce. The equation of geodesic deviation (25.34) says that, in a local Lorentz frame or in linearized theory, the relative acceleration of two test particles, separated by the vector $\boldsymbol{\xi}$, is $\Delta a_i = -\mathcal{E}_{ij}\xi^j$. Similarly, a gyroscope at the tip of $\boldsymbol{\xi}$ precesses relative to an inertial frame at its tail with the differential frame-dragging angular velocity $\Delta\Omega_i = \mathcal{B}_{ij}\xi^j$ (Ex. 25.22). Not surprisingly, in linearized theory, $\boldsymbol{\mathcal{B}}$ is the symmetrized gradient of the angular velocity $\boldsymbol{\Omega}_{\rm prec}$ of precession relative to distant inertial frames: $\boldsymbol{\mathcal{B}} = \left(\nabla\boldsymbol{\Omega}_{\rm prec}\right)^{\mathrm{S}}$.

Just as the electric and magnetic fields can be visualized using field lines that are the integral curves of **E** and **B**, $\boldsymbol{\mathcal{E}}$ and $\boldsymbol{\mathcal{B}}$ can be visualized using the integral curves of their eigenvectors. Since each field, $\boldsymbol{\mathcal{E}}$ or $\boldsymbol{\mathcal{B}}$, has three orthogonal eigenvectors, each has a network of three orthogonal sets of field lines. The field lines of $\boldsymbol{\mathcal{B}}$ are called (Nichols et al., 2011) *frame-drag vortex lines* or simply "vortex lines"; those of $\boldsymbol{\mathcal{E}}$ are called *tidal tendex lines* or simply "tendex lines"

(continued)

T2

BOX 25.2. (continued)

(from the Latin *tendere,* meaning "to stretch" by analogy with vortex from *vertere,* meaning "to turn"). For a spinning point mass in linearized theory, $\mathcal{E}$ and $\mathcal{B}$ are given by Eqs. (25.101), and their tendex and vortex lines are as shown below (adapted from Nichols et al., 2011). The red tendex lines (left diagram) stretch; the blue squeeze. The red vortex lines (right) twist a gyroscope at a person's feet (or head) counterclockwise relative to inertial frames at her head (or feet). The blue vortex lines twist clockwise. We explore these concepts in greater detail in Box 26.3.

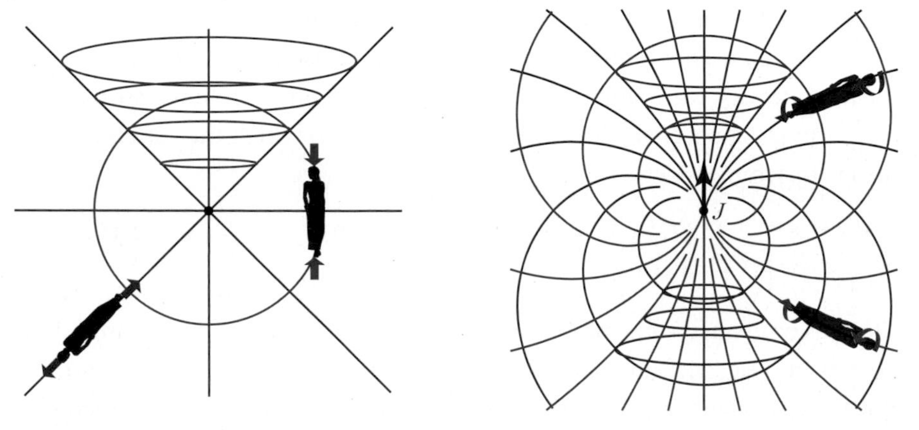

(b) Show that in linearized theory, $\mathcal{B}$ is the symmetrized gradient of the angular velocity $\mathbf{\Omega}_{\text{prec}}$ of precession of a gyroscope relative to distant inertial frames.

Exercise 25.23 *Problem: Spinning Particle in Linearized Theory: Tidal and Frame-Drag Fields* T2

Show that in linearized theory, for a spinning particle at the origin with mass M and with its spin J along the polar axis, the orthonormal-frame components of $\mathcal{E}$ and $\mathcal{B}$ are

$$\mathcal{E}_{\hat{r}\hat{r}} = -2\mathcal{E}_{\hat{\theta}\hat{\theta}} = -2\mathcal{E}_{\hat{\phi}\hat{\phi}} = -\frac{2M}{r^3}, \tag{25.101a}$$

$$\mathcal{B}_{\hat{r}\hat{r}} = -2\mathcal{B}_{\hat{\theta}\hat{\theta}} = -2\mathcal{B}_{\hat{\phi}\hat{\phi}} = -\frac{6J\cos\theta}{r^4}, \quad \mathcal{B}_{\hat{r}\hat{\theta}} = \mathcal{B}_{\hat{\theta}\hat{r}} = -\frac{3J\sin\theta}{r^4}. \tag{25.101b}$$

What are the eigenvectors of these fields? Convince yourself that these eigenvectors' integral curves (the tidal tendex lines and frame-drag vortex lines) are as depicted at the bottom of Box 25.2.

25.9.4 Conservation Laws for Mass, Momentum, and Angular Momentum in Linearized Theory T2

Consider a static (unmoving) sphere $\mathcal{S}$ surrounding our time-independent source of gravity, with such a large radius r that the $O(1/r^3)$ corrections in $\bar{h}_{\mu\nu}$ and in the metric [Eqs. (25.98)] can be ignored. Suppose that a small amount of mass-energy E (as measured in the sphere's and source's rest frame) is injected through the sphere, into the source. Then the special relativistic law of mass-energy conservation tells us that the source's mass $M = \int T^{00} dV$ will increase by $\Delta M = E$. Similarly, if an energy flux T^{0j} flows through the sphere, the source's mass will change by

for a stationary source of gravity perturbed by an infalling or outflowing stress-energy tensor:

$$\frac{dM}{dt} = -\int_{\mathcal{S}} T^{0j} d\Sigma_j, \tag{25.102}$$

mass conservation

where $d\Sigma_j$ is the sphere's outward-pointing surface-area element, and the minus sign is because $d\Sigma_j$ points outward, not inward. Since M is the mass that appears in the source's asymptotic gravitational field $\bar{h}_{\mu\nu}$ and metric $g_{\alpha\beta}$, this conservation law can be regarded as describing how the source's gravitating mass changes when energy is injected into it.

From the special relativistic law for angular momentum conservation (e.g., Misner, Thorne, and Wheeler, 1973, Box 5.6), we deduce a similar result. A flux $\epsilon_{ijk} x^j T^{km}$ of angular momentum through the sphere produces the following change in the angular momentum J_i that appears in the source's asymptotic field $\bar{h}_{\mu\nu}$ and metric:

$$\frac{dJ_i}{dt} = -\int_{\mathcal{S}} \epsilon_{ijk} x^j T^{km} d\Sigma_m. \tag{25.103}$$

angular-momentum conservation

There is also a conservation law for a gravitationally measured linear momentum. That linear momentum does not show up in the asymptotic field and metric that we wrote down above [Eqs. (25.98)], because our coordinates were chosen to be attached to the source's center of mass (i.e., they are the Lorentz coordinates of the source's rest frame). However, if linear momentum P_j is injected through our sphere $\mathcal{S}$ and becomes part of the source, then the source's center of mass will start moving, and the asymptotic metric will acquire a new term:

$$\delta g_{0j} = -4P_j/r, \tag{25.104}$$

where (after the injection)

$$P_j = P^j = \int T^{0j} dV \tag{25.105}$$

source's linear momentum

[see Eq. (25.91) with $\bar{h}^{0j} = -\bar{h}_{0j} = -h_{0j} = -\delta g_{0j}$; also see Ex. 25.21b]. More generally, the rate of change of the source's total linear momentum (the P_j term in the

T2

asymptotic g_{0j}) is the integral of the inward flux of momentum (inward component of the stress tensor) across the sphere:

momentum conservation

$$\frac{dP_j}{dt} = -\int_{\mathcal{S}} T^{jk} d\Sigma_k. \tag{25.106}$$

25.9.5 Conservation Laws for a Strong-Gravity Source T2

strong-gravity source surrounded by asymptotically flat spacetime:

For a time-independent source with strong internal gravity, not only does the asymptotic metric, far from the source, have the same form [Eqs. (25.98c), (25.98d), (25.104)] as for a weakly gravitating source, but also the conservation laws for its gravitationally measured mass, angular momentum, and linear momentum [Eqs. (25.102), (25.103), (25.106), respectively] continue to hold true. Of course, the sphere $\mathcal{S}$ must be placed far from the source, in a region where gravity is weak, so spacetime is *asymptotically flat*[6] and linearized theory is valid in the vicinity of $\mathcal{S}$. When this is done, then the special relativistic description of inflowing mass, angular momentum, and energy is valid at $\mathcal{S}$, and the linearized Einstein equation, applied in the vicinity of $\mathcal{S}$ (and not extended into the strong-gravity region), turn out to guarantee that the M, J_j, and P_j appearing in the asymptotic metric evolve in accord with the conservation laws (25.102), (25.103), and (25.106).

in asymptotically flat region, source's mass, momentum, angular momentum, and their conservation are the same as for a weak-gravity source

For strongly gravitating sources, these conservation laws owe their existence to the spacetime's asymptotic time-translation, rotation, and space-translation symmetries. In generic, strong-gravity regions of spacetime there are no such symmetries and correspondingly, no integral conservation laws for energy, angular momentum, or linear momentum.

If a strongly gravitating source is dynamical rather than static, it will emit gravitational waves (Chap. 27). The amplitudes of those waves, like the influence of the source's mass, die out as $1/r$ far from the source, so spacetime retains its asymptotic time-translation, rotation, and space-translation symmetries. These symmetries continue to enforce integral conservation laws on the gravitationally measured mass, angular momentum, and linear momentum [Eqs. (25.102), (25.103), and (25.106)], but with the new requirement that one include, in the fluxes through $\mathcal{S}$, contributions from the gravitational waves' energy, angular momentum, and linear momentum; see Chap. 27.

For a brief derivation and discussion of these asymptotic conservation laws, see Thorne (1983, Sec. 3.3.2); for far greater detail, see Misner, Thorne, and Wheeler (1973, Chaps. 18 and 19).

6. Our real universe, of course, is not asymptotically flat. However, nearly everywhere the distances between gravitating systems (e.g., between our solar system and the alpha centauri system) are so large that spacetime is very nearly asymptotically flat as one moves outward into the region between the systems. Correspondingly, to very high accuracy all the statements in this section remain true.

Bibliographic Note

For a superb, detailed historical account of Einstein's intellectual struggle to formulate the laws of general relativity, see Pais (1982). For Einstein's papers of that era, in the original German and in English translation, with detailed annotations and explanations by editors with strong backgrounds in both physics and the history of science, see Einstein (1989). For some key papers of that era by other major contributors besides Einstein, in English translation, see Lorentz et al. (1923).

This chapter's pedagogical approach to presenting the fundamental concepts of general relativity is strongly influenced by Misner, Thorne, and Wheeler (1973), where readers will find much greater detail. See, especially, Chap. 8 for the mathematics (differential geometry) of curved spacetime, or Chaps. 9–14 for far greater detail; Chap. 16 for the Einstein equivalence principle and how to lift laws of physics into curved spacetime; Chap. 17 for the Einstein field equations and many different ways to derive them; Chap. 18 for weak gravitational fields (the Newtonian limit and linearized theory); and Chaps. 19 and 20 for the metric in the asymptotically flat region outside a strongly gravitating source and for the source's conservation laws for mass, momentum, and angular momentum.

For an excellent, elementary introduction to the fundamental concepts of general relativity from a viewpoint that is somewhat less mathematical than this chapter or Misner, Thorne, and Wheeler (1973), see Hartle (2003). We also recommend, at a somewhat elementary level, Schutz (2009); and at a more advanced level, Carroll (2004), Straumann (2013), and Zee (2013). At a very advanced and mathematical level, we recommend Wald (1984). For a rather different approach to general relativity, one that emphasizes the connection to field theory over that to geometry, we recommend Weinberg (1972).

Our physicist's approach to differential geometry in this chapter lacks much of the rigor and beauty of a mathematician's approach, for which we recommend Spivak (1999).

CHAPTER TWENTY-SIX

Relativistic Stars and Black Holes

All light emitted from such a body would be made to return towards it by its own proper gravity.

JOHN MICHELL (1783)

26.1 Overview

26.1

Having sketched the fundamentals of Einstein's theory of gravity, general relativity, we now illustrate his theory by several concrete applications: stars and black holes in this chapter, gravitational waves in Chap. 27, and the large-scale structure and evolution of the universe in Chap. 28.

While stars and black holes are the central thread of this chapter, we study them less for their own intrinsic interest than for their roles as vehicles by which to understand general relativity. Using them, we elucidate some issues we have already met: the physical and geometric interpretations of spacetime metrics and of coordinate systems, the Newtonian limit of general relativity, the geodesic motion of freely falling particles and photons, local Lorentz frames and the tidal forces measured therein, proper reference frames, the Einstein field equation, the local law of conservation of 4-momentum, and the asymptotic structure of spacetime far from gravitating sources. Stars and black holes also serve to introduce several new physical phenomena that did not show up in our study of the foundations of general relativity: the "many-fingered" nature of time, event horizons, and spacetime singularities.

We begin this chapter, in Sec. 26.2, by studying the geometry of the curved spacetime outside any static star, as predicted by the Einstein field equation. In Sec. 26.3, we study general relativity's description of the interiors of static stars. In Sec. 26.4, we turn attention to the spherically symmetric gravitational implosion by which a nonrotating star is transformed into a black hole, and to the Schwarzschild spacetime geometry outside and inside the resulting static, spherical hole. In Sec. 26.5, we study the Kerr spacetime geometry of a spinning black hole. In Sec. 26.6, we elucidate the nature of time in the curved spacetimes of general relativity. And in Ex. 26.13, we explore the role of the vacuum Schwarzschild solution of the Einstein field equation as a wormhole.

BOX 26.1. READERS' GUIDE

- This chapter relies significantly on:
  - Chap. 2 on special relativity;
  - Chap. 24 on the transition from special relativity to general relativity; and
  - Chap. 25 on the fundamental concepts of general relativity.
- Portions of this chapter are a foundation for the applications of general relativity theory to gravitational waves (Chap. 27) and to cosmology (Chap. 28).

26.2

26.2 Schwarzschild's Spacetime Geometry

26.2.1

26.2.1 The Schwarzschild Metric, Its Connection Coefficients, and Its Curvature Tensors

On January 13, 1916, just 7 weeks after formulating the final version of his field equation, $\boldsymbol{G} = 8\pi \boldsymbol{T}$, Albert Einstein read to a meeting of the Prussian Academy of Sciences in Berlin a letter from the eminent German astrophysicist Karl Schwarzschild. Schwarzschild, as a member of the German army, had written from the World-War-I Russian front to tell Einstein of a mathematical discovery he had made: he had found the world's first exact solution to the Einstein field equation.

Written as a line element in a special coordinate system (coordinates $\{t, r, \theta, \phi\}$) that Schwarzschild invented for the purpose, Schwarzschild's solution takes the form (Schwarzschild, 1916a)

Schwarzschild metric

$$\boxed{ds^2 = -(1-2M/r)dt^2 + \frac{dr^2}{(1-2M/r)} + r^2(d\theta^2 + \sin^2\theta d\phi^2),} \qquad (26.1)$$

where M is a constant of integration. The connection coefficients, Riemann tensor, and Ricci and Einstein tensors for this metric can be computed by the methods of Chaps. 24 and 25; see Ex. 26.1. The results are tabulated in Box 26.2. The key bottom line is that the Einstein tensor vanishes. Therefore, the Schwarzschild metric (26.1) is a solution of the Einstein field equation with vanishing stress-energy tensor.

roles of the Schwarzschild metric: exterior of static star and imploding star; black hole and wormhole

Many readers know already the lore of this subject. The Schwarzschild metric is reputed to represent the vacuum exterior of a nonrotating, spherical star; and also the exterior of a spherical star as it implodes to form a black hole; and also the exterior and interior of a nonrotating, spherical black hole; and also a wormhole that connects two different universes or two widely separated regions of our own universe.

How does one discover these physical interpretations of the Schwarzschild metric (26.1)? The tools for discovering them—and, more generally, the tools for interpreting

BOX 26.2. CONNECTION COEFFICIENTS AND CURVATURE TENSORS FOR SCHWARZSCHILD SOLUTION

The coordinate basis vectors for the Schwarzschild solution of Einstein's equation are

$$\vec{e}_t = \frac{\partial}{\partial t}, \quad \vec{e}_r = \frac{\partial}{\partial r}, \quad \vec{e}_\theta = \frac{\partial}{\partial \theta}, \quad \vec{e}_\phi = \frac{\partial}{\partial \phi};$$

$$\vec{e}^{\,t} = \vec{\nabla} t, \quad \vec{e}^{\,r} = \vec{\nabla} r, \quad \vec{e}^{\,\theta} = \vec{\nabla} \theta, \quad \vec{e}^{\,\phi} = \vec{\nabla} \phi. \tag{1}$$

The covariant and contravariant metric coefficients in this coordinate basis are [cf. Eq. (26.1)]

$$g_{tt} = -\left(1 - \frac{2M}{r}\right), \quad g_{rr} = \frac{1}{(1 - 2M/r)}, \quad g_{\theta\theta} = r^2, \quad g_{\phi\phi} = r^2 \sin^2 \theta; \tag{2a}$$

$$g^{tt} = -\frac{1}{(1 - 2M/r)}, \quad g^{rr} = \left(1 - \frac{2M}{r}\right), \quad g^{\theta\theta} = \frac{1}{r^2} \quad g^{\phi\phi} = \frac{1}{r^2 \sin^2 \theta}. \tag{2b}$$

The nonzero connection coefficients in this coordinate basis are

$$\Gamma^t{}_{rt} = \Gamma^t{}_{tr} = \frac{M}{r^2} \frac{1}{(1 - 2M/r)}, \quad \Gamma^r{}_{tt} = \frac{M}{r^2}(1 - 2M/r), \quad \Gamma^r{}_{rr} = -\frac{M}{r^2} \frac{1}{(1 - 2M/r)},$$

$$\Gamma^r{}_{\theta\theta} = -r(1 - 2M/r), \quad \Gamma^\theta{}_{r\theta} = \Gamma^\theta{}_{\theta r} = \Gamma^\phi{}_{r\phi} = \Gamma^\phi{}_{\phi r} = \frac{1}{r}, \tag{3}$$

$$\Gamma^r{}_{\phi\phi} = -r \sin^2 \theta (1 - 2M/r), \quad \Gamma^\theta{}_{\phi\phi} = -\sin\theta \cos\theta, \quad \Gamma^\phi{}_{\theta\phi} = \Gamma^\phi{}_{\phi\theta} = \cot\theta.$$

The orthonormal basis associated with the above coordinate basis is

$$\vec{e}_{\hat{0}} = \frac{1}{\sqrt{1 - 2M/r}} \frac{\partial}{\partial t}, \quad \vec{e}_{\hat{r}} = \sqrt{1 - \frac{2M}{r}} \frac{\partial}{\partial r}, \quad \vec{e}_{\hat{\theta}} = \frac{1}{r} \frac{\partial}{\partial \theta}, \quad \vec{e}_{\hat{\phi}} = \frac{1}{r \sin\theta} \frac{\partial}{\partial \phi}. \tag{4}$$

The nonzero connection coefficients in this orthonormal basis are

$$\Gamma^{\hat{r}}{}_{\hat{0}\hat{0}} = \Gamma^{\hat{0}}{}_{\hat{r}\hat{0}} = \frac{M}{r^2 \sqrt{1 - 2M/r}}, \quad \Gamma^{\hat{\phi}}{}_{\hat{\theta}\hat{\phi}} = -\Gamma^{\hat{\theta}}{}_{\hat{\phi}\hat{\phi}} = \frac{\cot\theta}{r},$$

$$\Gamma^{\hat{\theta}}{}_{\hat{r}\hat{\theta}} = \Gamma^{\hat{\phi}}{}_{\hat{r}\hat{\phi}} = -\Gamma^{\hat{r}}{}_{\hat{\theta}\hat{\theta}} = -\Gamma^{\hat{r}}{}_{\hat{\phi}\hat{\phi}} = \frac{\sqrt{1 - 2M/r}}{r}. \tag{5}$$

(continued)

BOX 26.2. (continued)

The nonzero components of the Riemann tensor in this orthonormal basis are

$$R_{\hat{r}\hat{0}\hat{r}\hat{0}} = -R_{\hat{\theta}\hat{\phi}\hat{\theta}\hat{\phi}} = -\frac{2M}{r^3}, \quad R_{\hat{\theta}\hat{0}\hat{\theta}\hat{0}} = R_{\hat{\phi}\hat{0}\hat{\phi}\hat{0}} = -R_{\hat{r}\hat{\phi}\hat{r}\hat{\phi}} = -R_{\hat{r}\hat{\theta}\hat{r}\hat{\theta}} = \frac{M}{r^3}, \tag{6}$$

and those obtainable from these via the symmetries (25.45a) of Riemann. The Ricci tensor, curvature scalar, and Einstein tensor all vanish—which implies that the Schwarzschild metric is a solution of the vacuum Einstein field equation.

physically any spacetime metric that one encounters—are a central concern of this chapter.

EXERCISES

Exercise 26.1 *Practice: Connection Coefficients and the Riemann Tensor for the Schwarzschild Metric*

(a) Explain why, for the Schwarzschild metric (26.1), the metric coefficients in the coordinate basis have the values given in Eqs. (2a,b) of Box 26.2.

(b) Using tensor-analysis software on a computer,[1] derive the connection coefficients given in Eq. (3) of Box 26.2.

(c) Show that the basis vectors in Eqs. (4) of Box 26.2 are orthonormal.

(d) Using tensor-analysis software on a computer, derive the connection coefficients (5) and Riemann components (6) of Box 26.2 in the orthonormal basis.

26.2.2

26.2.2 The Nature of Schwarzschild's Coordinate System, and Symmetries of the Schwarzschild Spacetime

When presented with a line element such as Eq. (26.1), one of the first questions one is tempted to ask is "What is the nature of the coordinate system?" Since the metric coefficients will be different in some other coordinate system, surely one must know something about the coordinates to interpret the line element.

Remarkably, one need not go to the inventor of the coordinates to find out their nature. Instead, one can turn to the line element itself: the line element (or metric

1. Such as the simple Mathematica program in Hartle (2003, Appendix C), which is available on that book's website: http://web.physics.ucsb.edu/~gravitybook/mathematical.html.

coefficients) contain full information not only about the details of the spacetime geometry, but also about the nature of the coordinates. The line element (26.1) is a good example:

properties of the Schwarzschild metric and its Schwarzschild coordinates:

Look first at the 2-dimensional surfaces in spacetime that have constant values of t and r. We can regard $\{\theta, \phi\}$ as a coordinate system on each such 2-surface. The spacetime line element (26.1) tells us that the geometry of the 2-surface is given in terms of those coordinates by

$$^{(2)}ds^2 = r^2(d\theta^2 + \sin^2\theta d\phi^2) \tag{26.2}$$

(where the prefix $^{(2)}$ refers to the dimensionality of the surface). This is the line element (metric) of an ordinary, everyday 2-dimensional sphere expressed in standard spherical polar coordinates. Thus we have learned that the Schwarzschild spacetime is spherically symmetric, and moreover, θ and ϕ are standard spherical polar coordinates. This is an example of extracting from a metric information about both the coordinate-independent spacetime geometry and the coordinate system being used.

spherically symmetric

Furthermore, note from Eq. (26.2) that the circumferences and surface areas of the spheres $(t, r) = \text{const}$ in Schwarzschild spacetime are given by

radial coordinate r is circumference$/2\pi$, or square root of area$/4\pi$

$$\boxed{\text{circumference} = 2\pi r, \quad \text{area} = 4\pi r^2.} \tag{26.3}$$

This tells us one aspect of the geometric interpretation of the r coordinate: r is a radial coordinate in the sense that the circumferences and surface areas of the spheres in Schwarzschild spacetime are expressed in terms of r in the standard manner of Eq. (26.3). We must not go further, however, and assert that r is radius in the sense of being the proper distance from the center of one of the spheres to its surface. The center, and the line from center to surface, do not lie on the sphere itself, and they thus are not described by the spherical line element (26.2). Moreover, since we know that spacetime is curved, we have no right to expect that the proper distance from the center of a sphere to its surface will be given by distance = circumference$/2\pi = r$ as in flat spacetime.

26.2.3 Schwarzschild Spacetime at Radii $r \gg M$: The Asymptotically Flat Region

26.2.3

Returning to the Schwarzschild line element (26.1), let us examine several specific regions of spacetime: At "radii" r large compared to the integration constant M, the line element (26.1) takes the form

$$ds^2 = -dt^2 + dr^2 + r^2(d\theta^2 + \sin^2\theta d\phi^2). \tag{26.4}$$

This is the line element of flat spacetime, $ds^2 = -dt^2 + dx^2 + dy^2 + dz^2$ written in spherical polar coordinates $\{x = r \sin\theta \cos\phi,\ y = r \sin\theta \sin\phi,\ z = r \cos\theta\}$. Thus, Schwarzschild spacetime is asymptotically flat in the region of large radii $r/M \to \infty$. This is just what one might expect physically when one gets far away from all sources of

asymptotically flat

gravity. Thus, it is reasonable to presume that the Schwarzschild spacetime geometry is that of some sort of isolated, gravitating body that is located in the region $r \sim M$.

The large-r line element (26.4) not only reveals that Schwarzschild spacetime is asymptotically flat; it also shows that in the asymptotically flat region the Schwarzschild coordinate t is the time coordinate of a Lorentz reference frame. Notice that the region of strong spacetime curvature has a boundary (say, $r \sim 100M$) that remains forever fixed relative to the asymptotically Lorentz spatial coordinates $\{x = r \sin\theta \cos\phi,\ y = r \sin\theta \sin\phi,\ z = r \cos\theta\}$. This means that the asymptotic Lorentz frame can be regarded as the body's asymptotic rest frame. We conclude, then, that far from the body the Schwarzschild t coordinate becomes the Lorentz time of the body's asymptotic rest frame, and the Schwarzschild $\{r, \theta, \phi\}$ coordinates become spherical polar coordinates in the body's asymptotic rest frame.

asymptotic rest frame

As we move inward from $r = \infty$, we gradually begin to see spacetime curvature. That curvature shows up, at $r \gg M$, in slight deviations of the Schwarzschild metric coefficients from those of a Lorentz frame: to first order in M/r the line element (26.1) becomes

$$ds^2 = -\left(1 - \frac{2M}{r}\right) dt^2 + \left(1 + \frac{2M}{r}\right) dr^2 + r^2(d\theta^2 + \sin^2\theta d\phi^2), \quad (26.5)$$

or, equivalently, in Cartesian spatial coordinates:

$$\begin{aligned} ds^2 = &-\left(1 - \frac{2M}{\sqrt{x^2+y^2+z^2}}\right) dt^2 + dx^2 + dy^2 + dz^2 \\ &+ \frac{2M}{r}\left(\frac{x}{r}dx + \frac{y}{r}dy + \frac{z}{r}dz\right)^2 . \end{aligned} \quad (26.6)$$

It is reasonable to expect that, at these large radii where the curvature is weak, Newtonian gravity will be a good approximation to Einsteinian gravity. In Sec. 25.9.1, we studied in detail the transition from general relativity to Newtonian gravity, and found that in nearly Newtonian situations, if one uses a nearly globally Lorentz coordinate system (as we are doing), the line element should take the form [Eq. (25.79)]:

$$ds^2 = -(1 + 2\Phi)dt^2 + (\delta_{jk} + h_{jk})dx^j dx^k + 2h_{tj} dt\, dx^j, \quad (26.7)$$

where the $h_{\mu\nu}$ are metric corrections that are small compared to unity, and Φ (which shows up in the time-time part of the metric) is the Newtonian potential. Direct comparison of Eq. (26.7) with (26.6) shows that a Newtonian description of the body's distant gravitational field will entail a Newtonian potential given by

Newtonian limit of Schwarzschild metric

$$\Phi = -\frac{M}{r} \quad (26.8)$$

($\Phi = -GM/r$ in conventional units). This, of course, is the external Newtonian field of a body with mass M. Thus, the integration constant M in the Schwarzschild line element is the mass that characterizes the body's distant, nearly Newtonian gravitational field. This is an example of reading the mass of a body off the asymptotic form of the metric (last paragraph of Sec. 25.9.3).

M is the mass that characterizes the asymptotic Newtonian gravitational field

Notice that the asymptotic metric here [Eq. (26.5)] differs in its spatial part from that in Sec. 25.9.3 [Eq. (25.98d)]. This difference arises from the use of different radial coordinates here. If we define $\bar{r}$ by $r = \bar{r} + M$ at radii $r \gg M$, then to linear order in M/r, the asymptotic Schwarzschild metric (26.5) becomes

$$ds^2 = -\left(1 - \frac{2M}{\bar{r}}\right) dt^2 + \left(1 + \frac{2M}{\bar{r}}\right) [d\bar{r}^2 + \bar{r}^2(d\theta^2 + \sin^2\theta d\phi^2)], \tag{26.9}$$

which is the same as Eq. (25.98d) with vanishing angular momentum $J = 0$. This easy change of the spatial part of the metric reinforces the fact that one reads the asymptotic Newtonian potential and the source's mass M from the time-time components of the metric and not from the spatial part of the metric.

We can describe, in operational terms, the physical interpretation of M as the body's mass as follows. Suppose that a test particle (e.g., a small planet) moves around our central body in a circular orbit with radius $r \gg M$. A Newtonian analysis of the orbit predicts that, as measured using Newtonian time, the period of the orbit will be $P = 2\pi(r^3/M)^{\frac{1}{2}}$ (one of Kepler's laws). Moreover, since Newtonian time is very nearly equal to the time t of the nearly Lorentz coordinates used in Eq. (26.5) (cf. Sec. 25.9.1), and since that t is Lorentz time in the body's relativistic, asymptotic rest frame, the orbital period as measured by observers at rest in the asymptotic rest frame must be $P = 2\pi(r^3/M)^{\frac{1}{2}}$. Thus, M is the mass that appears in Kepler's laws for the orbits of test particles far from the central body. This quantity is sometimes called the body's "active gravitational mass," since it is the mass that characterizes the body's gravitational pull. It is also called the body's "total mass-energy," because it turns out to include all forms of mass and energy that the body possesses (rest mass; internal kinetic energy; and all forms of internal binding energy, including gravitational).

M is the mass in Kepler's laws for distant planets

We note in passing that one can use general relativity to deduce the Keplerian role of M without invoking the Newtonian limit: place a test particle in the body's equatorial plane $\theta = \pi/2$ at a radius $r \gg M$, and give it an initial velocity that lies in the equatorial plane. Then symmetry guarantees the particle remains in the equatorial plane: there is no way to prefer going toward north, $\theta < \pi/2$, or toward south, $\theta > \pi/2$. Furthermore, adjust the initial velocity so the particle remains always at a fixed radius. Then the only nonvanishing components $u^\alpha = dx^\alpha/d\tau$ of the particle's 4-velocity are $u^t = dt/d\tau$ and $u^\phi = d\phi/d\tau$. The particle's orbit is governed by the geodesic equation $\nabla_{\vec{u}} \vec{u} = 0$, where $\vec{u}$ is its 4-velocity. The radial component of this geodesic equation, computed in Schwarzschild coordinates, is [cf. Eq. (25.14) with a switch from affine parameter ζ to proper time $\tau = m\zeta$]

relativistic analysis of Keplerian orbital motion

$$\frac{d^2r}{d\tau^2} = -\Gamma^r{}_{\mu\nu}\frac{dx^\mu}{d\tau}\frac{dx^\nu}{d\tau} = -\Gamma^r{}_{tt}\frac{dt}{d\tau}\frac{dt}{d\tau} - \Gamma^r{}_{\phi\phi}\frac{d\phi}{d\tau}\frac{d\phi}{d\tau}. \tag{26.10}$$

(Here we have used the vanishing of all $dx^\alpha/d\tau$ except the t and ϕ components and have used the vanishing of $\Gamma^r{}_{t\phi} = \Gamma^r{}_{\phi t}$ [Eq. (3) of Box 26.2].) Since the orbit is circular, with fixed r, the left-hand side of Eq. (26.10) must vanish; correspondingly, the right-hand side gives

$$\frac{d\phi}{dt} = \frac{d\phi/d\tau}{dt/d\tau} = \left(-\frac{\Gamma^r{}_{tt}}{\Gamma^r{}_{\phi\phi}}\right)^{\frac{1}{2}} = \left(\frac{M}{r^3}\right)^{\frac{1}{2}}, \tag{26.11}$$

where we have used the values of the connection coefficients from Eq. (3) of Box 26.2, specialized to the equatorial plane $\theta = \pi/2$. Equation (26.11) tells us that the amount of coordinate time t required for the particle to circle the central body once, $0 \le \phi \le 2\pi$, is $\Delta t = 2\pi(r^3/M)^{\frac{1}{2}}$. Since t is the Lorentz time of the body's asymptotic rest frame, observers in the asymptotic rest frame will measure for the particle an orbital period $P = \Delta t = 2\pi(r^3/M)^{\frac{1}{2}}$. This, of course, is the same result as we obtained from the Newtonian limit—but our relativistic analysis shows it to be true for circular orbits of arbitrary radius r, not just for $r \gg M$.

26.2.4

26.2.4 Schwarzschild Spacetime at $r \sim M$

Next we move inward, from the asymptotically flat region of Schwarzschild spacetime toward smaller and smaller radii. As we do so, the spacetime geometry becomes more and more strongly curved, and the Schwarzschild coordinate system becomes less and less Lorentz. As an indication of extreme deviations from Lorentz, notice that the signs of the metric coefficients,

$$\frac{\partial}{\partial t}\cdot\frac{\partial}{\partial t} = g_{tt} = -\left(1 - \frac{2M}{r}\right), \quad \frac{\partial}{\partial r}\cdot\frac{\partial}{\partial r} = g_{rr} = \frac{1}{(1-2M/r)} \tag{26.12}$$

outside $r = 2M$, t is a time coordinate and r a space coordinate; inside, their roles are reversed

reverse as one moves from $r > 2M$ through $r = 2M$ and into the region $r < 2M$. Correspondingly, outside $r = 2M$, world lines of changing t but constant $\{r, \theta, \phi\}$ are timelike, while inside $r = 2M$, those world lines are spacelike. Similarly outside $r = 2M$, world lines of changing r but constant $\{t, \theta, \phi\}$ are spacelike, while inside they are timelike. In this sense, outside $r = 2M$, t plays the role of a time coordinate and r the role of a space coordinate; while inside $r = 2M$, t plays the role of a space coordinate and r the role of a time coordinate. Moreover, this role reversal occurs without any change in the role of r as $1/(2\pi)$ times the circumference of circles around the center [Eq. (26.3)].

For many decades this role reversal presented severe conceptual problems, even to the best experts in general relativity. We return to it in Sec. 26.4. Henceforth we refer to the location of role reversal, $r = 2M$, as the gravitational radius of the Schwarzschild spacetime, henceforth *gravitational radius*. It is also known as the *Schwarzschild radius* and, as we shall see, is the location of an absolute *event horizon*. In Sec. 26.4, we seek a clear understanding of the "interior" region, $r < 2M$; but until then, we confine attention to the region $r > 2M$, outside the gravitational radius.

Notice that the metric coefficients in the Schwarzschild line element (26.1) are all independent of the coordinate t. This means that the geometry of spacetime itself is invariant under the translation $t \rightarrow t + \text{const}$. At radii $r > 2M$, where t plays the role of a time coordinate, $t \rightarrow t + \text{const}$ is a time translation; correspondingly, the Schwarzschild spacetime geometry is time-translation-invariant (i.e., "static") outside the gravitational radius.

the spacetime geometry is static outside $r = 2M$

EXERCISES

Exercise 26.2 *Example: The Bertotti-Robinson Solution of the Einstein Field Equation*
Bruno Bertotti (1959) and Ivor Robinson (1959) independently solved the Einstein field equation to obtain the following metric for a universe endowed with a uniform magnetic field:

$$ds^2 = Q^2(-dt^2 + \sin^2 t\, dz^2 + d\theta^2 + \sin^2\theta\, d\phi^2). \tag{26.13}$$

Here

$$Q = \text{const}, \quad 0 \le t \le \pi, \quad -\infty < z < +\infty, \quad 0 \le \theta \le \pi, \quad 0 \le \phi \le 2\pi. \tag{26.14}$$

If one computes the Einstein tensor from the metric coefficients of the line element (26.13) and equates it to 8π times a stress-energy tensor, one finds a stress-energy tensor that is precisely the same as for an electromagnetic field [Eqs. (2.75) and (2.80)] lifted, unchanged, into general relativity. The electromagnetic field is one that, as measured in the local Lorentz frame of an observer with fixed $\{z, \theta, \phi\}$ (a "static" observer), has vanishing electric field and has a magnetic field directed along $\partial/\partial z$ with magnitude independent of where the observer is located in spacetime. In this sense, the spacetime metric (26.13) is that of a homogeneous magnetic universe. Discuss the geometry of this universe and the nature of the coordinates $\{t, z, \theta, \phi\}$. More specifically, do the following.

(a) Which coordinate increases in a timelike direction and which coordinates in spacelike directions?

(b) Is this universe spherically symmetric?

(c) Is this universe cylindrically symmetric?

(d) Is this universe asymptotically flat?

(e) How does the geometry of this universe change as t ranges from 0 to π? [Hint: Show that the curves $\{(z, \theta, \phi) = \text{const}, t = \tau/Q\}$ are timelike geodesics—the world lines of the static observers referred to above. Then argue from symmetry, or use the result of Ex. 25.4a.]

(f) Give as complete a characterization as you can of the coordinates $\{t, z, \theta, \phi\}$.

26.3

26.3 Static Stars

26.3.1

26.3.1 Birkhoff's Theorem

Birkhoff's theorem: uniqueness of Schwarzschild solution

In 1923, George Birkhoff, a professor of mathematics at Harvard, proved a remarkable theorem (Birkhoff, 1923). (For a textbook proof, see Misner, Thorne, and Wheeler, 1973, Sec. 32.2.) The Schwarzschild spacetime geometry is the unique spherically symmetric solution of the vacuum Einstein field equation $\boldsymbol{G} = 0$. This Birkhoff theorem can be restated in more operational terms as follows. Suppose that you find a solution of the vacuum Einstein field equation, written as a set of metric coefficients $g_{\bar{\alpha}\bar{\beta}}$ in some coordinate system $\{x^{\bar{\mu}}\}$. Suppose, further, that these $g_{\bar{\alpha}\bar{\beta}}(x^{\bar{\mu}})$ coefficients exhibit spherical symmetry but do not coincide with the Schwarzschild expressions [Eqs. (2a) of Box 26.2]. Then the Birkhoff theorem guarantees the existence of a coordinate transformation from your coordinates $x^{\bar{\mu}}$ to Schwarzschild's coordinates x^{ν} such that, when that transformation is performed, the resulting new metric components $g_{\alpha\beta}(x^{\nu})$ have precisely the Schwarzschild form [Eqs. (2a) of Box 26.2]. For an example, see Ex. 26.3. This implies that, thought of as a coordinate-independent spacetime geometry, the Schwarzschild solution is completely unique.

Now consider a static, spherically symmetric star (e.g., the Sun) residing alone in an otherwise empty universe (or, more realistically, residing in our own universe but so far from other gravitating matter that we can ignore all other sources of gravity when studying it). Since the star's interior is spherical, it is reasonable to presume that the exterior will be spherical; since the exterior is also vacuum ($\boldsymbol{T} = 0$), its spacetime geometry must be that of Schwarzschild. If the circumference of the star's surface is $2\pi R$ and its surface area is $4\pi R^2$, then that surface must reside at the location $r = R$ in the Schwarzschild coordinates of the exterior. In other words, the spacetime geometry will be described by the Schwarzschild line element (26.1) at radii $r > R$, but by something else inside the star, at $r < R$.

Since real atoms with finite rest masses reside on the star's surface, and since such atoms move along timelike world lines, it must be that the world lines $\{r = R$, $\theta =$ const, $\phi =$ const, t varying$\}$ are timelike. From the Schwarzschild invariant interval (26.1) we read off the squared proper time, $d\tau^2 = -ds^2 = (1 - 2M/R)dt^2$, along those world lines. This $d\tau^2$ is positive (timelike world line) if and only if $R > 2M$. Thus, a static star with total mass-energy (active gravitational mass) M can never have a circumference smaller than $2\pi R = 4\pi M$. Restated in conventional units:

a static star must have radius R greater than $2M$

$$\boxed{\begin{aligned}\frac{\text{circumference}}{2\pi} &= R \equiv \begin{pmatrix}\text{radius}\\ \text{of star}\end{pmatrix} > 2M = \frac{2GM}{c^2}\\ &= 2.953\ \text{km}\left(\frac{M}{M_\odot}\right) \equiv \begin{pmatrix}\text{gravitational}\\ \text{radius}\end{pmatrix}.\end{aligned}} \tag{26.15}$$

Here $M_\odot$ is the mass of the Sun. The Sun satisfies this constraint by a huge margin: $R = 7 \times 10^5$ km $\gg 2.953$ km. A 1-solar-mass white-dwarf star satisfies it by a smaller margin: $R \simeq 6 \times 10^3$ km. And a 1-solar-mass neutron star satisfies it by only a modest margin: $R \simeq 10$ km. For a pedagogical and detailed discussion see, for example, Shapiro and Teukolsky (1983).

EXERCISES

Exercise 26.3 *Problem: Schwarzschild Geometry in Isotropic Coordinates*

(a) It turns out that the following line element is a solution of the vacuum Einstein field equation $\boldsymbol{G} = 0$:

$$ds^2 = -\left(\frac{1 - M/(2\bar{r})}{1 + M/(2\bar{r})}\right)^2 dt^2 + \left(1 + \frac{M}{2\bar{r}}\right)^4 [d\bar{r}^2 + \bar{r}^2(d\theta^2 + \sin^2\theta d\phi^2)]. \tag{26.16}$$

Since this solution is spherically symmetric, Birkhoff's theorem guarantees it must represent the standard Schwarzschild spacetime geometry in a coordinate system that differs from Schwarzschild's. Show that this is so by exhibiting a coordinate transformation that converts this line element into Eq. (26.1). [Note: The $\{t, \bar{r}, \theta, \phi\}$ coordinates are called *isotropic*, because in them the spatial part of the line element is a function of $\bar{r}$ times the 3-dimensional Euclidean line element, and Euclidean geometry picks out at each point in space no preferred spatial directions (i.e., it is isotropic).]

(b) Show that at large radii $r \gg M$, the line element (26.16) takes the form (25.98c) discussed in Chap. 25, but with vanishing spin angular momentum $\mathbf{J} = 0$.

Exercise 26.4 ***Example: Gravitational Redshift of Light from a Star's Surface*

Consider a photon emitted by an atom at rest on the surface of a static star with mass M and radius R. Analyze the photon's motion in the Schwarzschild coordinate system of the star's exterior, $r \geq R > 2M$. In particular, compute the "gravitational redshift" of the photon by the following steps.

(a) Since the emitting atom is nearly an ideal clock, it gives the emitted photon nearly the same frequency ν_{em}, as measured in the emitting atom's proper reference frame (as it would give were it in an Earth laboratory or floating in free space). Thus the proper reference frame of the emitting atom is central to a discussion of the photon's properties and behavior. Show that the orthonormal basis vectors of that proper reference frame are

$$\vec{e}_{\hat{0}} = \frac{1}{\sqrt{1 - 2M/r}}\frac{\partial}{\partial t}, \quad \vec{e}_{\hat{r}} = \sqrt{1 - 2M/r}\frac{\partial}{\partial r}, \quad \vec{e}_{\hat{\theta}} = \frac{1}{r}\frac{\partial}{\partial \theta}, \quad \vec{e}_{\hat{\phi}} = \frac{1}{r\sin\theta}\frac{\partial}{\partial \phi}, \tag{26.17}$$

with $r = R$ (the star's radius).

(b) Explain why the photon's energy as measured in the emitter's proper reference frame is $\mathcal{E} = h\nu_{\text{em}} = -p_{\hat{0}} = -\vec{p} \cdot \vec{e}_{\hat{0}}$. (Here and below h is Planck's constant, and $\vec{p}$ is the photon's 4-momentum.)

(c) Show that the quantity $\mathcal{E}_\infty \equiv -p_t = -\vec{p} \cdot \partial/\partial t$ is conserved as the photon travels outward from the emitting atom to an observer at very large radius, which we idealize as $r = \infty$. [Hint: Recall the result of Ex. 25.4a.] Show, further, that $\mathcal{E}_\infty$ is the photon's energy, as measured by the observer at $r = \infty$—which is why it is called the photon's "energy-at-infinity" and denoted $\mathcal{E}_\infty$. The photon's frequency, as measured by that observer, is given, of course, by $h\nu_\infty = \mathcal{E}_\infty$.

(d) Show that $\mathcal{E}_\infty = \mathcal{E}\sqrt{1 - 2M/R}$ and thence that $\nu_\infty = \nu_{\rm em}\sqrt{1 - 2M/R}$, and that therefore the photon is redshifted by an amount

$$\boxed{\frac{\lambda_{\rm rec} - \lambda_{\rm em}}{\lambda_{\rm em}} = \frac{1}{\sqrt{1 - 2M/R}} - 1.} \tag{26.18}$$

Here $\lambda_{\rm rec}$ is the wavelength that the photon's spectral line exhibits at the receiver, and $\lambda_{\rm em}$ is the wavelength that the emitting kind of atom would produce in an Earth laboratory. Note that for a nearly Newtonian star (i.e., one with $R \gg M$), this redshift becomes $\simeq M/R = GM/Rc^2$.

(e) Evaluate this redshift for Earth, for the Sun, and for a 1.4-solar-mass, 10-km-radius neutron star.

26.3.2

26.3.2 Stellar Interior

We now take a temporary detour from our study of the Schwarzschild geometry to discuss the interior of a static, spherical star. We do so less because of an interest in stars than because the detour will illustrate the process of solving the Einstein field equation and the role of the contracted Bianchi identity in the solution process.

Since the star's spacetime geometry is to be static and spherically symmetric, we can introduce as coordinates in its interior: (i) spherical polar angular coordinates θ and ϕ, (ii) a radial coordinate r such that the circumferences of the spheres are $2\pi r$, and (iii) a time coordinate $\bar{t}$ such that the metric coefficients are independent of $\bar{t}$. By their geometrical definitions, these coordinates will produce a spacetime line element of the form

$$ds^2 = g_{\bar{t}\bar{t}} d\bar{t}^2 + 2 g_{\bar{t}r} d\bar{t} dr + g_{rr} dr^2 + r^2 (d\theta^2 + \sin^2\theta d\phi^2), \tag{26.19}$$

with $g_{\alpha\beta}$ independent of $\bar{t}$, θ, and ϕ. Metric coefficients $g_{\bar{t}\theta}$, $g_{r\theta}$, $g_{\bar{t}\phi}$, and $g_{r\phi}$ are absent from Eq. (26.19), because they would break the spherical symmetry: they would distinguish the $+\phi$ direction from $-\phi$ or $+\theta$ from $-\theta$, since they would give nonzero values for the scalar products of $\partial/\partial\phi$ or $\partial/\partial\theta$ with $\partial/\partial t$ or $\partial/\partial r$. [Recall that the metric coefficients in a coordinate basis are $g_{\alpha\beta} = \mathbf{g}(\partial/\partial x^\alpha, \partial/\partial x^\beta) = (\partial/\partial x^\alpha) \cdot (\partial/\partial x^\beta)$.] We can get rid of the off-diagonal $g_{\bar{t}r}$ term in the line element (26.19) by specializing the time coordinate. The coordinate transformation

$$\bar{t} = t - \int \left(\frac{g_{\bar{t}r}}{g_{\bar{t}\bar{t}}}\right) dr \tag{26.20}$$

brings the line element into the form

coordinates and line element for interior of a static star

$$\boxed{ds^2 = -e^{2\Phi}dt^2 + e^{2\Lambda}dr^2 + r^2(d\theta^2 + \sin^2\theta\, d\phi^2).} \tag{26.21}$$

Here, after the transformation (26.20), we have introduced the names $e^{2\Phi}$ and $e^{2\Lambda}$ for the time-time and radial-radial metric coefficients, respectively. The signs of these coefficients (negative for g_{tt} and positive for g_{rr}) are dictated by the fact that inside the star, as on its surface, real atoms move along world lines of constant $\{r, \theta, \phi\}$ and changing t, and thus those world lines must be timelike. The name $e^{2\Phi}$ is chosen because when gravity is nearly Newtonian, the time-time metric coefficient $-e^{2\Phi}$ must reduce to $-(1+2\Phi)$, with Φ the Newtonian potential [Eq. (25.79)]. Thus, the Φ used in Eq. (26.21) is a generalization of the Newtonian potential to relativistic, spherical, static gravitational situations.

To solve the Einstein field equation for the star's interior, we must specify the stress-energy tensor. Stellar material is excellently approximated by a perfect fluid, and since our star is static, at any point inside the star the fluid's rest frame has constant $\{r, \theta, \phi\}$. Correspondingly, the 4-velocity of the fluid is

fluid 4-velocity inside static star

$$\boxed{\vec{u} = e^{-\Phi}\frac{\partial}{\partial t}.} \tag{26.22}$$

Here the factor $e^{-\Phi}$ guarantees that the 4-velocity will have $\vec{u}^2 = -1$, as it must.

Of course, this fluid is not freely falling. Rather, for a fluid element to remain always at fixed $\{r, \theta, \phi\}$ it must accelerate relative to local freely falling observers with a 4-acceleration $\vec{a} \equiv \nabla_{\vec{u}}\,\vec{u} \neq 0$ (i.e., $a^\alpha = u^\alpha{}_{;\mu}u^\mu \neq 0$). Symmetry tells us that this 4-acceleration cannot have any θ or ϕ components, and orthogonality of the 4-acceleration to the 4-velocity tells us that it cannot have any t component. The r component, computed from $a^r = u^r{}_{;\mu}u^\mu = \Gamma^r{}_{00}u^0u^0$, is $a^r = e^{-2\Lambda}\Phi_{,r}$; and thus we have

fluid 4-acceleration inside static star

$$\boxed{\vec{a} = e^{-2\Lambda}\,\Phi_{,r}\frac{\partial}{\partial r}.} \tag{26.23}$$

Each fluid element can be thought of as carrying an orthonormal set of basis vectors:

orthonormal basis vectors of fluid's local rest frame (its proper reference frame)

$$\boxed{\vec{e}_{\hat{0}} = \vec{u} = e^{-\Phi}\frac{\partial}{\partial t}, \quad \vec{e}_{\hat{r}} = e^{-\Lambda}\frac{\partial}{\partial r}, \quad \vec{e}_{\hat{\theta}} = \frac{1}{r}\frac{\partial}{\partial\theta}, \quad \vec{e}_{\hat{\phi}} = \frac{1}{r\sin\theta}\frac{\partial}{\partial\phi};} \tag{26.24a}$$

$$\boxed{\vec{e}^{\hat{0}} = e^{\Phi}\vec{\nabla}t, \quad \vec{e}^{\hat{r}} = e^{\Lambda}\vec{\nabla}r, \quad \vec{e}^{\hat{\theta}} = r\vec{\nabla}\theta, \quad \vec{e}^{\hat{\phi}} = r\sin\theta\,\vec{\nabla}\phi.} \tag{26.24b}$$

These basis vectors play two independent roles. (i) One can regard the tangent space of each event in spacetime as being spanned by the basis (26.24), specialized to that event. From this viewpoint, Eqs. (26.24) constitute an orthonormal, noncoordinate basis that covers every tangent space of the star's spacetime. This basis is called the fluid's *orthonormal, local-rest-frame basis.* (ii) One can focus attention on a specific fluid element, which moves along the world line $r = r_o, \theta = \theta_o, \phi = \phi_o$; and one can construct the proper reference frame of that fluid element in the same manner as we constructed the proper reference frame of an accelerated observer in flat spacetime in Sec. 24.5. That proper reference frame is a coordinate system $\{x^{\hat{\alpha}}\}$ whose basis vectors on the fluid element's world line are equal to the basis vectors (26.24):

$$\frac{\partial}{\partial x^{\hat{\mu}}} = \vec{e}_{\hat{\mu}}, \quad \vec{\nabla} x^{\hat{\mu}} = \vec{e}^{\hat{\mu}} \text{ at } x^{\hat{j}} = 0, \quad \text{with } \hat{1} = \hat{r}, \hat{2} = \hat{\theta}, \hat{3} = \hat{\phi}. \tag{26.25a}$$

More specifically, the proper-reference-frame coordinates $x^{\hat{\mu}}$ are given, to second-order in spatial distance from the fluid element's world line, by

coordinates of fluid element's proper reference frame

$$x^{\hat{0}} = e^{\Phi_o} t, \quad x^{\hat{1}} = \int_{r_o}^{r} e^{\Lambda} dr - \frac{1}{2} e^{-\Lambda_o} r_o [(\theta - \theta_o)^2 + \sin^2 \theta_o (\phi - \phi_o)^2],$$
$$x^{\hat{2}} = r(\theta - \theta_o) - \frac{1}{2} r_o \sin \theta_o \cos \theta_o (\phi - \phi_o)^2, \quad x^{\hat{3}} = r \sin \theta (\phi - \phi_o), \tag{26.25b}$$

from which one can verify relation (26.25a) with the basis vectors given by Eqs. (26.24). [In Eqs. (26.25b) and throughout this discussion all quantities with subscripts $_o$ are evaluated on the fluid's world line.] In terms of the proper-reference-frame coordinates (26.25b), the line element (26.21) takes the following form, accurate to first order in distance from the fluid element's world line:

$$ds^2 = -[1 + 2\Phi_{,r}(r - r_o)](dx^{\hat{0}})^2 + \delta_{ij} dx^{\hat{i}} dx^{\hat{j}}. \tag{26.25c}$$

Notice that the quantity $\Phi_{,r}(r - r_o)$ is equal to the scalar product of (i) the spatial separation $\hat{\mathbf{x}} \equiv (r - r_o)\partial/\partial r + (\theta - \theta_o)\partial/\partial\theta + (\phi - \phi_o)\partial/\partial\phi$ of the "field point" (r, θ, ϕ) from the fluid element's world line, with (ii) the fluid's 4-acceleration (26.23), viewed as a spatial 3-vector $\mathbf{a} = e^{-2\Lambda_o}\Phi_{,r}\partial/\partial r$. Correspondingly, the spacetime line element (26.25c) in the fluid element's proper reference frame takes the standard proper-reference-frame form (24.60b):

$$ds^2 = -(1 + 2\mathbf{a} \cdot \hat{\mathbf{x}})(dx^{\hat{0}})^2 + \delta_{jk} dx^{\hat{j}} dx^{\hat{k}}, \tag{26.26}$$

accurate to first order in distance from the fluid element's world line. To second order, as discussed at the end of Sec. 25.3, there are corrections proportional to the spacetime curvature.

In the local rest frame of the fluid [i.e., when expanded on the fluid's orthonormal rest-frame basis vectors (26.24) or equally well (26.25a)], the components $T^{\hat{\alpha}\hat{\beta}} = (\rho + P)u^{\hat{\alpha}}u^{\hat{\beta}} + P g^{\hat{\alpha}\hat{\beta}}$ of the fluid's stress-energy tensor take on the standard form [Eq. (24.50)]:

stress-energy tensor in fluid's proper reference frame: fluid density and pressure

$$\boxed{T^{\hat{0}\hat{0}} = \rho, \quad T^{\hat{r}\hat{r}} = T^{\hat{\theta}\hat{\theta}} = T^{\hat{\phi}\hat{\phi}} = P,} \tag{26.27}$$

corresponding to a rest-frame mass-energy density ρ and isotropic pressure P. By contrast with the simplicity of these local-rest-frame components, the contravariant components $T^{\alpha\beta} = (\rho + P)u^\alpha u^\beta + P g^{\alpha\beta}$ in the $\{t, r, \theta, \phi\}$ coordinate basis are rather more complicated:

$$T^{tt} = e^{-2\Phi}\rho, \quad T^{rr} = e^{-2\Lambda}P, \quad T^{\theta\theta} = r^{-2}P, \quad T^{\phi\phi} = (r\,\sin\theta)^{-2}\,P. \tag{26.28}$$

This shows one advantage of using orthonormal bases: the components of vectors and tensors are generally simpler in an orthonormal basis than in a coordinate basis. A second advantage occurs when one seeks the physical interpretation of formulas. Because every orthonormal basis is the proper-reference-frame basis of some local observer (the observer with 4-velocity $\vec{u} = \vec{e}_{\hat{0}}$), components measured in such a basis have an immediate physical interpretation in terms of measurements by that observer. For example, $T^{\hat{0}\hat{0}}$ is the total density of mass-energy measured by the local observer. By contrast, components in a coordinate basis typically do not have a simple physical interpretation.

EXERCISES

Exercise 26.5 *Derivation: Proper-Reference-Frame Coordinates*
Show that in the coordinate system $\{x^{\hat{0}}, x^{\hat{1}}, x^{\hat{2}}, x^{\hat{3}}\}$ of Eqs. (26.25b), the coordinate basis vectors at $x^{\hat{j}} = 0$ are Eqs. (26.24), and, accurate through first order in distance from $x^{\hat{j}} = 0$, the spacetime line element is Eq. (26.26); that is, errors are no larger than second order.

26.3.3 Local Conservation of Energy and Momentum

26.3.3

Before inserting the perfect-fluid stress-energy tensor (26.27) into the Einstein field equation, we impose on it the local law of conservation of 4-momentum: $\vec{\nabla} \cdot \boldsymbol{T} = 0$. In doing so we require from the outset that, since the star is to be static and spherical, its density ρ and pressure P must be independent of t, θ, and ϕ (i.e., like the metric coefficients Φ and Λ, they must be functions of radius r only).

The most straightforward way to impose 4-momentum conservation is to equate to zero the quantities

$$T^{\alpha\beta}{}_{;\beta} = \frac{\partial T^{\alpha\beta}}{\partial x^\beta} + \Gamma^\beta{}_{\mu\beta} T^{\alpha\mu} + \Gamma^\alpha{}_{\mu\beta} T^{\mu\beta} = 0 \tag{26.29}$$

in our coordinate basis, making use of expressions (26.28) for the contravariant components of the stress-energy tensor, and the connection coefficients and metric components given in Box 26.2.

This straightforward calculation requires a lot of work. Much better is an analysis based on the local proper reference frame of the fluid. The temporal component of $\vec{\nabla} \cdot \boldsymbol{T} = 0$ in that reference frame [i.e., the projection $\vec{u} \cdot (\vec{\nabla} \cdot \boldsymbol{T}) = 0$ of this conservation law onto the time basis vector $\vec{e}_{\hat{0}} = e^{-\Phi}\partial/\partial t = \vec{u}$] represents energy conservation as seen by the fluid—the first law of thermodynamics:

first law of thermodynamics

$$\frac{d(\rho V)}{d\tau} = -P\frac{dV}{d\tau}. \tag{26.30}$$

Here τ is proper time as measured by the fluid element we are following, and V is the fluid element's volume. (This equation is derived in Ex. 2.26b, in a special relativistic context; but since it involves only one derivative, there is no danger of curvature coupling, so that derivation and the result can be lifted without change into general relativity, i.e., into the star's curved spacetime; cf. Ex. 26.6a.) Now, inside this static star, the fluid element sees and feels no changes. Its density ρ, pressure P, and volume V always remain constant along the fluid element's world line, and energy conservation is therefore guaranteed to be satisfied. Equation (26.30) tells us nothing new.

The spatial part of $\vec{\nabla} \cdot \boldsymbol{T} = 0$ in the fluid's local rest frame can be written in geometric form as $\boldsymbol{P} \cdot (\vec{\nabla} \cdot \boldsymbol{T}) = 0$. Here $\boldsymbol{P} \equiv \boldsymbol{g} + \vec{u} \otimes \vec{u}$ is the tensor that projects all vectors into the 3-surface orthogonal to $\vec{u}$, that is, into the fluid's local 3-surface of simultaneity (Exs. 2.10 and 25.1b). By inserting the perfect-fluid stress-energy tensor $\boldsymbol{T} = (\rho + P)\vec{u} \otimes \vec{u} + P\boldsymbol{g} = \rho\vec{u} \otimes \vec{u} + P\boldsymbol{P}$ into $\boldsymbol{P} \cdot (\vec{\nabla} \cdot \boldsymbol{T}) = 0$, reexpressing the result in slot-naming index notation, and carrying out some index gymnastics, we must obtain the same result as in special relativity (Ex. 2.26c):

force balance inside the fluid

$$(\rho + P)\vec{a} = -\boldsymbol{P} \cdot \vec{\nabla} P \tag{26.31}$$

(cf. Ex. 26.6b). Here $\vec{a}$ is the fluid's 4-acceleration. Recall from Ex. 2.27 that for a perfect fluid, $\rho + P$ is the inertial mass per unit volume. Therefore, Eq. (26.31) says that the fluid's inertial mass per unit volume times its 4-acceleration is equal to the negative of its pressure gradient, projected orthogonally to its 4-velocity. Since both sides of Eq. (26.31) are purely spatially directed as seen in the fluid's local proper reference frame, we can rewrite this equation in 3-dimensional language as

$$\boxed{(\rho + P)\mathbf{a} = -\nabla P.} \tag{26.32}$$

A Newtonian physicist, in the proper reference frame, would identify $-\mathbf{a}$ as the local gravitational acceleration, $\mathbf{g}$, and correspondingly, would rewrite Eq. (26.31) as

Newtonian viewpoint on force balance

$$\nabla P = (\rho + P)\mathbf{g}. \tag{26.33}$$

This is the standard equation of hydrostatic equilibrium for a fluid in an Earthbound laboratory (or swimming pool or lake or ocean), except for the presence of the pressure P in the inertial mass per unit volume (Ex. 2.27). On Earth the typical pressures of fluids, even deep in the ocean, are only $P \lesssim 10^9$ dyne/cm$^2 \simeq 10^{-12}$ g cm$^{-3} \lesssim 10^{-12}\rho$. Thus, to extremely good accuracy one can ignore the contribution of pressure to the inertial mass density. However, deep inside a neutron star, P may be within a factor 2 of ρ, so the contribution of P cannot be ignored.

We can convert the law of force balance (26.31) into an ordinary differential equation for the pressure P by evaluating its components in the fluid's proper reference frame. The 4-acceleration (26.23) is purely radial; its radial component is $a^{\hat{r}} = e^{-\Lambda}\Phi_{,r} = \Phi_{,\hat{r}}$. The gradient of the pressure is also purely radial, and its radial component is $P_{;\hat{r}} = P_{,\hat{r}} = e^{-\Lambda}P_{,r}$. Therefore, the law of force balance reduces to

$$\frac{dP}{dr} = -(\rho + P)\frac{d\Phi}{dr}. \tag{26.34}$$

force balance rewritten

EXERCISES

Exercise 26.6 *Practice and Derivation: Local Conservation of Energy and Momentum for a Perfect Fluid*

(a) Use index manipulations to show that in general (not just inside a static star), for a perfect fluid with $T^{\alpha\beta} = (\rho + P)u^\alpha u^\beta + Pg^{\alpha\beta}$, the law of energy conservation $u_\alpha\, T^{\alpha\beta}{}_{;\beta} = 0$ reduces to the first law of thermodynamics (26.30). [Hint: You will need the relation $u^\mu{}_{;\mu} = (1/V)(dV/d\tau)$; cf. Ex. 2.24.]

(b) Similarly, show that $P_{\mu\alpha}T^{\alpha\beta}{}_{;\beta} = 0$ reduces to the force-balance law (26.31).

26.3.4 The Einstein Field Equation

26.3.4

Turn, now, to the Einstein field equation inside a static, spherical star with isotropic pressure. To impose it, we must first compute, in our $\{t, r, \theta, \phi\}$ coordinate system, the components of the Einstein tensor $G_{\alpha\beta}$. In general, the Einstein tensor has 10 independent components. However, the symmetries of the line element (26.21) impose identical symmetries on the Einstein tensor computed from it: The only nonzero components in the fluid's proper reference frame will be $G^{\hat{0}\hat{0}}$, $G^{\hat{r}\hat{r}}$, and $G^{\hat{\theta}\hat{\theta}} = G^{\hat{\phi}\hat{\phi}}$; and these three independent components will be functions of radius r only. Correspondingly, the Einstein equation will produce three independent differential equations for our four unknowns: the metric coefficients ("gravitational potentials") Φ and Λ [Eq. (26.21)], and the radial distribution of density ρ and pressure P.

These three independent components of the Einstein equation will actually be redundant with the law of hydrostatic equilibrium (26.34). One can see this as follows. If we had not yet imposed the law of 4-momentum conservation, then the Einstein equation $\boldsymbol{G} = 8\pi \boldsymbol{T}$, together with the Bianchi identity $\vec{\nabla} \cdot \boldsymbol{G} \equiv 0$ [Eq. (25.69)], would enforce $\vec{\nabla} \cdot \boldsymbol{T} = 0$. More explicitly, our three independent components of the Einstein equation together would imply the law of radial force balance [i.e., of hydrostatic equilibrium (26.34)]. Since we have already imposed Eq. (26.34), we need evaluate only two of the three independent components of the Einstein equation; they will give us full information.

with force balance imposed, Einstein's equation provides just two additional constraints on the stellar structure

A long and rather tedious calculation (best done on a computer), based on the metric coefficients of Eq. (26.21) and on Eqs. (24.38), (25.50), (25.46), (25.49), and (25.68), produces the following for the time-time and radial-radial components of the Einstein tensor, and thence of the Einstein field equation:

$$G^{\hat{0}\hat{0}} = \frac{1}{r^2}\frac{d}{dr}[r(1-e^{-2\Lambda})] = 8\pi T^{\hat{0}\hat{0}} = 8\pi\rho, \tag{26.35}$$

$$G^{\hat{r}\hat{r}} = -\frac{1}{r^2}(1-e^{-2\Lambda}) + \frac{2}{r}e^{-2\Lambda}\frac{d\Phi}{dr} = 8\pi T^{\hat{r}\hat{r}} = 8\pi P. \tag{26.36}$$

We can bring these components of the field equation into simpler form by defining a new metric coefficient $m(r)$ by

$$\boxed{e^{2\Lambda} \equiv \frac{1}{1-2m/r}.} \tag{26.37}$$

Note [cf. Eqs. (26.1), (26.21), and (26.37)] that outside the star, m is equal to the star's total mass-energy M. This, plus the fact that in terms of m the time-time component of the field equation (26.35) takes the form

Einstein equation for the mass inside radius r, $m(r)$

$$\boxed{\frac{dm}{dr} = 4\pi r^2\rho,} \tag{26.38a}$$

motivates the name *mass inside radius r* for the quantity $m(r)$. In terms of m the radial-radial component (26.36) of the field equation becomes

Einstein equation for $\Phi(r)$

$$\boxed{\frac{d\Phi}{dr} = \frac{m+4\pi r^3 P}{r(r-2m)};} \tag{26.38b}$$

combining this with Eq. (26.34), we obtain an alternative form of the equation of hydrostatic equilibrium:

TOV equation of hydrostatic equilibrium

$$\boxed{\frac{dP}{dr} = -\frac{(\rho+P)(m+4\pi r^3 P)}{r(r-2m)}.} \tag{26.38c}$$

[This form is called the Tolman-Oppenheimer-Volkoff (or TOV) equation, because it was first derived by Tolman (1939) and first used in a practical calculation by Oppenheimer and Volkoff (1939).] Equations (26.38a), (26.38b), (26.38c) plus an equation of state for the pressure of the stellar material P in terms of its density of total mass-energy ρ,

$$\boxed{P = P(\rho),} \tag{26.38d}$$

summary: the relativistic equations of stellar structure, Eqs. (26.38)

determine the four quantities Φ, m, ρ, and P as functions of radius. In other words, Eqs. (26.38) are the relativistic equations of stellar structure.

Actually, for full determination, one also needs boundary conditions. Just as the surface of a sphere is everywhere locally Euclidean (i.e., is arbitrarily close to Euclidean in arbitrarily small regions), so also spacetime must be everywhere locally Lorentz; cf. Eqs. (25.9). For spacetime to be locally Lorentz at the star's center (in particular, for circumferences of tiny circles around the center to be equal to 2π times their radii), it is necessary that m vanish at the center:

boundary conditions for relativistic equations of stellar structure

$$m = 0 \text{ at } r = 0, \quad \text{and thus } \boxed{m(r) = \int_0^r 4\pi r^2 \rho dr} \tag{26.39}$$

[cf. Eqs. (26.21) and (26.37)]. At the star's surface the interior spacetime geometry (26.21) must join smoothly to the exterior Schwarzschild geometry (26.1):

$$\boxed{m = M \quad \text{and} \quad e^{2\Phi} = 1 - 2M/r \text{ at } r = R.} \tag{26.40}$$

26.3.5 Stellar Models and Their Properties

26.3.5

procedure for constructing a relativistic stellar model

A little thought now reveals a straightforward method of producing a relativistic stellar model.

1. Specify an equation of state for the stellar material $P = P(\rho)$, and specify a central density ρ_c or central pressure P_c for the star.
2. Integrate the coupled hydrostatic-equilibrium equation (26.38c) and "mass equation" (26.38a) outward from the center, beginning with the initial conditions $m = 0$ and $P = P_c$ at the center.
3. Terminate the integration when the pressure falls to zero; this is the surface of the star.
4. At the surface read off the value of m; it is the star's total mass-energy M, which appears in the star's external, Schwarzschild line element (26.1).
5. From this M and the radius $r \equiv R$ of the star's surface, read off the value of the gravitational potential Φ at the surface [Eq. (26.40)].
6. Integrate the Einstein field equation (26.38b) inward from the surface toward the center to determine Φ as a function of radius inside the star.

Just 6 weeks after reading to the Prussian Academy of Science the letter in which Karl Schwarzschild derived his vacuum solution (26.1) of the field equation, Albert Einstein again presented the Academy with results from Schwarzschild's fertile mind: an exact solution for the structure of the interior of a star that has constant density ρ. [And just 4 months after that, on June 29, 1916, Einstein had the sad task of announcing to the Academy that Schwarzschild had died of an illness contracted on the World-War-I Russian front.]

details of a relativistic star with constant density ρ

In our notation, Schwarzschild's solution for the interior of a star is characterized by its uniform density ρ, its total mass M, and its radius R, which is given in terms of ρ and M by

$$M = \frac{4\pi}{3}\rho R^3 \tag{26.41}$$

[Eq. (26.39)]. In terms of these quantities, the mass M inside radius r, the pressure P, and the gravitational potential Φ as functions of r are (Schwarzschild, 1916b)

$$m = \frac{4\pi}{3}\rho r^3, \quad P = \rho\left[\frac{(1-2Mr^2/R^3)^{\frac{1}{2}} - (1-2M/R)^{\frac{1}{2}}}{3(1-2M/R)^{\frac{1}{2}} - (1-2Mr^2/R^3)^{\frac{1}{2}}}\right], \tag{26.42a}$$

$$e^{\Phi} = \frac{3}{2}\left(1 - \frac{2M}{R}\right)^{\frac{1}{2}} - \frac{1}{2}\left(1 - \frac{2Mr^2}{R^3}\right)^{\frac{1}{2}}. \tag{26.42b}$$

We present these details less for their specific physical content than to illustrate the solution of the Einstein field equation in a realistic, astrophysically interesting situation. For discussions of the application of this formalism to neutron stars, where relativistic deviations from Newtonian theory can be rather strong, see, for example, Shapiro and Teukolsky (1983). For the seminal work on the theory of neutron-star structure, see Oppenheimer and Volkoff (1939).

upper limit on mass of a neutron star, and how it comes about

Among the remarkable consequences of the TOV equation of hydrostatic equilibrium (26.38c) for neutron-star structure are the following. (i) If the mass m inside radius r ever gets close to $r/2$, the "gravitational pull" [right-hand side of (26.38c)] becomes divergently large, forcing the pressure gradient that counterbalances it to be divergently large, and thereby driving the pressure quickly to zero as one integrates outward. This protects the static star from having M greater than $R/2$ (i.e., from having its surface inside its gravitational radius). (ii) Although the density of matter near the center of a neutron star is above that of an atomic nucleus (2.3×10^{17} kg m^{-3}), where the equation of state is ill-understood, we can be confident that there is an upper limit on the masses of neutron stars, a limit in the range $2M_\odot \lesssim M_{\rm max} \lesssim 3M_\odot$.[2] This mass limit cannot be avoided by postulating that a more massive neutron star develops an arbitrarily large central pressure and thereby supports itself against gravitational implosion. The reason is that an arbitrarily large central pressure is self-defeating: The "gravitational pull" that appears on the right-hand side of Eq. (26.38c) is quadratic in the pressure at very high pressures (whereas it would be independent of pressure in Newtonian theory). This purely relativistic feature guarantees that, if a star develops too high a central pressure, it will be unable to support itself against the resulting "quadratically too high" gravitational pull.

EXERCISES

Exercise 26.7 *Problem: Mass-Radius Relation for Neutron Stars*

The equation of state of a neutron star is very hard to calculate at the supra-nuclear densities required, because the calculation is a complex, many-body problem and the particle interactions are poorly understood and poorly measured. Observations of

2. Measured neutron star masses range from ~1.2 solar masses to more than 2.0 solar masses.

neutron stars' masses and radii can therefore provide valuable constraints on fundamental nuclear physics. As we discuss briefly in the following chapter, various candidate equations of state can already be excluded on these observational grounds.

We need an equation of state for when the density $\rho = 10^{18}\rho_{18}$ kg m^{-3} is close to nuclear density ($\rho_{18} \sim 0.27$). A simple functional form, which interpolates between the equation appropriate for a low density, non-relativistic, degenerate neutron gas with $P \propto \rho^{5/3}$ [cf. Eq. (3.43)] and the equation for ultrarelativistic particles, $P = 1/3\rho$, [Eq. (3.54a)], is given by

$$P_{18} = \frac{0.06\rho_{18}^{5/3} + \rho_{18}^4}{3(1+\rho_{18}^3)} \tag{26.43}$$

For this equation of state, use the equations of stellar structure (26.38a) and (26.38c) to find the masses and radii of stars with a range of central pressures, and hence deduce a mass-radius relation, $M(R)$. You should discover that, as the central pressure is increased, the mass passes through a maximum, while the radius continues to decrease. (Stars with radii smaller than that at the maximum mass are unstable to radial perturbations.)

26.3.6 Embedding Diagrams

26.3.6

We conclude our discussion of static stars by using them to illustrate a useful technique for visualizing the curvature of spacetime: the embedding of the curved spacetime, or a piece of it, in a flat space of higher dimensionality.

the problem of embedding a curved manifold inside a flat higher-dimensional manifold (the embedding space); number of dimensions needed in the embedding space

The geometry of a curved, n-dimensional manifold is characterized by $\frac{1}{2}n(n+1)$ metric components (since those components form a symmetric $n \times n$ matrix), of which only $\frac{1}{2}n(n+1) - n = \frac{1}{2}n(n-1)$ are of coordinate-independent significance (since we are free to choose arbitrarily the n coordinates of our coordinate system and can thereby force n of the metric components to take on any desired values, e.g., zero). If this n-dimensional manifold is embedded in a flat N-dimensional manifold, that embedding will be described by expressing $N - n$ of the embedding manifold's Euclidean (or Lorentz) coordinates in terms of the other n. Thus, the embedding is characterized by $N - n$ functions of n variables. For the embedding to be possible, in general, this number of choosable functions must be at least as large as the number of significant metric coefficients $\frac{1}{2}n(n-1)$. From this argument we conclude that the dimensionality of the embedding space must be $N \geq \frac{1}{2}n(n+1)$. Actually, this argument analyzes only the local features of the embedding. If one also wants to preserve the global topology of the n-dimensional manifold, one must in general go to an embedding space of even higher dimensionality.

Curved spacetime has $n = 4$ dimensions and thus requires for its local embedding a flat space with at least $N = \frac{1}{2}n(n+1) = 10$ dimensions. This is a bit much for 3-dimensional beings like us to visualize. If, as a sop to our visual limitations, we reduce our ambitions and seek only to extract a 3-surface from curved spacetime and visualize it by embedding it in a flat space, we will require a flat space of $N = 6$ dimensions. This is still a bit much. In frustration, we are driven to extract from spacetime $n = 2$ dimensional surfaces and visualize them by embedding in flat spaces with $N = 3$ dimensions. This is doable—and, indeed, instructive.

embedding the equatorial plane of a static, relativistic star in a flat 3-dimensional embedding space

As a nice example, consider the equatorial "plane" through the spacetime of a static spherical star, at a specific "moment" of coordinate time t [i.e., consider the 2-surface $t = \text{const}$, $\theta = \pi/2$ in the spacetime of Eqs. (26.21), (26.37)]. The line element on this equatorial 2-surface is

$$ {}^{(2)}ds^2 = \frac{dr^2}{1 - 2m/r} + r^2 d\phi^2, \quad \text{where } m = m(r) = \int_0^r 4\pi r^2 \rho dr \tag{26.44} $$

[cf. Eq. (26.39)]. We seek to construct in a 3-dimensional Euclidean space a 2-dimensional surface with precisely this same 2-geometry. As an aid, we introduce in the Euclidean embedding space a cylindrical coordinate system $\{r, z, \phi\}$, in terms of which the space's 3-dimensional line element is

$$ {}^{(3)}ds^2 = dr^2 + dz^2 + r^2 d\phi^2. \tag{26.45} $$

The surface we seek to embed is axially symmetric, so we can describe its embedding by the value of z on it as a function of radius r: $z = z(r)$. Inserting this (unknown) embedding function into Eq. (26.45), we obtain for the surface's 2-geometry,

$$ {}^{(2)}ds^2 = [1 + (dz/dr)^2]\, dr^2 + r^2 d\phi^2. \tag{26.46} $$

Comparing with our original expression (26.44) for the 2-geometry, we obtain a differential equation for the embedding function:

$$ \frac{dz}{dr} = \left(\frac{1}{1 - 2m/r} - 1 \right)^{\frac{1}{2}}. \tag{26.47} $$

If we set $z = 0$ at the star's center, then the solution of this differential equation is

shape of embedded equatorial plane in general

$$ \boxed{z = \int_0^r \frac{dr'}{[r'/(2m) - 1]^{\frac{1}{2}}}.} \tag{26.48} $$

shape outside the star where the spacetime geometry is Schwarzschild

Near the star's center $m(r)$ is given by $m = (4\pi/3)\rho_c r^3$, where ρ_c is the star's central density; and outside the star $m(r)$ is equal to the star's r-independent total mass M. Correspondingly, in these two regions Eq. (26.48) reduces to

$$ z = \sqrt{(2\pi/3)\rho_c}\, r^2 \quad \text{at } r \text{ very near zero.} $$

$$ z = \sqrt{8M(r - 2M)} + \text{const at } r > R, \quad \text{i.e., outside the star.} \tag{26.49} $$

Figure 26.1 shows the embedded 2-surface $z(r)$ for a star of uniform density $\rho = \text{const}$ (Ex. 26.8). For any other star the embedding diagram will be qualitatively similar, though quantitatively different.

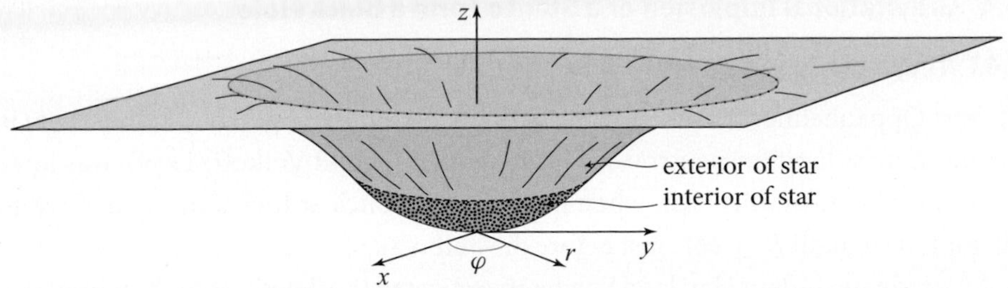

FIGURE 26.1 Embedding diagram depicting an equatorial, 2-dimensional slice $t = \text{const}, \theta = \pi/2$ through the spacetime of a spherical star with uniform density ρ and with radius R equal to 2.5 times the gravitational radius $2M$. See Ex. 26.8 for details.

The most important feature of this embedding diagram is its illustration of the fact [also clear in the original line element (26.44)] that, as one moves outward from the star's center, its circumference $2\pi r$ increases less rapidly than the proper radial distance traveled, $l = \int_0^r (1 - 2m/r)^{-\frac{1}{2}} dr$. As a specific example, the distance from the center of Earth to a perfect circle near Earth's surface is more than the circumference/2π by about 1.5 mm—a number whose smallness compared to the actual radius, 6.4×10^8 cm, is a measure of the weakness of the curvature of spacetime near Earth. As a more extreme example, the distance from the center of a massive neutron star to its surface is about 1 km greater than its circumference/2π—greater by an amount that is roughly 10% of the ~10-km circumference/2π. Correspondingly, in the embedding diagram for Earth the embedded surface would be so nearly flat that its downward dip at the center would be imperceptible, whereas the diagram for a neutron star would show a downward dip about like that of Fig. 26.1.

EXERCISES

Exercise 26.8 *Example: Embedding Diagram for Star with Uniform Density*

(a) Show that the embedding surface of Eq. (26.48) is a paraboloid of revolution everywhere outside the star.

(b) Show that in the interior of a uniform-density star, the embedding surface is a segment of a sphere.

(c) Show that the match of the interior to the exterior is done in such a way that, in the embedding space, the embedded surface shows no kink (no bend) at $r = R$.

(d) Show that, in general, the circumference/(2π) for a star is less than the distance from the center to the surface by an amount of order one sixth the star's gravitational radius, $M/3$. Evaluate this amount analytically for a star of uniform density, and numerically (approximately) for Earth and for a neutron star.

26.4 ## 26.4 Gravitational Implosion of a Star to Form a Black Hole

26.4.1 ### 26.4.1 The Implosion Analyzed in Schwarzschild Coordinates

J. Robert Oppenheimer, on discovering with his student George Volkoff that there is a maximum mass limit for neutron stars (Oppenheimer and Volkoff, 1939), was forced to consider the possibility that, when it exhausts its nuclear fuel, a more massive star will implode to radii $R \le 2M$. Just before the outbreak of World War II, Oppenheimer and his graduate student Hartland Snyder investigated the details of such an implosion for the idealized case of a perfectly spherical star in which all the internal pressure is suddenly extinguished (Oppenheimer and Snyder, 1939). In this section, we repeat their analysis, though from a more modern viewpoint and using somewhat different arguments.

By Birkhoff's theorem, the spacetime geometry outside an imploding, spherical star must be that of Schwarzschild. This means, in particular, that an imploding, spherical star cannot produce any gravitational waves; such waves would break the spherical symmetry. By contrast, a star that implodes nonspherically can produce a burst of gravitational waves (Chap. 27).

Since the spacetime geometry outside an imploding, spherical star is that of Schwarzschild, we can depict the motion of the star's surface by a world line in a 2-dimensional spacetime diagram with Schwarzschild coordinate time t plotted upward and Schwarzschild coordinate radius r plotted rightward (Fig. 26.2). The world line of the star's surface is an ingoing curve. The region to the left of the world line must be discarded and replaced by the spacetime of the star's interior, while the region to the right, $r > R(t)$, is correctly described by Schwarzschild.

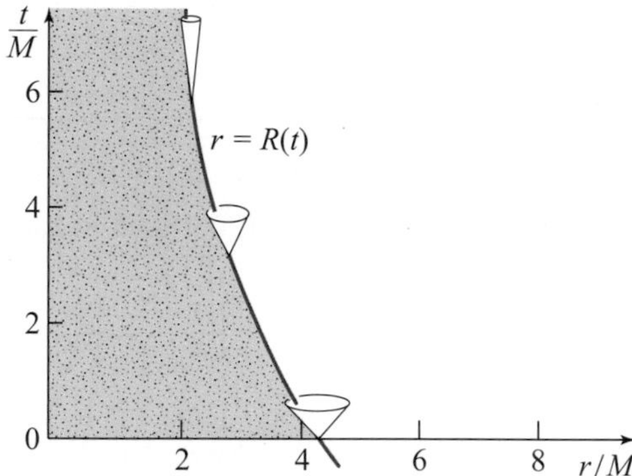

FIGURE 26.2 Spacetime diagram depicting the gravitationally induced implosion of a star in Schwarzschild coordinates. The thick solid curve is the world line of the star's surface, $r = R(t)$, in the external Schwarzschild coordinates. The stippled region to the left of that world line is not correctly described by the Schwarzschild line element (26.1); it requires for its description the spacetime metric of the star's interior. The surface's world line $r = R(t)$ is constrained to lie inside the light cones.

As for a static star, so also for an imploding one: because real atoms with finite rest masses live on the star's surface, the world line of that surface, $\{r = R(t), \theta \text{ and } \phi \text{ constant}\}$, must be timelike. Consequently, at each point along the world line it must lie within the local light cones that are depicted in Fig. 26.2.

The radial edges of the light cones are lines along which the Schwarzschild line element, the ds^2 of Eq. (26.1), vanishes with θ and ϕ held fixed:

$$0 = ds^2 = -(1 - 2M/R)dt^2 + \frac{dr^2}{1 - 2M/R}; \quad \text{that is,} \quad \frac{dt}{dr} = \pm\frac{1}{1 - 2M/R}. \tag{26.50}$$

Therefore, instead of having 45° opening angles $dt/dr = \pm 1$ as they do in a Lorentz frame of flat spacetime, the light cones "squeeze down" toward $dt/dr = \infty$ as the star's surface $r = R(t)$ approaches the gravitational radius: $R \to 2M$. This is a peculiarity due not to spacetime curvature, but rather to the nature of the Schwarzschild coordinates: If, at any chosen event of the Schwarzschild spacetime, we were to introduce a local Lorentz frame, then in that frame the light cones would have 45° opening angles.

Since the world line of the star's surface is confined to the interiors of the local light cones, the squeezing down of the light cones near $r = 2M$ prevents the star's world line $r = R(t)$ from ever, in any finite coordinate time t, reaching the gravitational radius, $r = 2M$.

This conclusion is completely general; it relies in no way on the details of what is going on inside the star or at its surface. It is just as valid for completely realistic stellar implosion (with finite pressure and shock waves) as for the idealized, Oppenheimer-Snyder case of zero-pressure implosion. In the special case of zero pressure, one can explore the details further:

Because no pressure forces act on the atoms at the star's surface, those atoms must move inward along radial geodesic world lines. Correspondingly, the world line of the star's surface in the external Schwarzschild spacetime must be a timelike geodesic of constant (θ, ϕ). In Ex. 26.9, the geodesic equation is solved to determine that world line $R(t)$, with a conclusion that agrees with the above argument: only after a lapse of infinite coordinate time t does the star's surface reach the gravitational radius $r = 2M$. A byproduct of that calculation is equally remarkable. Although the implosion to $R = 2M$ requires infinite Schwarzschild coordinate time t, it requires only a finite proper time τ as measured by an observer who rides inward on the star's surface. In fact, the proper time is

the surface of an imploding star reaches the gravitational radius $r = 2M$ in finite proper time τ, but infinite coordinate time t

$$\boxed{\tau \simeq \frac{\pi}{2}\left(\frac{R_o^3}{2M}\right)^{\frac{1}{2}} = 15\,\mu\text{s}\left(\frac{R_o}{2M}\right)^{3/2}\frac{M}{M_\odot} \quad \text{if } R_o \gg 2M,} \tag{26.51}$$

where R_o is the star's initial radius when it first begins to implode freely, $M_\odot$ denotes the mass of the Sun, and proper time τ is measured from the start of implosion. Note that this implosion time is equal to $1/(4\sqrt{2})$ times the orbital period of a test particle at the radius of the star's initial surface. For a star with mass and initial radius equal to those of the Sun, τ is about 30 minutes; for a neutron star that has been pushed over the maximum mass limit by accretion of matter from its surroundings, τ is about 50 μs. For a hypothetical supermassive star with $M = 10^9 M_\odot$ and $R_o \sim$ a few M, τ would be about a day.

26.4.2

26.4.2 Tidal Forces at the Gravitational Radius

What happens to the star's surface, and an imagined observer on it, when—after infinite coordinate time but a brief proper time—it reaches the gravitational radius? There are two possibilities: (i) the tidal gravitational forces there might be so strong that they destroy the star's surface and any observers on it; or (ii) the tidal forces are not that strong, and so the star and observers must continue to exist, moving into a region of spacetime (presumably $r < 2M$) that is not smoothly joined onto $r > 2M$ in the Schwarzschild coordinate system. In the latter case, the pathology is all due to poor properties of Schwarzschild's coordinates. In the former case, it is due to an intrinsic, coordinate-independent singularity of the tide-producing Riemann curvature.

To see which is the case, we must evaluate the tidal forces felt by observers on the surface of the imploding star. Those tidal forces are produced by the Riemann curvature tensor. More specifically, if an observer's feet and head have a vector separation $\boldsymbol{\xi}$ at time τ as measured by the observer's clock, then the curvature of spacetime will exert on them a relative gravitational acceleration given by the equation of geodesic deviation in the form appropriate to a local Lorentz frame:

$$\frac{d^2\xi^{\bar{j}}}{d\tau^2} = -R^{\bar{j}}{}_{\bar{0}\bar{k}\bar{0}}\xi^{\bar{k}} \tag{26.52}$$

[Eq. (25.34)]. Here the barred indices denote components in the observer's local Lorentz frame. The tidal forces will become infinite and will thereby destroy the observer and all forms of matter on the star's surface, if and only if the local Lorentz Riemann components $R_{\bar{j}\bar{0}\bar{k}\bar{0}}$ diverge as the star's surface approaches the gravitational radius. Thus, to test whether the observer and star survive, we must compute the components of the Riemann curvature tensor in the local Lorentz frame of the star's imploding surface.

The easiest way to compute those components is by a transformation from components as measured in the proper reference frames of observers who are "at rest" (fixed r, θ, ϕ) in the Schwarzschild spacetime. At each event on the world tube of the star's surface, then, we have two orthonormal frames: one (barred indices) a local Lorentz frame imploding with the star; the other (hatted indices) a proper reference frame at rest. Since the metric coefficients in these two bases have the standard flat-space form $g_{\bar{\alpha}\bar{\beta}} = \eta_{\alpha\beta}$, $g_{\hat{\alpha}\hat{\beta}} = \eta_{\alpha\beta}$, the bases must be related by a Lorentz transformation [cf. Eq. (2.35b) and associated discussion]. A little thought makes it clear that the required transformation matrix is that for a pure boost [Eq. (2.37a)]:

$$L^{\hat{0}}{}_{\bar{0}} = L^{\hat{r}}{}_{\bar{r}} = \gamma, \quad L^{\hat{0}}{}_{\bar{r}} = L^{\hat{r}}{}_{\bar{0}} = -\beta\gamma, \quad L^{\hat{\theta}}{}_{\bar{\theta}} = L^{\hat{\phi}}{}_{\bar{\phi}} = 1; \quad \gamma = \frac{1}{\sqrt{1-\beta^2}}, \tag{26.53}$$

with β the speed of implosion of the star's surface, as measured in the proper reference frame of the static observer when the surface flies by. The transformation law for the components of the Riemann tensor has, of course, the standard form for any fourth-rank tensor:

$$R_{\bar{\alpha}\bar{\beta}\bar{\gamma}\bar{\delta}} = L^{\hat{\mu}}{}_{\bar{\alpha}} L^{\hat{\nu}}{}_{\bar{\beta}} L^{\hat{\lambda}}{}_{\bar{\gamma}} L^{\hat{\sigma}}{}_{\bar{\delta}} R_{\hat{\mu}\hat{\nu}\hat{\lambda}\hat{\sigma}}. \tag{26.54}$$

The basis vectors of the proper reference frame are given by Eqs. (4) of Box 26.2, and from that box we learn that the components of Riemann in this basis are

$$R_{\hat{0}\hat{r}\hat{0}\hat{r}} = -\frac{2M}{R^3}, \quad R_{\hat{0}\hat{\theta}\hat{0}\hat{\theta}} = R_{\hat{0}\hat{\phi}\hat{0}\hat{\phi}} = +\frac{M}{R^3},$$
$$R_{\hat{\theta}\hat{\phi}\hat{\theta}\hat{\phi}} = \frac{2M}{R^3}, \quad R_{\hat{r}\hat{\theta}\hat{r}\hat{\theta}} = R_{\hat{r}\hat{\phi}\hat{r}\hat{\phi}} = -\frac{M}{R^3}. \tag{26.55}$$

These are the components measured by static observers.

By inserting these static-observer components and the Lorentz-transformation matrix (26.53) into the transformation law (26.54) we reach our goal: the following components of Riemann in the local Lorentz frame of the star's freely imploding surface:

$$R_{\bar{0}\bar{r}\bar{0}\bar{r}} = -\frac{2M}{R^3}, \quad R_{\bar{0}\bar{\theta}\bar{0}\bar{\theta}} = R_{\bar{0}\bar{\phi}\bar{0}\bar{\phi}} = +\frac{M}{R^3},$$
$$R_{\bar{\theta}\bar{\phi}\bar{\theta}\bar{\phi}} = \frac{2M}{R^3}, \quad R_{\bar{r}\bar{\theta}\bar{r}\bar{\theta}} = R_{\bar{r}\bar{\phi}\bar{r}\bar{\phi}} = -\frac{M}{R^3}. \tag{26.56}$$

These components are remarkable in two ways. First, they remain perfectly finite as the star's surface approaches the gravitational radius, $R \to 2M$; correspondingly, tidal gravity cannot destroy the star or the observers on its surface. Second, the components of Riemann are identically the same in the two orthonormal frames, hatted and barred, which move radially at finite speed β with respect to each other [expressions (26.56) are independent of β and are the same as Eqs. (26.55)]. This is a result of the very special algebraic structure that Riemann's components have for the Schwarzschild spacetime; it will not be true in typical spacetimes.

the Riemann tensor (tidal field), as measured on the imploding star's surface, remains finite as the surface approaches radius $r = 2M$

26.4.3 Stellar Implosion in Eddington-Finkelstein Coordinates

26.4.3

From the finiteness of the components of Riemann in the local Lorentz frame of the star's surface, we conclude that something must be wrong with Schwarzschild's $\{t, r, \theta, \phi\}$ coordinate system in the vicinity of the gravitational radius $r = 2M$: although nothing catastrophic happens to the star's surface as it approaches $2M$, those coordinates refuse to describe passage through $r = 2M$ in a reasonable, smooth, finite way. Thus to study the implosion as it passes through the gravitational radius and beyond, we need a new, improved coordinate system.

Several coordinate systems have been devised for this purpose. For a study and comparison of them see, for example, Misner, Thorne, and Wheeler (1973, Chap. 31). In this chapter we confine ourselves to one of them: a coordinate system devised for

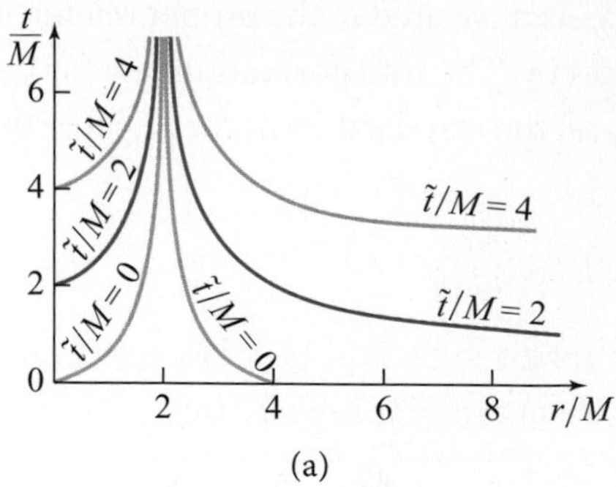

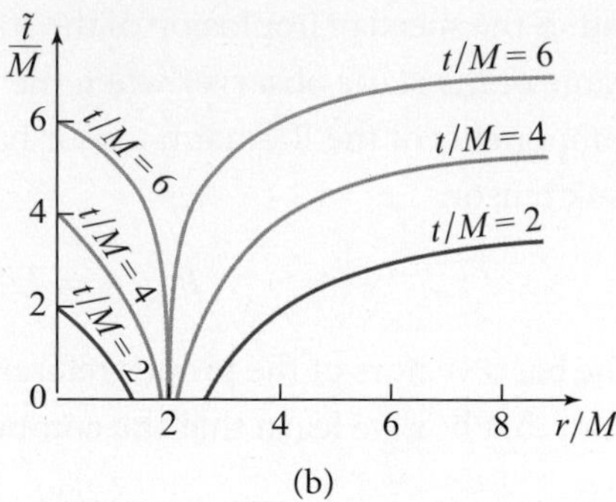

FIGURE 26.3 (a) The 3-surfaces of constant Eddington-Finkelstein time coordinate $\tilde{t}$ drawn in a Schwarzschild spacetime diagram, with the angular coordinates $\{\theta, \phi\}$ suppressed. (b) The 3-surfaces of constant Schwarzschild time coordinate t drawn in an Eddington-Finkelstein spacetime diagram, with angular coordinates suppressed.

other purposes by Arthur Eddington (1922), then long forgotten and only rediscovered independently and used for this purpose by Finkelstein (1958). Yevgeny Lifshitz, of Landau-Lifshitz fame, told one of the authors many years later what an enormous impact Finkelstein's coordinate system had on peoples' understanding of the implosion of stars. "You cannot appreciate how difficult it was for the human mind before Finkelstein to understand [the Oppenheimer-Snyder analysis of stellar implosion]," Lifshitz said. When, 19 years after Oppenheimer and Snyder, the issue of *Physical Review* containing Finkelstein's paper arrived in Moscow, suddenly everything was clear.

Finkelstein, a postdoctoral fellow at the Stevens Institute of Technology in Hoboken, New Jersey, found the following simple transformation, which moves the region $\{t = \infty,\ r = 2M\}$ of Schwarzschild coordinates in to a finite location. His transformation involves introducing a new time coordinate

Eddington-Finkelstein time coordinate

$$\boxed{\tilde{t} = t + 2M \ln\big|[r/(2M)] - 1\big|,} \tag{26.57}$$

but leaving unchanged the radial and angular coordinates. Figure 26.3 shows the surfaces of constant Eddington-Finkelstein time[3] $\tilde{t}$ in Schwarzschild coordinates, and the surfaces of constant Schwarzschild time t in Eddington-Finkelstein coordinates. Notice, as advertised, that $\{t = \infty, r = 2M\}$ is moved to a finite Eddington-Finkelstein location.

By inserting the coordinate transformation (26.57) into the Schwarzschild line element (26.1), we obtain the following line element for Schwarzschild spacetime written in Eddington-Finkelstein coordinates:

3. Sometimes $\tilde{t}$ is also called "ingoing Eddington-Finkelstein time," because it enables one to analyze infall through the gravitational radius.

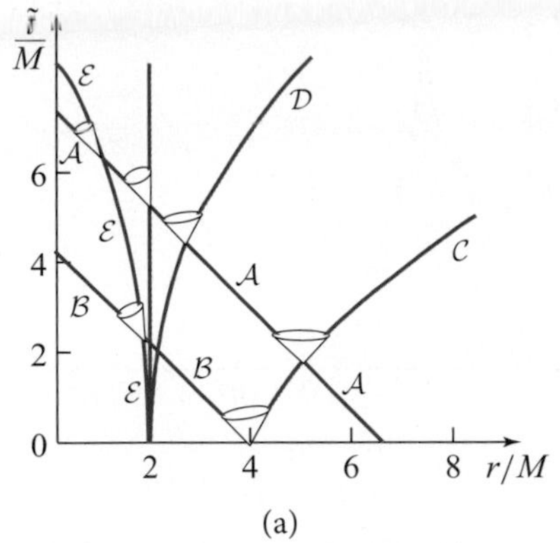

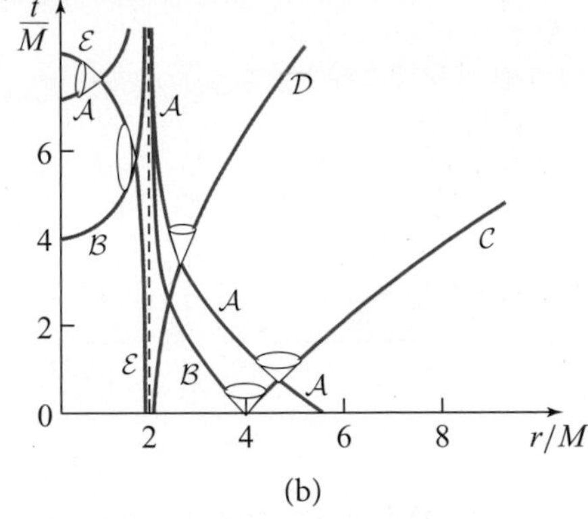

FIGURE 26.4 (a) Radial light rays and light cones for the Schwarzschild spacetime as depicted in Eddington-Finkelstein coordinates [Eq. (26.59)]. (b) The same light rays and light cones as depicted in Schwarzschild coordinates [cf. Fig. 26.2]. $\mathcal{A}$ and $\mathcal{B}$ are ingoing light rays that start far outside $r = 2M$; $\mathcal{C}$ and $\mathcal{D}$ are outgoing rays that start near $r = 2M$; $\mathcal{E}$ is an outgoing ray that is trapped inside the gravitational radius.

Schwarzschild metric in Eddington-Finkelstein coordinates

$$\boxed{ds^2 = -\left(1 - \frac{2M}{r}\right) d\tilde{t}^2 + \frac{4M}{r} d\tilde{t}\, dr + \left(1 + \frac{2M}{r}\right) dr^2 + r^2(d\theta^2 + \sin^2\theta d\phi^2).} \tag{26.58}$$

Notice that, by contrast with the line element in Schwarzschild coordinates, none of the metric coefficients diverge as r approaches $2M$. Moreover, in an Eddington-Finkelstein spacetime diagram, by contrast with Schwarzschild, the light cones do not pinch down to slivers at $r = 2M$ (compare Figs. 26.4a and 26.4b): The world lines of radial light rays are computable in Eddington-Finkelstein, as in Schwarzschild, by setting $ds^2 = 0$ (null world lines) and $d\theta = d\phi = 0$ (radial world lines) in the line element. The result, depicted in Fig. 26.4a, is

edges of radial light cone in Eddington-Finkelstein coordinates

$$\frac{d\tilde{t}}{dr} = -1 \text{ for ingoing rays; and } \frac{d\tilde{t}}{dr} = \left(\frac{1 + 2M/r}{1 - 2M/r}\right) \text{ for outgoing rays.} \tag{26.59}$$

Note that, in the Eddington-Finkelstein coordinate system, the ingoing light rays plunge unimpeded through $r = 2M$ and onward into $r = 0$ along 45° lines. The outgoing light rays, by contrast, are never able to escape outward through $r = 2M$: because of the inward tilt of the outer edge of the light cone, all light rays that begin inside $r = 2M$ are forced forever to remain inside, and in fact are drawn inexorably into $r = 0$, whereas light rays initially outside $r = 2M$ can escape to $r = \infty$.

imploding star passes through the gravitational radius and onward to $r = 0$ in finite Eddington-Finkelstein coordinate time

Now return to the implosion of a star. The world line of the star's surface, which became asymptotically frozen at the gravitational radius when studied in Schwarzschild coordinates, plunges unimpeded through $r = 2M$ and to $r = 0$ when studied in

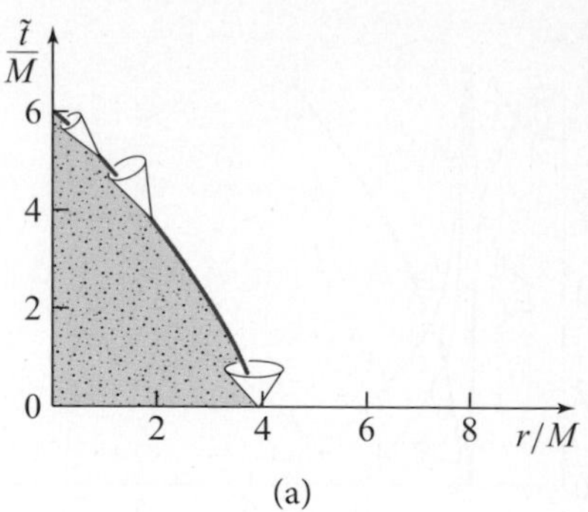

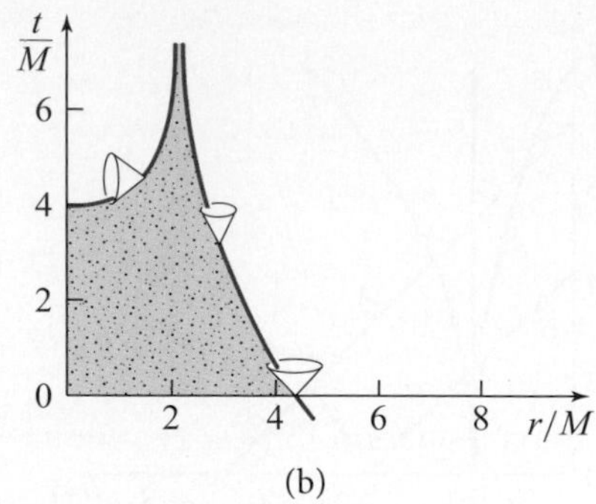

FIGURE 26.5 World line of an observer on the surface of an imploding star, as depicted (a) in an Eddington-Finkelstein spacetime diagram, and (b) in a Schwarzschild spacetime diagram; see Ex. 26.9.

Eddington-Finkelstein coordinates; see Ex. 26.9 and compare Figs. 26.5b and 26.5a. Thus to understand the star's ultimate fate, we must study the region $r = 0$.

EXERCISES

Exercise 26.9 *Example: Implosion of the Surface of a Zero-Pressure Star Analyzed in Schwarzschild and in Eddington-Finkelstein Coordinates*

Consider the surface of a zero-pressure star, which implodes along a timelike geodesic $r = R(t)$ in the Schwarzschild spacetime of its exterior. Analyze that implosion using Schwarzschild coordinates $\{t, r, \theta, \phi\}$ and the exterior metric (26.1) in those coordinates, and then repeat your analysis in Eddington-Finkelstein coordinates. More specifically, do the following.

(a) Using Schwarzschild coordinates, show that the covariant time component u_t of the 4-velocity $\vec{u}$ of a particle on the star's surface is conserved along its world line (cf. Ex. 25.4a). Evaluate this conserved quantity in terms of the star's mass M and the radius $r = R_o$ at which it begins to implode.

(b) Use the normalization of the 4-velocity to show that the star's radius R as a function of the proper time τ since implosion began (proper time as measured on its surface) satisfies the differential equation

$$\frac{dR}{d\tau} = -[\text{const} + 2M/R]^{\frac{1}{2}}, \tag{26.60}$$

and evaluate the constant. Compare this with the equation of motion for the surface as predicted by Newtonian gravity, with proper time τ replaced by Newtonian time. (It is a coincidence that the two equations are identical.)

(c) Show from the equation of motion (26.60) that the star implodes through the gravitational radius $R = 2M$ and onward to $R = 0$ in a finite proper time given by Eq. (26.51). Show that this proper time has the magnitudes cited in Eq. (26.51) and the sentences following it.

(d) Show that the Schwarzschild coordinate time t required for the star to reach its gravitational radius, $R \to 2M$, is infinite.

(e) Show, further, that when studied in Eddington-Finkelstein coordinates, the surface's implosion to $R = 2M$ requires only finite coordinate time $\tilde{t}$; in fact, a time of the same order of magnitude as the proper time (26.51). [Hint: Derive a differential equation for $d\tilde{t}/d\tau$ along the world line of the star's surface, and use it to examine the behavior of $d\tilde{t}/d\tau$ near $R = 2M$.]

(f) Show that the world line of the star's surface as depicted in an Eddington-Finkelstein spacetime diagram has the form shown in Fig. 26.5a, and that in a Schwarzschild spacetime diagram it has the form shown in Fig. 26.5b.

26.4.4 Tidal Forces at $r = 0$—The Central Singularity

26.4.4

As with $r \to 2M$, there are two possibilities: either the tidal forces as measured on the star's surface remain finite as $r \to 0$, in which case something must be going wrong with the coordinate system; or else the tidal forces diverge, destroying the star. The tidal forces are computed in Ex. 26.10, with a remarkable result: they diverge. Thus, the region $r = 0$ is a *spacetime singularity*: a region where tidal gravity becomes infinitely large, destroying everything that falls into it.

$r = 0$ is a spacetime singularity; tidal gravity becomes infinite there

This conclusion, of course, is very unsatisfying. It is hard to believe that the correct laws of physics will predict such total destruction. In fact, they probably do not. As we will find in Sec. 28.7.1, in discussing the origin of the universe, when the radius of curvature of spacetime becomes as small as $L_P \equiv (G\hbar/c^3)^{\frac{1}{2}} \simeq 10^{-33}$ cm, space and time must cease to exist as classical entities; they and the spacetime geometry must then become quantized. Correspondingly, general relativity must then break down and be replaced by a quantum theory of the structure of spacetime—a quantum theory of gravity. That quantum theory will describe and govern the classically singular region at $r = 0$. Since, however, only rough hints of the structure of that quantum theory are in hand at this time, it is not known what that theory will say about the endpoint of stellar implosion.

the singularity is governed by the laws of quantum gravity

EXERCISES

Exercise 26.10 *Example: Gore at the Singularity*

(a) Show that, as the surface of an imploding star approaches $R = 0$, its world line in Schwarzschild coordinates asymptotes to the curve $\{(t, \theta, \phi) = \text{const}, r \text{ variable}\}$.

(b) Show that this curve to which it asymptotes [part (a)] is a timelike geodesic. [Hint: Use the result of Ex. 25.4a.]

(c) Show that the basis vectors of the infalling observer's local Lorentz frame near $r = 0$ are related to the Schwarzschild coordinate basis by

$$\vec{e}_{\hat{0}} = -\left(\frac{2M}{r} - 1\right)^{\frac{1}{2}} \frac{\partial}{\partial r}, \quad \vec{e}_{\hat{1}} = \left(\frac{2M}{r} - 1\right)^{-\frac{1}{2}} \frac{\partial}{\partial t},$$

$$\vec{e}_{\hat{2}} = \frac{1}{r}\frac{\partial}{\partial \theta}, \quad \vec{e}_{\hat{3}} = \frac{1}{r \sin\theta}\frac{\partial}{\partial \phi}. \tag{26.61}$$

What are the components of the Riemann tensor in that local Lorentz frame?

(d) Show that the tidal forces produced by the Riemann tensor stretch an infalling observer in the radial, $\vec{e}_{\hat{1}}$, direction and squeeze the observer in the tangential, $\vec{e}_{\hat{2}}$ and $\vec{e}_{\hat{3}}$, directions. Show that the stretching and squeezing forces become infinitely strong as the observer approaches $r = 0$.

(e) Idealize the body of an infalling observer to consist of a head of mass $\mu \simeq 20$ kg and feet of mass $\mu \simeq 20$ kg separated by a distance $h \simeq 2$ m, as measured in the observer's local Lorentz frame, and with the separation direction radial. Compute the stretching force between head and feet, as a function of proper time τ, as the observer falls into the singularity. Assume that the hole has the mass $M = 7 \times 10^9 M_\odot$, which has been measured by astronomical observations for the black hole at the center of the supergiant elliptical galaxy M87. How long before hitting the singularity (at what proper time τ) does the observer die, if he or she is a human being made of flesh, bone, and blood?

26.4.5

26.4.5 Schwarzschild Black Hole

Unfortunately, the singularity and its quantum mechanical structure are totally invisible to observers in the external universe: the only way the singularity can possibly be seen is by means of light rays, or other signals, that emerge from its vicinity. However, because the future light cones are all directed into the singularity (Fig. 26.5), no light-speed or sub-light-speed signals can ever emerge from it. In fact, because the outer edge of the light cone is tilted inward at every event inside the gravitational radius (Figs. 26.4 and 26.5), no signal can emerge from inside the gravitational radius to tell external observers what is going on there. In effect, the gravitational radius is an *absolute event horizon* for our universe, a horizon beyond which we cannot see—except by plunging through it, and paying the ultimate price for our momentary exploration of the hole's interior: we cannot publish the results of our observations.

gravitational radius as an absolute event horizon

As most readers are aware, the region of strong, vacuum gravity left behind by the implosion of the star is called a *black hole*. The horizon, $r = 2M$, is the surface of the hole, and the region $r < 2M$ is its interior. The spacetime geometry of the black hole, outside and at the surface of the star that creates it by implosion, is that of

black hole, its interior, surface (horizon), and exterior

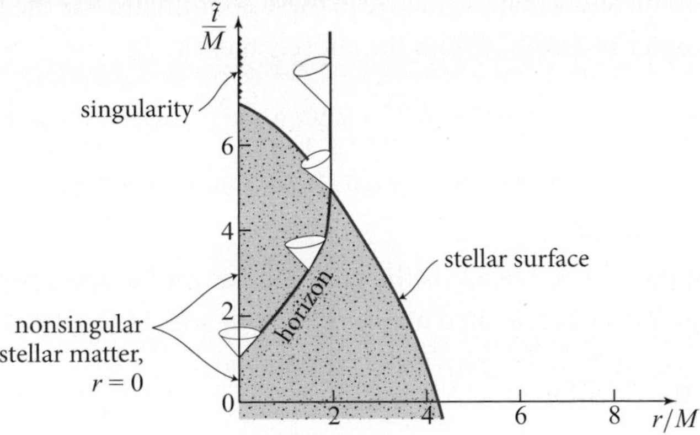

FIGURE 26.6 Spacetime diagram depicting the formation and evolution of the horizon of a black hole. The coordinates outside the surface of the imploding star are those of Eddington and Finkelstein; those inside are a smooth continuation of Eddington and Finkelstein (not explored in this book). Note that the horizon is the boundary of the region that is unable to send outgoing null geodesics to radial infinity.

Schwarzschild—though, of course, Karl Schwarzschild had no way of knowing this in the few brief months left to him after his discovery of the Schwarzschild line element.

evolution of the horizon as black hole is formed

The horizon—defined as the boundary between spacetime regions that can and cannot communicate with the external universe—actually forms initially at the star's center and then expands to encompass the star's surface at the precise moment when the surface penetrates the gravitational radius. This evolution of the horizon is depicted in an Eddington-Finkelstein-type spacetime diagram in Fig. 26.6.

genericity of singularities inside black holes

Our discussion here has been confined to spherically symmetric, nonrotating black holes created by the gravitational implosion of a spherically symmetric star. Of course, real stars are not spherical, and it was widely believed—perhaps we should say, hoped—in the 1950s and 1960s that black-hole horizons and singularities would be so unstable that small nonsphericities or small rotations of the imploding star would save it from the black-hole fate. However, elegant and very general analyses carried out in the 1960s, largely by the British physicists Roger Penrose and Stephen Hawking, showed otherwise. More recent numerical simulations on supercomputers have confirmed those analyses: singularities are a generic outcome of stellar implosion, as are the black-hole horizons that clothe them.

EXERCISES

Exercise 26.11 *Example: Rindler Approximation near the Horizon of a Schwarzschild Black Hole*

(a) Near the event $\{r = 2M, \theta = \theta_o, \phi = \phi_o, t \text{ finite}\}$, on the horizon of a black hole, introduce locally Cartesian spatial coordinates $\{x = 2M \sin\theta_o(\phi - \phi_o), y = 2M(\theta - \theta_o), z = \int_{2M}^{r} dr/\sqrt{1 - 2M/r}\}$, accurate to first order in distance from that event. Show that the metric in these coordinates has the form (accurate to leading order in distance from the chosen event):

$$ds^2 = -(g_H\, z)^2 dt^2 + dx^2 + dy^2 + dz^2, \quad \text{where} \quad g_H = \frac{1}{4M} \tag{26.62}$$

is the horizon's so-called *surface gravity,* to which we shall return, for a rotating black hole, in Eq. (26.90).

(b) Notice that the metric (26.62) is the same as that for flat spacetime as seen by a family of uniformly accelerated observers (i.e., as seen in the Rindler coordinates of Sec. 24.5.4). Why is this physically reasonable?

Exercise 26.12 ***Example: Orbits around a Schwarzschild Black Hole*

Around a Schwarzschild black hole, spherical symmetry dictates that every geodesic orbit lies in a plane that bifurcates the $t = \text{const}$ 3-volume. We are free to orient our coordinate system, for any chosen geodesic, so its orbital plane is equatorial: $\theta = \pi/2$. Then the geodesic has three conserved quantities: the orbiting particle's rest mass μ, energy-at-infinity $\mathcal{E}_\infty$, and angular momentum L, which are given by

$$\mu^2 = -\vec{p}^{\,2} = -g_{\alpha\beta}\frac{dx^\alpha}{d\zeta}\frac{dx^\beta}{d\zeta}, \quad \mathcal{E}_\infty = -p_t = -g_{tt}\frac{dt}{d\zeta}, \quad L = p_\phi = g_{\phi\phi}\frac{d\phi}{d\zeta}, \tag{26.63}$$

respectively. In this exercise we focus on particles with finite rest mass. Zero-rest-mass particles can be analyzed similarly; see the references at the end of this exercise.

(a) Set the rest mass μ to unity; equivalently, switch from 4-momentum to 4-velocity for the geodesic's tangent vector. Then, by algebraic manipulation of the constants of motion (26.63), derive the following orbital equations:

$$\left(\frac{dr}{d\zeta}\right)^2 + V^2(r) = \mathcal{E}_\infty^2, \quad \text{where} \quad V^2(r) = \left(1 - \frac{2M}{r}\right)\left(1 + \frac{L^2}{r^2}\right), \tag{26.64a}$$

$$\frac{d\phi}{d\zeta} = \frac{L}{r^2}, \quad \frac{dt}{d\zeta} = \frac{\mathcal{E}_\infty}{1 - 2M/r}. \tag{26.64b}$$

(b) We use a device that we have also encountered in our treatment of ion acoustic solitons in Sec. 23.6. We think of Eq. (26.64a) as an equivalent nonrelativistic energy equation with $\frac{1}{2}V^2$ being the effective potential energy and $\frac{1}{2}\mathcal{E}_\infty^2$ the effective total energy. As the energy we actually care about is $\mathcal{E}_\infty$, it is more direct to refer to $V(r)$ as our potential and investigate its properties. This $V(r)$ is plotted in Fig. 26.7a for several values of the particle's angular momentum L. Explain why: (i) Circular geodesic orbits are at extrema of $V(r)$—the large dots in the figure. (ii) Each bound orbit can be described by a horizontal line, such as the red one in the figure, with height equal to the orbit's $\mathcal{E}_\infty$; and the particle's radial motion is back and forth between the points at which the horizontal line intersects the potential.

(c) Show that the innermost stable circular orbit (often abbreviated as ISCO) is at $r = 6M$, and it occurs at a saddle point of the potential, for $L = 2\sqrt{3}\,M$. Show that all inward-moving particles with $L < 2\sqrt{3}\,M$ are doomed to fall into the black hole.

(d) Show that the innermost unstable circular orbit is at $r = 3M$ and has infinite energy for finite rest mass. From this infer that there should be an unstable circular orbit for photons at $r = 3M$.

(e) The geodesic equations of motion in the form (26.64) are not very suitable for numerical integration: at each radial turning point, where $V(r) = \mathcal{E}_\infty$ and $dr/d\zeta = 0$, the accuracy of straightforward integrations goes bad, and one must switch signs of $dr/d\zeta$ by hand, unless one is sophisticated. For these reasons and others, it is preferable in numerical integrations to use the super-hamiltonian form of the geodesic equation (Ex. 25.7), or to convert Eq. (26.64a) into a second-order differential equation before integrating. Show that Hamilton's equations, for the super-hamiltonian, are

$$\frac{dr}{d\zeta} = \left(1 - \frac{2M}{r}\right) p_r, \quad \frac{dp_r}{d\zeta} = \frac{L^2}{r^3} - \frac{M}{r^2} p_r^2 - \frac{M}{(r-2M)^2}\mathcal{E}_\infty^2, \quad \frac{d\phi}{d\zeta} = \frac{L}{r^2}. \tag{26.65}$$

When integrating these equations, one must make sure that the initial value of p_r satisfies $g^{\alpha\beta} p_\alpha p_\beta = -1$ (for our unit-rest-mass particle). Show that this condition reduces to

$$p_r = \pm\sqrt{\frac{\mathcal{E}_\infty{}^2/(1-2M/r) - L^2/r^2 - 1}{1 - 2M/r}}. \tag{26.66}$$

(f) Integrate the super-hamiltonian equations (26.65) numerically for the orbit described by the red horizontal line in Fig. 26.7a, which has $L = 3.75M$ and

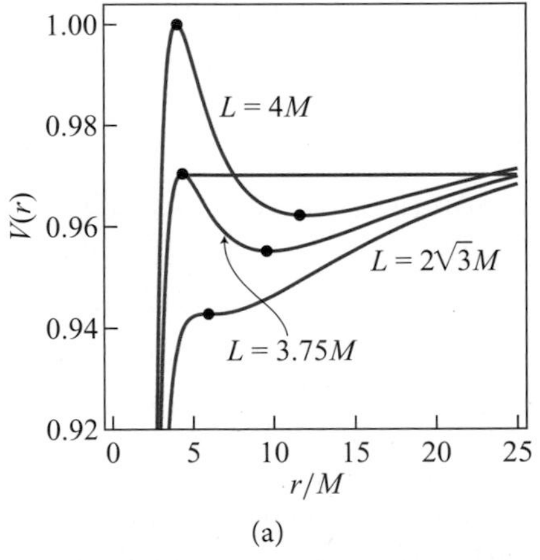

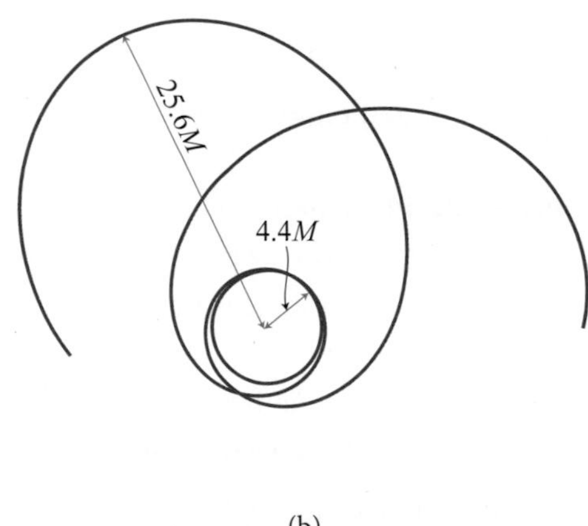

FIGURE 26.7 (a) The potential $V(r)$ for the geodesic radial motion $r(\zeta)$ of finite-rest-mass particles around a Schwarzschild black hole. Each curve is labeled by the particle's orbital angular momentum L. (b) The orbit $r(\phi)$ corresponding to the horizontal red line in (a); it has $L = 3.75M$ and $\mathcal{E}_\infty = 0.9704$. It is called a *zoom-whirl* orbit, because its particle zooms inward from a large radius, whirls around the black hole several times, then zooms back out—and then repeats.

$\mathcal{E}_\infty = 0.9704$. The result should be the zoom-whirl orbit depicted in Fig. 26.7b. (The initial conditions used in that figure were $r = 25M$, $\phi = 0$, and [from Eq. (26.66)] $p_r = 0.0339604$.)

(g) Carry out other numerical integrations to explore the variety of shapes of finite-rest-mass orbits around a Schwarzschild black hole.

Orbits around a Schwarzschild black hole are treated in most general relativity textbooks; see, for example, Misner, Thorne, and Wheeler (1973, Chap. 25) and Hartle (2003); see also Frolov and Novikov (1998), Sec. 2.8. Analytic solutions to the geodesic equation, expressed in terms of elliptic functions, are given by Darwin (1959).

Exercise 26.13 *Example: Schwarzschild Wormhole* T2

Our study of the Schwarzschild solution of Einstein's equations in this chapter has been confined to situations where, at small radii, the Schwarzschild geometry joins onto that of a star—either a static star or one that implodes to form a black hole. Suppose, by contrast, that there is no matter anywhere in the Schwarzschild spacetime. To get insight into this situation, construct an embedding diagram for the equatorial 2-surfaces $\{t = \text{const}, \theta = \pi/2\}$ of the vacuum Schwarzschild spacetime, using as the starting point the line element of such a 2-surface written in isotropic coordinates (Ex. 26.3):

$$^{(2)}ds^2 = \left(1 + \frac{M}{2\bar{r}}\right)^4 (d\bar{r}^2 + \bar{r}^2 d\phi^2). \tag{26.67}$$

Show that the region $0 < \bar{r} \ll M/2$ is an asymptotically flat space, that the region $\bar{r} \gg M/2$ is another asymptotically flat space, and that these two spaces are connected by a *wormhole* ("bridge," "tunnel") through the embedding space. This exercise reveals that the pure vacuum Schwarzschild spacetime represents a wormhole that connects two different universes—or, with a change of topology, a wormhole that connects two widely separated regions of one universe.

Exercise 26.14 *Example: Dynamical Evolution of Schwarzschild Wormhole* T2

The isotropic-coordinate line element (26.16) describing the spacetime geometry of a Schwarzschild wormhole is independent of the time coordinate t. However, because $g_{tt} = 0$ at the wormhole's throat, $\bar{r} = M/2$, the proper time $d\tau = \sqrt{-ds^2}$ measured by an observer at rest appears to vanish, which cannot be true. Evidently, the isotropic coordinates are ill behaved at the throat.

(a) Martin Kruskal (1960) and George Szekeres (1960) independently introduced a coordinate system that covers the wormhole's entire spacetime and elucidates its dynamics in a nonsingular manner. The Kruskal-Szekeres time and radial coordinates v and u are related to the Schwarzschild t and r by

$$[r/(2M) - 1]\, e^{r/(2M)} = u^2 - v^2, \tag{26.68}$$

$$t = 4M \tanh^{-1}(v/u) \quad \text{at } r > 2M, \quad t = 4M \tanh^{-1}(u/v) \quad \text{at } r < 2M.$$

Show that the metric of Schwarzschild spacetime written in these Kruskal-Szekeres coordinates is T2

$$ds^2 = (32M^3/r)e^{-r/(2M)}(-dv^2 + du^2) + r^2(d\theta^2 + \sin^2\theta d\phi^2), \quad (26.69)$$

where $r(u, v)$ is given by Eq. (26.68).

(b) Draw a spacetime diagram with v increasing upward and u increasing horizontally and rightward. Show that the radial light cones are 45° lines everywhere. Show that there are two $r = 0$ singularities, one on the past hyperbola, $v = -\sqrt{u^2+1}$, and the other on the future hyperbola, $v = +\sqrt{u^2+1}$. Show that the gravitational radius, $r = 2M$, is at $v = \pm u$. Show that our universe, outside the wormhole, is at $u \gg 1$, and there is another universe at $u \ll -1$.

(c) Draw embedding diagrams for a sequence of spacelike hypersurfaces, the first of which hits the past singularity and the last of which hits the future singularity. Thereby show that the metric (26.69) represents a wormhole that is created in the past, expands to maximum throat circumference $4\pi M$, then pinches off in the future to create a pair of singularities, one in each universe.

(d) Show that nothing can pass through the wormhole from one universe to the other; anything that tries gets crushed in the wormhole's pinch off.

For a solution, see Fuller and Wheeler (1962), or Misner, Thorne, and Wheeler (1973, Chap. 31). For discussions of what is required to hold a wormhole open so it can be traversed, see Morris and Thorne (1988); and for discussions of whether arbitrarily advanced civilizations can create wormholes and hold them open for interstellar travel, see the nontechnical discussions and technical references in Everett and Roman (2011). For a discussion of the possible use of traversable wormholes for backward time travel, see Sec. 2.9, and also Everett and Roman (2011) and references therein. For the visual appearance of wormholes and for guidance in constructing wormhole images by propagating light rays through and around them, see Thorne (2014) and James et al. (2015b).

26.5 Spinning Black Holes: The Kerr Spacetime T2 26.5

26.5.1 The Kerr Metric for a Spinning Black Hole T2 26.5.1

Consider a star that implodes to form a black hole, and assume for pedagogical simplicity that during the implosion no energy, momentum, or angular momentum flows through a large sphere surrounding the star. Then the asymptotic conservation laws discussed in Secs. 25.9.4 and 25.9.5 guarantee that the mass M, linear momentum P_j, and angular momentum J_j of the newborn hole, as encoded in its asymptotic metric, will be identical to those of its parent star. If (as we shall assume) our asymptotic coordinates are those of the star's rest frame ($P_j = 0$), then the hole will also be at rest in those coordinates (i.e., it will also have $P_j = 0$).

T2

If the star was nonspinning ($J_j = 0$), then the hole will also have $J_j = 0$, and a powerful theorem due to Werner Israel (reviewed in Carter 1979) guarantees that—after it has settled down into a quiescent state—the hole's spacetime geometry will be that of Schwarzschild.

If, instead, the star was spinning ($J_j \neq 0$), then the final, quiescent hole cannot be that of Schwarzschild. Instead, according to powerful theorems due to Stephen Hawking, Brandon Carter, David Robinson, and others (also reviewed in Carter 1979), its spacetime geometry will be that described by the following exact, vacuum solution to the Einstein field equation:

the Kerr metric for a spinning black hole in Boyer-Lindquist coordinates

$$\boxed{ds^2 = -\alpha^2 dt^2 + \frac{\rho^2}{\Delta} dr^2 + \rho^2 d\theta^2 + \varpi^2 (d\phi - \omega dt)^2,} \tag{26.70a}$$

where

$$\boxed{\Delta = r^2 + a^2 - 2Mr, \quad \rho^2 = r^2 + a^2 \cos^2\theta, \quad \Sigma^2 = (r^2 + a^2)^2 - a^2 \Delta \sin^2\theta,}$$

$$\boxed{\alpha^2 = \frac{\rho^2}{\Sigma^2}\Delta, \quad \varpi^2 = \frac{\Sigma^2}{\rho^2}\sin^2\theta, \quad \omega = \frac{2aMr}{\Sigma^2}.} \tag{26.70b}$$

This is called the *Kerr solution*, because it was discovered by the New Zealand mathematician Roy Kerr (1963).

In this line element, $\{t, r, \theta, \phi\}$ are the coordinates, and there are two constants, M and a. The physical meanings of M and a can be deduced from the asymptotic form of the Kerr metric (26.70) at large radii:

$$\begin{aligned} ds^2 = &-\left(1 - \frac{2M}{r}\right) dt^2 - \frac{4Ma}{r}\sin^2\theta d\phi dt \\ &+ \left[1 + O\left(\frac{M}{r}\right)\right] [dr^2 + r^2(d\theta^2 + \sin^2\theta d\phi^2)]. \end{aligned} \tag{26.71}$$

the black hole's mass M and spin angular momentum Ma

By comparing with the standard asymptotic metric in spherical coordinates [Eq. (25.98d)], we see that M is the mass of the black hole, $Ma \equiv J_H$ is the magnitude of its spin angular momentum, and its spin points along the polar axis, $\theta = 0$. Evidently, then, the constant a is the hole's angular momentum per unit mass; it has the same dimensions as M: length (in geometrized units).

It is easy to verify that, in the limit $a \to 0$, the Kerr metric (26.70) reduces to the Schwarzschild metric (26.1), and the coordinates $\{t, r, \theta, \phi\}$ in which we have written it (called "Boyer-Lindquist coordinates") reduce to Schwarzschild's coordinates.

Just as it is convenient to read the covariant metric components $g_{\alpha\beta}$ off the line element (26.70a) via $ds^2 = g_{\alpha\beta} dx^\alpha dx^\beta$, so also it is convenient to read the contravariant metric components $g^{\alpha\beta}$ off an expression for the wave operator: $\Box \equiv \vec{\nabla} \cdot \vec{\nabla} = g^{\alpha\beta} \nabla_\alpha \nabla_\beta$. (Here $\nabla_\alpha \equiv \nabla_{\vec{e}_\alpha}$ is the directional derivative along the basis vector $\vec{e}_\alpha$.) For the Kerr metric (26.70), a straightforward inversion of the matrix

$||g_{\alpha\beta}||$ gives the $||g^{\alpha\beta}||$ embodied in the following equation:

T2

the wave operator and contravariant components of the Kerr metric

$$\Box = -\frac{1}{\alpha^2}(\nabla_t + \omega\nabla_\phi)^2 + \frac{\Delta}{\rho^2}{\nabla_r}^2 + \frac{1}{\rho^2}{\nabla_\theta}^2 + \frac{1}{\varpi^2}{\nabla_\phi}^2. \tag{26.72}$$

26.5.2 Dragging of Inertial Frames T2

26.5.2

As we saw in Sec. 25.9.3, the angular momentum of a weakly gravitating body can be measured by its frame-dragging, precessional influence on the orientation of gyroscopes. Because the asymptotic metric (26.71) of a Kerr black hole is identical to the weak-gravity metric used to study gyroscopic precession in Sec. 25.9.3, the black hole's spin angular momentum J_{BH} can also be measured via frame-dragging gyroscopic precession.

This frame dragging also shows up in the geodesic trajectories of freely falling particles. For concreteness, consider a particle dropped from rest far outside the black hole. Its initial 4-velocity will be $\vec{u} = \partial/\partial t$, and correspondingly, in the distant, flat region of spacetime, the covariant components of $\vec{u}$ will be $u_t = -1$, $u_r = u_\theta = u_\phi = 0$.

Now, the Kerr metric coefficients $g_{\alpha\beta}$, like those of Schwarzschild, are independent of t and ϕ: the Kerr metric is symmetric under time translation (it is stationary) and under rotation about the hole's spin axis (it is axially symmetric). These symmetries impose corresponding conservation laws on the infalling particle (Ex. 25.4a): u_t and u_ϕ are conserved (i.e., they retain their initial values $u_t = -1$ and $u_\phi = 0$ as the particle falls). By raising indices—$u^\alpha = g^{\alpha\beta}u_\beta$, using the metric coefficients embodied in Eq. (26.72)—we learn the evolution of the contravariant 4-velocity components: $u^t = -g^{tt} = 1/\alpha^2$, $u^\phi = -g^{t\phi} = \omega/\alpha^2$. These in turn imply that, as the particle falls, it acquires an angular velocity around the hole's spin axis given by

frame dragging: angular velocity of a particle that falls in from $r = \infty$

$$\Omega = \frac{d\phi}{dt} = \frac{d\phi/d\tau}{dt/d\tau} = \frac{u^\phi}{u^t} = \omega. \tag{26.73}$$

(The coordinates ϕ and t are tied to the rotational and time-translation symmetries of the spacetime, so they are very special; that is why we can use them to define a physically meaningful angular velocity.)

At large radii, $\omega = 2aM/r^3 \to 0$ as $r \to \infty$. Therefore, when first dropped, the particle falls radially inward. However, as the particle nears the hole and picks up speed, it acquires a significant angular velocity around the hole's spin axis. The physical cause of this is *frame dragging*: The hole's spin drags inertial frames into rotation around the spin axis, and that inertial rotation drags the inertially falling particle into a circulatory orbital motion.

26.5.3 The Light-Cone Structure, and the Horizon T2

26.5.3

Just as for a Schwarzschild hole, so also for Kerr: the light-cone structure is a powerful tool for identifying the horizon and exploring the spacetime geometry near it.

T2

outer and inner edges of light cone

At any event in spacetime, the tangents to the light cone are those displacements $\{dt, dr, d\theta, d\phi\}$ along which $ds^2 = 0$. The outermost and innermost edges of the cone are those for which $(dr/dt)^2$ is maximal. By setting expression (26.70a) to zero, we see that dr^2 has its maximum value, for a given dt^2, when $d\phi = \omega dt$ and $d\theta = 0$. In other words, the photons that move radially outward or inward at the fastest possible rate are those whose angular motion is that of frame dragging [Eq. (26.73)]. For these extremal photons, the radial motion (along the outer and inner edges of the light cone) is

$$\frac{dr}{dt} = \pm \frac{\alpha\sqrt{\Delta}}{\rho} = \pm \frac{\Delta}{\Sigma}. \tag{26.74}$$

Now, Σ is positive definite, but Δ is not; it decreases monotonically with decreasing radius, reaching zero at

$$\boxed{r = r_H \equiv M + \sqrt{M^2 - a^2}} \tag{26.75}$$

[Eq. (26.70b)]. (We assume that $|a| < M$, so r_H is real; we justify this assumption below.) Correspondingly, the light cone closes up to a sliver and then pinches off as $r \to r_H$; it pinches onto a null curve (actually, a null geodesic) given by

$$r = r_H, \quad \theta = \text{const}, \quad \phi = \Omega_H t + \text{const}, \tag{26.76}$$

where

$$\boxed{\Omega_H = \omega(r = r_H) = \frac{a}{2Mr_H}}. \tag{26.77}$$

the horizon, $r = r_H$, and its rotational angular velocity, Ω_H

This light-cone structure is depicted in Fig. 26.8a,b. The light-cone pinch-off as shown there is the same as that for Schwarzschild spacetime (Fig. 26.2) except for the light cones' frame-dragging-induced angular tilt $d\phi/dt = \omega$. In the Schwarzschild case, as $r \to 2M$, the light cones pinch onto the geodesic world lines $\{r = 2M, \theta = \text{const}, \phi = \text{const}\}$ of photons that travel along the horizon. These null world lines are called the horizon's *generators*. In the Kerr case, the light-cone pinch-off reveals that the horizon is at $r = r_H$, and the horizon generators are null geodesics that travel around and around the horizon with angular velocity Ω_H. This motivates us to regard the horizon itself as having the rotational angular velocity Ω_H.

in Boyer-Lindquist coordinates, infalling particles asymptote to horizon and to its angular velocity

When a finite-rest-mass particle falls into a spinning black hole, its world line, as it nears the horizon, is constrained always to lie inside the light cone. The light-cone pinch-off then constrains its motion asymptotically to approach the horizon generators. Therefore, as seen in Boyer-Lindquist coordinates, the particle is dragged into an orbital motion, just above the horizon, with asymptotic angular velocity $d\phi/dt = \Omega_H$, and it travels around and around the horizon "forever" (for infinite Boyer-Lindquist coordinate time t), and never (as $t \to \infty$) manages to cross through the horizon.

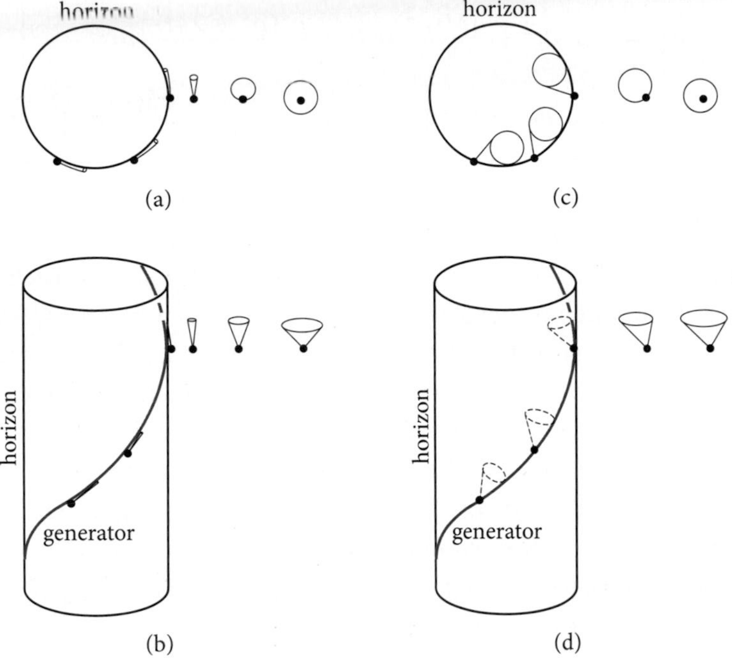

FIGURE 26.8 (a), (b): Light-cone structure of Kerr spacetime depicted in Boyer-Lindquist coordinates. Drawing (b) is a spacetime diagram; drawing (a) is the same diagram as viewed from above. (c), (d): The same light-cone structure in Kerr coordinates.

As in the Schwarzschild case, so also in Kerr: this infall to $r = r_H$ requires only finite proper time τ as measured by the particle, and the particle feels only finite tidal forces (only finite values of the components of Riemann in its proper reference frame). Therefore, as for Schwarzschild spacetime, the "barrier" to infall through $r = r_H$ must be an illusion produced by a pathology of the Boyer-Lindquist coordinates at $r = r_H$.

This coordinate pathology can be removed by a variety of different coordinate transformations. One is the following change of the time and angular coordinates:

Kerr coordinates

$$\boxed{\tilde{t} = t + \int \frac{2Mr}{\Delta} dr, \quad \tilde{\phi} = \phi + \int \frac{a}{\Delta} dr.} \tag{26.78}$$

Kerr metric in Kerr coordinates

The new (tilded) coordinates are a variant of a coordinate system originally introduced by Kerr, so we call them "Kerr coordinates."[4] By inserting the coordinate transformation (26.78) into the line element (26.70a), we obtain the following form of the Kerr metric in Kerr coordinates:

4. They are often called "ingoing Kerr coordinates," because they facilitate analyzing infall through the horizon.

T2

$$
\boxed{\begin{aligned} ds^2 = -\alpha^2 d\tilde{t}^2 + \frac{4Mr\rho^2}{\Sigma^2} dr d\tilde{t} + \frac{\rho^2(\rho^2+2Mr)}{\Sigma^2} dr^2 \\ + \rho^2 d\theta^2 + \varpi^2 \left[d\tilde{\phi} - \omega d\tilde{t} - \frac{a(\rho^2+2Mr)}{\Sigma^2} dr \right]^2 \end{aligned}} . \tag{26.79}
$$

It is easy to verify that when $a \to 0$ (so Kerr spacetime becomes Schwarzschild), the Kerr coordinates (26.78) become those of Eddington and Finkelstein [Eq. (26.57)], and the Kerr line element (26.79) becomes the Eddington-Finkelstein one [Eq. (26.58)]. Similarly, when one explores the light-cone structure for a spinning black hole in the Kerr coordinates (Fig. 26.8c,d), one finds a structure like that of Eddington-Finkelstein (Fig. 26.4a): At large radii, $r \gg M$, the light cones have their usual 45° form, but as one moves inward toward the horizon, they begin to tilt inward. In addition to the inward tilt, there is a frame-dragging-induced tilt in the direction of the hole's rotation, $+\phi$. At the horizon, the outermost edge of the light cone is tangent to the horizon generators; in Kerr coordinates, as in Boyer-Lindquist, these generators rotate around the horizon with angular velocity $d\tilde{\phi}/d\tilde{t} = \Omega_H$ [cf. Eq. (26.78), which says that at fixed r, $\tilde{t} = t +$ const and $\tilde{\phi} = \phi +$ const].

in Kerr coordinates, infalling particles cross the horizon and are pulled on inward

This light-cone structure demonstrates graphically that the horizon is at the radius $r = r_H$. Outside there, the outer edge of the light cone tilts toward increasing r, and so it is possible to escape to radial infinity. Inside r_H the outer edge of the light cone tilts inward, and all forms of matter and energy are forced to move inward, toward a singularity whose structure, presumably, is governed by the laws of quantum gravity.[5]

26.5.4

26.5.4 Evolution of Black Holes—Rotational Energy and Its Extraction T2

newborn black holes settle down into Kerr form

When a spinning star collapses to form a black hole, its centrifugal forces will flatten it, and the dynamical growth of flattening will produce gravitational radiation (Chap. 27). The newborn hole will also be flattened and will not have the Kerr shape; but rather quickly, within a time $\Delta t \sim 100M \sim 0.5$ ms $(M/M_\odot)$, the deformed hole will shake off its deformations as gravitational waves and settle down into the Kerr shape. This is the conclusion of extensive analyses, both analytic and numerical.

Many black holes are in binary orbits with stellar companions and pull gas off their companions and swallow it. Other black holes accrete gas from interstellar space. Any such accretion causes a hole's mass and spin to evolve in accord with the conservation laws (25.102) and (25.103). One might have thought that by accreting a large amount of angular momentum, a hole's angular momentum per unit mass a could grow larger than its mass M. If this were to happen, then $r_H = M + \sqrt{M^2 - a^2}$ would cease to

5. Much hoopla has been made of the fact that in the Kerr spacetime it is possible to travel inward, through a "Cauchy horizon" and then into another universe. However, the Cauchy horizon, located at $r = M - \sqrt{M^2 - a^2}$, is highly unstable against perturbations, which convert it into a singularity with infinite spacetime curvature. For details of this instability and the singularity, see, e.g., Brady, Droz, and Morsink (1998), Marolf and Ori (2013), and references therein.

T2

be a real radius—a fact that signals the destruction of the hole's horizon: as a grows to exceed M, the inward light-cone tilt gets reduced, so that everywhere the outer edge of the cone points toward increasing r, which means that light, particles, and information are no longer trapped.

accretion of matter onto a black hole pushes its spin parameter a up to a maximum value that is a little less than $a = M$

Remarkably, however, it appears that the laws of general relativity forbid a ever to grow larger than M. As accretion pushes a/M upward toward unity, the increasing drag of inertial frames causes a big increase of the hole's cross section to capture material with negative angular momentum (which will spin the hole down) and a growing resistance to capturing any further material with large positive angular momentum. Infalling particles that might try to push a/M over the limit get flung back out by huge centrifugal forces, before they can reach the horizon. A black hole, it appears, is totally resistant to having its horizon destroyed.

In 1969, Roger Penrose discovered that a large fraction of the mass of a spinning black hole is in the form of rotational energy, stored in the whirling spacetime curvature outside the hole's horizon. Although this rotational energy cannot be localized in any precise manner, it nevertheless can be extracted. Penrose discovered this by the following thought experiment, which is called the *Penrose process.*

Penrose process for extracting spin energy from a black hole

From far outside the hole, you throw a massive particle into the vicinity of the hole's horizon. Assuming you are at rest with respect to the hole, your 4-velocity is $\vec{U} = \partial/\partial t$. Denote by $\mathcal{E}_\infty^{\text{in}} = -\vec{p}^{\,\text{in}} \cdot \vec{U} = -\vec{p}^{\,\text{in}} \cdot (\partial/\partial t) = -p_t^{\text{in}}$ the energy of the particle (rest mass plus kinetic), as measured by you—its *energy-at-infinity* in the language of Ex. 26.4 and Sec. 4.10.2. As the particle falls, $\mathcal{E}_\infty^{\text{in}} = -p_t^{\text{in}}$ is conserved because of the Kerr metric's time-translation symmetry. Arrange that, as the particle nears the horizon, it splits into two particles; one (labeled "plunge") plunges through the horizon and the other (labeled "out") flies back out to large radii, where you catch it. Denote by $\mathcal{E}_\infty^{\text{plunge}} \equiv -p_t^{\text{plunge}}$ the conserved energy-at-infinity of the plunging particle and by $\mathcal{E}_\infty^{\text{out}} \equiv -p_t^{\text{out}}$ that of the out-flying particle. Four-momentum conservation at the event of the split dictates that $\vec{p}^{\,\text{in}} = \vec{p}^{\,\text{plunge}} + \vec{p}^{\,\text{out}}$, which implies this same conservation law for all components of the 4-momenta, in particular:

$$\mathcal{E}_\infty^{\text{out}} = \mathcal{E}_\infty^{\text{in}} - \mathcal{E}_\infty^{\text{plunge}}. \tag{26.80}$$

Now, it is a remarkable fact that the Boyer-Lindquist time basis vector $\partial/\partial t$ has a squared length $\partial/\partial t \cdot \partial/\partial t = g_{tt} = -\alpha^2 + \varpi^2\omega^2$ that becomes positive (so $\partial/\partial t$ becomes spacelike) at radii

$$\boxed{r < r_{\text{ergo}} \equiv M + \sqrt{M^2 - a^2 \cos^2\theta}.} \tag{26.81}$$

black hole's ergosphere

Notice that r_{ergo} is larger than r_H everywhere except on the hole's spin axis: $\theta = 0, \pi$. The region $r_H < r < r_{\text{ergo}}$ is called the hole's *ergosphere*. If the split into two particles occurs in the ergosphere, then it is possible to arrange the split such that the scalar product of the *timelike* vector $\vec{p}^{\,\text{plunge}}$ with the *spacelike* vector $\partial/\partial t$ is posi-

T2

tive, which means that the plunging particle's conserved energy-at-infinity, $\mathcal{E}_\infty^{\text{plunge}} = -\vec{p}^{\text{plunge}} \cdot (\partial/\partial t)$, is negative; whence [by Eq. (26.80)]

$$\mathcal{E}_\infty^{\text{out}} > \mathcal{E}_\infty^{\text{in}}. \tag{26.82}$$

See Ex. 26.16a.

When the outflying particle reaches your location, $r \gg M$, its conserved energy is equal to its physically measured total energy (rest-mass plus kinetic), and the fact that $\mathcal{E}_\infty^{\text{out}} > \mathcal{E}_\infty^{\text{in}}$ means that you get back more energy (rest-mass plus kinetic) than you put in. The hole's asymptotic energy-conservation law (25.102) implies that the hole's mass has decreased by precisely the amount of energy that you have extracted:

$$\Delta M = -(\mathcal{E}_\infty^{\text{out}} - \mathcal{E}_\infty^{\text{in}}) = \mathcal{E}_\infty^{\text{plunge}} < 0. \tag{26.83}$$

A closer scrutiny of this process (Ex. 26.16f) reveals that the plunging particle must have had negative angular momentum, so it has spun the hole down a bit. The energy you extracted, in fact, came from the hole's enormous store of rotational energy, which makes up part of its mass M. Your extraction of energy has reduced that rotational energy.

the second law of black-hole mechanics

Stephen Hawking has used sophisticated mathematical techniques to prove that, independently of how you carry out this thought experiment, and, indeed, independently of what is done to a black hole, general relativity requires that the horizon's surface area A_H never decrease. This is called the *second law of black-hole mechanics,* and it actually turns out to be a variant of the second law of thermodynamics in disguise (Ex. 26.16g). A straightforward calculation (Ex. 26.15) reveals that the horizon surface area is given by

$$\boxed{A_H = 4\pi(r_H^2 + a^2) = 8\pi M r_H \quad \text{for a spinning hole,}} \tag{26.84a}$$

$$\boxed{A_H = 16\pi M^2 \quad \text{for a nonspinning hole, } a = 0.} \tag{26.84b}$$

Demetrios Christodoulou has shown (cf. Ex. 26.16) that, in the Penrose process described here, the nondecrease of A_H is the only constraint on how much energy one can extract, so by a sequence of optimally designed particle injections and splits that keep A_H unchanged, one can reduce the mass of the hole to

$$\boxed{M_{\text{irr}} = \sqrt{\frac{A_H}{16\pi}} = \sqrt{\frac{M(M + \sqrt{M^2 - a^2})}{2}},} \tag{26.85}$$

black hole's irreducible mass and rotational energy

but no smaller. This is called the hole's irreducible mass. The hole's total mass is the sum of its irreducible mass and its rotational energy M_{rot}; so the rotational energy is

$$\boxed{M_{\text{rot}} = M - M_{\text{irr}} = M\left[1 - \sqrt{\frac{1}{2}\left(1 + \sqrt{1 - a^2/M^2}\right)}\right].} \tag{26.86}$$

For the fastest possible spin, $a = M$, this gives $M_{\rm rot} = M(1 - 1/\sqrt{2}) \simeq 0.2929M$. This is the maximum amount of energy that can be extracted, and it is enormous compared to the energy $\sim 0.005M$ that can be released by thermonuclear burning in a star of mass M. T2

The Penrose process of throwing in particles and splitting them in two is highly idealized, and of little or no importance in Nature. However, Nature seems to have found a very effective alternative method for extracting rotational energy from spinning black holes (Blandford and Znajek, 1977; Thorne, Price, and MacDonald, 1986; McKinney, Tchekhovskoy, and Blandford, 2012; Ex. 26.21) in which magnetic fields, threading through a black hole and held on the hole by a surrounding disk of hot plasma, extract energy electromagnetically. This process is thought to power the gigantic jets that shoot out of the nuclei of some active galaxies. It might also be the engine for some powerful gamma-ray bursts.

Blandford-Znajek process by which magnetic fields extract energy from black holes in Nature

EXERCISES

Exercise 26.15 *Derivation: Surface Area of a Spinning Black Hole* T2

From the Kerr metric (26.70) derive Eqs. (26.84) for the surface area of a spinning black hole's horizon—that is, the surface area of the 2-dimensional surface $\{r = r_H, t = \text{constant}\}$.

Exercise 26.16 ***Example: Penrose Process, Hawking Radiation, and Thermodynamics of Black Holes* T2

This exercise is a foundation for the discussion of black-hole thermodynamics in Sec. 4.10.2.

(a) Consider the Penrose process, described in the text, in which a particle flying inward toward a spinning hole's horizon splits in two inside the ergosphere, and one piece plunges into the hole while the other flies back out. Show that it is always possible to arrange this process so the plunging particle has negative energy-at-infinity: $\mathcal{E}_\infty^{\rm plunge} = -\vec{p}^{\,\rm plunge} \cdot \partial/\partial t < 0$. [Hint: Perform a calculation in a local Lorentz frame in which $\partial/\partial t$ points along a spatial basis vector, $\vec{e}_{\hat{1}}$. Why is it possible to find such a local Lorentz frame?]

(b) Around a spinning black hole consider the vector field

$$\vec{\xi}_H \equiv \partial/\partial t + \Omega_H \partial/\partial\phi, \tag{26.87}$$

where Ω_H is the horizon's angular velocity. Show that at the horizon (at radius $r = r_H$) this vector field is null and is tangent to the horizon generators. Show that all other vectors in the horizon are spacelike.

(c) In the Penrose process the plunging particle changes the hole's mass by an amount ΔM and its spin angular momentum by an amount ΔJ_H. Show that

$$\Delta M - \Omega_H \Delta J_H = -\vec{p}^{\,\rm plunge} \cdot \vec{\xi}_H. \tag{26.88}$$

T2

Here $\vec{p}^{\text{plunge}}$ and $\vec{\xi}_H$ are to be evaluated at the event where the particle plunges through the horizon, so they both reside in the same tangent space. [Hint: The angular momentum carried into the horizon is the quantity $p_\phi^{\text{plunge}} = \vec{p}^{\text{plunge}} \cdot \partial/\partial\phi$. Why? This quantity is conserved along the plunging particle's world line. Why?] Note that in Sec. 4.10.2, p_ϕ is denoted $\mathbf{j} \cdot \hat{\boldsymbol{\Omega}}_H$—the projection of the particle's orbital angular momentum on the black hole's spin axis.

(d) Show that if $\vec{A}$ is any future-directed timelike vector and $\vec{K}$ is any null vector, both living in the tangent space at the same event in spacetime, then $\vec{A} \cdot \vec{K} < 0$. [Hint: Perform a calculation in a specially chosen local Lorentz frame.] Thereby conclude that $-\vec{p}^{\text{plunge}} \cdot \vec{\xi}_H$ is positive, whatever may be the world line and rest mass of the plunging particle.

(e) Show that for the plunging particle to decrease the hole's mass, it must also decrease the hole's angular momentum (i.e., it must spin the hole down a bit).

(f) Hawking's second law of black-hole mechanics says that, whatever may be the particle's world line and rest mass, when the particle plunges through the horizon, it causes the horizon's surface area A_H to increase. This suggests that the always positive quantity $\Delta M - \Omega_H \Delta J_H = -\vec{p}^{\text{plunge}} \cdot \vec{\xi}_H$ might be a multiple of the increase ΔA_H of the horizon area. Show that this is indeed the case:

$$\boxed{\Delta M = \Omega_H \Delta J_H + \frac{g_H}{8\pi} \Delta A_H,} \tag{26.89}$$

where g_H is given in terms of the hole's mass M and the radius r_H of its horizon by

$$\boxed{g_H = \frac{r_H - M}{2M r_H}.} \tag{26.90}$$

(You might want to do the algebra, based on Kerr-metric formulas, on a computer.) The quantity g_H (which we have met previously in the Rindler approximation; Ex. 26.11) is called the hole's "surface gravity" for a variety of reasons. One reason is that an observer who hovers just above a horizon generator, blasting his or her rocket engines to avoid falling into the hole, has a 4-acceleration with magnitude g_H/α and thus feels a "gravitational acceleration" of this magnitude; here $\alpha = g^{tt}$ is a component of the Kerr metric called the *lapse function* [Eqs. (26.70) and (26.72)]. This gravitational acceleration is arbitrarily large for an observer arbitrarily close to the horizon (where Δ and hence α are arbitrarily close to zero); when renormalized by α to make it finite, the acceleration is g_H. Equation (26.89) is called the "first law of black-hole mechanics" because of its resemblance to the first law of thermodynamics.

(g) Stephen Hawking has shown, using quantum field theory, that a black hole's horizon emits thermal (blackbody) radiation. The temperature of this "Hawking radiation," as measured by the observer who hovers just above the horizon, is proportional to the gravitational acceleration g_H/α that the observer measures, with a proportionality constant $\hbar/(2\pi k_B)$, where $\hbar$ is Planck's reduced constant and k_B is Boltzmann's constant. As this thermal radiation climbs out of the horizon's vicinity and flies off to large radii, its frequencies and temperature get redshifted by the factor α, so as measured by distant observers the temperature is

T2

$$T_H = \frac{\hbar}{2\pi k_B} g_H. \tag{26.91}$$

This suggests a reinterpretation of the first law of black-hole mechanics (26.89) as the first law of thermodynamics for a black hole:

$$\Delta M = \Omega_H \Delta J_H + T_H \Delta S_H, \tag{26.92}$$

where S_H is the hole's entropy [cf. Eq. (4.62)]. Show that this entropy is related to the horizon's surface area by

$$S_H = k_B \frac{A_H}{4{L_P}^2}, \tag{26.93}$$

where $L_P = \sqrt{\hbar G/c^3} = 1.616 \times 10^{-33}$ cm is the Planck length (with G Newton's gravitation constant and c the speed of light). Because $S_H \propto A_H$, the second law of black-hole mechanics (nondecreasing A_H) is actually the second law of thermodynamics (nondecreasing S_H) in disguise.[6]

(h) For a 10-solar-mass, nonspinning black hole, what is the temperature of the Hawking radiation in Kelvins, and what is the hole's entropy in units of the Boltzmann constant?

(i) Reread the discussions of black-hole thermodynamics and entropy in the expanding universe in Secs. 4.10.2 and 4.10.3, which rely on the results of this exercise.

Exercise 26.17 *Problem: Thin Accretion Disks: Circular, Equatorial, Geodesic Orbits around a Kerr Black Hole* T2

Astronomers find and observe black holes primarily through the radiation emitted by infalling gas. Usually this gas has sufficient angular momentum to form an *accretion disk* around the black hole, lying in its equatorial plane. When the gas can cool efficiently, the disk is physically thin. The first step in describing accretion disks is to compute their equilibrium structure, which is usually approximated by assuming that the fluid elements follow circular geodesic orbits and pressure can be ignored.

6. Actually, the emission of Hawking radiation decreases the hole's entropy and surface area; but general relativity is oblivious to this, because general relativity is a classical theory, and Hawking's prediction of the thermal radiation is based on quantum theory. Thus, the Hawking radiation violates the second law of black-hole mechanics. It does not, however, violate the second law of thermodynamics, because the entropy carried into the surrounding universe by the Hawking radiation exceeds the magnitude of the decrease of the hole's entropy. The total entropy of hole plus universe increases.

T2

The next step is to invoke some form of magnetic viscosity that catalyzes the outward transport of angular momentum through the disk and the slow inspiral of the disk's gas. The final step is to compute the spectrum of the emitted radiation. In this problem, we consider only the first step: the gas's circular geodesic orbits.

(a) Following Ex. 26.12, we can write the 4-velocity in covariant form as $u_\alpha = \{-\mathcal{E}_\infty, u_r, 0, L\}$, where $\mathcal{E}_\infty$ and L are the conserved energy and angular momentum per unit mass, respectively, and u_r describes the radial motion (which we eventually set to zero). Use $u^\alpha u_\alpha = -1$ and $u^r = dr/d\tau$ to show that, for any equatorial geodesic orbit:

$$\frac{1}{2}\left(\frac{dr}{d\tau}\right)^2 + V_{\rm eff} = 0, \text{ where } V_{\rm eff} = \frac{1 + u^t u_t + u^\phi u_\phi}{2 g_{rr}} \tag{26.94a}$$

is an effective potential that includes $\mathcal{E}_\infty$ and L inside itself [by contrast with the effective potential we used for a Schwarzschild black hole, Eq. (26.64a), from which $\mathcal{E}_\infty$ was pulled out].

(b) Explain why a circular orbit is defined by the twin conditions, $V_{\rm eff} = 0$, $\partial_r V_{\rm eff} = 0$, and use the Boyer-Lindquist metric coefficients plus computer algebra to solve for the conserved energy and angular momentum:

$$\mathcal{E}_\infty = \frac{1 - 2M/r \pm aM^{1/2}/r^{3/2}}{(1 - 3M/r \pm 2aM^{1/2}/r^{3/2})^{1/2}}, \quad L = \frac{\pm M^{1/2} r^{1/2} - 2aM/r \pm M^{1/2}a^2/r^{3/2}}{(1 - 3M/r \pm 2aM^{1/2}/r^{3/2})^{1/2}}, \tag{26.94b}$$

where, in $\pm$, the $+$ is for a prograde orbit (same direction as hole spins) and $-$ is for retrograde.

(c) Explain mathematically and physically why the angular velocity $\Omega = u^\phi/u^t$ satisfies

$$\frac{\partial \mathcal{E}_\infty}{\partial r} = \Omega \frac{\partial L}{\partial r}, \tag{26.94c}$$

and show that it evaluates to

$$\Omega = \frac{\pm M^{1/2}}{r^{3/2} \pm aM^{1/2}}. \tag{26.94d}$$

(d) Modify the argument from Ex. 26.12 to show that the binding energy of a gas particle in its smallest, stable, circular, equatorial orbit around a maximally rotating ($a = M$) Kerr hole is $1 - 3^{-1/2} = 0.423$ per unit mass for prograde orbits and $1 - 5/3^{3/2} = 0.0377$ per unit mass for retrograde orbits. The former is a measure of the maximum power that can be released as gas accretes onto a hole through a thin disk in a cosmic object like a quasar. Note that it is two orders of magnitude larger than the energy typically recoverable from nuclear reactions.

For the visual appearance of a thin accretion disk around a black hole, distorted by gravitational lensing (bending of light rays in the Kerr metric), and for other aspects of a black hole's gravitational lensing, see, for example, James et al. (2015a) and references therein.

Exercise 26.18 *Problem: Thick Accretion Disks* T2

A quite different type of accretion disk forms when the gas is unable to cool. This can occur when its gas supply rate is either very large or very small. In the former case, the photons are trapped by the inflowing gas; in the latter, the radiative cooling timescale exceeds the inflow timescale. (In practice such inflows are likely to produce simultaneous outflows to carry off the energy released.) Either way, pressure and gravity are of comparable importance. There is an elegant description of gas flow close to the black hole as a sort of toroidal star, where the metric is associated with the hole and not the gas. In this problem we make simple assumptions to solve for the equilibrium flow.

(a) Treat the gas as a perfect fluid, and use as thermodynamic variables its pressure P and enthalpy density w. When the gas supply rate is large, the pressure is generally dominated by radiation. Show that in this case $w = \rho_o + 4P$, with ρ_o the rest-mass density of the plasma. Show, further, that $w = \rho_o + \frac{5}{2}P$ when the pressure is due primarily to nonrelativistic ions and their electrons.

(b) Assume that the motion is purely azimuthal, so the only nonzero covariant components of the gas's 4-velocity, in the Boyer-Lindquist coordinate system, are u^t and u^ϕ. (In practice there will also be slow poloidal circulation and slow inflow.) Show that the equation of hydrostatic equilibrium (26.31) can be written in the form

$$\frac{P_{,a}}{w} = u^t u_{t,a} + u^\phi u_{\phi,a}, \tag{26.95a}$$

where $a = r, \theta$, and the commas denote partial derivatives.

(c) Define for the gas the specific energy at infinity, $\mathcal{E}_\infty = -u_t$; the *specific fluid angular momentum*, $\ell = -u_\phi/u_t$ (which is $L/\mathcal{E}_\infty$ in the notation of the previous exercise); and the angular velocity, $\Omega = d\phi/dt = u^\phi/u^t$. Show that Eq. (26.95a) can be rewritten as

$$-\frac{\vec{\nabla} P}{w} = \vec{\nabla} \ln \mathcal{E}_\infty - \frac{\Omega \vec{\nabla} \ell}{1 - \Omega \ell}, \tag{26.95b}$$

where the spacetime gradient $\vec{\nabla}$ has components only in the r and θ directions, and ℓ and $\mathcal{E}_\infty$ are regarded as scalar fields.

(d) Now make the first simplifying assumption: the gas obeys a *barotropic* equation of state, $P = P(w)$ (cf. Sec. 14.2.2). Show that this assumption implies that the specific angular momentum ℓ is a function of the angular velocity.

(e) Show that the nonrelativistic limit of this result—applicable to stars like white dwarfs—is that the angular velocity is constant on cylindrical surfaces, which is von Zeipel's theorem (Ex. 13.8).

T2

(f) Compute the shape of the surfaces on which Ω and ℓ are constant in the r-θ "plane" (surface of constant t and ϕ) for a spinning black hole with $a = 0.9$ M.

(g) Now make the second simplifying assumption: the specific angular momentum ℓ is constant. Compute the shape of the isobars, also in the r-θ plane, and show that they exhibit a cusp along a circle in the equatorial plane, whose radius shrinks as ℓ increases.

(h) Compute the specific energy at infinity, $\mathcal{E}_\infty$, of the isobar that passes through the cusp, and show that it can vanish if ℓ is large enough. Interpret your answer physically.

Of course it is possible to deal with more realistic assumptions about the equation of state and the angular momentum distribution, but this relatively simple model brings out some salient features of the equilibrium flow.

Exercise 26.19 *Problem: Geodetic and Lense-Thirring Precession* T2

(a) Consider a pulsar in a circular, equatorial orbit around a massive Kerr black hole, with the pulsar's spin vector $\vec{S}$ lying in the hole's equatorial plane. Using the fact that the spin vector is orthogonal to the pulsar's 4-velocity (Sec. 25.7), show that its only nonzero covariant components are S_r, S_ϕ, and $S_t = -\Omega S_\phi$, where $\Omega = d\phi/dt$ is the pulsar's orbital angular velocity [Eq. (26.94d)]. Now, neglecting usually negligible quadrupole-curvature coupling forces (Sec. 25.7), the spin is parallel-transported along the pulsar's geodesic, so $\nabla_{\vec{u}}\vec{S} = 0$. Show that this implies

$$\frac{d^2 S_\alpha}{d\tau^2} = \Gamma^\delta{}_{\alpha\gamma}\Gamma^\beta{}_{\delta\epsilon}u^\gamma u^\epsilon S_\beta. \tag{26.96a}$$

(b) Next use Boyer-Lindquist coordinates and the results from Ex. (26.17) plus computer algebra to show that the spatial part of Eq. (26.96a) takes the remarkably simple form

$$\frac{d^2 S_i}{d\tau^2} = -\frac{M S_i}{r^3}, \tag{26.96b}$$

where $i = r, \phi$. Interpret this equation geometrically. In particular, comment on the absence of the hole's spin parameter a in the context of the dragging of inertial frames, and consider how two counter-orbiting stars would precess relative to each other.

(c) Explain why the rate of precession of the pulsar's spin as measured by a distant observer is given by

$$\Omega_p = \Omega - \frac{M^{1/2}}{r^{3/2}u^t}, \tag{26.96c}$$

where $u^t = dt/d\tau$ is the contravariant time component of the pulsar's 4-velocity.

(d) Use computer algebra to evaluate Ω_p, and show that a Taylor expansion for $r \gg M$ gives

$$\Omega_p = \frac{3M^{3/2}}{2r^{5/2}} - \frac{aM}{r^3} + \ldots . \tag{26.96d}$$

The first term on the right-hand side is known as *geodetic* precession; the second as *Lense-Thirring* precession.

Exercise 26.20 *Challenge: General Orbits around a Kerr Black Hole* T2

By combining the techniques used for general orbits around a Schwarzschild black hole (Ex. 26.12) with those for equatorial orbits around a Kerr black hole (Ex. 26.17), explore the properties and shapes of general, geodesic orbits around a Kerr hole. To make progress on this exercise, note that because the Kerr spacetime is not spherically symmetric, the orbits will not, in general, lie in "planes" (2- or 3-dimensional surfaces) of any sort. Correspondingly, one must explore the orbits in 4 spacetime dimensions rather than 3. In addition to the fairly obvious three constants of motion that the Kerr hole shares with Schwarzschild—μ, $\mathcal{E}_\infty$, and L [Eqs. (26.63)]—there is a fourth called the *Carter constant* after Brandon Carter, who discovered it:

$$\mathcal{Q} = p_\theta^2 + \cos^2\theta [a^2(\mu^2 - \mathcal{E}_\infty^2) + \sin^{-2}\theta L^2]. \tag{26.97}$$

The numerical integrations are best carried out using Hamilton's equations for the super-hamiltonian (Ex. 25.7).

Formulas for the geodesics and some of their properties are given in most general relativity textbooks, for example, Misner, Thorne, and Wheeler (1973, Sec. 33.5) and Straumann (2013, Sec. 8.4). For an extensive numerical exploration of the orbital shapes, based on the super-hamiltonian, see Levin and Perez-Giz (2008)—who present and discuss the Hamilton equations in their Appendix A.

Exercise 26.21 *Challenge: Electromagnetic Extraction of Energy from a Spinning Black Hole* T2

As discussed in the text, spinning black holes contain a considerable amount of rotational energy [Eq. (26.86)]. This exercise sketches how this energy may be extracted by an electromagnetic field: the Blandford-Znajek process.

(a) Suppose that a Kerr black hole, described using Boyer-Lindquist coordinates $\{t, r, \theta, \phi\}$, is orbited by a thick accretion disk (Ex. 26.18), whose surface near the hole consists of two funnels, axisymmetric around the hole's rotation axis $\theta = 0$ and $\theta = \pi$. Suppose, further, that the funnels' interiors contain a stationary, axisymmetric electromagnetic field described by a vector potential $A_\alpha(r, \theta)$. The surface of each funnel contains surface current and charge that keep the disk's interior free of electromagnetic field, and the disk's gas supplies pressure across the

T2

funnel's surface to balance the electromagnetic stress. Whatever plasma there may be inside the funnel has such low density that the electromagnetic contribution to the stress-energy tensor is dominant, and so we can write $F_{\alpha\beta}J^{\beta}=0$ [cf. Eq. (2.81b)]. Show that the electromagnetic field tensor [Eq. (25.61)] in the Boyer-Lindquist coordinate basis can be written $F_{\alpha\beta}=A_{\beta,\alpha}-A_{\alpha,\beta}$, where the commas denote partial derivatives.

(b) Write $A_{\phi}=\Phi/(2\pi)$, $A_t=-V$, and by considering the electromagnetic field for $r\gg M$, interpret Φ and V as the magnetic flux contained in a circle of fixed r,θ and the electrical potential of that circle, respectively. This also remains true near the black hole, where we define the magnetic and electric fields as those measured by observers who move orthogonally to the hypersurfaces of constant Boyer-Lindquist time t, so $\vec{E}$ and $\vec{B}$ can be regarded as spatial vectors **E** and **B** that lie in those hypersurfaces of constant t. Can you prove all this?

(c) Explain why it is reasonable to expect the electric field **E** to be orthogonal to the magnetic field **B** near the hole as well as far away, and use Faraday's law to show that its toroidal component $E_{\hat{\phi}}$ must vanish. Thereby conclude that the equipotential surfaces $V=$ constant coincide with the magnetic surfaces $\Phi=$ constant.

(d) Define an angular velocity by $\Omega=-2\pi V_{,\theta}/\Phi_{,\theta}$, show that it, too, is constant on magnetic surfaces, and show that an observer who moves with this angular velocity $d\phi/dt=\Omega$ measures vanishing electric field. Thereby conclude that the magnetic field lines rotate rigidly with this angular velocity, and deduce an expression for the electric field **E** in terms of Φ and Ω. (You might want to compare with Ex. 19.11.)

(e) Use the inhomogeneous Maxwell equations to show that the current density J^{α} describes a flow of charge along the magnetic field lines. Hence calculate the current I flowing inside a magnetic surface. Express **B** in terms of I and Φ.

(f) Now sketch the variation of the electromagnetic field in the funnel both near the horizon and at a large distance from it, assuming that there is an outward flow of energy and angular momentum.

(g) Use the fact that $\partial/\partial t$ and $\partial/\partial\phi$ are Killing vectors (Ex. 25.5) to confirm that electromagnetic energy and angular momentum are conserved in the funnels.

(h) We have derived these general principles without giving an explicit solution that demonstrates energy extraction. In order to do this, we must also specify boundary conditions at the horizon of the black hole and at infinity. The former is essentially that the electromagnetic field be nonsingular when measured by an infalling observer or when expressed in Kerr coordinates, for example. The latter describes the flow of energy and angular momentum as outward. Stable solutions can be exhibited numerically and, typically, they have $\Omega\sim0.3$ to $0.5\Omega_H$ (McKinney, Tchekhovskoy, and Blandford, 2012). These solutions demonstrate that energy and angular momentum can flow outward across the event horizon so the mass of the black hole gradually decreases. How can this be?

T2

26.6 The Many-Fingered Nature of Time T2

26.6

We conclude this chapter with a discussion of a concept that John Archibald Wheeler (the person who has most clarified the conceptual underpinnings of general relativity) calls the *many-fingered nature of time*.

In the flat spacetime of special relativity there are preferred families of observers: each such family lives in a global Lorentz reference frame and uses that frame to split spacetime into space plus time. The hypersurfaces of constant time (slices of simultaneity) that result from that split are flat hypersurfaces slicing through all of spacetime (Fig. 26.9a). Of course, different preferred families live in different global Lorentz frames and thus split up spacetime into space plus time in different manners (e.g., the dotted and dashed slices of constant time in Fig. 26.9a in contrast to the solid ones). As a result, no universal concept of time exists in special relativity. But at least there are some strong restrictions on time: each inertial family of observers will agree that another family's slices of simultaneity are flat slices.

in special relativity, different inertial observers slice spacetime into space plus time differently, but all slices are flat

In general relativity (i.e., in curved spacetime), even this restriction is gone. In a generic curved spacetime there are no flat hypersurfaces, and hence no candidates for flat slices of simultaneity. In addition, no global Lorentz frames and thus no preferred families of observers exist in a generic curved spacetime. A family of observers who are all initially at rest with respect to one another, and each of whom moves freely (inertially), will soon acquire relative motion because of tidal forces. As a result, their

in general relativity, the slices of constant time have arbitrary shapes; hence, the "many-fingered nature of time"

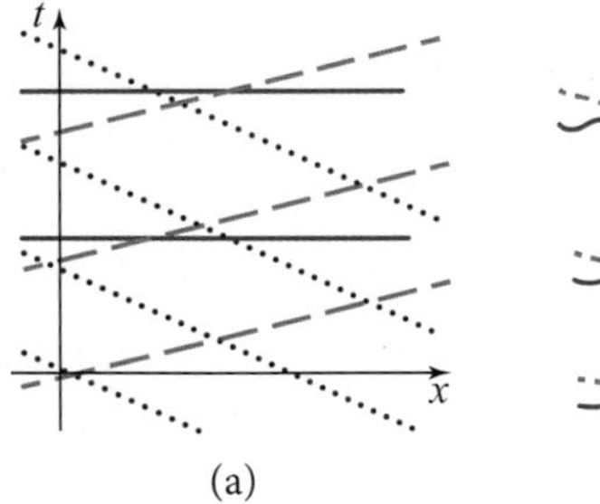

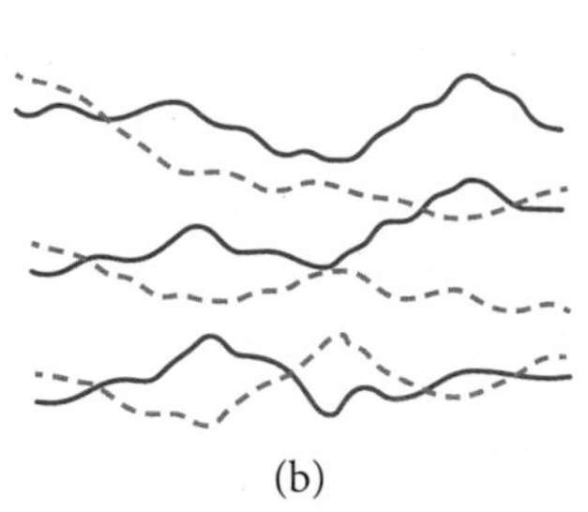

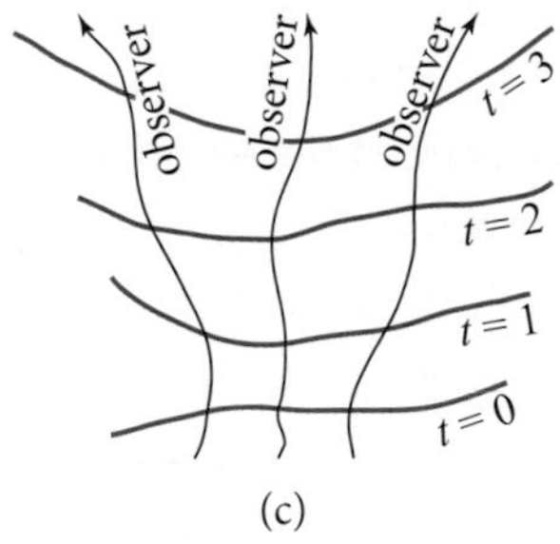

FIGURE 26.9 Spacetime diagrams showing the slices of simultaneity as defined by various families of observers. (a) Flat spacetime. The three families (those with solid slices, those with dashed, and those with dotted) are inertial, so their slices of constant time are those of global Lorentz frames. (b) Curved spacetime. The two families' slices of simultaneity illustrate the "many-fingered" nature of time. (c) Curved spacetime. The selection of an arbitrary foliation of spacelike hypersurfaces of simultaneity, and the subsequent construction of the world lines of observers who move orthogonally to those hypersurfaces (i.e., for whom light-ray synchronization will define those hypersurfaces as simultaneities).

T2

slices of simultaneity (defined locally by Einstein light-ray synchronization, and then defined globally by patching together the little local bits of slices) may soon become rather contorted. Correspondingly, as is shown in Fig. 26.9b, different families of observers will slice spacetime up into space plus time in manners that can be quite distorted, relative to one another—with "fingers" of one family's time slices pushing forward, ahead of the other family's here, and lagging behind there, and pushing ahead in some other place.

In curved spacetime it is best to not even restrict oneself to inertial (freely falling) observers. For example, in the spacetime of a static star, or of the exterior of a Schwarzschild black hole, the family of static observers [observers whose world lines are $\{(r, \theta, \phi) = \text{const}, t \text{ varying}\}$] are particularly simple; their world lines mold themselves to the static structure of spacetime in a simple, static manner. However, these observers are not inertial; they do not fall freely. This need not prevent us from using them to split up spacetime into space plus time, however. Their proper reference frames produce a perfectly good split. When one uses that split, in the case of a black hole, one obtains a 3-dimensional-space version of the laws of black-hole physics that is a useful tool in astrophysical research; see Thorne, Price, and MacDonald (1986).

an observer's local slice of physical simultaneity is orthogonal to the observer's world line

For any family of observers, accelerated or inertial, the slices of simultaneity as defined by Einstein light-ray synchronization over small distances (or equivalently by the space slices of the observer's proper reference frames) are the 3-surfaces orthogonal to the observers' world lines (cf. Fig. 26.9c). To see this most easily, pick a specific event along a specific observer's world line, and study the slice of simultaneity there from the viewpoint of a local Lorentz frame in which the observer is momentarily at rest. Light-ray synchronization guarantees that locally, the observer's slice of simultaneity will be the same as that of this local Lorentz frame. Since the frame's slice is orthogonal to its own time direction and that time direction is the same as the direction of the observer's world line, the slice is orthogonal to the observer's world line. By the discussion in Sec. 24.5, the slice is also locally the same (to first order in distance away from the world line) as a slice of constant time in the observer's proper reference frame.

if observers rotate around one another, their local slices of simultaneity cannot mesh; so best to choose time slices first and then compute observers' orthogonal world lines

If the observers rotate around one another (in curved spacetime or in flat), it is not possible to mesh their local slices of simultaneity, defined in this manner, into global slices of simultaneity (i.e., there are no global 3-dimensional hypersurfaces orthogonal to their world lines). We can protect against this eventuality by choosing the slices first: select any family of nonintersecting spacelike slices through the curved spacetime (Fig. 26.9c). Then there will be a family of timelike world lines that are everywhere orthogonal to these hypersurfaces. A family of observers who move along those orthogonal world lines and who define their 3-spaces of simultaneity by local light-ray synchronization will thereby identify the orthogonal hypersurfaces as their simultaneities. Exercise 26.22 illustrates these ideas using Schwarzschild spacetime, and Box 26.3 uses these ideas to visualize a black hole's spacetime curvature.

T2

BOX 26.3. TENDEX AND VORTEX LINES OUTSIDE A BLACK HOLE T2

When one uses a family of spacelike slices (a *foliation*) with unit normals (4-velocities of orthogonally moving observers) $\vec{w}$ to split spacetime up into space plus time, the electromagnetic field tensor $\boldsymbol{F}$ splits into the electric field $E^\alpha = F^{\alpha\beta} w_\beta$ and the magnetic field $B^\beta = \frac{1}{2}\epsilon^{\alpha\beta\gamma\delta} F_{\gamma\delta} w_\alpha$. These are 3-vectors lying in the spacelike slices. In terms of components in the observers' proper reference frames, they are $E^{\hat{j}} = F^{\hat{0}\hat{j}}$ and $B^{\hat{i}} = \epsilon^{\hat{i}\hat{j}\hat{k}} F_{\hat{j}\hat{k}}$ (see Sec. 2.11 and especially Fig. 2.9). Similarly, the foliation splits the vacuum Riemann tensor into the symmetric, trace-free tidal field $\mathcal{E}_{\hat{i}\hat{j}} = R_{\hat{i}\hat{0}\hat{j}\hat{0}}$ (which produces relative accelerations $\Delta a_{\hat{i}} = -\mathcal{E}_{\hat{i}\hat{j}}\xi^{\hat{j}}$ of particles separated by $\xi^{\hat{j}}$), and the frame-drag field $\mathcal{B}_{\hat{i}\hat{j}} = \frac{1}{2}\epsilon_{\hat{i}\hat{p}\hat{q}} R^{\hat{p}\hat{q}}{}_{\hat{j}\hat{0}}$ (which produces differential frame dragging $\Delta\Omega_{\hat{i}} = \mathcal{B}_{\hat{i}\hat{j}}\xi^{\hat{j}}$). See Box 25.2.

Just as the electromagnetic field can be visualized using electric and magnetic field lines that live in the spacelike slices, so also the tidal field $\mathcal{E}_{\hat{i}\hat{j}}$ and frame-drag field $\mathcal{B}_{\hat{i}\hat{j}}$ can each be visualized using integral curves of its three eigenvector fields: the tidal field's tendex lines and the frame-drag field's vortex lines (Box 25.2). These lines lie in the space slices and are often color coded by their eigenvalues, which are called the lines' *tendicities* and *(frame-drag) vorticities*.

For a nonspinning (Schwarzschild) black hole, the frame-drag field vanishes (no spin implies no frame dragging); and the tidal field (with slicing either via Schwarzschild coordinate time t or Eddington-Finkelstein coordinate time $\tilde{t}$) has proper-reference-frame components $\mathcal{E}_{\hat{r}\hat{r}} = -2M/r^3$, $\mathcal{E}_{\hat{\theta}\hat{\theta}} = \mathcal{E}_{\hat{\phi}\hat{\phi}} = +M/r^3$ [Eq. (6) of Box 26.2]. This tidal field's eigenvectors are $\vec{e}_{\hat{r}}$, $\vec{e}_{\hat{\theta}}$, and $\vec{e}_{\hat{\phi}}$, and their integral curves (the tendex lines) are identical to those shown for a weakly gravitating, spherical body in the figure at the bottom left of Box 25.2.

For a black hole with large spin, $a/M = 0.95$ for example, it is convenient to choose as our space slices, hypersurfaces of constant Kerr coordinate time $\tilde{t}$ [Eq. (26.78)], since these penetrate smoothly through the horizon. The tendex and vortex lines that live in those slices have the following forms (Zhang et al., 2012).

The tendex lines are shown on the left and in the central inset; the vortex lines are shown on the right. The horizon is color coded by the sign of the tendicity (left) and vorticity (right) of the lines that emerge from it: blue for positive and red for negative.

The blue tendex lines have positive tendicity, and the tidal-acceleration equation $\Delta a_{\hat{i}} = -\mathcal{E}_{\hat{i}\hat{j}}\xi^{\hat{j}}$ says that a woman whose body is oriented along them

(continued)

T2

BOX 26.3. (continued)

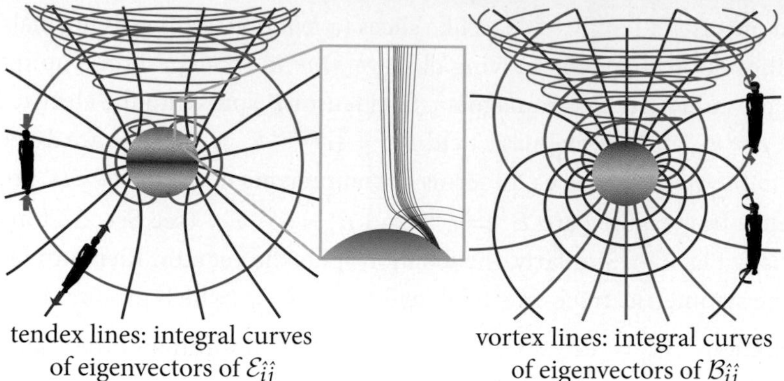

tendex lines: integral curves of eigenvectors of $\mathcal{E}_{\hat{i}\hat{j}}$

vortex lines: integral curves of eigenvectors of $\mathcal{B}_{\hat{i}\hat{j}}$

gets squeezed with relative “gravitational” acceleration between head and foot equal to the lines’ tendicity times her body length. The red lines have negative tendicity, so they stretch a man with a head-to-foot relative acceleration equal to the magnitude of their tendicity times his body length. Notice that near the poles of a fast-spinning black hole, the radial tendex lines are blue, and in the equatorial region they are red. Therefore, a man falling into the polar regions of a fast-spinning hole gets squeezed radially, and one falling into the equatorial regions gets stretched radially.

For a woman with her body oriented along a vortex line, the differential frame-drag equation $\Delta\Omega_{\hat{i}} = \mathcal{B}_{\hat{i}\hat{j}}\xi^{\hat{j}}$ says that a gyroscope at her feet precesses clockwise (blue line, positive vorticity) or counterclockwise (red line, negative vorticity) relative to inertial frames at her head. And a gyroscope at her head precesses in that same direction relative to inertial frames at her feet. Thus, the vortex lines can be regarded as either counterclockwise (red) or clockwise (blue). The precessional angular velocity is equal to the line’s (frame-drag) vorticity times the woman’s body length.

Notice that counterclockwise (red) vortex lines emerge from the north polar region of the horizon, swing around the south pole, and descend back into the north polar region. Similarly, (blue) vortex lines emerge from the horizon’s south polar region, swing around the north pole, and descend back into the south. This is similar to the vortex lines for a spinning body in linearized theory (bottom right figure in Box 25.2). The differential precession is counterclockwise along the (red) near-horizon radial lines in the north polar region, because gyroscopes near the horizon are dragged into precession more strongly by the hole’s spin, the nearer one is to the horizon. This also explains the clockwise precession along the (blue) near-horizon radial lines in the south polar region.

For more details about the tendex and vortex lines of Kerr black holes, see Zhang et al. (2012).

Exercise 26.22 *Practice: Slices of Simultaneity in Schwarzschild Spacetime* T2

(a) One possible choice of slices of simultaneity for Schwarzschild spacetime is the set of 3-surfaces $\{t = \text{const}\}$, where t is the Schwarzschild time coordinate. Show that the unique family of observers for whom these are the simultaneities are the static observers, with world lines $\{(r, \theta, \phi) = \text{const}, t \text{ varying}\}$. Explain why these slices of simultaneity and families of observers exist only outside the horizon of a black hole and cannot be extended into the interior. Draw a picture of the world lines of these observers and their slices of simultaneity in an Eddington-Finkelstein spacetime diagram.

(b) A second possible choice of simultaneities is the set of 3-surfaces $\{\tilde{t} = \text{const}\}$, where $\tilde{t}$ is the Eddington-Finkelstein time coordinate. What are the world lines of the observers for whom these are the simultaneities? Draw a picture of those world lines in an Eddington-Finkelstein spacetime diagram. Note that they and their simultaneities cover the interior of the hole as well as its exterior.

Bibliographic Note

In our opinion, the best elementary textbook treatment of black holes and relativistic stars is that in Hartle (2003, Chaps. 12, 13, 15, 24); this treatment is also remarkably complete.

For the treatment of relativistic stars at an elementary level, we also recommend Schutz (2009, Chap. 10), and at a more advanced level (including stellar pulsations), Straumann (2013, Chaps. 4, 9) and Misner, Thorne, and Wheeler (1973, Chaps. 31–34).

For the study of black holes at an intermediate level, see Carroll (2004, Chaps. 5, 6), and Hobson, Efstathiou, and Lasenby (2006, Chaps. 9, 11, 13), and at more advanced levels, Wald (1984, Chap. 12), which is brief and highly mathematical, and Straumann (2013, Chap. 7) and Misner, Thorne, and Wheeler (1973, Chaps. 23, 24, 26), which are long and less mathematical.

The above are all portions of general relativity textbooks. There are a number of books and monographs devoted solely to the theory of black holes and/or relativistic stars. Among these, we particularly recommend the following. Shapiro and Teukolsky (1983) is an astrophysically oriented book at much the same level as this chapter, but with much greater detail and extensive applications; it deals with black holes, neutron stars, and white dwarf stars in astrophysical settings. Meier (2012) is a much more up-to-date and quite comprehensive treatment of the astrophysics and observations of black holes of all sizes. Frolov and Novikov (1998) is a thorough monograph on black holes, including their fundamental theory and their interactions with the rest of the universe; it includes extensive references to the original literature and readable summaries of all the important issues that had been treated by black-hole researchers as of 1997. Chandrasekhar (1983) is an idiosyncratic but elegant and complete monograph on the theory of black holes and especially small perturbations of them. Thorne, Price, and MacDonald (1986) is an equally idiosyncratic monograph that formulates the theory of black holes in 3+1 language, which facilitates physical understanding.

27

CHAPTER TWENTY-SEVEN

Gravitational Waves and Experimental Tests of General Relativity

A system of moving bodies emits gravitational waves. . . . We shall assume that the speeds of all the system's bodies are small compared to the speed of light [and that the system's gravity is weak]. . . . Because of the presence of matter, the equation for the radiated waves . . . will have the form $\frac{1}{2}\Box\psi_i^k = \kappa\tau_i^k$, . . . where τ_i^k . . . contain, along with the [material stress-energy tensor], also terms of quadratic order from [the Einstein tensor].[1]

LEV LANDAU AND EVGENY LIFSHITZ (1941)

27.1 Overview 27.1

In 1915, when Einstein formulated general relativity, human technology was inadequate for testing it definitively. Only a half century later did technology begin to catch up. In the years since then, the best experiments have improved from accuracies of a few tens of percent to a part in 10,000 or 100,000, and general relativity has passed the tests with flying colors. In Sec. 27.2, we describe some of these tests, derive general relativity's predictions for them, and discuss the experimental results.

Observations of gravitational waves are changing the character of research on general relativity. As of 2016, they have enabled observational studies of the large-amplitude, highly nonlinear vibrations of curved spacetime triggered when two black holes collide. Thereby, they have produced, for the first time, tests of general relativity when gravity is ultra strong and dynamical. They are enabling high-accuracy studies of relativistic effects in inspiraling black-hole binaries and soon should do the same for neutron stars—where they may also teach us about the equation of state of high-density nuclear matter. In the future, they will enable us to map the spacetime geometries of quiescent black holes with high precision. And they might provide a window into the physical conditions present during the first moments of the expansion of the universe (Sec. 28.7.1).

In this chapter, we develop the theory of gravitational waves in much detail and describe efforts to detect them and the sources that may be seen. More specifically, in Sec. 27.3, we develop the mathematics of gravitational waves, both classically and quantum mechanically (in the language of gravitons), and we study their propagation through flat spacetime. Then, in Sec. 27.4, we study their propagation through curved spacetime using the tools of geometric optics. In Sec. 27.5, we develop the simplest approximate method for computing the generation of gravitational waves, the "quadrupole-moment formalism." We also describe and present a few details of

1. By including the quadratic terms from the Einstein tensor, Landau and Lifshitz made their analysis of gravitational-wave generation valid for self-gravitating bodies such as binary stars—a remarkable achievement at this early date in the development of gravitational-wave theory.

BOX 27.1. READERS' GUIDE

- This chapter relies significantly on:
  - Chap. 2 on special relativity;
  - Chap. 24 on the transition from special relativity to general relativity;
  - Chap. 25 on the fundamental concepts of general relativity, especially Sec. 25.9 on weak, relativistic gravitational fields;
  - Chap. 26 on relativistic stars and black holes; and
  - Sec. 7.3 on geometric optics.
- In addition, Sec. 27.2.3 on Fermat's principle and gravitational lenses is closely linked to Sec. 7.3.6 on Fermat's principle, Sec. 7.6 on gravitational lenses, and Sec. 8.6 on diffraction at a caustic.
- Portions of this chapter are a foundation for Chap. 28 on cosmology.

other, more sophisticated and accurate methods based on multipolar expansions, post-Newtonian techniques, and numerical simulations on supercomputers (numerical relativity). In Sec. 27.6, we turn to gravitational-wave detection, focusing especially on the Laser Interferometer Gravitational wave Observatory (LIGO), pulsar timing arrays, and the proposed Laser Interferometer Space Antenna (LISA).

27.2

27.2 Experimental Tests of General Relativity

In this section, we describe briefly some of the most important experimental tests of general relativity. For greater detail and other tests, see Will (1993a, 2014).

27.2.1

27.2.1 Equivalence Principle, Gravitational Redshift, and Global Positioning System

weak equivalence principle (i.e., universality of free fall) and tests of it

A key aspect of the equivalence principle is the prediction that any object, whose size is extremely small compared to the radius of curvature of spacetime and on which no nongravitational forces act, should move on a geodesic. This means, in particular, that its trajectory through spacetime should be independent of its chemical composition. This is called the *weak equivalence principle*, or the *universality of free fall*.

Efforts to test the universality of free fall date back to Galileo's (perhaps apocryphal) experiment of dropping objects from the leaning tower of Pisa. Over the past century, a sequence of ever-improving experiments led by Roland von Eötvös (ca. 1920), Robert Dicke (ca. 1964), Vladimir Braginsky (ca. 1972), and Eric Adelberger (ca. 2008) have led to an accuracy $\Delta a/a < 2 \times 10^{-13}$ for the difference of gravitational acceleration toward the Sun for Earthbound bodies with very different chemical compositions (Schlamminger et al., 2008). A planned atom-interferometer experiment is designed to reach $\Delta a/a \lesssim 1 \times 10^{-15}$ (Biedermann et al., 2015). The MICROSCOPE satellite (Touboul et al., 2017) has achieved this accuracy and the proposed space experiment STEP promises a further improvement of a thousand (Overduin et al., 2012).

tests for self-gravitating bodies

General relativity predicts that bodies with significant self-gravity (even black holes) should also fall, in a nearly homogeneous external gravitational field, with the same acceleration as a body with negligible self-gravity. This prediction, sometimes called the *strong equivalence principle,* has been tested by comparing the gravitational accelerations of Earth and the Moon toward the Sun. Their fractional difference of acceleration (as determined by tracking the relative motions of the Moon and Earth using laser beams fired from Earth, reflected off mirrors that astronauts and cosmonauts have placed on the Moon, and received back at Earth) has been measured to be $\Delta a/a \lesssim 1 \times 10^{-13}$. Since Earth and the Moon have (gravitational potential energy)/(rest-mass energy) $\simeq -4 \times 10^{-10}$ and $\simeq -2 \times 10^{-11}$, respectively, this verifies that gravitational energy falls with the same acceleration as other forms of energy to within about 2.5 parts in 10,000. The pulsar J0337+1715 has a binding energy roughly equal to ten percent of its mass and is found in a triple system (Ransom et al., 2014). The acceleration of the pulsar and its white dwarf companion differs fractionally by less than 3×10^{-6}, which furnishes an even stronger test (Archibald et al., 2018). For further discussion, see Merkowitz (2010) and Will (1993a, 2014).

gravitational redshift, and tests of it

From the equivalence principle one can deduce that, for an emitter and absorber at rest in a Newtonian gravitational field Φ, light (or other electromagnetic waves) must be gravitationally redshifted by an amount $\Delta\lambda/\lambda = \Delta\Phi$, where $\Delta\Phi$ is the difference in Newtonian potential between the locations of the emitter and receiver. (See Ex. 26.4 for a general relativistic derivation when the field is that of a nonspinning, spherical central body with the emitter on the body's surface and the receiver far from the body; see Ex. 24.16 for a derivation when the emitter and receiver are on the floor and ceiling of an Earthbound laboratory.) Relativistic effects produce a correction to this shift of magnitude $\sim(\Delta\Phi)^2$ [cf. Eq. (26.18)], but for experiments performed in the solar system, the currently available precision is too poor to detect this correction, so such experiments test the equivalence principle and not the details of general relativity.

A famous measurement of this gravitational redshift was NASA's 1976 Gravity-Probe-A Project, in which several atomic clocks were flown to a height of about 10,000 km above Earth and were compared with atomic clocks on Earth via radio signals transmitted downward. After correcting for special relativistic effects, the measured gravitational redshift agreed with the prediction to within the experimental accuracy of about 2 parts in 10,000. An update, using satellites orbiting Jupiter, improves the accuracy by a factor of six (Delva et al., 2018).

influence of gravitational redshift on GPS

The Global Positioning System (GPS), by which one can routinely determine one's location on Earth to within an accuracy of about 10 m, is based on signals transmitted from a set of Earth-orbiting satellites. Each satellite's position is encoded on its transmitted signals, together with the time of transmission as measured by atomic clocks onboard the satellite. A person's GPS receiver contains a high-accuracy clock and a computer. It measures the signal arrival time and compares with the encoded transmission time to determine the distance from satellite to receiver. It uses those distances from several satellites, together with the encoded satellite positions, to determine (by triangulation) the receiver's location on Earth.

The transmission times encoded on the signals are corrected for the gravitational redshift before transmission. Without this redshift correction, the satellite clocks would quickly get out of synchronization with all the clocks on the ground, thereby eroding the GPS accuracy; see Ex. 27.1. Thus a clear understanding of general relativity was crucial to the design of GPS!

EXERCISES

Exercise 27.1 *Practice: Gravitational Redshift for GPS*
The GPS satellites are in circular orbits at a height of 20,200 km above Earth's surface, where their orbital period is 12 sidereal hours. If the ticking rates of the clocks on the satellites were not corrected for the gravitational redshift, roughly how long would it take them to accumulate a time shift, relative to clocks on Earth, large enough to degrade the GPS position accuracy by 10 m? by 1 km?

27.2.2

27.2.2 Perihelion Advance of Mercury

It was known at the end of the nineteenth century that the point in Mercury's orbit closest to the Sun, known as its perihelion, advances at a rate of about 575″ per century with respect to the fixed stars, of which about 532″ can be accounted for by Newtonian perturbations due to the other planets. The remaining ~43″ per century was a mystery until Einstein showed that it can be accounted for quantitatively by the general theory of relativity.

More specifically (as is demonstrated in Ex. 27.2), if we idealize the Sun as nonrotating and spherical (so its external gravitational field is Schwarzschild), we ignore the presence of the other planets, and we note that the radius of Mercury's orbit is very large compared to the Sun's mass (in geometrized units), then Mercury's orbit will be very nearly an ellipse; and the ellipse's perihelion will advance, from one orbit to the next, by an angle

predicted perihelion or periastron advance

$$\boxed{\Delta\phi = 6\pi M/p + O(M^2/p^2) \text{ radians.}} \tag{27.1}$$

Here M is the Sun's mass,[2] and p is the ellipse's *semi-latus rectum*, which is related to its semimajor axis a (half its major diameter) and its eccentricity e by $p = a(1 - e^2)$. For the parameters of Mercury's orbit ($M = M_\odot \simeq 1.4766$ km, $a = 5.79089 \times 10^7$ km, $e = 0.205628$), this advance is 0.10352″ per orbit. Since the orbital period is 0.24085 Earth years, this advance corresponds to 42.98″ per century.

The solar oblateness, inferred from helioseismology (Fig. 16.3), contributes about 0.03" per century to the perihelion advance. The frame dragging due to the Sun's

2. The same formula is true in a binary whose two masses are comparable, with M the sum of the masses.

rotational angular momentum is about another order of magnitude smaller. Modern observational data agree with the relativistic 42.98" to within the data's accuracy of about 1 part in 1,000 (Iorio, Ruggiero, and Corda, 2013). Further accuracy is promised by the BeppiColombo mission to Mercury.

measurements of perihelion advance of Mercury's orbit

The advance of periastron has also been measured in several binary pulsars. In the double pulsar, PSR J0737-3039 (see Sec. 27.2.4), the rate is $17° \text{ yr}^{-1}$. In practice this is used to measure the masses of the binary's neutron stars, but it also validates Eq. (27.1) for the advance rate at the same level, $\sim 10^{-3}$, as the Mercury measurement, with the important difference that it is testing the influence of having comparable masses.

measurements of periastron advance in binary pulsars

Gravity in the solar system is weak. Even at Mercury's orbit, the gravitational potential of the Sun is only $|\Phi| \sim 3 \times 10^{-8}$. Therefore, when one expands the spacetime metric in powers of Φ, current experiments with their fractional accuracies $\sim 10^{-5}$ or worse are able to see only the first-order terms beyond Newtonian theory (i.e., terms of *first post-Newtonian order*). To move on to second post-Newtonian order, $O(\Phi^2)$ beyond Newton, in our solar system will require major advances in technology. However, second- and higher-order effects are beginning to be measured via gravitational waves from inspiraling compact binaries, where $|\Phi|$ is $\gtrsim 0.1$.

higher-order post-Newtonian corrections to periastron advance

EXERCISES

Exercise 27.2 *Example: Perihelion Advance*

Consider a small satellite in a noncircular orbit about a spherical body with much larger mass M, for which the external gravitational field is Schwarzschild. The satellite will follow a timelike geodesic. Orient the Schwarzschild coordinates so the satellite's orbit is in the equatorial plane: $\theta = \pi/2$.

(a) Because the metric coefficients are independent of t and ϕ, the satellite's energy-at-infinity $\mathcal{E}_\infty = -p_t$ and angular momentum $L = p_\phi$ must be constants of the satellite's motion (Ex. 25.4a). Show that

$$\mathcal{E}_\infty = \left(1 - \frac{2M}{r}\right)\frac{dt}{d\tau}, \quad L = r^2\frac{d\phi}{d\tau}. \tag{27.2a}$$

See Ex. 26.12. Here and below we take the satellite to have unit mass, so its momentum and 4-velocity are the same, and its affine parameter ζ and proper time τ are the same.

(b) Introduce the coordinate $u = r^{-1}$ and use the normalization of the 4-velocity to derive the following differential equation for the orbit:

$$\left(\frac{du}{d\phi}\right)^2 = \frac{\mathcal{E}_\infty^2}{L^2} - \left(u^2 + \frac{1}{L^2}\right)(1 - 2Mu). \tag{27.2b}$$

(c) Differentiate this equation with respect to ϕ to obtain a second-order differential equation:

$$\frac{d^2u}{d\phi^2} + u - \frac{M}{L^2} = 3Mu^2. \tag{27.2c}$$

By reinstating the constants G, c, and comparing with the Newtonian orbital equation, argue that the right-hand side represents a relativistic perturbation to the Newtonian equation of motion.

(d) Henceforth in this exercise, assume that $r \gg M$ (i.e., $u \ll 1/M$), and solve the orbital equation (27.2c) by perturbation theory. More specifically, at zero order (i.e., setting the right-hand side to zero), show that the Kepler ellipse (Goldstein, Poole, and Safko, 2002, Sec. 3.7),

$$u_K = \left(\frac{M}{L^2}\right)(1 + e\cos\phi), \tag{27.2d}$$

is a solution. Here e (a constant of integration) is the ellipse's eccentricity, and L^2/M is the ellipse's *semi-latus rectum, p*. The orbit has its minimum radius at $\phi = 0$.

(e) By substituting u_K from part (d) into the right-hand side of the relativistic equation of motion (27.2c), show (to first-order in the relativistic perturbation) that in one orbit the angle ϕ at which the satellite is closest to the mass advances by $\Delta\phi \simeq 6\pi M^2/L^2$. [Hint: Try to write the differential equation in the form $d^2u/d\phi^2 + (1+\epsilon)^2 u \simeq \ldots$, where $\epsilon \ll 1$.]

(f) For the planet Mercury, using the parameter values given after Eq. (27.1), deduce that the relativistic contribution to the rate of advance of the perihelion (point of closest approach to the Sun) is 42.98″ per century.

Exercise 27.3 *Example: Gravitational Deflection of Light*

Repeat the analysis of Ex. 27.2 for a photon following a null geodesic. More specifically, do the following.

(a) Show that the photon trajectory $u(\phi)$ (with $u \equiv 1/r$) obeys the differential equation

$$\frac{d^2u}{d\phi^2} + u = 3Mu^2. \tag{27.3}$$

(b) Obtain the zeroth-order solution by ignoring the right-hand side:

$$u = \frac{\sin\phi}{b}, \tag{27.4}$$

where b is an integration constant. Show that this is just a straight line in the asymptotically flat region far from the body, and b is the impact parameter (projected distance of closest approach to the body).

(c) Substitute this solution (27.4) into the right-hand side of Eq. (27.3), and show that the perturbed trajectory satisfies

$$u = \frac{\sin\phi}{b} + \frac{M}{b^2}(1 - \cos\phi)^2. \tag{27.5}$$

(d) Hence show that a ray with impact parameter $b \gg M$ will be deflected through an angle

$$\Delta\phi = \frac{4M}{b} \tag{27.6}$$

[cf. Eq. (7.87) and associated discussion].

27.2.3 Gravitational Deflection of Light, Fermat's Principle, and Gravitational Lenses

27.2.3

Einstein not only explained the anomalous perihelion shift of Mercury. He also predicted (Ex. 27.3) that the null rays along which starlight propagates will be deflected, when passing through the curved spacetime near the Sun, by an angle

gravitational deflection of starlight

$$\boxed{\Delta\phi = 4M/b + \mathrm{O}(M^2/b^2)} \tag{27.7}$$

relative to their trajectories if spacetime were flat. Here M is the Sun's mass, and b is the ray's impact parameter (distance of closest approach to the Sun's center). For comparison, theories that incorporated a Newtonian-like gravitational field into special relativity (Sec 25.1 and Ex. 25.1) predicted no deflection of light rays; the corpuscular theory of light combined with Newtonian gravity predicted half the general relativistic deflection, as did a 1911 principle-of-equivalence argument by Einstein that was ignorant of the curvature of space. The deflection was measured to an accuracy $\sim$20% during the 1919 solar eclipse and agreed with general relativity rather than with the competing theories—a triumph that helped make Einstein famous. Modern experiments, based on the deflection of radio waves from distant quasars, as measured using very long baseline interferometry (interfering the waves arriving at radio telescopes with transcontinental or transworld separations; Sec. 9.3), have achieved accuracies of about 1 part in 10,000, and they agree completely with general relativity. Similar accuracies are now achievable using optical interferometers in space and may soon be achievable via optical interferometry on the ground.

measurements of gravitational light deflection

These accuracies are so great that, when astronomers make maps of the sky using either radio interferometers or optical interferometers, they must now correct for gravitational deflection of the rays not only when the rays pass near the Sun but also for rays coming in from nearly all directions. This correction is not quite as easy as Eq. (27.7) suggests, since that equation is valid only when the telescope is much farther from the Sun than the impact parameter. In the more general case, the correction is more complicated and must include aberration due to the telescope motion as well as the effects of spacetime curvature.

light deflection for large impact parameters: influences on astronomical observations

The gravitational deflection of light rays passing through or near a cluster of galaxies can produce a spectacular array of distorted images of the light source. In Sec. 7.6, we deduced the details of this gravitational lens effect using a model in which we treated spacetime as flat but endowed with a refractive index $\mathfrak{n}(\mathbf{x}) = 1 - 2\Phi(\mathbf{x})$, where $\Phi(\mathbf{x})$ is the Newtonian gravitational potential of the lensing system. This model can also be used to compute light deflection in the solar system. We now derive this model from general relativity.

gravitational lensing

The foundation for this model is the following general relativistic version of Fermat's principle [see Eq. (7.46) for the Newtonian version]: Consider any static spacetime geometry [i.e., one for which we can introduce a coordinate system in which $\partial g_{\alpha\beta}/\partial t = 0$ and $g_{jt} = 0$; so the only nonzero metric coefficients are $g_{00}(x^k)$ and $g_{ij}(x^k)$]. In such a spacetime the time coordinate t is special, since it is tied to the spacetime's temporal symmetry. An example is Schwarzschild spacetime and the Schwarzschild time coordinate t. Now, consider a light ray emitted from a spatial point $x^j = a^j$ in the static spacetime and received at a spatial point $x^j = b^j$. Assuming the spatial path along which the ray travels is $x^j(\eta)$, where η is any parameter with $x^j(0) = a^j$, $x^j(1) = b^j$, then the total coordinate time Δt required for the light's trip from a^j to b^j is

general relativistic Fermat's principle

$$\boxed{\Delta t = \int_0^1 \sqrt{\gamma_{jk}\frac{dx^j}{d\eta}\frac{dx^k}{d\eta}}\,d\eta, \quad \text{where } \gamma_{jk} \equiv \frac{g_{jk}}{-g_{00}}} \tag{27.8}$$

(computed using the fact that the ray must be null so $ds^2 = g_{00}dt^2 + g_{ij}dx^i dx^j = 0$). Fermat's principle says that *the actual spatial trajectory of the light path, in any static spacetime, is one that extremizes this coordinate time lapse.*

This principle can be proved (Ex. 27.4) by showing that the Euler-Lagrange equation for the action (27.8) is equivalent to the geodesic equation for a photon in the static spacetime with metric $g_{\mu\nu}(x^k)$.

Derivation of Index-of-Refraction Model. The index-of-refraction model used to study gravitational lenses in Sec. 7.6 is easily deduced as a special case of this Fermat principle. In a nearly Newtonian situation, the linearized-theory, Lorenz-gauge, trace-reversed metric perturbation has the form (25.91) with only the time-time component being significantly large: $\bar{h}_{00} = -4\Phi$, $\bar{h}_{0j} \simeq 0$, $\bar{h}_{jk} \simeq 0$. Correspondingly, the metric perturbation [obtained by inverting Eq. (25.85)] is $h_{00} = -2\Phi$, $h_{jk} = -2\Phi\delta_{jk}$, and the full spacetime metric $g_{\mu\nu} = \eta_{\mu\nu} + h_{\mu\nu}$ is

$$ds^2 = -(1 + 2\Phi)dt^2 + (1 - 2\Phi)\delta_{jk}dx^j dx^k. \tag{27.9}$$

This is the standard spacetime metric (25.79) in the Newtonian limit with a special choice of spatial coordinates, those of linearized-theory Lorenz gauge. The Newtonian limit includes the slow-motion constraint that time derivatives of the metric are small compared to spatial derivatives [Eq. (25.75b)], so on the timescale for light to travel through a lensing system, the Newtonian potential can be regarded as static: $\Phi = \Phi(x^j)$. Therefore, the Newtonian-limit metric (27.9) can be regarded as static, and the coordinate time lapse along a trajectory between two spatial points, Eq. (27.8), reduces to

Newtonian limit of Fermat's principle

$$\Delta t = \int_0^L (1 - 2\Phi)d\ell, \tag{27.10}$$

where $d\ell = \sqrt{\delta_{jk}dx^j dx^k}$ is distance traveled treating the coordinates as though they were Cartesian in flat space and L is that total distance between the two points.

According to the relativistic Fermat principle (27.8), this Δt is extremal for light rays. However, Eq. (27.10) is also the action for the Newtonian, nongravitational version of Fermat's principle [Eq. (7.47)], with index of refraction

$$\boxed{\mathfrak{n}(x^j) = 1 - 2\Phi(x^j).} \tag{27.11}$$

refractive-index model for computing gravitational lensing when gravity is weak

Therefore, the spatial trajectories of the light rays can be computed via the Newtonian Fermat principle, with the index of refraction (27.11). ■

Although this index-of-refraction model involves treating a special (Lorenz-gauge) coordinate system as though the spatial coordinates were Cartesian and space were flat (so $d\ell^2 = \delta_{jk}dx^j dx^k$)—which does not correspond to reality—nevertheless, this model predicts the correct gravitational lens images. The reason is that it predicts the correct rays through the Lorenz-gauge coordinates, and when the light reaches Earth, the cumulative lensing has become so great that the slight difference in the coordinates here from truly Cartesian has negligible influence on the images one sees.

EXERCISES

Exercise 27.4 *Derivation: Fermat's Principle for a Photon's Path in a Static Spacetime*
Show that the Euler-Lagrange equation for the action principle (27.8) is equivalent to the geodesic equation for a photon in the static spacetime metric $g_{00}(x^k)$, $g_{ij}(x^k)$. Specifically, do the following.

(a) The action (27.8) is the same as that for a geodesic in a 3-dimensional space with metric γ_{jk} and with t playing the role of proper distance traveled [Eq. (25.19) converted to a positive-definite, 3-dimensional metric]. Therefore, the Euler-Lagrange equation for Eq. (27.8) is the geodesic equation in that (fictitious) space [Eq. (25.14) with t the affine parameter]. Using Eq. (24.38c) for the connection coefficients, show that the geodesic equation can be written in the form

$$\gamma_{jk}\frac{d^2x^k}{dt^2} + \frac{1}{2}(\gamma_{jk,l} + \gamma_{jl,k} - \gamma_{kl,j})\frac{dx^k}{dt}\frac{dx^l}{dt} = 0. \tag{27.12a}$$

(b) Take the geodesic equation (25.14) for the light ray in the real spacetime, with spacetime affine parameter ζ, and change parameters to coordinate time t. Thereby obtain

$$g_{jk}\frac{d^2x^k}{dt^2} + \Gamma_{jkl}\frac{dx^k}{dt}\frac{dx^l}{dt} - \Gamma_{j00}\frac{g_{kl}}{g_{00}}\frac{dx^k}{dt}\frac{dx^l}{dt} + \frac{d^2t/d\zeta^2}{(dt/d\zeta)^2}g_{jk}\frac{dx^k}{dt} = 0,$$

$$\frac{d^2t/d\zeta^2}{(dt/d\zeta)^2} + 2\Gamma_{0k0}\frac{dx^k/dt}{g_{00}} = 0. \tag{27.12b}$$

(c) Insert the second of these equations into the first, and write the connection coefficients in terms of derivatives of the spacetime metric. With a little algebra, bring your result into the form Eq. (27.12a) of the Fermat-principle Euler-Lagrange equation.

27.2.4

27.2.4 Shapiro Time Delay

Shapiro time delay

In 1964, Irwin Shapiro proposed a new experiment to test general relativity: monitor the round-trip travel time for radio waves transmitted from Earth and bounced off Venus or some other planet, or transponded by a spacecraft. As the line-of-sight between Earth and the planet or spacecraft gradually moves nearer and then farther from the Sun, the waves' rays will pass through regions of greater and then smaller spacetime curvature, which will influence the round-trip travel time by greater and then smaller amounts. From the time evolution of the round-trip time, one can deduce the changing influence of the Sun's spacetime curvature.

One can compute the round-trip travel time with the aid of Fermat's (geometric-optics) principle. The round-trip proper time, as measured on Earth (neglecting, for simplicity, Earth's orbital motion; i.e., pretending Earth is at rest relative to the Sun while a radio-wave's rays go out and back), is $\Delta\tau_\oplus = \sqrt{1-2M/r_\oplus}\,\Delta t \simeq (1 - M/r_\oplus)\Delta t$, where M is the Sun's mass, $r_\oplus$ is Earth's distance from the Sun's center, Δt is the round-trip coordinate time in the static solar-system coordinates, and we have used $g_{00} = -(1-2M/r_\oplus)$ at Earth. Because Δt obeys Fermat's principle, it is stationary under small perturbations of the ray's spatial trajectory. This allows us to compute it using a straight-line trajectory through the spatial coordinate system. Letting b be the impact parameter (the ray's closest coordinate distance to the Sun) and ℓ be coordinate distance along the straight-line trajectory and neglecting the gravitational fields of the planets, we have $\Phi = -M/\sqrt{\ell^2+b^2}$, so the coordinate time lapse out and back is [Eq. (27.10)]

$$\Delta t = 2\int_{-\sqrt{r_\oplus^2-b^2}}^{\sqrt{r_{\rm refl}^2-b^2}} \left(1 + \frac{2M}{\sqrt{\ell^2+b^2}}\right) d\ell. \tag{27.13}$$

Here $r_{\rm refl}$ is the radius of the location at which the ray is reflected (or transponded) back to Earth. Performing the integral and multiplying by $\sqrt{g_{00}} \simeq 1 - M/r_\oplus$, we obtain for the round-trip travel time measured on Earth:

$$\Delta\tau_\oplus = 2\left(a_\oplus + a_{\rm refl}\right)\left(1 - \frac{M}{r_\oplus}\right) + 4M \ln\left[\frac{(a_\oplus + r_\oplus)(a_{\rm refl} + r_{\rm refl})}{b^2}\right], \tag{27.14}$$

where $a_\oplus = \sqrt{r_\oplus^2 - b^2}$, and $a_{\rm refl} = \sqrt{r_{\rm refl}^2 - b^2}$.

sharply varying term in Shapiro time delay

As Earth and the reflecting planet or transponding spacecraft move along their orbits, only one term in this round-trip time varies sharply: the term

$$\boxed{\Delta\tau_\oplus = 4M\ln(1/b^2) = -8M\ln b \simeq -40\ \mu\text{s}\ \ln b.} \tag{27.15}$$

When the planet or spacecraft passes nearly behind the Sun, as seen from Earth, b plunges to a minimum (on a timescale of hours or days) and then rises back up; correspondingly, the time delay shows a sharp blip. By comparing the observed blip with the theory in a measurement with the Cassini spacecraft, this Shapiro time delay has been verified to the remarkable precision of 2×10^{-5} (Bertotti, Iess, and Tortora, 2003).

measurements of Shapiro time delay in the solar system and in binary pulsars

The Shapiro effect has been seen in several binary pulsar systems. The two best examples are the double pulsar PSR J0737-3039 and PSR J1614-2230, where the lines of sight pass within 1° of the orbital plane, very close to the companions. The peak Shapiro delays are 51 μs and 21 μs, respectively. Remarkably, in the second example, the neutron star has a well-measured mass of 2.0 $M_\odot$, which is large enough to rule out several candidate equations of state for cold nuclear matter (DeMorest et al., 2010) (cf. Sec. 26.3.5).

27.2.5 Geodetic and Lense-Thirring Precession

27.2.5

As we have discussed in Secs. 25.9.3 and 26.5, the mass M and the angular momentum $\mathbf{J}$ of a gravitating body place their imprint on the body's asymptotic spacetime metric:

$$ds^2 = -\left(1 - \frac{2M}{r}\right) dt^2 - \frac{4\epsilon_{jkm} J^k x^m}{r^3} dt\, dx^j + \left[1 + \mathrm{O}\left(\frac{M}{r}\right)\right] \delta_{jk} dx^j dx^k \quad (27.16)$$

[Eq. (25.98c)]. The mass imprint can be deduced from measurements of the orbital angular velocity $\Omega = \sqrt{M/r^3}$ of an object in a circular orbit, and the angular-momentum imprint, from the precession it induces in an orbiting gyroscope [Eq. (25.100)].

geodetic precession of a gyroscope

As we deduced in Ex. 26.19, there are actually two precessions: geodetic precession and Lense-Thirring precession. The geodetic precession, like the orbital angular velocity, arises from the spherically symmetric part of the metric; it says that the spin of a gyroscope in a circular orbit of radius r will precess around the orbit's angular momentum vector at a rate $\frac{3}{2}(M/r)\Omega$. The Lense-Thirring precession is the average of Eq. (25.100) around the orbit, which gives $\mathbf{J}/(2r^3)$ for a polar orbit, $-\mathbf{J}/r^3$ for an equatorial orbit, and $(-J_i + \frac{3}{2}P_{ij}J_j)/r^3$ for a general circular orbit, where $\mathbf{J}$ is the central body's angular momentum, r is the orbital radius, and P_{ij} projects into the plane of the orbit.

Lense-Thirring precession of a gyroscope

measurements of precessions by GP-B and LAGEOS

The earth-orbiting experiment Gravity Probe B (GP-B), led by Francis Everitt, has measured these two precessions (Everitt et al., 2011). GP-B comprises four spinning spheres (gyroscopes) in one satellite on a polar orbit, and has verified the geodetic precession of $6.6''\ \mathrm{yr}^{-1}$ to a fractional accuracy of 0.003, and the Lense-Thirring precession of $0.040''\ \mathrm{yr}^{-1}$ to fractional accuracy 0.2. More recently, a combination of three satellites, LAGEOS, LAGEOS2, and LARES, has been used to measure the Lense-Thirring precession of an inclined equatorial orbit (rather than a gyroscope in orbit), where the prediction is $J/r^3 = 0.080''\ \mathrm{yr}^{-1}$. A fractional accuracy of 0.02 has been reported (Ciufolini et al., 2019).

Relativistic precession has also been observed in six binary pulsar systems. Here we must take account of the pulsar (p) and its compact companion (c). We can give a heuristic argument for the precession rate. When $m_p \ll m_c$, the dominant precession is geodetic in the gravitational field of the companion at a rate $\Omega_p = \frac{3}{2}(m_c/r)\Omega$. In the opposite limit, the companion undergoes orbital Lense-Thirring precession at a rate $\Omega_p = 2J_p/r^3$ and so the torque is $(2m_c\vec{J}_p/r) \times \vec{\Omega}$, and the reflex precession of the pulsar, about the total (orbital plus spin) angular momentum, will be at a rate $\Omega_p = (2m_c/r)\Omega$. The simplest interpolation between these two limiting cases is

precession in binary pulsars

$$\Omega_p = \Omega \frac{m_c}{r} \frac{(3m_c + 4m_p)}{2(m_c + m_p)}. \quad (27.17)$$

This turns out to be correct. If the orbit is elliptical, we (again) replace the radius r with the *semi-latus rectum* p.

The best measured example is the double pulsar PSR J0737-3039, where pulses from both pulsars were observed. One pulsar has a spin almost aligned with the orbital angular momentum, and so its precession is hard to measure. The other has an inclined spin, which has now precessed so far that the pulses are no longer detected. However, this precession changes the projected shape of the magnetosphere, which occults the pulses from the nonprecessing pulsar, allowing a measurement of the rate of precession of $4.7 \pm 0.6°\ \mathrm{yr}^{-1}$, which is consistent with the predicted value of $5.1°\ \mathrm{yr}^{-1}$ (Kramer et al., 2006).

measurements of precession in double pulsars

27.2.6

27.2.6 Gravitational Radiation Reaction

Radio observations of binary pulsars have already provided several indirect detections of gravitational waves, via radiation reaction in the binary.

Hulse-Taylor pulsar

The first binary pulsar (PSR B1913+16) was discovered in 1974 by Russell Hulse and Joseph Taylor. One star is a pulsar that emits radio pulses with a period of $\sim$59 ms at predictable times (allowing for the slowing down of the pulsar), and their arrival times can be determined to $\sim 15\ \mu$s. Its companion is almost certainly a neutron star, but it does not pulse. The orbital period is roughly 8 hours, and the orbit's eccentricity is $\sim$0.6. The pulses are received at Earth with Shapiro time delays due to crossing the binary orbit, and with other relativistic effects.

We do not know a priori the orbital inclination or the neutron-star masses. However, we obtain one relation between these three quantities by analyzing the Newtonian orbit. A second relation comes from measuring the consequences of the combined second-order Doppler shift and gravitational redshift as the pulsar moves in and out of its companion's gravitational field. A third relation comes from measuring the relativistic precession of the orbit's periastron. From these three relations, one can solve for the stars' masses and the orbital inclination, and as a check can verify that the Shapiro time delay comes out correctly. One can then use the system's parameters to predict the rate of orbital inspiral due to gravitational radiation reaction—a phenomenon with a magnitude of $\sim|\Phi|^{2.5}$ beyond Newton (i.e., 2.5 post-Newtonian order; Sec. 27.5.3). The prediction agrees with the measurements to an accuracy of $\sim 2 \times 10^{-3}$ (Weissberg, Nice, and Taylor, 2010)—a major triumph for general relativity! The agreement is even better, $<10^{-3}$, for the double pulsar PSR J0737-3039.

measurement of gravitational radiation reaction in binary pulsars

In LIGO's observations of gravitational waves from the inspiral of binary black holes, the radiation reaction is measured with lower accuracy than this, but because the binary is so compact—with M/a increasing from ~ 0.01 to ~ 0.3—higher-order corrections to the radiation reaction are readily measured.

For reviews of other tests of general relativity using binary pulsars, see Kaspi and Kramer (2016) and papers cited therein.

27.3 Gravitational Waves Propagating through Flat Spacetime

27.3

nature of gravitational waves

Gravitational waves are ripples in the curvature of spacetime that are emitted by violent astrophysical events and that propagate with the speed of light. It was clear to Einstein and others, even before general relativity was fully formulated, that his theory would have to predict gravitational waves, and within months after completing the theory, Einstein (1916b, 1918) worked out those waves' basic properties.

penetrating power of gravitational waves

It turns out that, after they have been emitted, gravitational waves propagate through matter with near impunity: they propagate as though in vacuum, even when other matter and fields are present. (For a proof and discussion see, e.g., Thorne, 1983, Sec. 2.4.3.) This justifies specializing our analysis to vacuum propagation.

27.3.1 Weak, Plane Waves in Linearized Theory

27.3.1

Once the waves are far from their source, the radii of curvature of their phase fronts are huge compared to a wavelength, as is the radius of curvature of the spacetime through which they propagate. Thus to high accuracy, we can idealize the waves as plane-fronted and as propagating through flat spacetime. The appropriate formalism for describing this is the linearized theory developed in Sec. 25.9.2.

linearized theory of gravitational-wave propagation:

We introduce coordinates that are as nearly Lorentz as possible, so the spacetime metric can be written as

metric perturbation

$$\boxed{g_{\alpha\beta} = \eta_{\alpha\beta} + h_{\alpha\beta}, \quad \text{with } |h_{\alpha\beta}| \ll 1} \tag{27.18a}$$

[Eq. (25.82)], and we call $h_{\alpha\beta}$ the waves' *metric perturbation*. We perform an "infinitesimal coordinate transformation" (gauge change):

$$x^\alpha_{\text{new}}(\mathcal{P}) = x^\alpha_{\text{old}}(\mathcal{P}) + \xi^\alpha(\mathcal{P}), \quad \text{which produces } h^{\text{new}}_{\mu\nu} = h^{\text{old}}_{\mu\nu} - \xi_{\mu,\nu} - \xi_{\nu,\mu} \tag{27.18b}$$

impose Lorenz gauge

[Eqs. (25.87) and (25.88)], with the gauge-change generators $\xi^\alpha(\mathcal{P})$ chosen to impose the Lorenz gauge condition

$$\boxed{\bar{h}_{\mu\nu}{}^{,\nu} = 0} \tag{27.18c}$$

trace-reversed metric perturbation

[Eq. (25.89)] on the new trace-reversed metric perturbation:

$$\boxed{\bar{h}_{\mu\nu} \equiv h_{\mu\nu} - \frac{1}{2} h\, \eta_{\mu\nu}, \quad h \equiv \eta^{\alpha\beta} h_{\alpha\beta}} \tag{27.18d}$$

[Eqs. (25.85) and (25.84)]. In this Lorenz gauge, the vacuum Einstein field equation becomes the flat-space wave equation for $\bar{h}^{\mu\nu}$ and so also for $h^{\mu\nu}$:

wave equation

$$\boxed{\bar{h}_{\mu\nu,\alpha}{}^{\alpha} = h_{\mu\nu,\alpha}{}^{\alpha} = 0} \tag{27.18e}$$

[Eq. (25.90)]. Here all indices after a comma are partial derivatives and they are raised with the flat metric $h_{\mu\nu,\alpha}{}^{\alpha} = h_{\mu\nu,\alpha\beta}\eta^{\beta\alpha}$.

This is as far as we went in vacuum (far from the waves' source) in Chap. 25. We now go further. We simplify the mathematics by orienting the axes of our nearly Lorentz coordinates so the waves are planar and propagate in the z direction. Then the obvious solution to the wave equation (27.18e) and the consequence of the Lorenz gauge condition (27.18c) are

$$\bar{h}_{\mu\nu} = \bar{h}_{\mu\nu}(t - z), \quad \bar{h}_{\mu 0} = -\bar{h}_{\mu z}. \tag{27.19}$$

There are now six independent components of the trace-reversed metric perturbation: the six spatial $\bar{h}_{ij}$; the second of Eqs. (27.19) fixes the time-space and time-time components in terms of them.

further specialization to TT gauge

Remarkably, these six independent components can be reduced to two by a further specialization of gauge. The original infinitesimal coordinate transformation (27.18b), which brought us into Lorenz gauge, relied on four functions $\xi_\mu(\mathcal{P}) = \xi_\mu(x^\alpha)$ of four spacetime coordinates. A more restricted set of gauge-change generators, $\xi_\mu(t - z)$, that are functions solely of retarded time (and thus satisfy the wave equation) will keep us in Lorenz gauge and can be used to annul the four components $\bar{h}_{xz}$, $\bar{h}_{yz}$, $\bar{h}_{zz}$, and $\bar{h} \equiv \eta^{\mu\nu}\bar{h}_{\mu\nu}$, whence (thanks to the Lorenz conditions $\bar{h}_{\mu 0} = -\bar{h}_{\mu z}$) all the $\bar{h}_{\mu 0}$ are also annulled. See Ex. 27.5. As a result, the trace-reversed metric perturbation $\bar{h}_{\mu\nu}$ and the metric perturbation $h_{\mu\nu}$ are now equal, and their only nonzero components are $h_{xx} = -h_{yy}$ and $h_{xy} = +h_{yx}$.

for locally plane waves propagating in z direction

This special new gauge has the name *transverse-traceless gauge* or *TT gauge*, because in it the metric perturbation is purely spatial, it is transverse to the waves' propagation direction (the z direction), and it is traceless. It is convenient to use the notation $h^{\rm TT}_{\mu\nu}$ for the metric perturbation in this TT gauge, and convenient to give the names h_+ and $h_\times$ to its two independent, nonzero components (which are associated with two polarization states for the waves, "+" and "×"):

the metric perturbation in TT gauge; gravitational-wave fields h_+ and $h_\times$

$$\boxed{h^{\rm TT}_{xx} = -h^{\rm TT}_{yy} = h_+(t - z), \quad h^{\rm TT}_{xy} = +h^{\rm TT}_{yx} = h_\times(t - z).} \tag{27.20}$$

The Riemann curvature tensor in this TT gauge, as in any gauge, can be expressed as

Riemann tensor

$$\boxed{R_{\alpha\beta\gamma\delta} = \frac{1}{2} h^{\rm TT}_{\{\alpha\beta,\gamma\delta\}} \equiv \frac{1}{2}(h^{\rm TT}_{\alpha\delta,\beta\gamma} + h^{\rm TT}_{\beta\gamma,\alpha\delta} - h^{\rm TT}_{\alpha\gamma,\beta\delta} - h^{\rm TT}_{\beta\delta,\alpha\gamma})} \tag{27.21}$$

[Eq. (25.80)]. Here the subscript symbol $\{\cdot\}$, analogous to $[\cdot]$ for antisymmetrization and $(\cdot)$ for symmetrization, means the combination of four terms on the right side of the $\equiv$ sign. Of particular interest for physical measurements is the relativistic tidal field $\mathcal{E}_{ij} = R_{i0j0}$, which produces a relative acceleration of freely falling particles [geodesic deviation; Eq. (25.34)]. Since the temporal components of $h^{\rm TT}_{\mu\nu}$ vanish, the only nonzero term in Eq. (27.21) for R_{i0j0} is the third one, in which the temporal components are derivatives, whence

$$\boxed{\mathcal{E}_{ij} = R_{i0j0} = -\frac{1}{2}\ddot{h}^{\rm TT}_{ij};} \quad \text{or}$$

tidal fields

$$\boxed{\mathcal{E}_{xx} = -\mathcal{E}_{yy} = -\frac{1}{2}\ddot{h}_+(t-z), \quad \mathcal{E}_{xy} = +\mathcal{E}_{yx} = -\frac{1}{2}\ddot{h}_\times(t-z).} \tag{27.22}$$

Here the dots mean time derivatives: $\ddot{h}_+(t-x) \equiv \partial^2 h_+/\partial t^2$. A useful index-free way to write these equations is

$$\boxed{\boldsymbol{\mathcal{E}} = -\frac{1}{2}\ddot{\mathbf{h}}^{\rm TT} = -\frac{1}{2}\ddot{h}_+\mathbf{e}^+ - \frac{1}{2}\ddot{h}_\times\mathbf{e}^\times,} \tag{27.23a}$$

where

$$\boxed{\mathbf{e}^+ = \vec{e}_x \otimes \vec{e}_x - \vec{e}_y \otimes \vec{e}_y,} \quad \boxed{\mathbf{e}^\times = \vec{e}_x \otimes \vec{e}_y + \vec{e}_y \otimes \vec{e}_x} \tag{27.23b}$$

polarization tensors

are the *polarization tensors* associated with the $+$ and $\times$ polarizations, respectively.

gauge invariance of Riemann tensor

It is a very important fact that the Riemann curvature tensor is gauge invariant. An infinitesimal coordinate transformation $x^\alpha_{\rm new}(\mathcal{P}) = x^\alpha_{\rm old}(\mathcal{P}) + \xi^\alpha(\mathcal{P})$ changes it by tiny fractional amounts of order ξ^α, by contrast with the metric perturbation, which is changed by amounts of order itself: $\delta h_{\mu\nu} = -2\xi_{(\mu,\nu)}$ (i.e., by fractional amounts of order unity; Ex. 25.19). This has two important consequences. (i) The gauge-invariant Riemann tensor (or its space-time-space-time part, the tidal field) is an optimal tool for discussing physical measurements (Sec. 27.3.2)—a much better tool than, for example, the gauge-dependent metric perturbation. (ii) The gauge invariance of Riemann motivates us to change our viewpoint on $h^{\rm TT}_{ij}$ in the following way.

alternative viewpoint: gravitational wave field defined in terms of Riemann tensor's tidal field

We define a dimensionless "gravitational-wave field" $h^{\rm TT}_{ij}$ to be minus twice the double time integral of the wave's tidal field:

$$\boxed{h^{\rm TT}_{ij} \equiv -2\int dt \int dt\, \mathcal{E}_{ij}.} \tag{27.24}$$

And we regard the computation that led to Eq. (27.22) as a demonstration that it is possible to find a gauge in which the metric perturbation is purely spatial and its spatial part is equal to this gravitational-wave field: $h_{0\mu} = 0$ and $h_{ij} = h^{\rm TT}_{ij}$.

In Box 27.2, we show that, if we have found a gauge in which the metric perturbation propagates as a plane wave at the speed of light, then we can compute the gravitational-wave field $h^{\rm TT}_{ij}$ from that gauge's $h_{\alpha\beta}$ or $\bar{h}_{\alpha\beta}$ by a simple projection process. This result is useful in the theory of gravitational-wave generation (see, e.g., Sec. 27.5.2).

BOX 27.2. PROJECTING OUT THE GRAVITATIONAL-WAVE FIELD h_{ij}^{TT}

Suppose that, for some gravitational wave, we have found a gauge (not necessarily TT) in which $h_{\mu\nu} = h_{\mu\nu}(t - z)$. Then a simple calculation with Eq. (25.80) reveals that the only nonzero components of this wave's tidal field are $\mathcal{E}_{ab} = -\frac{1}{2}\ddot{h}_{ab}$, where a and b run over x and y. But by definition, $\mathcal{E}_{ab} = -\frac{1}{2}\ddot{h}_{ab}^{\mathrm{TT}}$. Therefore, in this gauge we can compute the gravitational-wave field by simply throwing away all parts of $h_{\mu\nu}$ except the spatial, transverse parts: $h_{xx}^{\mathrm{TT}} = h_{xx}$, $h_{xy}^{\mathrm{TT}} = h_{xy}$, and $h_{yy}^{\mathrm{TT}} = h_{yy}$.

When computing the generation of gravitational waves, it is often easier to evaluate the trace-reversed metric perturbation $\bar{h}_{\alpha\beta}$ than the metric perturbation itself [e.g., Eq. (25.91)]. But $\bar{h}_{\alpha\beta}$ differs from $h_{\alpha\beta}$ by only a trace, and the gravitational-wave field h_{jk}^{TT} is traceless. Therefore, in any gauge where $\bar{h}_{\mu\nu} = \bar{h}_{\mu\nu}(t - z)$, we can compute the gravitational-wave field h_{jk}^{TT} from $\bar{h}_{\mu\nu}$ by throwing away everything except its spatial, transverse part, and by then removing its trace (i.e., by projecting out the spatial, transverse, traceless part):

$$\begin{aligned} &h_{jk}^{\mathrm{TT}} = \left(\bar{h}_{jk}\right)^{\mathrm{TT}}; \quad \text{or} \\ &h_+ = h_{xx}^{\mathrm{TT}} = \bar{h}_{xx} - \tfrac{1}{2}(\bar{h}_{xx} + \bar{h}_{yy}) = \tfrac{1}{2}(h_{xx} - h_{yy}), \\ &\quad h_\times = h_{xy}^{\mathrm{TT}} = \bar{h}_{xy}. \end{aligned} \tag{1}$$

Here the symbol $\left(\bar{h}_{jk}\right)^{\mathrm{TT}}$ means "project out the spatial, transverse, traceless part."

If we rotate the spatial axes so the waves propagate along the unit spatial vector $\mathbf{n}$ instead of along $\mathbf{e}_z$, then the "speed-of-light-propagation" forms of the metric perturbation and its trace reversal become $h_{\alpha\beta} = h_{\alpha\beta}(t - \mathbf{n}\cdot\mathbf{x})$ and $\bar{h}_{\alpha\beta} = \bar{h}_{\alpha\beta}(t - \mathbf{n}\cdot\mathbf{x})$, respectively, and the TT projection can be achieved with the aid of the transverse projection tensor:

$$P^{jk} \equiv \delta^{jk} - n^j n^k. \tag{2}$$

Specifically, we have

$$h_{jk}^{\mathrm{TT}} = (\bar{h}_{jk})^{\mathrm{TT}} = P_j{}^l P_k{}^m \bar{h}_{lm} - \frac{1}{2} P_{jk} P^{lm} \bar{h}_{lm}. \tag{3}$$

Here the notation is that of Cartesian coordinates with $P_j{}^k = P^{jk} = P_{jk}$.

EXERCISES

Exercise 27.5 *Derivation: Bringing $h_{\mu\nu}$ into TT Gauge*

Consider a weak, planar gravitational wave propagating in the z direction, written in a general Lorenz gauge [Eqs. (27.19)]. Show that by appropriate choices of new gauge-change generators that have the plane-wave form $\xi_\mu(t-z)$, one can (i) keep the metric perturbation in Lorenz gauge; (ii) annul $\bar{h}_{xz}$, $\bar{h}_{yz}$, $\bar{h}_{zz}$, and $\bar{h} \equiv \eta^{\mu\nu}\bar{h}_{\mu\nu}$; and (iii) thereby make the only nonzero components of the metric perturbation be $h_{xx} = -h_{yy}$ and $h_{xy} = +h_{yx}$. [Hint: Show that a gauge change (27.18b) produces $\bar{h}^{\rm new}_{\mu\nu} = \bar{h}^{\rm old}_{\mu\nu} - \xi_{\mu,\nu} - \xi_{\nu,\mu} + \eta_{\mu\nu}\xi_\alpha{}^{,\alpha}$, and use this in your computations.]

27.3.2 Measuring a Gravitational Wave by Its Tidal Forces

27.3.2

We seek physical insight into gravitational waves by studying the following idealized problem. Consider a cloud of test particles that floats freely in space and is static and spherical before the waves pass. Study the wave-induced deformations of the cloud as viewed in the nearest thing there is to a rigid, orthonormal coordinate system: the local Lorentz frame (in the physical spacetime) of a "fiducial particle" that sits at the cloud's center. In that frame the displacement vector ζ^j between the fiducial particle and some other particle has components $\zeta^j = x^j + \delta x^j$, where x^j is the other particle's spatial coordinate before the waves pass, and δx^j is its coordinate displacement produced by the waves. By inserting this into the local-Lorentz-frame variant of the equation of geodesic deviation [Eq. (25.34)], and neglecting the tiny δx^k compared to x^k on the right-hand side, we obtain

gravitational-wave tidal acceleration

$$\frac{d^2\delta x^j}{dt^2} = -R_{j0k0}x^k = -\mathcal{E}_{jk}x^k = \frac{1}{2}\ddot{h}^{\rm TT}_{jk}x^k, \qquad (27.25)$$

which can be integrated twice to give

gravitational-wave tidal displacement

$$\boxed{\delta x^j = \frac{1}{2}h^{\rm TT}_{jk}x^k.} \qquad (27.26)$$

Expression (27.25) is the *gravitational-wave tidal acceleration* that moves the particles back and forth relative to one another. It is completely analogous to the Newtonian tidal acceleration, $-\mathcal{E}_{jk}x^k = -(\partial^2\Phi/\partial x^j\partial x^k)x^k$, by which the Moon raises tides on Earth's oceans (Sec. 25.5.1).

Now specialize to a wave with + polarization (for which $h_\times = 0$). By inserting expression (27.20) into (27.26), we obtain

tidal displacements produced by h_+

$$\delta x = \frac{1}{2}h_+ x, \quad \delta y = -\frac{1}{2}h_+ y, \quad \delta z = 0. \qquad (27.27)$$

This displacement is shown in Fig. 27.1a,b. Notice that as the gravitational-wave field h_+ oscillates at the spherical cloud's location, the cloud is left undisturbed in the z-direction (propagation direction), and in transverse planes it gets deformed into an ellipse elongated first along the x-axis (when $h_+ > 0$), then along the y-axis (when

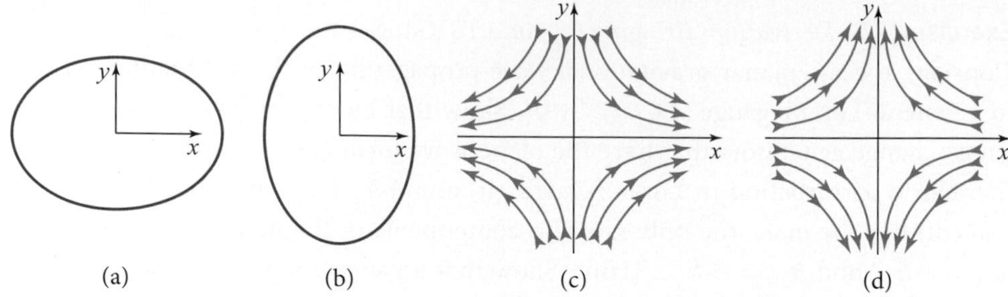

FIGURE 27.1 Physical manifestations, in a particle's local Lorentz frame, of h_+ gravitational waves. (a) Transverse deformation of an initially spherical cloud of test particles in a transeverse plane at a phase of the wave when $h_+ > 0$. (b) Deformation of the cloud when $h_+ < 0$. (c) Field lines representing the acceleration field $\delta\ddot{\mathbf{x}}$ that produces the cloud's deformation, at a phase when $\ddot{h}_+ > 0$. (d) Acceleration field lines when $\ddot{h}_+ < 0$.

$h_+ < 0$). Because $\mathcal{E}_{xx} = -\mathcal{E}_{yy}$ (i.e., because $\mathcal{E}_{jk}$ is traceless), the ellipse is squashed along one axis by the same amount as it is stretched along the other (i.e., the area of the ellipse is preserved during the oscillations).

h_+ tidal acceleration described by lines of force

The effects of the h_+ polarization state can also be described in terms of the *tidal acceleration field* that it produces in the central particle's local Lorentz frame:

$$\frac{d^2}{dt^2}\delta\mathbf{x} = -\boldsymbol{\mathcal{E}}_+ \cdot \mathbf{x} = \frac{1}{2}\ddot{h}_+(x\mathbf{e}_x - y\mathbf{e}_y), \tag{27.28}$$

where $\ddot{h}_+ \equiv \partial^2 h_+/\partial t^2$. Notice that this acceleration vector field $\delta\ddot{\mathbf{x}}$ is divergence free. Because it is divergence free, it can be represented by lines of force analogous to electric field lines, which point along the field and have a density of lines proportional to the magnitude of the field. When this is done, the field lines never end. Figure 27.1c,d shows this acceleration field at the phases of oscillation when $\ddot{h}_+$ is positive and when it is negative. Notice that the field is quadrupolar in shape, with a field strength (density of lines) that increases linearly with distance from the origin of the local Lorentz frame. The elliptical deformations of the spherical cloud of test particles shown in Fig. 27.1a,b are the responses of that cloud to this quadrupolar acceleration field. The polarization state that produces these accelerations and deformations is called the + state because of the orientation of the axes of the quadrupolar acceleration field (Fig. 27.1c,d).

tidal effects of $h_\times$

Next consider the × polarization state. In this state the deformations of the initially spherical cloud are described by

$$\delta x = \frac{1}{2}h_\times y, \quad \delta y = \frac{1}{2}h_\times x, \quad \delta z = 0. \tag{27.29}$$

These deformations, like those for the + state, are purely transverse; they are depicted in Fig. 27.2a,b. The acceleration field that produces these deformations is

$$\frac{d^2}{dt^2}\delta\mathbf{x} = -\boldsymbol{\mathcal{E}}_\times \cdot \mathbf{x} = \frac{1}{2}\ddot{h}_\times(y\mathbf{e}_x + x\mathbf{e}_y). \tag{27.30}$$

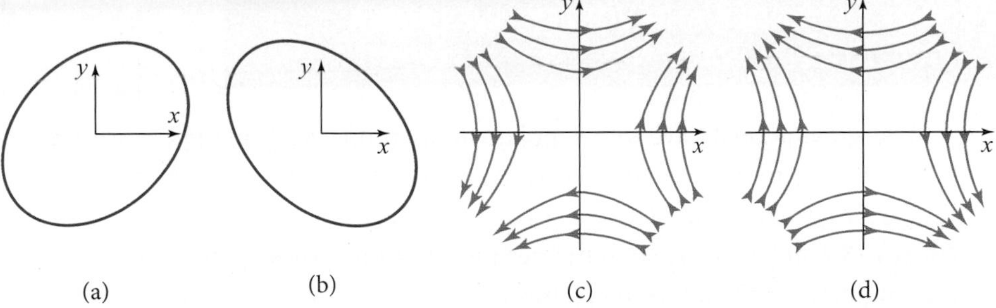

FIGURE 27.2 Physical manifestations, in a particle's local Lorentz frame, of $h_\times$ gravitational waves. (a) Deformation of an initially spherical cloud of test particles in a transverse plane at a phase of the wave when $h_\times > 0$. (b) Deformation of the sphere when $h_\times < 0$. (c) Field lines representing the acceleration field $\delta\ddot{\mathbf{x}}$ that produces the sphere's deformation, at a phase of the wave when $\ddot{h}_\times > 0$. (d) Acceleration field lines when $\ddot{h}_\times < 0$.

This acceleration field, like the one for the + polarization state, is divergence free and quadrupolar; the field lines describing it are depicted in Fig. 27.2c,d. The name "× polarization state" comes from the orientation of the axes of this quadrupolar acceleration field.

Planar gravitational waves can also be depicted in terms of the tendex and vortex lines associated with their tidal tensor field $\boldsymbol{\mathcal{E}}$ and frame-drag tensor field $\boldsymbol{\mathcal{B}}$; see Box 27.3.

When defining the gravitational-wave fields h_+ and $h_\times$, we have relied on a choice of (local Lorentz) reference frame (i.e., a choice of local Lorentz basis vectors $\vec{e}_\alpha$). Exercise 27.6 explores how these fields change when the basis is changed. The conclusions are simple. (i) When one rotates the transverse basis vectors $\vec{e}_x$ and $\vec{e}_y$ through an angle ψ, then h_+ and $h_\times$ rotate through 2ψ in the sense that:

behavior of gravitational wave fields under rotations and boosts

$$(h_+ + ih_\times)_{\text{new}} = (h_+ + ih_\times)_{\text{old}} e^{2i\psi}, \quad \text{when } (\vec{e}_x + i\vec{e}_y)_{\text{new}} = (\vec{e}_x + i\vec{e}_y)_{\text{old}} e^{i\psi}. \tag{27.31}$$

(ii) When one boosts from an old frame to a new one moving at some other speed but chooses the old and new spatial bases such that (a) the waves propagate in the z direction in both frames and (b) the plane spanned by $\vec{e}_x$ and $\vec{\kappa} \equiv \vec{e}_0 + \vec{e}_z =$ (propagation direction in spacetime) is the same in both frames, then h_+ and $h_\times$ are the same in the two frames—they are scalars under such a boost! The same is true of the transverse components of the vector potential $\mathbf{A}$ for an electromagnetic wave.

EXERCISES

Exercise 27.6 *Derivation: Behavior of h_+ and $h_\times$ under Rotations and Boosts*

(a) Derive the behavior [Eq. (27.31)] of h_+ and $h_\times$ under rotations in the transverse plane. [Hint: Write the gravitational-wave field, viewed as a geometric object, as $\boldsymbol{h}^{\text{TT}} = \Re\left[(h_+ + ih_\times)(\boldsymbol{e}^+ - i\,\boldsymbol{e}^\times)\right]$, where $\boldsymbol{e}^+ = (\vec{e}_x \otimes \vec{e}_x - \vec{e}_y \otimes \vec{e}_y)$ and

BOX 27.3. TENDEX AND VORTEX LINES FOR A GRAVITATIONAL WAVE T2

A plane gravitational wave with + polarization, propagating in the z direction, has as its only nonzero tidal-field components $\mathcal{E}_{xx} = -\mathcal{E}_{yy} = -\frac{1}{2}\ddot{h}_+(t - z)$ [Eq. (27.22)]. This tidal field's eigenvectors are $\mathbf{e}_x$ and $\mathbf{e}_y$, so its tendex lines (Boxes 25.2 and 26.3) are straight lines pointing along these basis vectors (i.e., the solid lines in the following picture).

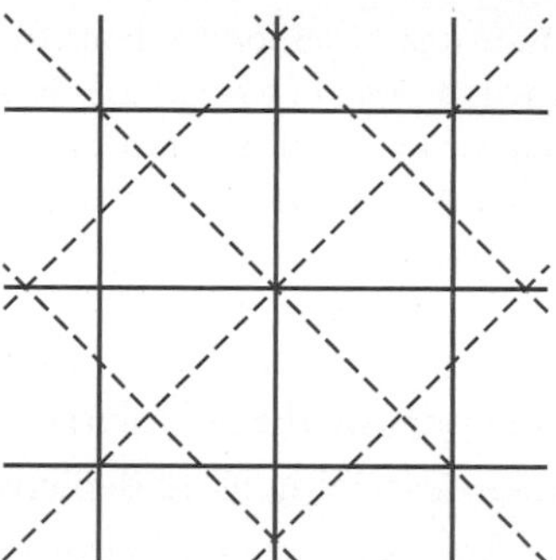

These lines' tendicities $\mathcal{E}_{xx}$ and $\mathcal{E}_{yy}$ are equal and opposite, so one set of lines stretches (red) and the other squeezes (blue). As the wave propagates, each line's tendicity oscillates as seen at fixed z, so its color oscillates between red and blue.

From the Maxwell-like Bianchi identity $\partial\boldsymbol{\mathcal{B}}/\partial t = -(\nabla \times \boldsymbol{\mathcal{E}})^{\mathrm{S}}$ (Box 25.2)—with $\boldsymbol{\mathcal{E}}$ a function of $t - \mathbf{n} \cdot \mathbf{x}$, and $\mathbf{n} = \mathbf{e}_z$ the wave's propagation direction—we infer that the wave's frame-drag field and tidal field are related by $\boldsymbol{\mathcal{B}} = (\mathbf{n} \times \boldsymbol{\mathcal{E}})^{\mathrm{S}}$. This means that the nonzero components of $\boldsymbol{\mathcal{B}}$ are $\mathcal{B}_{xy} = \mathcal{B}_{yx} = \mathcal{E}_{xx} = -\mathcal{E}_{yy} = -\frac{1}{2}\ddot{h}_+(t - z)$. Therefore, the gravitational wave's vortex lines are the dashed lines in the figure above (where the propagation direction, $\mathbf{n} = \mathbf{e}_z$, is out of the screen or paper, toward you).

Electric and magnetic field lines are generally drawn with line densities proportional to the magnitude of the field—a convention motivated by flux conservation. Not so for tendex and vortex lines, which have no corresponding conservation law. Instead, their field strengths (tendicities and vorticities) are usually indicated by color coding (see, e.g., Nichols et al., 2011).

Most discussions of gravitational waves (including the text of this chapter) focus on their tidal field $\boldsymbol{\mathcal{E}}$ and its physical stretch and squeeze; they ignore the frame-drag field with its differential precession (twisting) of gyroscopes. The reason is that modern technology is able to detect and monitor the stretch and squeeze, but the precession is far too small to be detected.

$\mathbf{e}^{\times} = (\vec{e}_x \otimes \vec{e}_y + \vec{e}_y \otimes \vec{e}_x)$ are the polarization tensors associated with + and × polarized waves, respectively [Eqs. (27.23b)]. Then show that $\mathbf{e}_+ - i\mathbf{e}_\times$ rotates through -2ψ, and use this to infer the desired result.]

(b) Show that, with the orientations of spatial basis vectors described after Eq. (27.31), h_+ and $h_\times$ are unchanged by boosts.

27.3.3 Gravitons and Their Spin and Rest Mass

27.3.3

Most of the abovementioned features of gravitational waves (though not expressed in this language) were clear to Einstein in 1918. Two decades later, as part of the effort to understand quantum fields, Markus Fierz and Wolfgang Pauli (1939), at the Eidgenössische Technische Hochschule (ETH) in Zurich, Switzerland, formulated a classical theory of linear fields of arbitrary spin so designed that the fields would be quantizable by canonical methods. Remarkably, their canonical theory for a field of spin two and zero rest mass is identical to general relativity with nonlinear effects removed, and the plane waves of that spin-two theory are identical to the waves described above. When quantized by canonical techniques, these waves are carried by zero-rest-mass, spin-two gravitons.

quantization of gravitational waves: gravitons with zero rest mass and spin two

One can see by simple arguments that the gravitons that carry gravitational waves must have zero rest mass and spin two. First, fundamental principles of quantum theory guarantee that any wave that propagates in vacuum with the speed of light must be carried by particles that have that same speed (i.e., particles whose 4-momenta are null, which means particles with zero rest mass). General relativity predicts that gravitational waves propagate with the speed of light. Therefore, its gravitons must have zero rest mass.

elementary explanations of the graviton rest mass and spin

Second, consider any plane-wave field (neutrino, electromagnetic, gravitational, etc.) that propagates at the speed of light in the z-direction of a (local) Lorentz frame. At any moment of time examine any physical manifestation of that field (e.g., the acceleration field it produces on test particles). Rotate that manifestation of the field around the z-axis, and ask what minimum angle of rotation is required to bring the field back to its original configuration. Call that minimum angle, $\theta_{\rm ret}$, the waves' *return angle*. The spin S of the particles that carry the wave will necessarily be related to that return angle by[3]

$$\boxed{S = \frac{360°}{\theta_{\rm ret}}.} \qquad (27.32)$$

This simple formula corresponds to the elegant mathematical statement that "the waves generate an irreducible representation of order $S = 360°/\theta_{\rm ret}$ of that subgroup of the Lorentz group that leaves their propagation vector unchanged (the 'Little group'

3. For spin 0 this formula fails. Spin 0 corresponds to circular symmetry around the spin axis.

of the propagation vector)." For electromagnetic waves, a physical manifestation is the electric field, which is described by a vector lying in the x-y plane; if one rotates that vector about the z-axis (propagation axis), it returns to its original orientation after a return angle $\theta_{\rm ret} = 360°$. Correspondingly, the spin of the particle that carries the electromagnetic waves (the photon) is one. For neutrinos, the return angle is $\theta_{\rm ret} = 720°$; and correspondingly, the spin of a neutrino is $\frac{1}{2}$. For gravitational waves, the physical manifestations include the deformation of a sphere of test particles (Figs. 27.1a,b and 27.2a,b) and the acceleration fields (Figs. 27.1c,d and 27.2c,d). Both the deformed, ellipsoidal spheres and the quadrupolar lines of force return to their original orientations after rotation through $\theta_{\rm ret} = 180°$; correspondingly, the graviton must have spin two. This spin two also shows up in the rotation factor $e^{i2\psi}$ of Eq. (27.31).

Although Fierz and Pauli (1939) showed us how to quantize linearized general relativity, the quantization of full, nonlinear general relativity remains a difficult subject of current research.

27.4

27.4 Gravitational Waves Propagating through Curved Spacetime

Richard Isaacson (1968a,b) has developed a geometric-optics formulation of the theory of gravitational waves propagating through curved spacetime, and as a by-product he has given a rigorous mathematical description of the waves' stress-energy tensor and thence the energy and momentum carried by the waves. In this section, we sketch the main ideas and results of Isaacson's analysis.[4]

two-lengthscale expansion for gravitational waves in curved spacetime

The foundation for the analysis is a two-lengthscale expansion $\lambdabar/\mathcal{L}$ like we used in Sec. 7.3 when formulating geometric optics. For any physical quantity, we identify the wave contribution as the portion that varies on some short lengthscale $\lambdabar = \lambda/(2\pi)$ (the waves' reduced wavelength), and the background as the portion that varies on a far longer lengthscale $\mathcal{L}$ (which is less than or of order the background's spacetime radius of curvature $\mathcal{R}$); see Fig. 27.3.

steady coordinates

To make this idea work, we must use "steady" coordinates (i.e., coordinates that are smooth to as great an extent as the waves permit, on lengthscales shorter than $\mathcal{L}$). In such coordinates, components of the spacetime metric $g_{\alpha\beta}$ and of the Riemann curvature tensor $R_{\alpha\beta\gamma\delta}$ split into background (B) plus gravitational waves (GW),

metric and Riemann tensor split into background and gravitational-wave parts

$$g_{\alpha\beta} = g^{\rm B}_{\alpha\beta} + h_{\alpha\beta}, \quad R_{\alpha\beta\gamma\delta} = R^{\rm B}_{\alpha\beta\gamma\delta} + R^{\rm GW}_{\alpha\beta\gamma\delta}, \tag{27.33a}$$

where the background quantities are defined as the averages (denoted $\langle\cdot\rangle$) of the full quantities over lengthscales long compared to $\lambdabar$ and short compared to $\mathcal{L}$:

$$g^{\rm B}_{\alpha\beta} \equiv \langle g_{\alpha\beta}\rangle, \quad R^{\rm B}_{\alpha\beta\gamma\delta} \equiv \langle R_{\alpha\beta\gamma\delta}\rangle. \tag{27.33b}$$

4. In the 1980s and 1990s, Isaacson, as the Program Director for Gravitational Physics at the U.S. National Science Foundation (NSF), played a crucial role in the creation of the LIGO Project for detecting gravitational waves and in moving LIGO toward fruition.

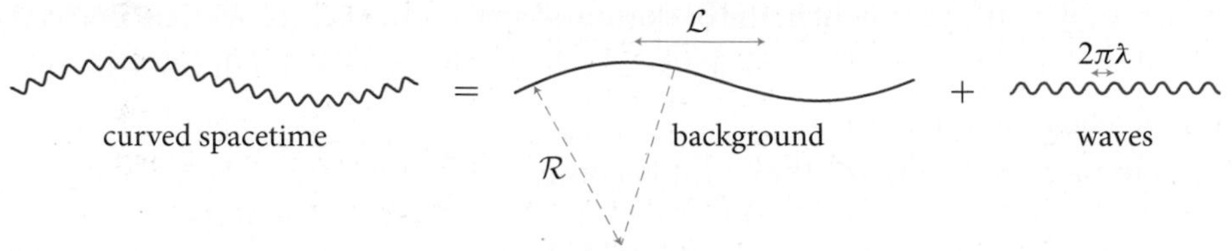

curved spacetime = background + waves

$\mathcal{L}$ $\mathcal{R}$ $2\pi\lambda\!\!\!^{-}$

FIGURE 27.3 Heuristic embedding diagram for the decomposition of curved spacetime into a background spacetime plus gravitational waves.

To assist us in solving the Einstein equation, we treat the Einstein tensor $G_{\alpha\beta}$ a bit differently from the metric and Riemann. We begin by expanding $G_{\alpha\beta}$ as a power series in the metric perturbation $h_{\alpha\beta}$: $G_{\alpha\beta} = G^{\mathrm{B}}_{\alpha\beta} + G^{(1)}_{\alpha\beta} + G^{(2)}_{\alpha\beta} + \cdots$. Here $G^{\mathrm{B}}_{\alpha\beta}$ is the Einstein tensor computed from the background metric $g^{\mathrm{B}}_{\alpha\beta}$, $G^{(1)}_{\alpha\beta}$ is linear in $h_{\alpha\beta}$, $G^{(2)}_{\alpha\beta}$ is quadratic, and so forth. We then split $G_{\alpha\beta}$ into its rapidly varying part, which is simply $G^{(1)}_{\alpha\beta}$ to leading order, and its smoothly varying part, which through quadratic order is $\langle G_{\alpha\beta}\rangle = G^{\mathrm{B}}_{\alpha\beta} + \langle G^{(2)}_{\alpha\beta}\rangle$. The vacuum Einstein equation $G_{\alpha\beta} = 0$ will be satisfied only if the rapidly and smoothly varying parts both vanish.

In Sec. 27.4.1, by setting the fast-varying part $G^{(1)}_{\alpha\beta}$ to zero, we obtain a wave equation in the background curved spacetime for $h_{\alpha\beta}$ (the gravitational waves), which we can solve (Sec. 27.4.2) using the geometric-optics approximation that underlies this analysis. Then, in Sec. 27.4.3, by setting the slowly varying part $G^{B}_{\alpha\beta} + \langle G^{(2)}_{\alpha\beta}\rangle$ to zero, we obtain Isaacson's description of gravitational-wave energy and momentum.

27.4.1 Gravitational Wave Equation in Curved Spacetime

27.4.1

The metric perturbation $h_{\alpha\beta}$ can be regarded as a tensor field that lives in the background spacetime.[5] The rapidly varying part of the Einstein equation, $G^{(1)}_{\alpha\beta} = 0$, gives rise to a wave equation for this tensorial metric perturbation (Isaacson, 1968a; Misner, Thorne, and Wheeler, 1973, Secs. 35.13, 35.14). We can infer this wave equation most easily from a knowledge of the form it takes in any local Lorentz frame of the background (with size $\gg \lambda\!\!\!^{-}$ but $\ll \mathcal{L}$). In such a frame, $G^{(1)}_{\alpha\beta} = 0$ must reduce to the field equation of linearized theory [the vacuum version of Eq. (25.83)]. And if we introduce Lorenz gauge [Eq. (27.18c)], then $G^{(1)}_{\alpha\beta} = 0$ must become, in a local Lorentz frame, the vacuum wave equation (27.18e). The frame-invariant versions of these local-Lorentz-frame equations, in the background spacetime, should be obvious. The trace-reversed metric perturbation (27.18d) in frame-invariant form must become

$$\bar{h}_{\mu\nu} \equiv h_{\mu\nu} - \frac{1}{2} h\, g^{\mathrm{B}}_{\mu\nu}, \quad h \equiv g^{\alpha\beta}_{\mathrm{B}} h_{\alpha\beta}. \tag{27.34a}$$

The Lorenz-gauge condition (27.18c) must become

Lorenz-gauge condition

$$\bar{h}_{\mu\nu|}{}^{\nu} = 0, \tag{27.34b}$$

5. Actually, this characterization requires that we restrict the coordinates to be steady.

where the | denotes a gradient in the background spacetime (i.e., a covariant derivative computed using connection coefficients constructed from $g^{\rm B}_{\mu\nu}$) and the index is raised with the background metric, $\bar{h}_{\mu\nu|}{}^{\nu} = \bar{h}_{\mu\nu|\alpha} g_B^{\alpha\nu}$. And the gravitational wave equation (Einstein field equation) (27.18e) must become

gravitational wave equation

$$\bar{h}_{\mu\nu|\alpha}{}^{\alpha} = 0 \tag{27.34c}$$

curvature coupling is negligible

plus curvature coupling terms, such as $R^{\rm B}_{\alpha\mu\beta\nu}\bar{h}^{\alpha\beta}$, that result from the noncommutation of the double gradients. The curvature coupling terms have magnitude $h/\mathcal{R}^2 \lesssim h/\mathcal{L}^2$ (where $\mathcal{R}$ is the radius of curvature of the background spacetime; cf. Fig. 27.3), while the terms kept in Eq. (27.34c) have the far-larger magnitude $h/\lambda\!\!\!^{-2}$, so the curvature coupling terms can be (and are) neglected.

27.4.2

27.4.2 Geometric-Optics Propagation of Gravitational Waves

geometric-optics wave propagation

When one solves Eqs. (27.34) using the geometric-optics techniques developed in Sec. 7.3, one obtains precisely the results that one should expect, knowing the solution (27.19) for weak, planar gravitational waves in flat spacetime (linearized theory).

1. If we split $h_{\mu\nu}$ up into two polarization pieces + and ×, each with its own rapidly varying phase φ and slowly varying amplitude $\mathsf{A}_{\mu\nu}$, then the Lorenz-gauge, trace-reversed metric perturbation for each piece takes the standard geometric-optics form [eikonal approximation; Eq. (7.20)]:

eikonal approximation

$$h_{\mu\nu} = \Re(\mathsf{A}_{\mu\nu} e^{i\varphi}). \tag{27.35a}$$

2. Because the linearized-theory waves propagate in a straight line (z direction) and travel at the speed of light, the geometric-optics waves propagate through curved spacetime on rays that are null geodesics. More specifically, the wave vector $\vec{k} = \vec{\nabla}\varphi$ is tangent to the null-ray geodesics, and φ is constant along a ray and hence is a rapidly varying function of the retarded time τ_r at which the source (in its own reference frame) emitted the ray:

rays are null geodesics

$$\varphi = \varphi(\tau_r), \quad \vec{k} = \vec{\nabla}\varphi, \quad \vec{k}\cdot\vec{k} = 0, \quad \nabla_{\vec{k}}\vec{k} = 0, \quad \nabla_{\vec{k}}\varphi = 0. \tag{27.35b}$$

3. Because the x- and y-axes that define the two polarizations in linearized theory remain fixed as the wave propagates, for each polarization we can split the amplitude $\mathsf{A}_{\mu\nu}$ up into a scalar amplitude A_+ or $\mathsf{A}_\times$, and a polarization tensor $\boldsymbol{e}^+$ or $\boldsymbol{e}^\times$ [like those of Ex. 27.6 and Eqs. (27.23b)], and the polarization tensors are parallel-transported along the rays:

polarization tensors are parallel propagated along rays

$$\mathsf{A}_{\mu\nu} = \mathsf{A}\, e_{\mu\nu}, \quad \nabla_{\vec{k}}\, \boldsymbol{e} = 0. \tag{27.35c}$$

graviton conservation implies amplitude scales inversely with square root of cross section of a bundle of rays

4. Because gravitons are conserved (cf. the conservation of quanta in our general treatment of geometric optics, Sec. 7.3.2), the flux of gravitons (which is proportional to the square A^2 of the scalar amplitude) times the cross sectional area $\mathcal{A}$ of a bundle of rays that are carrying the gravitons must be

constant. Therefore, the scalar wave amplitude A must die out as 1/(square root of cross sectional area $\mathcal{A}$ of a bundle of rays):

$$\mathsf{A} \propto 1/\sqrt{\mathcal{A}}. \tag{27.35d}$$

Now, just as the volume of a 3-dimensional fluid element, for a perfect fluid, changes at the rate $d \ln V/d\tau = \vec{\nabla} \cdot \vec{u}$, where $\vec{u}$ is the 4-velocity of the fluid [Eq. (2.65)], so (it turns out) the cross sectional area of a bundle of rays increases as $\nabla_{\vec{k}} \mathcal{A} = \vec{\nabla} \cdot \vec{k}$. Therefore, the transport law for the wave amplitude, $\mathsf{A} \propto 1/\sqrt{\mathcal{A}}$, becomes

$$\nabla_{\vec{k}} \mathsf{A} = -\frac{1}{2}(\vec{\nabla} \cdot \vec{k})\mathsf{A}. \tag{27.35e}$$

transport law for amplitude

Equations (27.35) are derived more rigorously in Isaacson (1968a) and Misner, Thorne, and Wheeler (1973, Sec. 35.14). They can be used to compute the influence of the cosmological curvature of our universe, and the gravitational fields of intervening bodies, on the propagation of gravitational waves from their sources to Earth. Once the waves have reached Earth, we can compute the measured gravitational-wave fields h_+ and $h_\times$ by projecting out the spatial, transverse-traceless parts of $h_{\mu\nu} = \Re(\mathsf{A} e_{\mu\nu} e^{i\varphi})$, as discussed in Box 27.2.

The geometric-optics propagation of gravitational waves, as embodied in Eqs. (27.35), is essentially identical to that of electromagnetic waves. Both waves, gravitational and electromagnetic, propagate along rays that are null geodesics. Both parallel-transport their polarizations. Both are carried by quanta that are conserved as they propagate along bundles of rays and as a result both have scalar amplitudes that vary as $\mathsf{A} \propto 1/\sqrt{\mathcal{A}}$, where $\mathcal{A}$ is the cross sectional area of a ray bundle.

Therefore, gravitational waves must exhibit exactly the same vacuum propagation phenomena as electromagnetic waves: Doppler shifts, cosmological redshifts, gravitational redshifts, gravitational deflection of rays, and gravitational lensing!

gravitational waves exhibit same vacuum propagation phenomena as electromagnetic waves: redshifts, ray deflection, gravitational lensing

In Ex. 27.14, we illustrate this geometric optics propagation of gravitational waves by applying it to the waves from a binary system, which travel outward through our expanding universe.

Exercise 27.7 explores an application where geometric optics breaks down due to diffraction.

EXERCISES

Exercise 27.7 ***Example: Gravitational Lensing of Gravitational Waves by the Sun*
Gravitational waves from a distant source travel through the Sun with impunity (negligible absorption and scattering), and their rays are gravitationally deflected. The Sun is quite centrally condensed, so most of the deflection is produced by a central region with mass $M_c \simeq 0.3 M_\odot$ and radius $R_c \simeq 10^5$ km $\simeq R_\odot/7$, and the maximum deflection angle is therefore $\Delta\phi \simeq 4M_c/R_c$ [Eq. (27.7)]. A few of the deflected rays, along which the waves propagate according to geometric optics, are shown in Fig. 27.4.

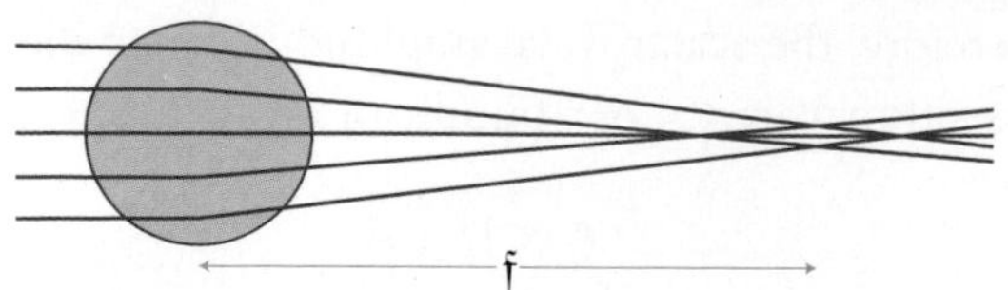

FIGURE 27.4 Some gravitational-wave rays that pass through the Sun are brought to an imperfect focus at a distance $\mathfrak{f}$, the focal length.

(a) Show that the rays are brought to an imperfect focus and thence produce caustics (Sec. 7.5) at a distance from the Sun (the focal length) $\mathfrak{f} \sim R_c^2/(4M_c) \sim 38$ AU. A calculation with a more accurate solar model gives $\mathfrak{f} \sim 20$ AU, which is near the orbit of Uranus.

(b) If the waves were to have arbitrarily small wavelength λ, then at the caustics, their wave fields h_+ and $h_\times$ would become divergently large (Sec. 7.5). Finite wavelength causes diffraction near the caustics (Sec. 8.6). Explain why the focused field thereby is smeared out over a region with transverse size $\sigma \sim [\lambdabar/(2R_c)]\mathfrak{f} \sim [\lambdabar/(8M_c)]R_c$. [Hint: See Eq. (8.9) and associated discussion.]

(c) Explain why, if $\sigma \ll R_c$ (i.e., if $\lambdabar \ll 8M_c \sim 3M_\odot$), substantial focusing occurs, and the field near the caustics is strongly amplified; but if $\sigma \gtrsim R_c$ (i.e., $\lambdabar \gtrsim 3M_\odot$), there is only slight or no focusing. Explain why it is unlikely that any discrete, strong gravitational-wave sources in the universe emit wavelengths shorter than $3M_\odot \sim 5$ km and therefore are strongly lensed by the Sun.

27.4.3

27.4.3 Energy and Momentum in Gravitational Waves

Now turn from the rapidly varying piece of the vacuum Einstein equation, $G^{(1)}_{\mu\nu} = 0$, to the piece that is averaged over scales long compared to $\lambdabar$ and short compared to $\mathcal{L}$:

$$G^{\mathrm{B}}_{\alpha\beta} + \langle G^{(2)}_{\alpha\beta}\rangle = 0. \tag{27.36}$$

(Recall that $G^{\mathrm{B}}_{\alpha\beta}$ is the Einstein tensor constructed from the slowly varying background metric, and $G^{(2)}_{\alpha\beta}$ is the piece of the full Einstein tensor that is quadratic in the rapidly varying metric perturbation $h_{\mu\nu}$ and that therefore does not average to zero.)

Notice that Eq. (27.36) can be brought into the standard form for Einstein's equation in the background spacetime,

Einstein equation for background metric in vacuum

$$\boxed{G^{\mathrm{B}}_{\alpha\beta} = 8\pi T^{\mathrm{GW}}_{\alpha\beta},} \tag{27.37}$$

by moving $\langle G^{(2)}_{\alpha\beta}\rangle$ to the right-hand side and then attributing to the waves a stress-energy tensor defined by

stress-energy tensor for gravitational waves: formal expression

$$T^{\mathrm{GW}}_{\alpha\beta} = -\frac{1}{8\pi}\langle G^{(2)}_{\alpha\beta}\rangle. \tag{27.38}$$

Because this stress-energy tensor involves an average over a few wavelengths, its energy density, momentum density, energy flux, and momentum flux are not defined on lengthscales shorter than a wavelength. One cannot say how much energy or momentum resides in the troughs of the waves and how much in the crests. One can only say how much total energy there is in a region containing a few or more wavelengths. However, once reconciled to this amount of nonlocality, one finds that $T^{\rm GW}_{\alpha\beta}$ has all the other properties expected of any good stress-energy tensor. Most especially, in the absence of coupling of the waves to matter (the situation we are treating), it obeys the standard conservation law:

conservation law for gravitational-wave energy and momentum

$$\boxed{T^{\rm GW\,\alpha\beta}{}_{|\beta} = 0,} \tag{27.39}$$

where, as above, the symbol $_|$ denotes the covariant derivative in the background spacetime (i.e., the derivative using the connection coefficients of $g^B_{\alpha\beta}$). This law is a direct consequence of the averaged field equation (27.37) and the contracted Bianchi identity for the background spacetime: $G^{\rm B\,\alpha\beta}{}_{|\beta} = 0$.

By grinding out the second-order perturbation of the Einstein tensor and inserting it into Eq. (27.38), performing several integrations by parts in the average $\langle\cdot\rangle$, and expressing the result in terms of h_+ and $h_\times$, one arrives at the following simple expression for $T^{\rm GW}_{\alpha\beta}$ in terms of the wave fields h_+ and $h_\times$:

stress-energy tensor in terms of gravitational-wave fields

$$\boxed{T^{\rm GW}_{\alpha\beta} = \frac{1}{16\pi}\langle h_{+,\alpha}h_{+,\beta} + h_{\times,\alpha}h_{\times,\beta}\rangle.} \tag{27.40}$$

[For details of the derivation, see Isaacson (1968b) or Misner, Thorne, and Wheeler (1973, Secs. 35.13, 35.15).]

Let us examine this stress-energy tensor in a local Lorentz frame of the background spacetime where the waves are locally plane and are propagating in the z direction—the kind of frame we used in Sec. 27.3.2 when exploring the properties of gravitational waves. Because in this frame, $h_+ = h_+(t-z)$ and $h_\times = h_\times(t-z)$, the only nonzero components of Eq. (27.40) are

$$\boxed{T^{\rm GW\,00} = T^{\rm GW\,0z} = T^{\rm GW\,z0} = T^{\rm GW\,zz} = \frac{1}{16\pi}\langle \dot h_+^2 + \dot h_\times^2\rangle.} \tag{27.41}$$

This has the same form as the stress-energy tensor for a plane electromagnetic wave propagating in the z direction, and the same form as the stress-energy tensor for any collection of zero-rest-mass particles moving in the z-direction [cf. Eq. (3.32d)], as it must, since the gravitational waves are carried by zero-rest-mass gravitons just as electromagnetic waves are carried by zero-rest-mass photons.

Suppose that the waves have frequency $\sim f$ and that the amplitudes of oscillation of h_+ and $h_\times$ are $\sim h_{\rm amp}$. Then by inserting factors of G and c into Eq. (27.41) (i.e., by switching from geometrized units to conventional units) and by setting

$\langle(\partial h_+/\partial t)^2\rangle \simeq \frac{1}{2}(2\pi f h_{\rm amp})^2$ and similarly for $h_\times$, we obtain the following approximate expression for the energy flux in the waves:

magnitude of gravitational-wave energy flux

$$T^{\rm GW\, 0z} \simeq \frac{\pi}{4}\frac{c^3}{G} f^2 h_{\rm amp}^2 \simeq \frac{0.01\,{\rm W}}{{\rm m}^2}\left(\frac{f}{200\ {\rm Hz}}\right)^2 \left(\frac{h_{\rm amp}}{10^{-21}}\right)^2. \quad (27.42)$$

The numbers in this equation are those for the first gravitational waves ever detected: LIGO's GW150914 at its peak brightness. Those waves' observed energy flux, $\sim 0.01\ {\rm W\, m^{-2}}$, was several times higher than that from the full moon as seen on Earth, but the waves' source, two $\sim 30 M_\odot$ colliding black holes, is $\sim$1.2 billion light years away compared to the Moon's distance of 1 light second. Of course, the moon shines steadily, while the holes' collision and this enormous flux lasted for only $\sim$20 ms.

For a short gravitational wave burst such as GW150914 (only a few wave cycles, which we shall approximate as just one), the enormous energy flux (27.42) corresponds to a huge mean occupation number for the quantum states of the gravitational-wave field (i.e., a huge value for the number of spin-two, zero-rest-mass gravitons in each quantum state). To compute that occupation number, we evaluate the volume in phase space occupied by the waves and then divide by the volume occupied by each quantum state (cf. Sec. 3.2.5). At a time when the waves have reached a distance r from the source, they occupy a spherical shell of area $4\pi r^2$ and thickness $\sim c/f = 2\pi \lambda\!\!\!^{-}$, where $\lambda\!\!\!^{-} = 1/(2\pi f)$ is their reduced wavelength, so their volume in physical space is $\mathcal{V}_x \sim 8\pi^2 r^2 \lambda\!\!\!^{-}$. As seen by observers whom the waves are passing, they come from a solid angle $\Delta\Omega \sim (2\lambda\!\!\!^{-}/r)^2$ centered on the source, and they have a spread of angular frequencies $\Delta\omega \sim \omega = c/\lambda\!\!\!^{-}$. Since each graviton carries an energy $\hbar\omega = \hbar c/\lambda\!\!\!^{-}$ and a momentum $\hbar\omega/c = \hbar/\lambda\!\!\!^{-}$, the volume that they occupy in momentum space is $\mathcal{V}_p \sim (\hbar/\lambda\!\!\!^{-})^3 \Delta\Omega$, or $\mathcal{V}_p \sim 4\hbar^3/(\lambda\!\!\!^{-} r^2)$. The gravitons' volume in phase space, then, is

$$\mathcal{V}_x \mathcal{V}_p \sim 32\pi^2\hbar^3 \sim (2\pi\hbar)^3. \quad (27.43)$$

Since each quantum state for a zero rest-mass particle occupies a volume $(2\pi\hbar)^3$ in phase space [Eq. (3.17) with $g_s = 1$], the total number of quantum states occupied by the gravitons is of order unity! Correspondingly, for a total energy radiated like that of GW150914, $\sim M_\odot c^2 \sim 10^{47} J$ with each graviton carrying an energy $\hbar c/\lambda\!\!\!^{-} \sim 10^{-31} J$, the mean occupation number of each occupied state is of order the total number of gravitons emitted:

mean occupation number for quantum states of gravitational waves from a strong gravitational wave burst

$$\eta \sim \frac{M_\odot c^2}{\hbar\omega} \sim 10^{78}. \quad (27.44a)$$

This is the mean occupation number from the viewpoint of the emitter.

A detector on Earth has available to it only those gravitons that pass through a region with transverse size of order their wavelength λ—which means a fraction $(2\pi\lambda\!\!\!^{-})^2/(4\pi r^2)$ of the emitted waves' volume. We can think of the detector as collapsing the gravitons' wave function into that volume. The number of available quantum states is still of order unity (demonstrate this!), but the number of gravitons occupying them is reduced by this factor, so from the detector's viewpoint, the mean occupation number is

mean occupation number from viewpoint of a detector on Earth

$$\eta_{\text{collapsed}} \sim \frac{M_\odot c^2}{\hbar\omega}\frac{(2\pi\,\bar\lambda)^2}{4\pi r^2} \sim \frac{T^{\text{GW}0z}}{\hbar\omega c}(2\pi\,\bar\lambda)^3 \sim 10^{39}. \tag{27.44b}$$

Notice that this is the number of gravitons in a cubic wavelength at the wave burst.

Whichever viewpoint one takes, the occupation number is enormous. It guarantees that the waves behave exceedingly classically; quantum-mechanical corrections to the classical theory have fractional magnitude $1/\sqrt{\eta} \sim 10^{-39}$, or $\sim 10^{-20}$.

27.5 The Generation of Gravitational Waves

27.5

When analyzing the generation of gravitational waves, it is useful to divide space around the source (in the source's rest frame) into the regions shown in Fig. 27.5.

If the source has size $L \lesssim M$, where M is its mass, then spacetime is strongly curved inside and near it, and we refer to it as a *strong-gravity source.* The region with radii (measured from its center of mass) $r \lesssim 10M$ is called the source's *strong-field region.* Examples of strong-gravity sources are vibrating or spinning neutron stars, and merging binary black holes. The region with radii $10M \lesssim r \lesssim \bar\lambda =$ (the reduced wavelength of the emitted waves) is called the source's *weak-field near zone.* In this region, the source's gravity is fairly well approximated by Newtonian theory and a Newtonian gravitational potential Φ. As in electromagnetic theory, the region $\bar\lambda \lesssim r \lesssim \lambda$ is called the *induction zone* or the *intermediate zone.* The *wave zone* begins at $r \sim \lambda = 2\pi\,\bar\lambda$.

regions of space around a gravitational-wave source: Fig. 27.5

It is useful to divide the wave zone into two parts: a part near the source ($r \lesssim r_o$ for some r_o) called the *local wave zone,* in which the spacetime curvatures of external bodies and of the universe as a whole are unimportant, and the *distant wave zone* ($r \gtrsim r_o$), in which the emitted waves are significantly affected by external bodies and the external universe (i.e., by background spacetime curvature). The theory of

theory of wave generation predicts waves in source's local wave zone; geometric optics carries them onward through distant wave zone

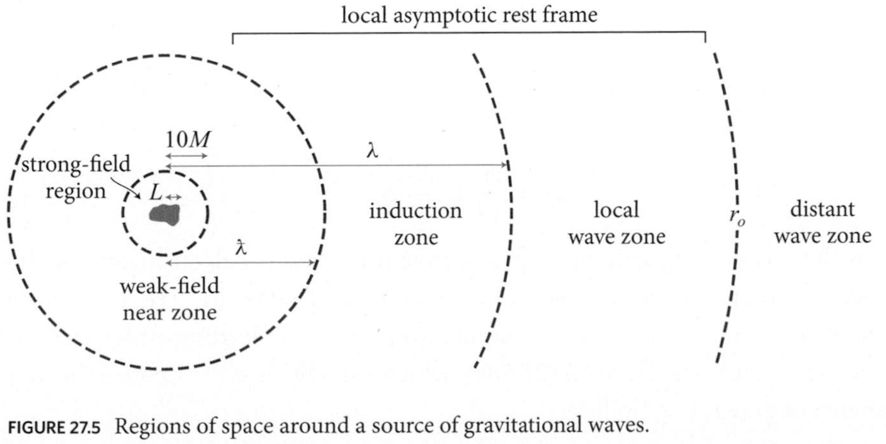

FIGURE 27.5 Regions of space around a source of gravitational waves.

gravitational-wave generation deals with computing, from the source's dynamics, the gravitational waves in the local wave zone. Propagation of the waves to Earth is dealt with by using geometric optics (or other techniques) to carry the waves from the local wave zone outward, through the distant wave zone, to Earth.

source's local asymptotic rest frame

The entire region in which gravity is weak and the spacetime curvatures of external bodies and the universe are unimportant ($10M \lesssim r \lesssim r_o$)—when viewed in nearly Lorentz coordinates in which the source is at rest—is called the source's *local asymptotic rest frame.*

27.5.1

27.5.1 Multipole-Moment Expansion

The electromagnetic waves emitted by a dynamical charge distribution are usually expressed as a sum over the source's multipole moments. There are two families of moments: the electric moments (moments of the electric-charge distribution) and the magnetic moments (moments of the electric-current distribution).

source's mass moments and current moments

Similarly, the gravitational waves emitted by a dynamical distribution of mass-energy and momentum can be expressed, in the local wave zone, as a sum over multipole moments. Again there are two families of moments: the *mass moments* (moments of the mass-energy distribution) and the *current moments* (moments of the mass-current distribution, i.e., the momentum distribution). The multipolar expansion of gravitational waves is developed in great detail in Blanchet (2014) and Thorne (1980). In this section, we sketch and explain its qualitative and order-of-magnitude features.

In the source's weak-gravity near zone (if it has one), the mass moments show up in the time-time part of the metric in a form familiar from Newtonian theory:

multipolar expansion of metric outside a source

$$g_{00} = -(1+2\Phi) = -1 \,\&\, \frac{\mathcal{I}_0}{r} \,\&\, \frac{\mathcal{I}_1}{r^2} \,\&\, \frac{\mathcal{I}_2}{r^3} \,\&\, \cdots \qquad (27.45)$$

[cf. Eq. (25.79)]. Here r is radius, $\mathcal{I}_\ell$ is the mass moment of order ℓ, and "&" means "plus a term with the form" (i.e., a term whose magnitude and parameter dependence are shown but whose multiplicative numerical coefficients do not interest us, at least not for the moment). The mass monopole moment $\mathcal{I}_0$ is the source's mass, and the mass dipole moment $\mathcal{I}_1$ can be made to vanish by placing the origin of coordinates at the center of mass [Eq. (25.96) and Ex. 25.21].

In the source's weak-gravity near zone, its current moments $\mathcal{S}_\ell$ similarly show up in the space-time part of the metric:

$$g_{0j} = \frac{\mathcal{S}_1}{r^2} \,\&\, \frac{\mathcal{S}_2}{r^3} \,\&\, \cdots . \qquad (27.46)$$

Just as there is no magnetic monopole moment in classical electromagnetic theory, so there is no current monopole moment in general relativity. The current dipole moment $\mathcal{S}_1$ is the source's angular momentum J_k, so the leading-order term in the expansion (27.46) has the form (25.98c), which we have used to deduce the angular momenta of gravitating bodies.

If the source has mass M, size L, and internal velocities $\sim v$, then the magnitudes of its moments are

$$\mathcal{I}_\ell \sim M L^\ell, \quad \mathcal{S}_\ell \sim M v L^\ell. \tag{27.47}$$

magnitudes of moments

These formulas guarantee that the near-zone fields g_{00} and g_{0j}, as given by Eqs. (27.45) and (27.46), are dimensionless.

As the source's moments oscillate dynamically, they produce gravitational waves. Mass-energy conservation [Eq. (25.102)] prevents the mass monopole moment $\mathcal{I}_0 = M$ from oscillating; angular-momentum conservation [Eq. (25.103)] prevents the current dipole moment $\mathcal{S}_1 =$ (angular momentum) from oscillating; and because the time derivative of the mass dipole moment $\mathcal{I}_1$ is the source's linear momentum, momentum conservation [Eq. (25.106)] prevents the mass dipole moment from oscillating. Therefore, the lowest-order moments that can oscillate and thereby contribute to the waves are the quadrupolar ones. The wave fields h_+ and $h_\times$ in the source's local wave zone must (i) be dimensionless, (ii) die out as $1/r$, and (iii) be expressed as a sum over derivatives of the multipole moments. These considerations guarantee that the waves will have the following form:

lowest-order moments that produce gravitational waves are quadrupolar

$$\boxed{h_+ \sim h_\times \sim \frac{\partial^2 \mathcal{I}_2/\partial t^2}{r} \;\&\; \frac{\partial^3 \mathcal{I}_3/\partial t^3}{r} \;\&\; \ldots \;\&\; \frac{\partial^2 \mathcal{S}_2/\partial t^2}{r} \;\&\; \frac{\partial^3 \mathcal{S}_3/\partial t^3}{r} \;\&\; \cdots} \tag{27.48}$$

multipolar expansion of gravitational waves

(Ex. 27.8).

The timescale on which the moments oscillate is $T \sim L/v$, so each time derivative produces a factor v/L. Correspondingly, the ℓ-pole contributions to the waves have magnitudes

$$\frac{\partial^\ell \mathcal{I}_\ell/\partial t^\ell}{r} \sim \frac{M}{r} v^\ell, \quad \frac{\partial^\ell \mathcal{S}_\ell/\partial t^\ell}{r} \sim \frac{M}{r} v^{(\ell+1)}. \tag{27.49}$$

This means that, for a *slow-motion source* (one with internal velocities v small compared to light, so the reduced wavelength $\lambda\!\!\!^{-} \sim L/v$ is large compared to the source size L), the mass quadrupole moment $\mathcal{I}_2$ will produce the strongest waves. The mass octupole (3-pole) waves and current quadrupole waves will be weaker by $\sim v \sim L/\lambda\!\!\!^{-}$; the mass 4-pole and current octupole waves will be weaker by $\sim v^2 \sim L^2/\lambda\!\!\!^{-2}$, and so forth. This is analogous to the electromagnetic case, where the electric dipole waves are the strongest, the electric quadrupole and magnetic dipole are smaller by $\sim L/\lambda\!\!\!^{-}$, and so on.

for slow-motion source: relative magnitudes of gravitational waves' multipolar components

In the next section, we develop the theory of mass quadrupole gravitational waves. For the corresponding theory of higher-order multipoles, see, for example, Thorne (1980, Secs. IV and VIII) and Blanchet (2014). In Sec. 27.5.3, we will see that a source's mass quadruopole waves cannot, by themselves, carry net linear momentum. Net wave momentum and the corresponding recoil of the source require a beating of mass quadrupole waves against current quadrupole or mass octupole waves.

EXERCISES

Exercise 27.8 *Derivation: Multipolar Expansion of Gravitational Waves*
Show that conditions (i), (ii), and (iii) preceding Eq. (27.48) guarantee that the multipolar expansion of the gravitational-wave fields will have the form (27.48).

27.5.2

27.5.2 Quadrupole-Moment Formalism

Consider a weakly gravitating, nearly Newtonian system (which is guaranteed to be a slow-motion gravitational-wave source, since Newtonian theory requires internal velocities $v \ll 1$). An example is a binary star system. Write the system's Newtonian potential (in its near zone) in the usual way:

$$\Phi(\mathbf{x}) = -\int \frac{\rho(\mathbf{x}')}{|\mathbf{x}-\mathbf{x}'|} dV_{x'}. \tag{27.50}$$

By using Cartesian coordinates, placing the origin of coordinates at the center of mass so $\int \rho x^j dV_x = 0$, and expanding:

$$\frac{1}{|\mathbf{x}-\mathbf{x}'|} = \frac{1}{r} + \frac{x^j x^{j'}}{r^3} + \frac{x^j x^k (3x^{j'}x^{k'} - r'^2 \delta_{jk})}{2r^5} + \cdots, \tag{27.51}$$

we obtain the multipolar expansion of the Newtonian potential:

for nearly Newtonian source: multipolar expansion of Newtonian potential

$$\boxed{\Phi(\mathbf{x}) = -\frac{M}{r} - \frac{3\mathcal{I}_{jk}x^j x^k}{2r^5} + \cdots.} \tag{27.52}$$

Here

$$\boxed{M = \int \rho dV_x, \quad \mathcal{I}_{jk} = \int \rho \left(x^j x^k - \frac{1}{3} r^2 \delta_{jk}\right) dV_x} \tag{27.53}$$

are the system's mass and mass quadrupole moment. Note that the mass quadrupole moment is equal to the second moment of the mass distribution with its trace removed.

As we have discussed, dynamical oscillations of the quadrupole moment generate the source's strongest gravitational waves. Those waves must be describable, in the source's near zone and local wave zone, by an outgoing-wave solution to the Lorenz-gauge, linearized Einstein equation,

$$\bar{h}_{\mu\nu,}{}^{\nu} = 0, \quad \bar{h}_{\mu\nu,\alpha}{}^{\alpha} = 0 \tag{27.54}$$

[Eqs. (25.89) and (25.90)], that has the near-zone Newtonian limit:

$$\frac{1}{2}(\bar{h}_{00} + \bar{h}_{xx} + \bar{h}_{yy} + \bar{h}_{zz}) = h_{00} = -(\text{quadrupole part of } 2\Phi) = \frac{3\mathcal{I}_{jk}x^j x^k}{r^5} \tag{27.55}$$

[cf. Eq. (25.79)].

The desired solution can be written in the form

Newtonian potential transitions into this linearized-theory gravitational field in transition and local wave zones

$$\bar{h}_{00} = 2\left[\frac{\mathcal{I}_{jk}(t-r)}{r}\right]_{,jk}, \quad \bar{h}_{0j} = 2\left[\frac{\dot{\mathcal{I}}_{jk}(t-r)}{r}\right]_{,k}, \quad \bar{h}_{jk} = 2\frac{\ddot{\mathcal{I}}_{jk}(t-r)}{r}, \tag{27.56}$$

where the coordinates are Cartesian, $r \equiv \sqrt{\delta_{jk}x^j x^k}$, and the dots denote time derivatives. To verify that this is the desired solution: (i) Compute its divergence $\bar{h}_{\alpha\beta,}{}^{\beta}$ and obtain zero almost trivially. (ii) Notice that each Lorentz-frame component of $\bar{h}_{\alpha\beta}$ has the form $f(t-r)/r$ aside from some derivatives that commute with the wave operator, which implies that it satisfies the wave equation. (iii) Notice that in the near zone, the slow-motion assumption inherent in the Newtonian limit makes the time derivatives negligible, so $\bar{h}_{jk} \simeq 0$ and $\bar{h}_{00}$ is twice the right-hand side of Eq. (27.55), as desired.

Because the trace-reversed metric perturbation (27.56) in the local wave zone has the speed-of-light-propagation form, aside from its very slow decay as $1/r$, we can compute the gravitational-wave field $h^{\rm TT}_{jk}$ from it by transverse-traceless projection [Eq. (3) of Box 27.2 with $\mathbf{n} = \mathbf{e}_r$]:

resulting quadrupolar gravitational-wave field

$$h^{\rm TT}_{jk} = 2\left[\frac{\ddot{\mathcal{I}}_{jk}(t-r)}{r}\right]^{\rm TT}. \tag{27.57}$$

This is called the *quadrupole-moment formula for gravitational-wave generation*. Our derivation shows that it is valid for any nearly Newtonian source.

Looking back more carefully at the derivation, one can see that, in fact, it relies only on the linearized Einstein equation and the Newtonian potential in the source's local asymptotic rest frame. Therefore, *this quadrupole formula is also valid for slow-motion sources that have strong internal gravity (e.g., slowly spinning neutron stars), so long as we read the quadrupole moment $\mathcal{I}_{jk}(t-r)$ off the source's weak-field, near-zone Newtonian potential (27.52) and don't try to compute it via the Newtonian volume integral (27.53).*

validity for strong-gravity slow-motion sources

When the source is nearly Newtonian, so the volume integral (27.53) can be used to compute the quadrupole moment, the computation of the waves is simplified by computing instead the second moment of the mass distribution:

$$I_{jk} = \int \rho x^j x^k dV_x, \tag{27.58}$$

which differs from the quadrupole moment solely in its trace. Then, because the TT projection is insensitive to the trace, the gravitational-wave field (27.57) can be computed as

for Newtonian source: gravitational-wave field in terms of second moment of mass distribution

$$h^{\rm TT}_{jk} = 2\left[\frac{\ddot{I}_{jk}(t-r)}{r}\right]^{\rm TT}. \tag{27.59}$$

27.5.3

27.5.3 Quadrupolar Wave Strength, Energy, Angular Momentum, and Radiation Reaction

To get an order-of-magnitude feel for the strength of the gravitational waves, notice that the second time derivative of the quadrupole moment, in order of magnitude, is the nonspherical part of the source's internal kinetic energy, $E^{\rm ns}_{\rm kin}$:

magnitude of slow-motion source's quadrupolar gravitational-wave field

$$\boxed{h_+ \sim h_\times \sim \frac{E^{\rm ns}_{\rm kin}}{r} = G\frac{E^{\rm ns}_{\rm kin}}{c^4 r},} \tag{27.60}$$

where the second expression is written in conventional units. Although this estimate is based on the slow-motion assumption of source size small compared to reduced wavelength, $L \ll \lambda\!\!\!^{-}$, it remains valid in order of magnitude when extrapolated into the realm of the strongest of all realistic astrophysical sources, which have $L \sim \lambda\!\!\!^{-}$. In Ex. 27.17 we use Eq. (27.60) to estimate the strongest gravitational waves that might be seen by ground-based gravitational-wave detectors.

Because the gravitational stress-energy tensor $T^{\rm GW}_{\mu\nu}$ produces background curvature via the Einstein equation $G^{\rm B}_{\mu\nu} = 8\pi T^{\rm GW}_{\mu\nu}$, just like nongravitational stress-energy tensors, it must contribute to the rate of change of the source's mass M, linear momentum P_j, and angular momentum J_i [Eqs. (25.102)–(25.106)] just like other stress-energies. When one inserts the quadrupolar $T^{\rm GW}_{\mu\nu}$ into Eqs. (25.102)–(25.106) and integrates over a sphere in the wave zone of the source's local asymptotic rest frame, one finds that (Ex. 27.11):

source's rate of change of mass, momentum, and angular momentum due to mass quadrupolar gravitational-wave emission

$$\boxed{\frac{dM}{dt} = -\frac{1}{5}\left\langle \frac{\partial^3 \mathcal{I}_{jk}}{\partial t^3}\frac{\partial^3 \mathcal{I}_{jk}}{\partial t^3}\right\rangle,} \tag{27.61}$$

$$\boxed{\frac{dJ_i}{dt} = -\frac{2}{5}\epsilon_{ijk}\left\langle \frac{\partial^2 \mathcal{I}_{jm}}{\partial t^2}\frac{\partial^3 \mathcal{I}_{km}}{\partial t^3}\right\rangle,} \tag{27.62}$$

and $dP_j/dt = 0$. It turns out (cf. Thorne, 1980, Sec. IV) that the dominant linear-momentum change (i.e., the dominant radiation-reaction "kick") arises from a beating of the mass quadrupole moment against the mass octupole moment, and mass quadrupole against current quadrupole:

$$\boxed{\frac{dP_i}{dt} = -\frac{2}{63}\left\langle \frac{\partial^3 \mathcal{I}_{jk}}{\partial t^3}\frac{\partial^4 \mathcal{I}_{jki}}{\partial t^4}\right\rangle - \frac{16}{45}\epsilon_{ijk}\left\langle \frac{\partial^3 \mathcal{I}_{jl}}{\partial t^3}\frac{\partial^3 \mathcal{S}_{kl}}{\partial t^3}\right\rangle.} \tag{27.63}$$

Here the mass octupole moment $\mathcal{I}_{jki}$ is the trace-free part of the third moment of the mass distribution, and the current quadrupole moment $\mathcal{S}_{kp}$ is the symmetric, trace-free part of the first moment of the vectorial angular momentum distribution. (See, e.g., Thorne, 1980, Secs. IV.C, V.C; Thorne, 1983, Sec. 3.)

The back reaction of the emitted waves on their source shows up not only in changes of the source's mass, momentum, and angular momentum, but also in accompanying changes of the source's internal structure. These structure changes can be deduced fully, in many cases, from dM/dt, dJ_j/dt, and dP_j/dt. A nearly Newtonian binary system is an example (Sec. 27.5.4). However, in other cases (e.g., a compact body orbiting near the horizon of a massive black hole), the only way to compute the structure changes is via a gravitational-radiation-reaction force that acts back on the system.

The simplest example of such a force is one derived by William Burke (1971) for quadrupole waves emitted by a nearly Newtonian system. Burke's quadrupolar radiation-reaction force can be incorporated into Newtonian gravitation theory by simply augmenting the system's near-zone Newtonian potential with a radiation-reaction term, computed from the fifth time derivative of the system's quadrupole moment:

gravitational radiation-reaction potential

$$\boxed{\Phi^{\rm react} = \frac{1}{5}\frac{\partial^5 \mathcal{I}_{jk}}{\partial t^5} x^j x^k.} \tag{27.64}$$

This potential satisfies the vacuum Newtonian field equation $\nabla^2\Phi \equiv \delta^{jk}\Phi_{,jk} = 0$, because $\mathcal{I}_{jk}$ is traceless.

This augmentation of the Newtonian potential arises as a result of general relativity's outgoing-wave condition. If one were to switch to an ingoing-wave condition, $\Phi^{\rm react}$ would change sign, and if the system's oscillating quadrupole moment were joined onto standing gravitational waves, $\Phi^{\rm react}$ would go away. In Ex. 27.12, it is shown that the radiation-reaction force density $-\rho\nabla\Phi^{\rm react}$ saps energy from the system at the same rate as the gravitational waves carry it away.

radiation-reaction force density in source

Burke's gravitational radiation-reaction potential $\Phi^{\rm react}$ and force density $-\rho\nabla\Phi^{\rm react}$ are close analogs of the radiation-reaction potential [last term in Eq. (16.79)] and acceleration [right-hand side of Eq. (16.82)] that act on an oscillating ball that emits sound waves into a surrounding fluid. Moreover, Burke's derivation of his gravitational radiation-reaction potential is conceptually the same as the derivation, in Sec. 16.5.3, of the sound-wave reaction potential.

EXERCISES

Exercise 27.9 *Problem: Gravitational Waves from Arm Waving*
Wave your arms rapidly, and thereby try to generate gravitational waves.

(a) Using classical general relativity, compute in order of magnitude the wavelength of the waves you generate and their dimensionless amplitude at a distance of one wavelength away from you.

(b) How many gravitons do you produce per second? Discuss the implications of your result.

Exercise 27.10 *Example: Quadrupolar Wave Generation in Linearized Theory*
Derive the quadrupolar wave-generation formula (27.59) for a slow-motion, weak-gravity source in linearized theory using Lorenz gauge and beginning with the retarded-integral formula:

$$\bar{h}_{\mu\nu}(t, \mathbf{x}) = \int \frac{4T_{\mu\nu}(t - |\mathbf{x} - \mathbf{x}'|, \mathbf{x}')}{|\mathbf{x} - \mathbf{x}'|} dV_{x'} \tag{27.65}$$

[Eq. (25.91)]. Your derivation might proceed as follows.

(a) Show that for a slow-motion source, the retarded integral gives for the $1/r \equiv 1/|\mathbf{x}|$ (radiative) part of $\bar{h}_{jk}$:

$$\bar{h}_{jk}(t, \mathbf{x}) = \frac{4}{r} \int T_{jk}(t - r, \mathbf{x}') dV_{x'}. \tag{27.66}$$

(b) Show that in linearized theory using Lorenz gauge, the vacuum Einstein equation $-\bar{h}_{\mu\nu,\alpha}{}^{\alpha} = 16\pi T_{\mu\nu}$ [Eq. (25.90)] and the Lorenz gauge condition $\bar{h}_{\mu\nu,}{}^{\nu} = 0$ [Eq. (25.89)] together imply that the stress-energy tensor that generates the waves must have vanishing coordinate divergence: $T^{\mu\nu}{}_{,\nu} = 0$. This means that linearized theory is ignorant of the influence of self-gravity on the gravitating $T^{\mu\nu}$!

(c) Show that this vanishing divergence implies $[T^{00}x^j x^k]_{,00} = [T^{lm}x^j x^k]_{,ml}$ $-2[T^{lj}x^k + T^{lk}x^j]_{,l} + 2T^{jk}$.

(d) By combining the results of parts (a) and (c), deduce that

$$\bar{h}_{jk}(t, \mathbf{x}) = \frac{2}{r}\frac{d^2 I_{jk}(t - r)}{dt^2}, \tag{27.67}$$

where I_{jk} is the second moment of the source's (Newtonian) mass-energy distribution $T^{00} = \rho$ [Eq. (27.58)].

(e) Noticing that the trace-reversed metric perturbation (27.67) has the "speed-of-light-propagation" form, deduce that the gravitational-wave field h^{TT}_{jk} can be computed from Eq. (27.67) by a transverse-traceless projection [Box 27.2].

Part (b) shows that this linearized-theory analysis is incapable of deducing the gravitational waves emitted by a source whose dynamics is controlled by its self-gravity (e.g., a nearly Newtonian binary star system). By contrast, the derivation of the quadrupole formula given in Sec. 27.5.2 is valid for any slow-motion source, regardless of the strength and roles of its internal gravity; see the discussion following Eq. (27.57).

Exercise 27.11 *Problem: Energy and Angular Momentum Carried by Gravitational Waves*

(a) Compute the net rate at which the quadrupolar waves (27.57) carry energy away from their source, by carrying out the surface integral (25.102) with T^{0j} being Isaacson's gravitational-wave energy flux (27.40). Your answer should be Eq. (27.61). [Hint: Perform the TT projection in Cartesian coordinates using the projection tensor, Eq. (2) of Box 27.2, and make use of the following integrals over the solid angle on the unit sphere:

$$\frac{1}{4\pi}\int n_i d\Omega = 0, \quad \frac{1}{4\pi}\int n_i n_j d\Omega = \frac{1}{3}\delta_{ij}, \quad \frac{1}{4\pi}\int n_i n_j n_k d\Omega = 0,$$
$$\frac{1}{4\pi}\int n_i n_j n_k n_l d\Omega = \frac{1}{15}(\delta_{ij}\delta_{kl} + \delta_{ik}\delta_{jl} + \delta_{il}\delta_{jk}). \tag{27.68}$$

These integrals should be obvious by symmetry, aside from the numerical factors out in front. Those factors are most easily deduced by computing the z components (i.e., by setting $i = j = k = l = z$ and using $n_z = \cos\theta$).]

(b) The computation of the waves' angular momentum can be carried out in the same way, but is somewhat delicate, because a tiny nonradial component of the energy flux, that dies out as $1/r^3$, gives rise to the $O(1/r^2)$ angular momentum flux (see Thorne, 1980, Sec. IV.D).

Exercise 27.12 *Problem: Energy Removed by Gravitational Radiation Reaction*

Burke's radiation-reaction potential (27.64) produces a force per unit volume $-\rho\nabla\Phi^{\rm react}$ on its nearly Newtonian source. If we multiply this force per unit volume by the velocity $\mathbf{v} = d\mathbf{x}/dt$ of the source's material, we obtain thereby a rate of change of energy per unit volume. Correspondingly, the net rate of change of the system's mass-energy must be

$$\frac{dM}{dt} = -\int \rho\mathbf{v}\cdot\nabla\Phi^{\rm react} dV_x. \tag{27.69}$$

Show that, when averaged over a few gravitational-wave periods, this formula agrees with the rate of change of mass (27.61) that we derived in Ex. 27.11 by integrating the outgoing waves' energy flux.

27.5.4 Gravitational Waves from a Binary Star System

27.5.4

A very important application of the quadrupole formalism is to wave emission by a nearly Newtonian binary star system. Denote the stars by indices A and B and their masses by M_A and M_B, so their total and reduced mass are (as usual)

$$\boxed{M = M_A + M_B, \quad \mu = \frac{M_A M_B}{M};} \tag{27.70a}$$

and for simplicity, let the binary's orbit be circular, with separation a between the stars' centers of mass. Then Newtonian force balance dictates that the orbital angular velocity Ω is given by Kepler's law:

dynamics of a Newtonian binary in a circular orbit

$$\boxed{\Omega = \sqrt{M/a^3},} \tag{27.70b}$$

and the orbits of the two stars are

$$x_A = \frac{M_B}{M} a \cos \Omega t, \quad y_A = \frac{M_B}{M} a \sin \Omega t,$$
$$x_B = -\frac{M_A}{M} a \cos \Omega t, \quad y_B = -\frac{M_A}{M} a \sin \Omega\, t. \tag{27.70c}$$

The second moment of the mass distribution [Eq. (27.58)] is $I_{jk} = M_A x_A^j x_A^k + M_B x_B^j x_B^k$. Inserting the stars' time-dependent positions (27.70c), we obtain as the only nonzero components:

$$I_{xx} = \mu a^2 \cos^2 \Omega t, \quad I_{yy} = \mu a^2 \sin^2 \Omega t, \quad I_{xy} = I_{yx} = \mu a^2 \cos \Omega\, t \sin \Omega\, t. \tag{27.70d}$$

Noting that $\cos^2 \Omega t = \frac{1}{2}(1 + \cos 2\Omega t)$, $\sin^2 \Omega t = \frac{1}{2}(1 - \cos 2\Omega t)$, and $\cos \Omega t \sin \Omega t = \frac{1}{2} \sin 2\Omega t$ and evaluating the double time derivative, we obtain:

$$\ddot{I}_{xx} = -2\mu (M\Omega)^{2/3} \cos 2\Omega t, \quad \ddot{I}_{yy} = 2\mu (M\Omega)^{2/3} \cos 2\Omega t,$$
$$\ddot{I}_{xy} = \ddot{I}_{yx} = -2\mu (M\Omega)^{2/3} \sin 2\Omega\, t. \tag{27.70e}$$

We express these components in terms of Ω rather than a, because Ω is a direct gravitational-wave observable: the waves' angular frequency is 2Ω.

To compute the gravitational-wave field (27.59), we must project out the transverse traceless part of this $\ddot{I}_{jk}$. The projection is most easily performed in an orthonormal spherical basis, since there the transverse part is just the projection onto the plane spanned by $\vec{e}_{\hat{\theta}}$ and $\vec{e}_{\hat{\phi}}$, and the transverse-traceless part has components

$$(\ddot{I}_{\hat{\theta}\hat{\theta}})^{\rm TT} = -(\ddot{I}_{\hat{\phi}\hat{\phi}})^{\rm TT} = \frac{1}{2}(\ddot{I}_{\hat{\theta}\hat{\theta}} - \ddot{I}_{\hat{\phi}\hat{\phi}}), \quad (\ddot{I}_{\hat{\theta}\hat{\phi}})^{\rm TT} = \ddot{I}_{\hat{\theta}\hat{\phi}} \tag{27.70f}$$

[cf. Eq. (1) of Box 27.2]. Now, a little thought will save us much work: We need only compute these quantities at $\phi = 0$ (i.e., in the x-z plane), since their circular motion guarantees that their dependence on t and ϕ must be solely through the quantity $\Omega t - \phi$. At $\phi = 0$, $\vec{e}_{\hat{\theta}} = \vec{e}_x \cos \theta - \vec{e}_z \sin \theta$ and $\vec{e}_{\hat{\phi}} = \vec{e}_y$, so the only nonzero components of the transformation matrices from the Cartesian basis to the transverse part of the spherical basis are $L^x{}_{\hat{\theta}} = \cos \theta$, $L^z{}_{\hat{\theta}} = -\sin \theta$, and $L^y{}_{\hat{\phi}} = 1$. Using this transformation matrix at $\phi = 0$, we obtain: $\ddot{I}_{\hat{\theta}\hat{\theta}} = \ddot{I}_{xx} \cos^2 \theta$, $\ddot{I}_{\hat{\phi}\hat{\phi}} = \ddot{I}_{yy}$, and $\ddot{I}_{\hat{\theta}\hat{\phi}} = \ddot{I}_{xy} \cos \theta$. Inserting these and expressions (27.70e) into Eq. (27.70f), and setting $\Omega t \to \Omega t - \phi$ to make the formulas valid away from $\phi = 0$, we obtain:

$$(\ddot{I}_{\hat{\theta}\hat{\theta}})^{\rm TT} = -(\ddot{I}_{\hat{\phi}\hat{\phi}})^{\rm TT} = -(1 + \cos^2 \theta)\, \mu (M\Omega)^{2/3} \cos[2(\Omega t - \phi)],$$
$$(\ddot{I}_{\hat{\theta}\hat{\phi}})^{\rm TT} = +(\ddot{I}_{\hat{\phi}\hat{\theta}})^{\rm TT} = -2 \cos \theta\, \mu (M\Omega)^{2/3} \sin[2(\Omega t - \phi)]. \tag{27.70g}$$

The gravitational-wave field (27.59) is $2/r$ times this quantity evaluated at the retarded time $t - r$.

We make the conventional choice for the polarization tensors:

$$\boxed{\mathbf{e}^{+} = (\vec{e}_{\hat{\theta}} \otimes \vec{e}_{\hat{\theta}} - \vec{e}_{\hat{\phi}} \otimes \vec{e}_{\hat{\phi}}), \quad \mathbf{e}^{\times} = (\vec{e}_{\hat{\theta}} \otimes \vec{e}_{\hat{\phi}} + \vec{e}_{\hat{\phi}} \otimes \vec{e}_{\hat{\theta}});}$$

$$\boxed{\vec{e}_{\hat{\theta}} = \frac{1}{r}\frac{\partial}{\partial\theta}, \quad \vec{e}_{\hat{\phi}} = \frac{1}{r\sin\theta}\frac{\partial}{\partial\phi}.} \tag{27.71a}$$

Then Eqs. (27.59) and (27.70g) tell us that the gravitational-wave field is, in slot-naming index notation:

gravitational waves from Newtonian binary

$$\boxed{h^{\mathrm{TT}}_{\mu\nu} = h_{+} e^{+}_{\mu\nu} + h_{\times} e^{\times}_{\mu\nu},} \tag{27.71b}$$

where

$$\boxed{h_{+} = h^{\mathrm{TT}}_{\hat{\theta}\hat{\theta}} = \frac{2}{r}[\ddot{I}_{\hat{\theta}\hat{\theta}}(t-r)]^{\mathrm{TT}} = -2(1+\cos^2\theta)\,\frac{\mu(M\Omega)^{2/3}}{r}\,\cos[2(\Omega t - \Omega r - \phi)],} \tag{27.71c}$$

$$\boxed{h_{\times} = h^{\mathrm{TT}}_{\hat{\theta}\hat{\phi}} = \frac{2}{r}[\ddot{I}_{\hat{\theta}\hat{\phi}}(t-r)]^{\mathrm{TT}} = -4\cos\theta\,\frac{\mu(M\Omega)^{2/3}}{r}\,\sin[2(\Omega t - \Omega r - \phi)].} \tag{27.71d}$$

We have expressed the amplitudes of these waves in terms of the dimensionless quantity $(M\Omega)^{2/3} = M/a = v^2$, where v is the relative velocity of the two stars.

Notice that, as viewed from the polar axis $\theta = 0$, h_{+} and $h_{\times}$ are identical except for a $\pi/2$ phase delay, which means that the net stretch-squeeze ellipse (the combination of those in Figs. 27.1 and 27.2) rotates with angular velocity Ω. This is the gravitational-wave variant of circular polarization, and it arises because the binary motion as viewed from the polar axis looks circular. By contrast, as viewed by an observer in the equatorial plane $\theta = \pi/2$, $h_{\times}$ vanishes, so the net stretch-squeeze ellipse just oscillates along the $+$ axes, and the waves have linear polarization. This is natural, since the orbital motion as viewed by an equatorial observer is just a linear, horizontal, back-and-forth oscillation. Notice also that the gravitational-wave frequency is twice the orbital frequency:

binary's gravitational-wave frequency is twice its orbital frequency

$$\boxed{f = 2\frac{\Omega}{2\pi} = \frac{\Omega}{\pi}.} \tag{27.72}$$

To compute, via Eqs. (27.61) and (27.62), the rate at which energy and angular momentum are lost from the binary, we need to know the double and triple time derivatives of its quadrupole moment $\mathcal{I}_{jk}$. The double time derivative is just $\ddot{I}_{jk}$ with its trace removed, but Eq. (27.70e) shows that $\ddot{I}_{jk}$ is already traceless so $\ddot{\mathcal{I}}_{jk} = \ddot{I}_{jk}$. Inserting Eq. (27.70e) for this quantity into Eqs. (27.61) and (27.62) and performing the average over a gravitational-wave period, we find that

$$\frac{dM}{dt} = -\frac{32}{5}\frac{\mu^2}{M^2}(M\Omega)^{10/3}, \quad \frac{dJ_z}{dt} = -\frac{1}{\Omega}\frac{dM}{dt}, \quad \frac{dJ_x}{dt} = \frac{dJ_y}{dt} = 0. \tag{27.73}$$

binary's orbital inspiral due to gravitational-wave emission

This loss of energy and angular momentum causes the binary to spiral inward, decreasing the stars' separation a and increasing their orbital angular velocity Ω. By comparing Eqs. (27.73) with the standard equations for the binary's orbital energy and angular momentum [$M -$ (sum of rest masses of stars) $= E = -\frac{1}{2}\mu M/a = -\frac{1}{2}\mu(M\Omega)^{2/3}$, and $J_z = \mu a^2\Omega = \mu(M\Omega)^{2/3}/\Omega$], we obtain an equation for $d\Omega/dt$, which we can integrate to give

$$\Omega = \pi f = \left(\frac{5}{256}\frac{1}{\mu M^{2/3}}\frac{1}{t_o - t}\right)^{3/8}. \tag{27.74}$$

Here t_o (an integration constant) is the time remaining until the two stars merge, if the stars are thought of as point masses so their surfaces do not collide sooner. This equation can be inverted to read off the time until merger as a function of gravitational-wave frequency.

These results for a binary's waves and radiation-reaction-induced inspiral are of great importance for gravitational-wave detection (see, e.g., Cutler and Thorne, 2002; Sathyaprakash and Schutz, 2009).

EXERCISES

Exercise 27.13 *Problem: Gravitational Waves Emitted by a Linear Oscillator*
Consider a mass m attached to a spring, so it oscillates along the z-axis of a Cartesian coordinate system, moving along the world line $z = a\cos\Omega t$, $y = x = 0$. Use the quadrupole-moment formalism to compute the gravitational waves $h_+(t, r, \theta, \phi)$ and $h_\times(t, r, \theta, \phi)$ emitted by this oscillator, with the polarization tensors chosen as in Eqs. (27.71a). Pattern your analysis after the computation of waves from a binary in Sec. 27.5.4.

Exercise 27.14 ***Example: Propagation of a Binary's Waves Through an Expanding Universe*
As we shall see in Sec. 28.3, the following line element is a possible model for the large-scale structure of our universe:

$$ds^2 = b^2[-d\eta^2 + d\chi^2 + \chi^2(d\theta^2 + \sin^2\theta d\phi^2)], \quad \text{where } b = b_o\eta^2, \tag{27.75}$$

and b_o is a constant with dimensions of length. This is an expanding universe with flat spatial slices $\eta =$ constant. Notice that the proper time measured by observers at rest in the spatial coordinate system is $t = b_o \int \eta^2 d\eta = (b_o/3)\eta^3$.

A nearly Newtonian, circular binary is at rest at $\chi = 0$ in an epoch when $\eta \simeq \eta_o$. The coordinates of the binary's local asymptotic rest frame are $\{t, r, \theta, \phi\}$, where $r = b\chi$ and the coordinates cover only a tiny region of the universe: $\chi \lesssim \chi_o \ll \eta_o$. The gravitational waves in this local asymptotic rest frame are described by Eqs. (27.71). Use geometric optics (Sec. 27.4.2) to propagate these waves out through the expanding universe. In particular, do the following.

(a) Show that the null rays along which the waves propagate are the curves of constant θ, ϕ, and $\eta - \chi$.

(b) Each ray can be characterized by the retarded time τ_r at which the source emitted it. Show that

$$\tau_r = \frac{1}{3} b_o (\eta - \chi)^3. \tag{27.76a}$$

(c) Show that in the source's local asymptotic rest frame, this retarded time is $\tau_r = t - r$, and the phase of the wave is $\varphi = 2(\Omega\, \tau_r + \phi)$ [cf. Eqs. (27.71c) and (27.71d)]. Because the frequency Ω varies with time due to the binary inspiral, a more accurate formula for the wave's phase is $\varphi = 2(\int \Omega\, d\tau_r + \phi)$. Using Eq. (27.74), show that

$$\varphi = 2\phi - \left(\frac{t_o - \tau_r}{5\mathcal{M}}\right)^{5/8}, \quad \Omega = \frac{d\varphi}{d\tau_r} = \left(\frac{5}{256} \frac{1}{\mathcal{M}^{5/3}} \frac{1}{t_o - \tau_r}\right)^{3/8}, \tag{27.76b}$$

where

$$\mathcal{M} \equiv \mu^{3/5} M^{2/5} \tag{27.76c}$$

(with μ the reduced mass and M the total mass) is called the binary's *chirp mass,* because, as Eqs. (27.76b) show, it controls the rate at which the binary's orbital angular frequency Ω and the gravitational-wave angular frequency 2Ω "chirp upward" as time passes. The quantity τ_r as given by Eq. (27.76a) is constant along rays when they travel out of the local wave zone and into and through the universe. Correspondingly, if we continue to write φ in terms of τ_r on those rays using Eqs. (27.76b), this φ will be conserved along the rays in the external universe and therefore will satisfy the geometric-optics equation: $\nabla_{\vec{k}}\, \varphi = 0$ [Eqs. (27.35b)].

(d) Show that the orthonormal basis vectors and polarization tensors

$$\vec{e}_{\hat\theta} = \frac{1}{b\chi} \frac{\partial}{\partial\theta}, \quad \vec{e}_{\hat\phi} = \frac{1}{b\chi \sin\theta} \frac{\partial}{\partial\phi},$$

$$\mathbf{e}^{+} = (\vec{e}_{\hat\theta} \otimes \vec{e}_{\hat\theta} - \vec{e}_{\hat\phi} \otimes \vec{e}_{\hat\phi}), \quad \mathbf{e}^{\times} = (\vec{e}_{\hat\theta} \otimes \vec{e}_{\hat\phi} + \vec{e}_{\hat\phi} \otimes \vec{e}_{\hat\theta}) \tag{27.76d}$$

in the external universe: (i) are parallel-transported along rays and (ii) when carried backward on rays into the local asymptotic rest frame, become the basis vectors and tensors used in that frame's solution (27.71) for the gravitational waves. Therefore, these $\mathbf{e}^{+}_{\mu\nu}$ and $\mathbf{e}^{\times}_{\mu\nu}$ are the polarization tensors needed for our geometric-optics waves.

(e) Consider a bundle of rays that, at the source, extends from ϕ to $\phi + \Delta\phi$ and from θ to $\theta + \Delta\theta$. Show that this bundle's cross sectional area, as it moves outward to larger and larger χ, is $\mathcal{A} = r^2 \sin\theta\,\Delta\theta\,\Delta\phi$, where r is a function of η and χ given by

$$r = b\chi = b_o\eta^2\chi. \tag{27.76e}$$

Show that in the source's local asymptotic rest frame, this r is the same as the distance r from the source that appears in Eqs. (27.71c) and (27.71d) for h_+ and $h_\times$.

(f) By putting together all the pieces from parts (a) through (e), show that the solution to the equations of geometric optics (27.35) for the gravitational-wave field as it travels outward through the universe is

$$\mathbf{h}^{\mathrm{TT}} = h_+\mathbf{e}^+ + h_\times\mathbf{e}^\times, \tag{27.76f}$$

with $\mathbf{e}^+$ and $\mathbf{e}^\times$ given by Eqs. (27.76d), with h_+ and $h_\times$ given by

$$h_+ = -2(1+\cos^2\theta)\frac{\mathcal{M}^{5/3}\Omega^{2/3}}{r}\cos\varphi, \quad h_\times = -4\cos\theta\frac{\mathcal{M}^{5/3}\Omega^{2/3}}{r}\sin\varphi, \tag{27.76g}$$

and with Ω, φ, and r given by Eqs. (27.76b), (27.76a), and (27.76e). [Hint: Note that all quantities in this solution except r are constant along rays, and r varies as $1/\sqrt{\mathcal{A}}$, where $\mathcal{A}$ is the area of a bundle of rays.]

(g) The angular frequency of the waves that are emitted at retarded time τ_r is $\omega_e = 2\Omega$. When received at Earth these waves have a cosmologically redshifted frequency $\omega_r = \partial\varphi/\partial t$, where $t = (b_o/3)\eta^3$ is proper time measured at Earth, and in the derivative we must hold fixed the spatial coordinates of Earth: $\{\chi, \theta, \phi\}$. The ratio of these frequencies is $\omega_e/\omega_r = 1 + z$, where z is the so-called cosmological redshift of the waves. Show that $1 + z = (\partial\tau_r/\partial t)^{-1} = \eta^2/(\eta - \chi)^2$.

(h) Show that the information carried by the binary's waves is the following. (i) From the ratio of the amplitudes of the two polarizations, one can read off the inclination angle θ of the binary's spin axis to the line of sight to the binary. (ii) From the waves' measured angular frequency ω and its time rate of change $d\omega/dt$, one can read off $(1+z)\mathcal{M}$, the binary's redshifted chirp mass. (iii) From the amplitude of the waves, with θ and $(1+z)\mathcal{M}$ known, one can read off $(1+z)r$, a quantity known to cosmologists as the binary's *luminosity distance*. [Note: It is remarkable that gravitational waves by themselves reveal the source's luminosity distance but not its redshift, while electromagnetic observations reveal the redshift but not the luminosity distance. This complementarity illustrates the importance and power of combined gravitational-wave and electromagnetic observations.]

27.5.5 Gravitational Waves from Binaries Made of Black Holes, Neutron Stars, or Both: Numerical Relativity T2

27.5.5

Among the most interesting sources of gravitational waves are binary systems made of two black holes, a black hole and a neutron star, or two neutron stars—so-called *compact binaries.* When the two bodies are far apart, their motion and waves can be described accurately by Newtonian gravity and the quadrupole-moment formalism: the formulas in Sec. 27.5.4. As the bodies spiral inward, $(M\Omega)^{2/3} = M/a = v^2$ grows larger, h_+ and $h_\times$ grow larger, and relativistic corrections to our Newtonian, quadrupole analysis grow larger. Those relativistic corrections (including current-quadrupole waves, mass-octupole waves, etc.) can be computed using a post-Newtonian expansion of the Einstein field equations (i.e., an expansion in $M/a \sim v^2$). The accuracies of ground-based detectors such as LIGO require that, for compact binaries, the expansion be carried at least to order v^7 beyond our Newtonian, quadrupole analysis! (See Blanchet, 2014.)

compact binaries analyzed by a post-Newtonian expansion

At the end of the inspiral, the binary's bodies come crashing together. To compute the waves from this final merger with an accuracy comparable to the observations, it is necessary to solve the Einstein field equation on a computer. The techniques for doing this are called *numerical relativity* (Baumgarte and Shapiro, 2010; Shibata, 2016) and were pioneered by Bryce DeWitt, Larry Smarr, Saul Teukolsky, Frans Pretorius, and others.

merger analyzed by numerical relativity

For binary black holes with approximately equal masses, simulations using numerical relativity reveal that the total energy radiated in gravitational waves is $\Delta E \sim 0.1Mc^2$, where M is the binary's total mass. Most of this energy is emitted in the last $\sim$5 to 10 cycles of waves, at wave periods $P \sim (10 \text{ to } 20)GM/c^3$ [i.e., frequencies $f = 1/P \sim 1{,}000$ Hz $(10M_\odot/M)$]. The gravitational-wave power output in these last 5–10 cycles is $dE/dt \sim 0.1Mc^2/(100GM/c^3) = 0.001c^5/G$, which is roughly 10^{23} times the luminosity of the Sun, and 100 times the luminosity of all the stars in the observable universe put together! If the holes have masses $\sim 10M_\odot$, this enormous luminosity lasts for only $\sim$0.1 s, and the total energy emitted is the rest-mass energy of the Sun. If the holes have masses $\sim 10^9 M_\odot$, the enormous luminosity lasts for $\sim$1 yr, and the energy emitted is the rest-mass energy of $\sim 10^8$ Suns.

gravitational-wave power output in black hole mergers: 100 universe luminosities

For the simplest case of two identical, nonspinning black holes that spiral together in a circular orbit, both waveforms (both wave shapes) have the simple time evolution shown in Fig. 27.6. As the holes spiral together, their amplitude and phase increase in accord with the Newtonian-quadrupole formulas (27.71), (27.72), and (27.74) but by the time of this figure, post-Newtonian corrections are producing noticeable differences from those formulas. When the holes merge, the gravitational-wave amplitude

T2

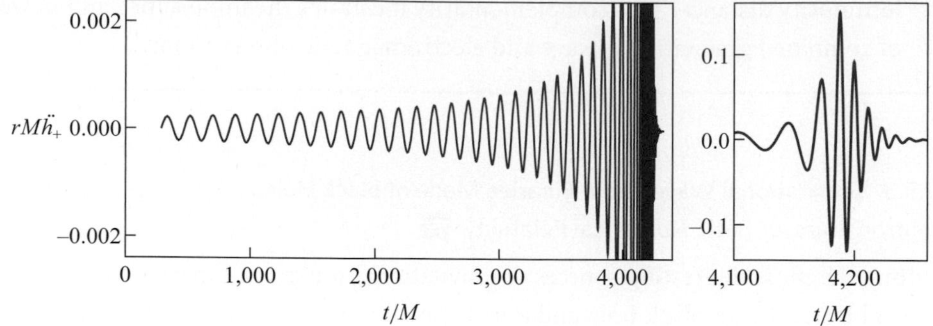

FIGURE 27.6 For a binary made of two identical, nonspinning black holes that spiral together and merge: the time evolution of the gravitational-wave tidal field $\mathcal{E}_{ij} \propto \ddot{h}_+$ [Eq. (27.22)] for the + polarization. The × polarization waveform is the same as this but with a phase shift. The right panel shows the detailed signal just prior to merger. Based on simulations performed by the Caltech/Cornell/CITA numerical relativity group (Mroué et. al. 2013). (CITA is the Canadian Institute for Theoretical Astrophysics.)

reaches a maximum amplitude. The merged hole then vibrates and the waves "ring down" with exponentially decaying amplitude.

Much more interesting are binaries made of black holes that spin. In this case, the angular momentum of each spinning hole drags inertial frames, as does the binary's orbital angular momentum. This frame-dragging causes the spins and the orbital plane to precess, and those precessions modulate the waves. Figure 27.7 depicts a generic example: a binary whose holes have a mass ratio 6:1, dimensionless spins $a_A/M_A = 0.91$, $a_B/M_B = 0.30$, and randomly chosen initial spin axes and orbital plane. Frame dragging causes the orbital motion to be rather complex, and correspondingly, the two waveforms are much richer than in the nonspinning case. The waveforms carry detailed information about the binary's masses, spins, and orbital evolution, and also about the geometrodynamics of its merger (Box 27.4).

information in gravitational waves

EXERCISES

Exercise 27.15 *Problem: Maximum Gravitational-Wave Amplitude* T2

Extrapolating Eqs. (27.71)–(27.73) into the strong-gravity regime, estimate the maximum gravitational-wave amplitude and emitted power for a nonspinning binary black hole with equal masses and with unequal masses. Compare with the results from numerical relativity discussed in the text.

Exercise 27.16 *Problem: Gravitational Radiation from Binary Pulsars in Elliptical Orbits* T2

Many precision tests of general relativity are associated with binary pulsars in elliptical orbits (Sec. 27.2.6).

(a) Verify that the radius of the relative orbit of the pulsars can be written as $r = p/(1 + e \cos\phi)$, where p is the *semi-latus rectum*, e is the eccentricity, and $d\phi/dt = (Mp)^{1/2}/r^2$ with M the total mass.

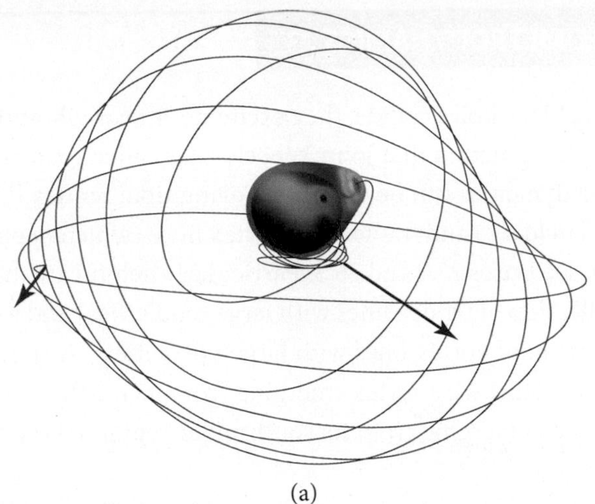

(a)

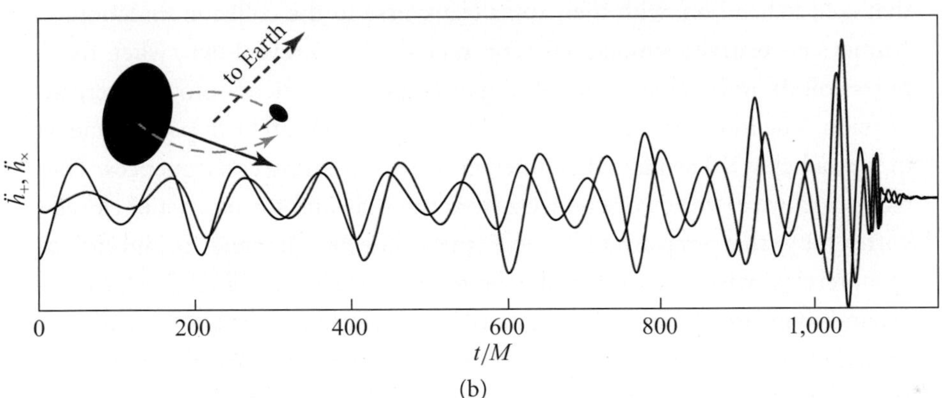

(b)

FIGURE 27.7 (a) The orbital motion of a small black hole around a larger black hole (mass ratio $M_B/M_A = 1/6$), when the spins are $a_B/M_B = 0.30$ and $a_A/M_A = 0.91$ and the initial spin axes and orbital plane are as shown in panel b. (b) The two gravitational waveforms emitted in the direction toward Earth (blue dashed line). These waveforms are from a catalog of simulations of 174 different binary-black-hole mergers, carried out by the Caltech/Cornell/CITA numerical relativity group (Mroué et al., 2013).

(b) Show that the traceless mass quadrupole moment, Eq. (27.53), in suitable coordinates, is [cf. Eq. (27.70d)]

$$\mathcal{I}_{jk} = \frac{\mu p^2}{(1 + e\cos\phi)^2} \begin{pmatrix} \cos^2\phi - \frac{1}{3} & \cos\phi\sin\phi & 0 \\ \cos\phi\sin\phi & \sin^2\phi - \frac{1}{3} & 0 \\ 0 & 0 & -\frac{1}{3} \end{pmatrix}. \quad (27.77a)$$

(c) Use computer algebra to evaluate the second and third time derivatives of this tensor, and then use Eqs. (27.61) and (27.62) to calculate the orbit-averaged energy and angular momentum emitted in gravitational waves, per orbit.

T2

BOX 27.4. GEOMETRODYNAMICS T2

When spinning black holes collide, they excite nonlinear vibrations of curved spacetime—a phenomenon that John Wheeler has called *geometrodynamics.* This nonlinear dynamics can be visualized using tidal tendex lines (which depict the tidal field $\mathcal{E}_{ij}$) and frame-drag vortex lines (which depict the frame-drag field $\mathcal{B}_{ij}$); see Boxes 25.2 and 26.3. Particularly helpful are the concepts of a *tendex* (a collection of tendex lines with large tendicities) and a *(frame-drag) vortex* (a collection of vortex lines with large vorticities). A spinning black hole has a counterclockwise vortex emerging from its north polar region, and a clockwise vortex emerging from its south polar region (right diagram in Box 26.3).

As an example of geometrodynamics, consider two identical black holes that collide head on, with their spins transverse to the collision direction. Numerical-relativity simulations (Owen et al., 2011) reveal that, when the holes collide and merge, each hole deposits its two vortices onto the merged horizon. The four vortices dynamically attach to each other in pairs (panel a in figure below). The pairs then interact, with surprising consequences. The blue (clockwise) vortex disconnects from the hole and forms a set of closed vortex loops that wrap around a torus (thick blue lines in panel c), and the red (counterclockwise) vortex does the same (thin red lines in panel c). This torus expands outward at the speed of light, while energy temporarily stored in near-horizon tendices (not shown) regenerates the new pair of horizon-penetrating vortices shown in panel b, with reversed vorticities (reversed directions of twist). As the torus expands outward, its motion, via the Maxwell-like Bianchi identity $\partial\mathcal{E}/\partial t = (\nabla\times\mathcal{B})^{\rm S}$ (Box 25.2), generates a set of tendex lines that wrap around the torus at 45° angles to the vortex lines (dashed lines in panel c). The torus's interleaved vortex and tendex lines have become a gravitational wave, which locally looks like the plane wave discussed in Box 27.3. This process repeats, with amplitude damping, generating a sequence of expanding tori. (Figure adapted from Owen et al., 2011.)

(continued)

Your answer should be

$$\Delta E = \frac{64\pi\mu^2 M^{5/2}}{5p^{7/2}}\left(1+\frac{73}{24}e^2+\frac{37}{96}e^4\right), \quad \Delta J = \frac{64\pi\mu^2 M^2}{5p^2}\left(1+\frac{7}{8}e^2\right). \tag{27.77b}$$

(d) Combine the results from part (c) with Kepler's laws to calculate an expression for the rate of increase of pulse period and decrease of the eccentricity (cf. Sec. 27.2.6).

T2

BOX 27.4. (continued)

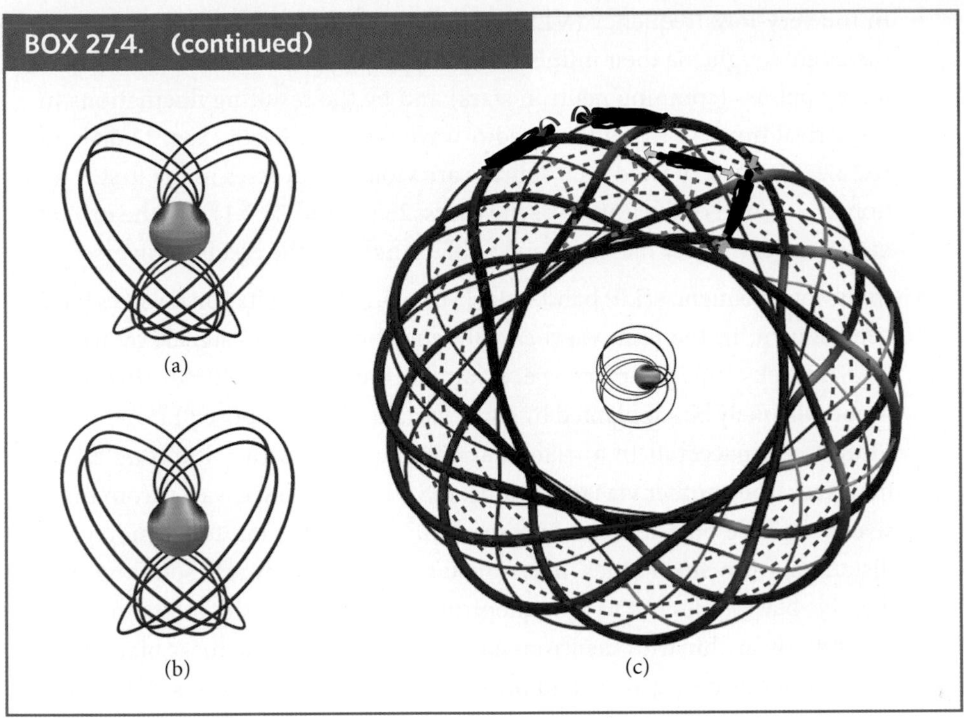

(e) Consider a parabolic encounter between two stars, and show that the energy and angular momentum radiated are, respectively,

$$\Delta E = \frac{170\pi\,\mu^2 M^{5/2}}{3p^{7/2}}, \quad \Delta L = \frac{24\pi\,\mu^2 M^2}{p^2}. \tag{27.78}$$

27.6 The Detection of Gravitational Waves

27.6

27.6.1 Frequency Bands and Detection Techniques

27.6.1

Physicists and astronomers are searching for gravitational waves in four different frequency bands using four different techniques:

gravitational-wave frequency bands: ELF, VLF, LF, and HF; sources and detection techniques in each band

- In the extremely low-frequency (ELF) band, $\sim 10^{-18}$ to $\sim 10^{-15}$ Hz, gravitational waves are sought via their imprint on the polarization of the cosmic microwave background (CMB) radiation. There is only one expected ELF source of gravitational waves, but it is a very interesting one: quantum fluctuations in the gravitational field (spacetime curvature) that emerge from the big bang's quantum-gravity regime, the *Planck era,* and that are subsequently amplified to classical, detectable sizes by the universe's early inflationary expansion. We shall study this amplification and the resulting ELF gravitational waves in Sec. 28.7.1 and shall see these waves' great potential for probing the physics of inflation.

- In the very-low-frequency (VLF) band, $\sim 10^{-9}$ to $\sim 10^{-7}$ Hz, gravitational waves are sought via their influence on the propagation of radio waves emitted by pulsars (spinning neutron stars) and by the resulting fluctuations in the arrival times of the pulsars' radio-wave pulses at Earth (Sec. 27.6.6 and Ex. 27.20). The expected VLF sources are violent processes in the first fraction of a second of the universe's life (Secs. 28.4.1 and 28.7.1) and the orbital motion of extremely massive pairs of black holes in the distant universe.
- In the low-frequency (LF) band, $\sim 10^{-4}$ to ~ 0.1 Hz, gravitational waves have been sought, in the past, via their influence on the radio signals by which NASA tracks interplanetary spacecraft. In the 2020s or 2030s, this technique will likely be supplanted by some variant of the proposed LISA—three "drag-free" spacecraft in a triangular configuration with 5-km-long arms, that track one another via laser beams. LISA is likely to see waves from massive black-hole binaries (hole masses $\sim 10^5$ to $10^7 M_\odot$) out to cosmological distances; from small holes, neutron stars, and white dwarfs spiraling into massive black holes out to cosmological distances; from the orbital motion of white-dwarf binaries, neutron-star binaries, and stellar-mass black-hole binaries in our own galaxy; and possibly from violent processes in the very early universe.
- The high-frequency (HF) band, ~ 10 to $\sim 10^3$ Hz, is where Earth-based detectors operate: laser interferometer gravitational-wave detectors, such as LIGO, and resonant-mass detectors in which a gravitational wave alters the amplitude and phase of vibrations of a normal mode of a large, solid cylinder or sphere. On September 14, 2015, the advanced LIGO gravitational wave detectors made their first detection: a wave burst named GW150914 with amplitude 1.0×10^{-21}, duration ~ 150 ms, and frequency chirping upward from ~ 50 Hz (when it entered the LIGO band) to 240 Hz (Abbott et al., 2016). By comparing the observed waveform with those from numerical relativity simulations, the LIGO-VIRGO scientists deduced that the waves came from the merger of a $29M_\odot$ black hole with a $36M_\odot$ black hole, 1.2 billion light years from Earth, to form a $62M_\odot$ black hole, with a release of $3M_\odot c^2$ of energy in gravitational waves. To date (spring, 2020), roughly 60 black hole binaries have been observed; also a spectacular neutron star binary, GW170817 (Abbott et al., 2017), and a source that, based on the measured masses, is probably a neutron star-black hole merger. As LIGO's sensitivity improves and additional interferometers come on line, the LIGO scientists expect to see other sources: waves from spinning, slightly deformed neutron stars (e.g., pulsars), supernovae, the triggers of gamma-ray bursts, and possibly waves from violent processes in the very early universe.

For detailed discussions of these gravitational-wave sources in all four frequency bands, and of prospects for their detection, see, for example, Cutler and Thorne (2002) and Sathyaprakash and Schutz (2009) and references therein. It is likely that waves will be seen in all four bands by about 2030.

EXERCISES

Exercise 27.17 *Example: Strongest Gravitational Waves in HF Band*

(a) Using an order-of-magnitude analysis based on Eq. (27.60), show that the strongest gravitational waves that are likely to occur each year in LIGO's HF band have $h_+ \sim h_\times \sim 10^{-21}$—which is the actual amplitude of LIGO's first observed wave burst, GW150914. [Hint: The highest nonspherical kinetic energy achievable must be for a highly deformed object (or two colliding objects), in which the internal velocities approach the speed of light—say, for realism, $v \sim 0.3c$. To achieve these velocities, the object's size L must be of order 2 or 3 Schwarzschild radii, $L \sim 5M$, where M is the source's total mass. The emitted waves must have $f \sim 200$ Hz (the frequency at the minimum of Advanced LIGO's noise curve—which is similar to initial LIGO, Fig. 6.7, but a factor ~10 lower). Using these considerations, estimate the internal angular frequency of the source's motion, and thence the source's mass, and finally the source's internal kinetic energy. Such a source will be very rare, so to see a few per year, its distance must be some substantial fraction of the Hubble distance. From this, estimate $h_+ \sim h_\times$.]

(b) As a concrete example, estimate the gravitational-wave strength from the final moments of inspiral and merger of two black holes, as described by Eqs. (27.71) and (27.70b) extrapolated into the highly relativistic domain.

27.6.2 Gravitational-Wave Interferometers: Overview and Elementary Treatment

27.6.2

idealized gravitational-wave interferometer

We briefly discussed Earth-based gravitational-wave interferometers such as LIGO in Sec. 9.5, focusing on optical interferometry issues. In this section we analyze the interaction of a gravitational wave with such an interferometer. This analysis will not only teach us much about gravitational waves, but it will also illustrate some central issues in the physical interpretation of general relativity theory.

To get quickly to the essentials, we examine a rather idealized interferometer: a Michelson interferometer (one without the input mirrors of Fig. 9.13) that floats freely in space, so there is no need to hang its mirrors by wires; see Fig. 27.8. In Sec. 27.6.5, we briefly discuss more realistic interferometers.

If we ignore delicate details, the operation of this idealized interferometer is simple. As seen in a local Lorentz frame of the beam splitter, the gravitational wave changes the length of the x arm by $\delta x = \frac{1}{2}h_+\ell_x$, where ℓ_x is the unperturbed length, and it changes that of the y arm by the opposite amount: $\delta y = -\frac{1}{2}h_+\ell_y$ [Eqs. (27.29)]. The interferometer is operated with unperturbed lengths ℓ_x and ℓ_y that are nearly but not quite equal, so there is a small amount of light going toward the photodetector. The wave-induced change of arm length causes a relative phase shift of the light

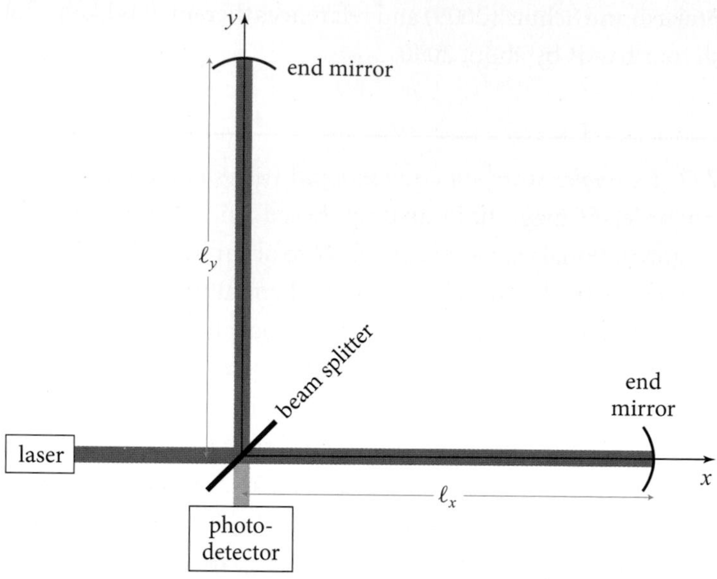

FIGURE 27.8 An idealized gravitational-wave interferometer.

returning down the two arms to the beam splitter given by $\Delta\varphi(t) = \omega_o(2\delta y - 2\delta x) = \omega_o(\ell_x + \ell_y)h_+(t)$, where ω_o is the light's angular frequency (and we have set the speed of light to unity); cf. Sec. 9.5. This oscillating phase shift modulates the intensity of the light going into the photodetector by $\Delta I_{\text{PD}}(t) \propto \Delta\varphi(t)$. Setting $\ell_x \simeq \ell_y = \ell$, this modulation is

interferometer's photodetector current output

$$\boxed{\Delta I_{\text{PD}}(t) \propto \Delta\varphi(t) = 2\omega_o \ell h_+(t).} \tag{27.79}$$

Therefore, the photodetector output tells us directly the gravitational waveform $h_+(t)$.

In the following two (Track-Two) subsections, we rederive this result much more carefully in two different coordinate systems (two different gauges). Our two analyses predict the same result (27.79) for the interferometer output, but they appear to attribute that result to two different mechanisms.

TT-gauge analysis attributes output signal to influence of waves on interferometer's light

In our first analysis (performed in TT gauge; Sec. 27.6.3), the interferometer's test mass remains always at rest in our chosen coordinate system, and the gravitational wave $h_+(t - z)$ interacts with the interferometer's light. The imprint that $h_+(t - z)$ leaves on the light causes a fluctuating light intensity $I_{\text{out}}(t) \propto h_+(t)$ to emerge from the interferometer's output port and be measured by the photodetector.

proper-reference-frame analysis attributes output signal to interaction of waves with interferometer mirrors

In our second analysis (a more rigorous version of the above quick analysis, performed in the proper reference frame of the interferometer's beam splitter; Sec. 27.6.4) the gravitational waves interact hardly at all with the light. Instead, they push the end mirrors back and forth relative to the coordinate system, thereby lengthening one arm while shortening the other. These changing arm lengths cause a changing interference of the light returning to the beam splitter from the two arms, and that changing interference produces the fluctuating light intensity $I_{\text{out}}(t) \propto h_+(t)$ measured by the photodetectors.

These differences of viewpoint are somewhat like the differences between the Heisenberg picture and the Schrödinger picture in quantum mechanics. The intuitive pictures associated with two viewpoints appear to be very different (Schrödinger's wave function versus Heisenberg's matrices; gravitational waves interacting with light versus gravitational waves pushing on mirrors). But when one computes the same physical observable from the two different viewpoints (probability for a quantum measurement outcome; light intensity measured by photodetector), the two viewpoints give the same answer.

analogy with Heisenberg and Schrödinger pictures in quantum mechanics

27.6.3 Interferometer Analyzed in TT Gauge T2

27.6.3

For our first analysis, we place the interferometer at rest in the x-y plane of a TT coordinate system, with its arms along the x- and y-axes and its beam splitter at the origin, as shown in Fig. 27.8. For simplicity, we assume that the gravitational wave propagates in the z direction and has + polarization, so the linearized spacetime metric has the TT-gauge form:

interferometer analyzed in TT gauge

spacetime metric

$$\boxed{ds^2 = -dt^2 + [1 + h_+(t - z)]\, dx^2 + [1 - h_+(t - z)]\, dy^2 + dz^2} \tag{27.80}$$

[Eq. (27.20)]. For ease of notation, we omit the subscript + from h_+ in the remainder of this section.

The beam splitter and end mirrors move freely and thus travel along geodesics of the metric (27.80). The splitter and mirrors are at rest in the TT coordinate system before the wave arrives, so initially, the spatial components of their 4-velocities vanish: $u_j = 0$. Because the metric coefficients $g_{\alpha\beta}$ are all independent of x and y, the geodesic equation dictates that the components u_x and u_y are conserved and thus remain zero as the wave passes, which implies (since the metric is diagonal) $u^x = dx/d\tau = 0$ and $u^y = dy/d\tau = 0$. One can also show (see Ex. 27.18) that $u^z = dz/d\tau = 0$ throughout the wave's passage. Thus, in terms of motion relative to the TT coordinate system, the gravitational wave has no influence at all on the beam splitter and mirrors; they all remain at rest (constant x, y, and z) as the waves pass.

mirrors and beam splitter do not move relative to TT coordinates

(Despite this lack of motion, the proper distances between the mirrors and the beam splitter—the interferometer's physically measured arm lengths—do change. If the unchanging coordinate lengths of the two arms are $\Delta x = \ell_x$ and $\Delta y = \ell_y$, then the metric (27.80) says that the physically measured arm lengths are

$$\boxed{L_x = \left[1 + \frac{1}{2}h(t)\right] \ell_x, \quad L_y = \left[1 - \frac{1}{2}h(t)\right] \ell_y.} \tag{27.81}$$

When h is positive, the x arm is lengthened and the y arm is shortened; when negative, L_x is shortened and L_y is lengthened.)

Next turn to the propagation of light in the interferometer. We assume, for simplicity, that the light beams have large enough transverse sizes that we can idealize them, on their optic axes, as plane electromagnetic waves. (In reality, they will be Gaussian

T2

beams, of the sort studied in Sec. 8.5.5.) The light's vector potential A^α satisfies the curved-spacetime vacuum wave equation $A^{\alpha;\mu}{}_{\mu} = 0$ [Eq. (25.60) with vanishing Ricci tensor]. We write the vector potential in geometric optics (eikonal-approximation) form as

$$A^\alpha = \Re(\mathsf{A}^\alpha e^{i\varphi}), \tag{27.82}$$

where A^α is a slowly varying amplitude, and φ is a rapidly varying phase [cf. Eq. (7.20)]. Because the wavefronts are (nearly) planar and the spacetime metric is nearly flat, the light's amplitude A^α will be nearly constant as it propagates down the arms, and we can ignore its variations. Not so the phase. It oscillates at the laser frequency $\omega_o/2\pi \sim 3 \times 10^{14}$ Hz [i.e., $\varphi^{\rm out}_{x\ \rm arm} \simeq \omega_o(x - t)$ for light propagating outward from the beam splitter along the x arm, and similarly for the returning light and the light in the y arm]. The gravitational wave places tiny deviations from this $\omega_o(x - t)$ onto the phase; we must compute those deviations.

In the spirit of geometric optics, we introduce the light's spacetime wave vector

$$\vec{k} \equiv \vec{\nabla}\varphi, \tag{27.83}$$

and we assume that $\vec{k}$ varies extremely slowly compared to the variations of φ. Then the wave equation $A^{\alpha;\mu}{}_{\mu} = 0$ reduces to the statement that the wave vector is null: $\vec{k} \cdot \vec{k} = \varphi_{,\alpha}\varphi_{,\beta}g^{\alpha\beta} = 0$. For light in the x arm the phase depends only on x and t; for that in the y arm it depends only on y and t. Combining this with the TT metric (27.80) and noting that the interferometer lies in the $z = 0$ plane, we obtain

influence of waves on phase of light in interferometer arms

$$\begin{aligned} -\left(\frac{\partial\varphi_{x\ \rm arm}}{\partial t}\right)^2 + [1 - h(t)]\left(\frac{\partial\varphi_{x\ \rm arm}}{\partial x}\right)^2 &= 0, \\ -\left(\frac{\partial\varphi_{y\ \rm arm}}{\partial t}\right)^2 + [1 + h(t)]\left(\frac{\partial\varphi_{y\ \rm arm}}{\partial y}\right)^2 &= 0. \end{aligned} \tag{27.84}$$

We idealize the laser as perfectly monochromatic, and we place it at rest in our TT coordinates, arbitrarily close to the beam splitter. Then the outgoing light frequency, as measured by the beam splitter, must be precisely ω_o and cannot vary with time. Since proper time as measured by the beam splitter is equal to coordinate time t [cf. the metric (27.80)], the frequency that the laser and beam splitter measure must be $\omega = -\partial\varphi/\partial t = -k_t$. This dictates the following boundary conditions (initial conditions) on the phase of the light that travels outward from the beam splitter:

$$\frac{\partial\varphi^{\rm out}_{x\ \rm arm}}{\partial t} = -\omega_o \text{ at } x = 0, \qquad \frac{\partial\varphi^{\rm out}_{y\ \rm arm}}{\partial t} = -\omega_o \text{ at } y = 0. \tag{27.85}$$

It is straightforward to verify that the solutions to Eq. (27.84) (and thence to the wave equation and thence to Maxwell's equations) that satisfy the boundary conditions (27.85) are

T2

$$\varphi_{x\,\text{arm}}^{\text{out}} = -\omega_o\left[t - x + \frac{1}{2}H(t-x) - \frac{1}{2}H(t)\right],$$

$$\varphi_{y\,\text{arm}}^{\text{out}} = -\omega_o\left[t - y - \frac{1}{2}H(t-y) + \frac{1}{2}H(t)\right], \tag{27.86}$$

where $H(t)$ is the first time integral of the gravitational waveform:

$$H(t) \equiv \int_0^t h(t')dt' \tag{27.87}$$

(cf. Ex. 27.19).

The outgoing light reflects off the mirrors, which are at rest in the TT coordinates at locations $x = \ell_x$ and $y = \ell_y$. As measured by observers at rest in these coordinates, there is no Doppler shift of the light, because the mirrors are not moving. Correspondingly, the phases of the reflected light, returning back along the two arms, have the following forms:

$$\varphi_{x\,\text{arm}}^{\text{back}} = -\omega_o\left[t + x - 2\ell_x + \frac{1}{2}H(t + x - 2\ell_x) - \frac{1}{2}H(t)\right],$$

$$\varphi_{y\,\text{arm}}^{\text{back}} = -\omega_o\left[t + y - 2\ell_y - \frac{1}{2}H(t + y - 2\ell_y) + \frac{1}{2}H(t)\right]. \tag{27.88}$$

The difference of the phases of the returning light, at the beam splitter ($x = y = 0$), is

difference between arms for output light's phase shift

$$\Delta\varphi \equiv \varphi_{x\,\text{arm}}^{\text{back}} - \varphi_{y\,\text{arm}}^{\text{back}} = -\omega_o[-2(\ell_x - \ell_y) + \frac{1}{2}H(t - 2\ell_x) + \frac{1}{2}H(t - 2\ell_y) - H(t)]$$

$$\simeq +2\omega_o[\ell_x - \ell_y + \ell h(t)] \quad \text{for Earth-based interferometers.} \tag{27.89}$$

In the final expression we have used the fact that for Earth-based interferometers operating in the HF band, the gravitational wavelength $\lambda_{GW} \sim c/(100\text{ Hz}) \sim 3{,}000$ km is long compared to the interferometers' ~4-km arms, and the arms have nearly the same length: $\ell_y \simeq \ell_x \equiv \ell$.

The beam splitter sends a light field $\propto e^{i\varphi_{x\,\text{arm}}^{\text{back}}} + e^{i\varphi_{y\,\text{arm}}^{\text{back}}}$ back toward the laser, and a field $\propto e^{i\varphi_{x\,\text{arm}}^{\text{back}}} - e^{i\varphi_{y\,\text{arm}}^{\text{back}}} = e^{i\varphi_{y\,\text{arm}}^{\text{back}}}(e^{i\Delta\varphi} - 1)$ toward the photodetector. The intensity of the light entering the photodetector is proportional to the squared amplitude of the field: $I_{PD} \propto |e^{i\Delta\varphi} - 1|^2$. We adjust the interferometer's arm lengths so their difference $\ell_x - \ell_y$ is small compared to the light's reduced wavelength $1/\omega_o = c/\omega_o$ but large compared to $|\ell h(t)|$. Correspondingly, $|\Delta\varphi| \ll 1$, so only a tiny fraction of the light goes toward the photodetector (it is the interferometer's "dark port"), and that dark-port light intensity is

photodiode output

$$\boxed{I_{PD} \propto |e^{i\Delta\varphi} - 1|^2 \simeq |\Delta\varphi|^2 \simeq 4\omega_o^2(\ell_x - \ell_y)^2 + 8\omega_o^2(\ell_x - \ell_y)\ell h_+(t).} \tag{27.90}$$

Here we have restored the subscript + onto h. The time-varying part of this intensity is proportional to the gravitational waveform $h_+(t)$ [in agreement with Eq. (27.79)]. It is this time-varying part that the photodetector reports as the interferometer output.

T2

EXERCISES

Exercise 27.18 *Derivation and Practice: Geodesic Motion in TT Coordinates* T2

Consider a particle that is at rest in the TT coordinate system of the gravitational-wave metric (27.80) before the gravitational wave arrives. In the text it is shown that the particle's 4-velocity has $u^x = u^y = 0$ as the wave passes. Show that $u^z = 0$ and $u^t = 1$ as the wave passes, so the components of the particle's 4-velocity are unaffected by the passing gravitational wave, and the particle remains at rest (constant x, y, and z) in the TT coordinate system.

Exercise 27.19 *Example: Light in an Interferometric Gravitational-Wave Detector in TT Gauge* T2

Consider the light propagating outward from the beam splitter, along the x arm of an interferometric gravitational-wave detector, as analyzed in TT gauge, so (suppressing the subscript "x arm" and superscript "out") the electromagnetic vector potential is $A^\alpha = \Re(\mathsf{A}^\alpha e^{i\varphi(x,t)})$, with A^α constant and with $\varphi = -\omega_o\left[t - x + \frac{1}{2}H(t-x) - \frac{1}{2}H(t)\right]$ [Eqs. (27.86) and (27.87)].

(a) Show that this φ satisfies the nullness equation (27.84), as claimed in the text—which implies that $A^\alpha = \Re(\mathsf{A}^\alpha e^{i\varphi(x,t)})$ satisfies Maxwell's equations in the geometric-optics limit.

(b) Show that this φ satisfies the initial condition (27.85), as claimed in the text.

(c) Show that, because the gradient $\vec{k} = \vec{\nabla}\varphi$ of this φ satisfies $\vec{k}\cdot\vec{k} = 0$, it also satisfies $\nabla_{\vec{k}}\vec{k} = 0$. Thus, the wave vector is the tangent vector to geometric-optics rays that are null geodesics in the gravitational-wave metric. Photons travel along these null geodesics and have 4-momenta $\vec{p} = \hbar\vec{k}$.

(d) Because the gravitational-wave metric (27.80) is independent of x, the p_x component of a photon's 4-momentum must be conserved along its geodesic world line. Compute $p_x = k_x = \partial\varphi/\partial x$, and thereby verify this conservation law.

(e) Explain why the photon's frequency, as measured by observers at rest in our TT coordinate system, is $\omega = -k_t = -\partial\varphi/\partial t$. Explain why the rate of change of this frequency, as computed moving with the photon, is $d\omega/dt \simeq (\partial/\partial t + \partial/\partial x)\omega$, and show that $d\omega/dt \simeq -\frac{1}{2}\omega_o dh/dt$.

27.6.4

27.6.4 Interferometer Analyzed in the Proper Reference Frame of the Beam Splitter T2

proper-reference-frame analysis

We now carefully reanalyze our idealized interferometer in the proper reference frame of its beam splitter, denoting that frame's coordinates by $\hat{x}^\alpha$. Because the beam splitter is freely falling (moving along a geodesic through the gravitational-wave spacetime), its proper reference frame is locally Lorentz (LL), and its metric coefficients have the form $g_{\hat{\alpha}\hat{\beta}} = \eta_{\alpha\beta} + O(\delta_{jk}\hat{x}^j\hat{x}^k/\mathcal{R}^2)$ [Eq. (25.9a)]. Here $\mathcal{R}$ is the radius of curvature of spacetime, and $1/\mathcal{R}^2$ is of order the components of the Riemann tensor, which have

T2

magnitude $\ddot{h}(\hat{t}-\hat{z})$ [Eq. (27.22) with t and z equal to $\hat{t}$ and $\hat{z}$, aside from fractional corrections of order h]. Thus we have:

$$g_{\hat{\alpha}\hat{\beta}} = \eta_{\alpha\beta} + \mathrm{O}[\ddot{h}(\hat{t}-\hat{z})\delta_{jk}\hat{x}^j\hat{x}^k]. \tag{27.91}$$

(Here and below we again omit the subscript + on h for ease of notation.)

The following coordinate transformation takes us from the TT coordinates x^α used in Sec. 27.6.3 to the beam splitter's LL coordinates:

$$x = \left[1 - \frac{1}{2}h(\hat{t}-\hat{z})\right]\hat{x}, \quad y = \left[1 + \frac{1}{2}h(\hat{t}-\hat{z})\right]\hat{y},$$
$$t = \hat{t} - \frac{1}{4}\dot{h}(\hat{t}-\hat{z})(\hat{x}^2 - \hat{y}^2), \quad z = \hat{z} - \frac{1}{4}\dot{h}(\hat{t}-\hat{z})(\hat{x}^2 - \hat{y}^2). \tag{27.92}$$

It is straightforward to insert this coordinate transformation into the TT-gauge metric (27.80) and thereby obtain, to linear order in h:

$$\boxed{ds^2 = -d\hat{t}^2 + d\hat{x}^2 + d\hat{y}^2 + d\hat{z}^2 + \frac{1}{2}(\hat{x}^2 - \hat{y}^2)\ddot{h}(\hat{t}-\hat{z})(d\hat{t} - d\hat{z})^2.} \tag{27.93}$$

spacetime metric in proper reference frame of beam splitter

This has the expected LL form (27.91) and, remarkably, it turns out not only to be a solution of the vacuum Einstein equation in linearized theory but also an exact solution to the full vacuum Einstein equation (cf. Misner, Thorne, and Wheeler, 1973, Ex. 35.8)!

Throughout our idealized interferometer, the magnitude of the metric perturbation in these LL coordinates is $|h_{\hat{\alpha}\hat{\beta}}| \lesssim (\ell/\lambdabar_{GW})^2 h$, where $\lambdabar_{GW} = \lambda_{GW}/(2\pi)$ is the waves' reduced wavelength, and h is the magnitude of $h(\hat{t}-\hat{z})$. For Earth-based interferometers operating in the HF band (~10 to ~1000 Hz), $\lambdabar_{GW}$ is of order 50–5,000 km, and the arm lengths are $\ell \le 4$ km, so $(L/\lambdabar)^2 \lesssim 10^{-6}$ to 10^{-2}. Thus, the metric coefficients $h_{\hat{\alpha}\hat{\beta}}$ are no larger than $h/100$. This has a valuable consequence for the analysis of the interferometer: up to fractional accuracy $\sim(\ell/\lambdabar_{GW})^2 h \lesssim h/100$, the LL coordinates are globally Lorentz throughout the interferometer (i.e., $\hat{t}$ measures proper time, and $\hat{x}^j$ are Cartesian and measure proper distance). In the rest of this section, we restrict attention to such Earth-based interferometers but continue to treat them as though they were freely falling. (See Sec. 27.6.5 for the influence of Earth's gravity.)

Being initially at rest at the origin of these LL coordinates, the beam splitter remains always at rest, but the mirrors move. Not surprisingly, the geodesic equation for the mirrors in the metric (27.93) dictates that their coordinate positions are, up to fractional errors of order $(\ell/\lambdabar_{GW})^2 h$,

T2

mirror motions in proper reference frame

$$\hat{x} = L_x = \left[1 + \frac{1}{2}h(\hat{t})\right]\ell_x, \quad \hat{y} = \hat{z} = 0 \quad \text{for mirror in } x \text{ arm},$$
$$\hat{y} = L_y = \left[1 - \frac{1}{2}h(\hat{t})\right]\ell_y, \quad \hat{x} = \hat{z} = 0 \quad \text{for mirror in } y \text{ arm}. \tag{27.94}$$

[Equations (27.94) can also be deduced from the gravitational-wave tidal acceleration $-\mathcal{E}_{\hat{j}\hat{k}}\hat{x}^k$, as in Eq. (27.25), and from the fact that to good accuracy $\hat{x}$ and $\hat{y}$ measure proper distance from the beam splitter.] So even though the mirrors do not move in TT coordinates, they do move in LL coordinates. The two coordinate systems predict the same time-varying physical arm lengths (the same proper distances from beam splitter to mirrors), L_x and L_y [Eqs. (27.81) and (27.94)].

As in TT coordinates, so also in LL coordinates, we can analyze the light propagation in the geometric-optics approximation, with $A^{\hat{\alpha}} = \Re(\mathcal{A}^{\hat{\alpha}} e^{i\varphi})$. Just as the wave equation for the vector potential dictates in TT coordinates that the rapidly varying phase of the outward light in the x arm has the form $\varphi^{\text{out}}_{x\text{ arm}} = -\omega_o(t - x) + \text{O}(\omega_o \ell h_{\mu\nu})$ [Eqs. (27.86) with $x \sim \ell \ll \lambda\!\!\!^{-}_{\text{GW}}$, so $H(t-x) - H(t) \simeq \dot{H}(t)x = h(t)x \sim h\ell \sim h_{\mu\nu}\ell$], so similarly the wave equation in LL coordinates turns out to dictate that

$$\varphi^{\text{out}}_{x\text{ arm}} = -\omega_o(\hat{t} - \hat{x}) + \text{O}(\omega_o \ell h_{\hat{\mu}\hat{\nu}}) = -\omega_o(\hat{t} - \hat{x}) + \text{O}\left(\omega_o \ell h \frac{\ell^2}{\lambda\!\!\!^{-2}_{\text{GW}}}\right), \tag{27.95}$$

and similarly for the returning light and the light in the y arm. The term $\text{O}(\omega_o \ell h\, \ell^2/\lambda\!\!\!^{-2}_{\text{GW}})$ is the influence of the direct interaction between the gravitational wave and the light, and it is negligible in the final answer (27.96) for the measured phase shift. Aside from this term, the analysis of the interferometer proceeds in exactly the same way as in flat space (because $\hat{t}$ measures proper time and $\hat{x}$ and $\hat{y}$ proper distance): the light travels a round trip distance L_x in one arm and L_y in the other, and therefore acquires a phase difference, on arriving back at the beam splitter, given by

difference in output phase shift between arms: same as in TT gauge

$$\Delta\varphi = -\omega_o[-2(L_x - L_y)] + \text{O}\left(\omega_o \ell h \frac{\ell^2}{\lambda\!\!\!^{-2}_{\text{GW}}}\right)$$
$$\simeq +2\omega_o[\ell_x - \ell_y + \ell h(\hat{t})] + \text{O}\left(\omega_o \ell h \frac{\ell^2}{\lambda\!\!\!^{-2}_{\text{GW}}}\right). \tag{27.96}$$

This net phase difference for the light returning from the two arms is the same as we deduced in TT coordinates [Eq. (27.89)], up to the negligible correction $\text{O}(\omega_o \ell h\, \ell^2/\lambda\!\!\!^{-2}_{\text{GW}})$, and therefore the time-varying intensity of the light into the photodetector will be the same [Eq. (27.90)].

In our TT analysis the phase shift $2\omega_o \ell h(t)$ arose from the interaction of the light with the gravitational waves. In the LL analysis, it is due to the displacements

T2

of the mirrors in the LL coordinates (i.e., the displacements as measured in terms of proper distance) that cause the light to travel different distances in the two arms. The direct LL interaction of the waves with the light produces only the tiny correction $O(\omega_o \ell h\, \ell^2/\lambda\!\!\!^{-}_{GW}^2)$ to the phase shift.

It should be evident that the LL description is much closer to elementary physics than the TT description is. This is always the case, when one's apparatus is sufficiently small that one can regard $\hat{t}$ as measuring proper time and $\hat{x}^j$ as Cartesian coordinates that measure proper distance throughout the apparatus. But for a large apparatus (e.g., planned space-based interferometers such as LISA, with arm lengths $\ell \gtrsim \lambda\!\!\!^{-}_{GW}$, and the pulsar timing arrays of Sec. 27.6.6) the LL analysis becomes quite complicated, as one must pay close attention to the $O(\omega_o \ell h\, \ell^2/\lambda\!\!\!^{-}_{GW}^2)$ corrections. In such a case, the TT analysis is much simpler.

27.6.5 Realistic Interferometers T2

27.6.5

For realistic, Earth-based interferometers, one must take account of the acceleration of gravity. Experimenters do this by hanging their beam splitters and test masses on wires or fibers. The simplest way to analyze such an interferometer is in the proper reference frame of the beam splitter, where the metric must now include the influence of the acceleration of gravity by adding a term $-2g\hat{z}$ to the metric coefficient $h_{\hat{0}\hat{0}}$ [cf. Eq. (24.60b)]. The resulting analysis, like that in the LL frame of our freely falling interferometer, will be identical to what one would do in an accelerated reference frame of flat spacetime, so long as one takes account of the motion of the test masses driven by the gravitational-wave tidal acceleration $-\mathcal{E}_{\hat{i}\hat{j}}\hat{x}^j$, and so long as one is willing to ignore the tiny effects of $O(\omega_o \ell h\, \ell^2/\lambda\!\!\!^{-}_{GW}^2)$.

To make the realistic interferometer achieve high sensitivity, the experimenters introduce a lot of clever complications, such as the input mirrors of Fig. 9.13, which turn the arms into Fabry-Perot cavities. All these complications can be analyzed, in the beam splitter's proper reference frame, using standard flat-spacetime techniques augmented by the gravitational-wave tidal acceleration. The direct coupling of the light to the gravitational waves can be neglected, as in our idealized interferometer.

27.6.6 Pulsar Timing Arrays T2

27.6.6

A rather different approach to direct detection of gravitational waves is by the influence of the waves on the timing of an array of radio pulsars (Hobbs et al., 2010).

gravitational-wave searches via pulsar timing arrays in VLF band

As we have discussed in Sec. 27.2.6, many pulsars, especially those with millisecond periods, have pulse arrival times at the radio telescope that can be predicted with high accuracy—30 to 100 ns in the best cases—after correcting for slowing down of the pulsar, propagation effects, and the motion of the telescope relative to the center of mass of the solar system. If the radio pulses travel through a gravitational wave, then they incur tiny variations in arrival time that can be used to detect the wave. This technique is most sensitive to waves with frequency $f \sim 30$ nHz (the gravitational

T2

VLF band; see Sec. 27.6.1), so a few periods are sampled over the duration of an observation, typically a few years. The most promising sources are supermassive binary black holes in the nuclei of distant galaxies.

Let us consider the idealized problem of radio pulses emitted at a steady rate by a stationary pulsar with position $\mathbf{z}$ relative to a stationary telescope on Earth and traveling at the speed of light. Now, suppose that there is a single, monochromatic, plane gravitational wave, $\boldsymbol{h} = \boldsymbol{h}_o \cos[2\pi f (t + \mathbf{x} \cdot \hat{\mathbf{r}})]$, where $\hat{\mathbf{r}}$ is a unit vector pointing toward the source. For the wave, adopt the metric (27.18a) and TT gauge. The pulsar and the telescope will remain at rest with respect to the coordinates (Ex. 27.18), and t measures their proper times.

The wave-induced time delay in the arrival of a pulse, called the *residual* $R(t)$, is then given by integrating along a past directed null geodesic ($ds = 0$) from the telescope to the pulsar:[6]

influence of gravitational wave on radio pulses' time delay (their "residual")

$$
\begin{aligned}
R(t) &= -\frac{1}{2}\int_0^z dz' \hat{\mathbf{z}} \cdot \boldsymbol{h}_o \cdot \hat{\mathbf{z}} \cos\big(2\pi f [t - z'(1 - \hat{\mathbf{z}} \cdot \hat{\mathbf{r}})]\big) \\
&= \frac{\hat{\mathbf{z}} \cdot \boldsymbol{h}_o \cdot \hat{\mathbf{z}}\big\{\sin(2\pi f [t - z(1 - \hat{\mathbf{z}} \cdot \hat{\mathbf{r}})]) - \sin(2\pi f t)\big\}}{4\pi f (1 - \hat{\mathbf{z}} \cdot \hat{\mathbf{r}})}
\end{aligned}
\tag{27.97}
$$

(Ex. 27.20a). Note that two terms contribute to the residual; the first is associated with the wave as it passes the pulsar, the second is local and associated with the wave passing Earth.

correlation of residuals between different pulsars

This search for gravitational waves is being prosecuted not just with a single pulsar but also with an array of tens (currently and hundreds in the future) of millisecond pulsars. Programs using different telescopes are being coordinated and combined through the International Pulsar Timing Array (IPTA) collaboration (http://www.ipta4gw.org). The local residual (27.97) is correlated between different lines of sight to different pulsars with differing polarization projections $\hat{\mathbf{z}} \cdot \boldsymbol{h}_o \cdot \hat{\mathbf{z}}$. This facilitates identifying the waves amidst different noises associated with different pulsars, and it allows a coherent addition of the data from many pulsars, which will be particularly valuable in searches for individual sources of gravitational radiation. If and when a source is located, the array can be used to fix its direction $\hat{\mathbf{r}}$ and the polarization of the waves. In addition to extracting astrophysics of the source, there should be a clear affirmation that gravitational waves have spin two and travel at the speed of light.

gravitational-wave sources in VLF band

However, unless we are very lucky, it is more likely that the first detection or the most prescriptive upper limit will relate to a stochastic background created by the superposed waves from many black hole binaries (Ex. 27.21). Although the bandwidth from an individual source will be of order the reciprocal of the time it takes the binary

6. Of course the actual null geodesic followed by the radio waves changes in the presence of the gravitational wave but, invoking Fermat's principle, the travel time is unchanged to $O(h)$ if we integrate along the unperturbed trajectory.

frequency to double—in the fHz to pHz range—the randomly phased signals are likely to overlap in frequency, ensuring that the gravitational radiation can be well described by a spectral energy density.

T2

Other more speculative sources of gravitational waves in the VLF frequency band have been proposed, especially cosmic strings. There could be many surprises waiting to be discovered.

EXERCISES

Exercise 27.20 *Example: Pulsar Timing Array* T2

Explore how pulsar timing can be used to detect a plane gravitational wave.

(a) Derive Eq. (27.97). [Hint: One way to derive this result is from the action

$$\frac{1}{2}\int g_{\alpha\beta}(dx^{\alpha}/d\zeta)(dx^{\beta}/d\zeta)d\zeta$$

that underlies the rays' super-hamiltonian; Ex. 25.7. The numerical value of the action is zero, and since it is an extremum along each true ray, if you evaluate it along a path that is a straight line in the TT coordinate system instead of along the true ray, you will still get zero at first order in $\boldsymbol{h}_o$.]

(b) Recognizing local and pulsar contributions to the timing residuals, explain how much information about the amplitude, direction, and polarization of a gravitational wave from a single black-hole binary can be obtained using accurate timing data from one, two, three, four, and many pulsars.

(c) Suppose, optimistically, that 30 pulsars will be monitored with timing accuracy ~100 ns, and that arrival times will be measured 30 times a year. Make an estimate of the minimum measurable amplitude of the sinusoidal residual created by a single binary as a function of observing duration and wave frequency f.

(d) Using the result from part (c) and using the predicted residual averaged over the direction and the orientation of the source from Eqs. (27.71c) and (27.71d), estimate the maximum distance to which an individual source could be detected as a function of the chirp mass $\mathcal{M}$ and the frequency f.

Exercise 27.21 *Problem: Stochastic Background from Binary Black Holes* T2

The most likely signal to be detected using a pulsar timing array is a stochastic background formed by perhaps billions of binary black holes in the nuclei of galaxies.

(a) For simplicity, suppose that these binaries are all on circular orbits and that they lose energy at a rate given by Eqs. (27.73). Show that for each binary the gravitational wave energy radiated per unit log wave frequency is

$$\frac{dE_{GW}}{d\ln f} = -\frac{dE}{d\ln f} = -\frac{2E}{3} = \frac{\pi^{2/3}}{3}\mathcal{M}^{5/3}f^{2/3}, \qquad (27.98\text{a})$$

where $\mathcal{M}$ is the chirp mass given by Eq. (27.76c) and E is its orbital energy given by $-\frac{1}{2}\mu M/a$.

T2

(b) Suppose that, on astrophysical grounds, binaries only radiate gravitational radiation efficiently if they merge in less than $t_{\rm merge} \sim 300$ Myr. Show that this requires

$$\mathcal{M} > \left(\frac{5}{256\, t_{\rm merge}}\right)^{3/5} (\pi f)^{-8/5}. \tag{27.98b}$$

This evaluates to $\sim$1 million solar masses for $f \sim 30$ nHz. By contrast, measured masses of the large black holes in the nuclei of galaxies range from about 4 million solar masses as in our own modest spiral galaxy to perhaps 20 billion solar masses in the largest galaxies observed.

(c) Now suppose that these black holes grew mostly through mergers of holes with very different masses. Show that the total energy radiated per unit log frequency over the course of the many mergers that led to a black hole with mass M is

$$\frac{dE_{GW}}{d\ln f} \sim \frac{\pi^{2/3}}{5} M^{5/3} f^{2/3}. \tag{27.98c}$$

(d) The number density per unit mass dn/dM today of these merging black holes has been estimated to be

$$\frac{dn}{d\ln M} = \frac{\rho_{BH} M^2}{2M_t^3} e^{-M/M_t}, \tag{27.98d}$$

where $\rho_{BH} \sim 2 \times 10^{-15}$ J m^{-3} is the contribution of black holes (i.e., of their masses) to the average energy density in the local universe and $M_t = 5 \times 10^7$ solar masses $\sim 10^{38}$ kg. Show that

$$\frac{d\rho_{GW}}{d\ln f} \sim \frac{\pi^{2/3}\,\Gamma(11/3)}{10} (M_t f)^{2/3} \rho_{BH} \sim 7 \times 10^{-19} \left(\frac{f}{30\ \text{nHz}}\right)^{2/3} \text{J m}^{-3}. \tag{27.99}$$

This is an overestimate, because most of the black holes were likely assembled in the past, when the universe was about three times smaller than it is today. If one thinks of the gravitational waves as gravitons that lose energy as the universe expands, then this energy density should be reduced by a factor of three. Furthermore, as the black holes are thought to grow through accretion of gas and not by mergers, this estimate should be considered an upper bound.

(e) Making the same assumptions as in the previous exercise, determine whether it will be possible to detect this background in 5 years of observation.

Bibliographic Note

For an elementary introduction to experimental tests of general relativity in the solar system, we recommend Hartle (2003, Chap. 10). For an enjoyable, popular-level book on experimental tests, see Will (1993b). For a very complete monograph on the theory underlying experimental tests, see Will (1993a), and for a more nearly up-to-date review of experimental tests, see Will (2014).

For elementary and fairly complete introductions to gravitational waves, we recommend Hartle (2003, Chaps. 16, 23) and Schutz (2009, Chap. 9). For more advanced treatments, we suggest Misner, Thorne, and Wheeler (1973, Sec. 18.2 and Chaps. 35, 36), Thorne (1983), and Straumann (2013, Secs. 5.3–5.7); but Misner, Thorne, and Wheeler (1973, Chap. 37) on gravitational-wave detection is terribly out of date and is not recommended. For fairly complete reviews of gravitational-wave sources for ground-based detectors (LIGO, etc.) and space-based detectors (LISA, etc.), see Cutler and Thorne (2002) and Sathyaprakash and Schutz (2009). For a lovely monograph on the physics of interferometric gravitational-wave detectors, see Saulson (1994), and for much greater detail, see Reitze, Saulson, and Grote (2019).

Because gravitational-wave science is a rapidly maturing and burgeoning field, there are long, in-depth treatments that include considerable experimental detail and much detail on data analysis techniques, as well as on wave sources and the fundamental theory: Maggiore (2007, 2018), and Creighton and Anderson (2011).

CHAPTER TWENTY-EIGHT

Cosmology

The expansion thus took place in three phases, a first period of rapid expansion in which the atom universe was broken into atomic stars, a period of slowing down, followed by a third period of accelerated expansion.

GEORGES LEMAÎTRE (1933)

28.1 Overview

28.1

The extragalactic sky is isotropic and dark at night. Distant galaxies accelerate away from us. Five-sixths of the matter in the universe is in a "dark" form that we can only detect through its gravitational effects. We are immersed in a bath of blackbody radiation—the *cosmic microwave background* (CMB)—with a temperature of $\sim$2.7 K that exhibits tiny fluctuations with fractional amplitude $\sim 10^{-5}$. These four profound observations lead inexorably to a description of the universe that began from a hot big bang nearly 14 billion years ago. Remarkable progress over recent years in making careful measurements of these features has led to a *standard cosmology* that involves mostly classical physics and incorporates many of the ideas and techniques we have discussed in the preceding 27 chapters.

Let us begin by describing the universe. We live on Earth, the "third rock" out from an undistinguished star located in the outskirts of a quite ordinary galaxy—the second largest in a loose federation of galaxies called the *local group* on the periphery of the *local supercluster.* Our neighbors are no more remarkable. Meanwhile, the distant galaxies and microwave background photons that we observe exhibit no preferred directions. In short, we do not appear to be special, and nowhere else that we can see appears to be special. It is therefore reasonable to assume that the universe has no "center" and that it is essentially homogeneous on the largest scales that we observe, so that all parts are equivalent at the same time. By extension, the average observed properties of the universe and its inferred history would be similar if we lived in any other galaxy.

However, it has long been known that the universe cannot also be infinite and static. If it were, then any line of sight would eventually intercept the surface of a star, and the night sky would be bright. Instead, as we observe directly, galaxies and their constituent stars are receding from us, and the speed of recession is now increasing with time, contrary to our Newtonian expectation. This expansion is observed out to distances that are so great that the recession velocities are relativistic, dimming the light from the distant stars and galaxies, and making the night sky between nearby

BOX 28.1. READERS' GUIDE

- This chapter draws on every one of the preceding chapters:
  - Chap. 1—geometry, tensors;
  - Chap. 2—stress energy tensor, Lorentz invariance;
  - Chap. 3—phase space, radiation thermodynamics, Liouville equation, stars, Boltzmann equation;
  - Chap. 4—entropy, Fermi-Dirac and Bose-Einstein distribution functions, statistical mechanics in the presence of gravity;
  - Chap. 5—chemical potential, Monte Carlo method;
  - Chap. 6—correlation functions and spectral density, Wiener-Khintchine theorem, Fokker-Planck formalism;
  - Chap. 7—paraxial optics, gravitational lenses, polarization;
  - Chap. 8—Fourier transforms, point spread function;
  - Chap. 9—coherence and random processes;
  - Chap. 10—radiation physics, parametric resonance;
  - Chap. 11—strain tensor, spherical coordinates;
  - Chap. 12—longitudinal modes, zero point fluctuations;
  - Chap. 13—relativistic fluids, stars;
  - Chap. 14—barotropic equation of state, viscosity;
  - Chap. 15—turbulence, power law spectrum;
  - Chap. 16—sound waves, nonlinear waves;
  - Chap. 17—compressible flow, 1-dimensional flow;
  - Chap. 18—heat conduction;
  - Chap. 19—nuclear reactions, magnetic stress tensor;
  - Chap. 20—Saha equation, Coulomb collisions;
  - Chap. 21—plasma oscillations, Debye screening;
  - Chap. 22—Jeans' theorem, collisionless particles, Landau damping;
  - Chap. 23—nonlinear wave dynamics, quantization of classical fields, collisionless shocks;
  - Chap. 24—differential geometry, local Lorentz frames;
  - Chap. 25—Riemann tensor, Einstein equation, geodesic deviation, geometrized units;
  - Chap. 26—horizons, pressure as source of gravitation; and
  - Chap. 27—gravitational time dilation, gravitational waves.

stars as dark as we observe. Furthermore, the amount of matter in the universe induces gravitational potential differences of order c^2 across the observed universe.

These features require a general relativistic description. However, even if we had tested it adequately in the strong field regime of black holes and neutron stars (which we have only begun to do), there is no more guarantee that general relativity describes the universe at large than that classical mechanics and electromagnetism suffice to describe atoms! Despite this, the best way to proceed is to continue to adopt general relativity and be alert to the possibility that it could fail us. What we shall discover is that it provides a sufficient and successful framework for describing the observations.

The discovery of the CMB in 1964 transformed cosmology. This radiation must have dominated the stress-energy tensor and, consequently, the geometry and dynamics of the universe in the past. More recently, it has been used to show that the spatial curvature of the universe is very small and that the CMB's tiny temperature fluctuations have simple behavior consistent with basic (mostly classical) physics. And the fluctuations can be used to argue that the universe exhibited a brief growth spurt, called *inflation,* when it was very young—a surge that established both the observed uniformity and the structure that we see around us today. No less important has been the quantitative description, mostly developed over the past 20 years, of the two invisible entities (popularly known as *dark matter* and the *cosmological constant* or, more generally, *dark energy*) that account for 95% of the modern universe. In fact, the measurements have become so good that most alternative descriptions of the universe have been ruled out and no longer need to be discussed. This is a good time to benefit from this simplification.

This chapter is longer, more ambitious, and less didactic—there are no practice and derivation exercises—than its predecessors, because we want to take advantage of the opportunity to exhibit modern classical physics in one of its most exciting and currently successful applications. The reader must expect to read slowly and carefully to "connect the dots." To keep the chapter at a manageable length, we eschew essentially all critical discussion of the observations and measurements, including errors. We also avoid the idiosyncratic terminology and conventions of astronomy. Finally, we stop short of describing the birth and development of galaxies, stars, and planets, as these phenomenological investigations would take us too far away from the direct application of general principles. Despite these limitations, it is striking how much of what is known about the universe can be calculated with passable accuracy using the ideas, principles, and techniques discussed in this book.

We begin in Sec. 28.2 by developing general relativistic cosmology, emphasizing the geometry and the kinematics before turning to the dynamics. In Sec. 28.3, we describe all the major constituents of the universe today and we follow this in Sec. 28.4 with a development of standard cosmology starting from when the temperature was roughly 10^{11} K, distinguishing seven distinct ages. In Sec. 28.5, we develop a theory to describe the growth of perturbations, which ultimately led to clustered galaxies. Most of what we know about the universe comes from observing photons, and so in

Sec. 28.6, we develop the cosmological optics that we need to draw inferences about the birth and growth of the universe. In Sec. 28.7, we conclude by discussing three foci of current research and incipient progress—attempts to understand the origin of the universe, notably inflation, theories of the creation of dark matter and baryons, and speculations about the fate of the universe involving the role of the cosmological constant.

28.2

28.2 General Relativistic Cosmology

28.2.1

28.2.1 Isotropy and Homogeneity

galaxies at a glance

We start by introducing *galaxies* and describing their distribution in space. A typical galaxy, like our own, comprises $\sim 10^{11}$ luminous stars (with a combined luminosity of $\sim 10^{37}$ W)[1] and gas, mostly concentrated in a sphere of radius about 3×10^{20} m $\sim$ 10 kpc located at the center of a sphere about ten times larger dominated by collisionless dark matter (Fig. 28.1). A typical galaxy mass is 10^{42} kg. There is a large range in galaxy masses from less than a millionth of the mass of our galaxy to more than a hundred times its mass.

distribution of galaxies

Roughly a trillion galaxies can be seen over the whole sky (Fig. 28.1; Beckwith et al., 2006). Their distribution appears to be quite isotropic. We can estimate their distances and study their spatial distribution. It is found that the galaxies are strongly clustered on scales of $\lesssim 3 \times 10^{23}$ m $\sim$ 10 Mpc. Especially prominent are clusters of roughly a thousand galaxies. The strength of their relative clustering diminishes with increasing linear scale to $|\delta N/N| \sim 10^{-5}$ when the scale size is comparable with the "size" of the universe (which we will define more carefully below as $\sim 10^{26}$ m $\sim$ 3 Gpc). Insofar as luminous galaxies fairly sample all material in the universe, it seems that on large enough scales, we can regard the matter distribution today as quite homogeneous.

cosmic microwave background

An even more impressive demonstration of this uniformity comes from observations of the CMB (Fig. 28.1; Penzias and Wilson, 1965). Not only does it retain a blackbody spectrum, the temperature fluctuations are also only $|\delta T/T| \lesssim 3 \times 10^{-5}$ on angular scales from radians to arcminutes. (There is a somewhat larger dipolar component, but this is attributable to a Doppler shift caused by the compounded motion of Earth, the Sun, and our galaxy.) This radiation has propagated to us from an epoch when the universe was a thousand times smaller than it is today and, as we shall demonstrate, the temperature was a thousand times greater, roughly 3,000 K. If the universe were significantly inhomogeneous at this time, then we would see far larger fluctuations in its temperature, spectrum, and polarization. We conclude that the young universe was quite homogeneous, just like the contemporary universe appears to be on the largest scales.

1. For calibration, the Sun has a mass of 2.0×10^{30} kg and a luminosity of 3.9×10^{26} W. The standard astronomical distance measure is 1 parsec (pc) $= 3.1 \times 10^{16}$ m.

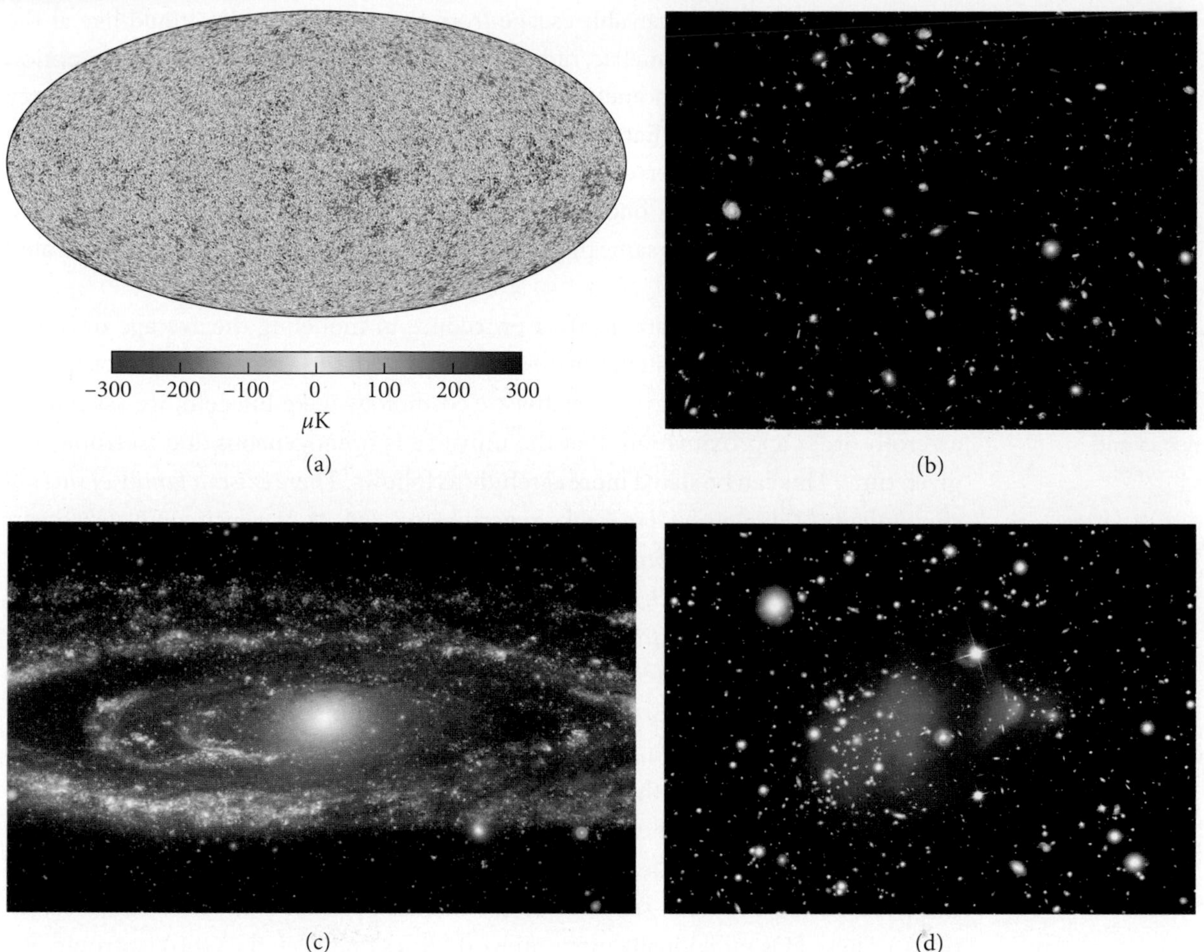

FIGURE 28.1 Four astronomical images that illustrate recent discoveries about the universe. (a) Image of the whole sky made by the Planck satellite, exhibiting $\sim 10\ \mu\text{K}$ fluctuations in the observed $\sim 3\ \text{K}$ temperature of the CMB (Sec. 28.6.1; credit: Planck Collaboration, 2016a). These fluctuations, observed when the universe was ~400,000 yr old, depict the seeds out of which grew the large structures we see around us today. (b) The deepest image of the sky taken by Hubble Space Telescope, roughly 2 arcmin across (Hubble eXtreme Deep Field, http://en.wikipedia.org/wiki/Hubble_eXtreme_Deep_Field; credit: NASA; ESA; G. Illingworth, and the HUDF09 Team, 2013). The light from the most distant galaxies in this image is estimated to have been emitted when the universe was only ~0.08 of its present size and less than 500 Myr old (Sec. 28.4.5). (c) Combined ultraviolet and infrared image of the Andromeda galaxy, the nearest large galaxy to our Milky Way galaxy. (Credit: NASA/JPL-Caltech.) Orbiting hydrogen gas can be seen out to about five times the size of this image, demonstrating that the stars we see form at the bottom of a large potential well of dark matter (Sec. 28.2.1). (d) The Bullet cluster of galaxies. (Credit: X-ray: NASA/CXC/CfA/M. Markevitch et al.; lensing map: NASA/STScI; ESO WFI; Magellan/University of Arizona/D. Clowe et al.; optical: NASA/STScI; Magellan/University of Arizona/D. Clowe et al.) Two clusters, each containing roughly several hundred galaxies, are in the process of merging. The hot gas, in red, can be traced by its X-ray emission and is separated from the dark matter, in blue, which can be located by the weak gravitational lensing distortion it imposes on the images of background galaxies (Sec. 28.6.2). The separation of these concentrations of matter demonstrates that dark matter is effectively collisionless (Sec. 28.3.2).

There is only one conceivable escape from this conclusion. We could live at the origin of a spherically symmetric, radially inhomogeneous universe. The assumption that this is not the case is sometimes known as the *Copernican Principle* by analogy with the proposition that Earth is not at the center of the solar system. According to this principle, the universe would look similarly isotropic from all other vantage points at the same time. If one accepts this hypothesis, then isotropy about us and about all other points at the same time implies homogeneity. This can be demonstrated formally.

These observational data justify a procedure in modeling the average universe, which was adopted by Einstein (1917) and others with little more than philosophical justification in the early days of relativistic cosmology. Like Einstein, we assume, as a zeroth order approximation, that the universe is homogeneous and isotropic at a given time. This can be stated more carefully as follows. *There exists a family of slices of simultaneity (3-dimensional spacelike hypersurfaces; Fig. 28.2), which completely covers spacetime, with the special property that on a given slice of simultaneity (i) no two points are distinguishable from each other (homogeneity) and (ii) at a given point no one spatial direction is distinguishable from any other (isotropy).*

homogeneous and isotropic universe

So, how do we assign these slices of simultaneity? Fortunately, the universe comes with a clock, the temperature of the CMB. (For the moment, ignore the tiny fluctuations.) We then introduce a set of imaginary *fundamental observers* (FOs; Fig. 28.2) and give them the velocity that removes the dipole anisotropy in the temperature distribution.[2] (Henceforth when we talk about observations at Earth, we imagine that these are observations made by an FO coincident with Earth today. To get the actual Earth-based observations, we just add a small Doppler shift to these FO observations.) These FOs individually move on world lines that keep the CMB isotropic. We regard the 3-dimensional hypersurfaces that they inhabit when they all measure the same CMB temperature as spaces in their own right, which we approximate as homogeneous. Isotropy guarantees that the FO world lines are orthogonal to these "slices of simultaneity." Let us focus on the hypersurface that we inhabit today, formed by freezing the action when everyone measures the CMB temperature to be 2.725 K.

fundamental observers

28.2.2

28.2.2 Geometry

METRIC TENSOR

We want to explore the geometry of this frozen, 3-dimensional space.[3] Our first task is to deduce the form of its metric tensor. Introduce a spherical coordinate system $\{\chi, \theta, \phi\}$, centered on us. Here χ is the radius, the proper distance (measured by a rather large number of meter rules) from us to a distant point; θ, ϕ are spherical polar angles, measured from Earth's north pole and the direction of

spatial coordinates

2. Roughly 360 km s^{-1} for Earth.
3. For a fuller treatment see, e.g., Misner, Thorne, and Wheeler (1973), Sec. 27.6 and Box 27.2.

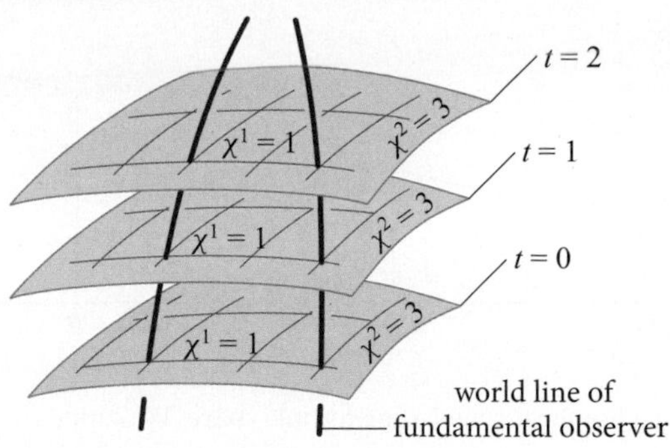

FIGURE 28.2 The slices of simultaneity, world lines of fundamental observers, and synchronous coordinate system for a homogeneous, isotropic model of the universe.

the Sun at the vernal equinox, for example.[4] We will also find it convenient to introduce an equivalent Cartesian coordinate system (Fig. 28.2): $\boldsymbol{\chi} \equiv \{\chi^1, \chi^2, \chi^3\} = \{\chi \sin\theta \cos\phi, \chi \sin\theta \sin\phi, \chi \cos\theta\}$. We know that the space around us is isotropic, which implies that the angular part of the metric is $d\theta^2 + \sin^2\theta d\phi^2$ at each radius.[5] However, we cannot assume that the space is globally flat and that the area of a sphere of constant radius χ is $4\pi\chi^2$. Therefore, we generalize the metric to become

$$^{(3)}ds^2 = d\chi^2 + \Sigma^2(d\theta^2 + \sin^2\theta d\phi^2), \tag{28.1}$$

where $\Sigma(\chi)$ is a function to be determined. Now we know that $\Sigma \simeq \chi$ for small χ to satisfy the requirement that space be locally flat.

HOMOGENEOUS 2-DIMENSIONAL SPACES

Restrict attention to the 2-dimensional subspace $\phi = \text{const}$, which has metric $^{(2)}ds^2 = d\chi^2 + \Sigma^2(\chi)d\theta^2$, and consider a curve with $\theta = 0$ emanating from us at point O. This is obviously a geodesic—the shortest distance between its start at O and any point we care to choose along it. Let this geodesic pass through two galaxies labeled A, B and end at a third galaxy C (Fig. 28.3). Let the proper distances from O to A, A to B, and B to C, be χ_1, χ_2, χ_3, respectively. Now add a second geodesic, also emanating from O, and inclined to the first geodesic by a small angle θ.[6] As θ is

4. This is how astronomers set up their equivalent "right ascension"–"declination" coordinate system.
5. The deep, underlying reason it is not possible to cover the surface of a sphere with a metric with constant coefficients is that the vector field $(\partial/\partial\phi)_\theta$ that generates rotations about the polar axis does not commute with the vector field that generates rotations about any other axis. In principle, we could rotate our angular coordinate system from one radius to the next, but this would hide the symmetry that we are so eager to exploit!
6. This is similar in spirit to the description of gravitational lensing in Sec. 7.6.

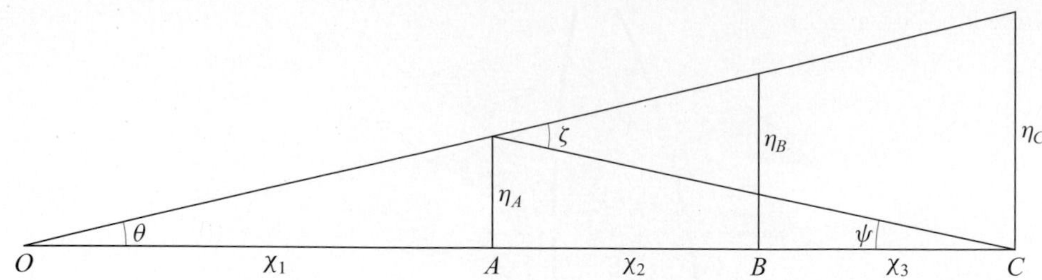

FIGURE 28.3 Geodesics in a homogeneous 2-dimensional space. We suppose that the angles θ, ψ, ζ are small and compute the transverse displacements η assuming the metric (28.1).

small, the proper separations of these geodesics (measured along short paths perpendicular to the first geodesic at A, B, and C) will be $\eta_A = \theta\Sigma(\chi_1)$, $\eta_B = \theta\Sigma(\chi_1+\chi_2)$, $\eta_C = \theta\Sigma(\chi_1+\chi_2+\chi_3)$, respectively, ignoring terms $O(\theta^3)$, which will vanish in the limit $\theta \to 0$. Next, introduce a third geodesic backward from C and intersecting with the second geodesic, at a point a perpendicular distance η_A from galaxy A. Denote the (small) angle it makes with the first geodesic at C to be ψ and with the second geodesic near A to be ζ.

derivation of metric tensor

Now, by assumption, the metric must have the same form, with the same metric function $\Sigma(\chi)$ when the origin is at O, A, or C. We can use this to derive a functional relationship that must be satisfied by $\Sigma(\chi)$ by evaluating η_B in two different ways:

$$\begin{aligned}\eta_B &= \theta\Sigma(\chi_1+\chi_2) = \psi\Sigma(\chi_3) + \zeta\Sigma(\chi_2)\\ &= \theta\left(\frac{\Sigma(\chi_3)\Sigma(\chi_1)}{\Sigma(\chi_2+\chi_3)} + \frac{\Sigma(\chi_2)\Sigma(\chi_1+\chi_2+\chi_3)}{\Sigma(\chi_2+\chi_3)}\right).\end{aligned} \tag{28.2}$$

(In the last equality, we have used elementary geometry to express ψ and ζ in terms of θ.) Rearranging, we obtain

$$\Sigma(\chi_1+\chi_2)\Sigma(\chi_2+\chi_3) = \Sigma(\chi_1)\Sigma(\chi_3) + \Sigma(\chi_1+\chi_2+\chi_3)\Sigma(\chi_2), \tag{28.3}$$

for all choices of χ_1, χ_2, χ_3.[7]

This sort of nonlocal, functional relationship is probably quite unfamiliar and, in general, such equations are hard to solve. However, in this instance we can use the device of solving it in a special case and then showing that the solution satisfies the original equation. If we assume that χ_1 and χ_3 are equal and small, and we expand Eq. (28.3) to second order in χ_1 with $\chi_2 \equiv \chi$ finite, we find that

$$\Sigma''\Sigma - \Sigma'^2 + 1 = 0. \tag{28.4}$$

Next, multiply the left-hand side by $2\Sigma'/\Sigma^3$, and note that it is the derivative of $(\Sigma'^2 - 1)/\Sigma^2$. Hence, we get

$$\Sigma'^2 + K\Sigma^2 - 1 = 0, \tag{28.5}$$

7. The well-educated reader may notice a similarity to Ptolemy's theorem. The comparison is instructive.

where the integration constant K is known as the *Gaussian curvature* of the 2-dimensional surface, or equivalently, half the scalar curvature (Sec. 25.6). Differentiating this equation leads to the simple harmonic equation

$$\Sigma'' + K\Sigma = 0. \tag{28.6}$$

Gaussian curvature K

We need solutions that will be locally Euclidean with $\Sigma(0) = 0$, $\Sigma'(0) = 1$.

There are three cases. If $K = 0$, then $\Sigma = \chi$. If K is positive, we write it as $K = R^{-2}$, where R is the *radius of curvature* and $\Sigma(\chi) = R \sin(\chi/R)$. If $K < 0$, we write it as $K = -R^{-2}$ and $\Sigma(\chi) = R \sinh(\chi/R)$. The final step is to verify that these solutions are valid for general χ_1, χ_2, χ_3, which they are. We have therefore found three general solutions to the functional equation (28.3). Suppose that there were an additional solution. We could subject it to the same limiting procedure and recover the same nonlinear differential equation (28.4) as $\chi_1 = \chi_3 \rightarrow 0$ for all $\chi_2 = \chi$. However, we know all the locally Euclidean solutions $\Sigma(\chi)$ to this equation, and our hypothetical fourth solution would not be among them. Therefore, it cannot exist, and so we have found all possible distance relations for homogeneous 2-dimensional spaces.

We summarize these three solutions by introducing a parameter $k = \pm 1$ or 0:

$$K = k/R^2; \quad \Sigma = R \sin(\chi/R) \text{ for } k = +1, \ \ \chi \text{ for } k = 0, \ \ R \sinh(\chi/R) \text{ for } k = -1. \tag{28.7}$$

NON-EUCLIDEAN GEOMETRIES

The $K = 0$ solution is instantly recognizable as describing the geometry of flat, Euclidean space. The $K > 0$ solution describes the 2-dimensional surface of a sphere embedded in 3-dimensional flat space. The circumference of a circle of radius χ is $2\pi R \sin(\chi/R) < 2\pi\chi$ and vanishes at an antipodal point, where $\chi = \pi R$. The space has a natural end—we call it *closed*. The associated area of this space is $2\pi \int_0^{\pi R} d\chi\, \Sigma = 4\pi R^2$. The geodesics we have been using are simply "great circles" on the 2-sphere.[8] The theorems of spherical trigonometry, which we could have gone on to prove, are familiar to navigators, astronomers, and crystallographers.

homogeneous positive curvature space

When $K < 0$, the circumference is $2\pi R \sinh(\chi/R) > 2\pi\chi$ and the area is unbounded, so the space is *open*. This 2-dimensional space *cannot* be embedded in 3-dimensional Euclidean space. This discovery, made independently in the nineteenth century by Bolyai, Gauss, and Lobachevsky, was a source of mathematical wonder and philosophical consternation.[9]

homogeneous negative curvature space

8. To go from the skinny triangles we have been considering so far to large values of η, we need to construct the geodesic $\chi(\theta)$ by minimizing the length $\int d\theta (\chi'^2 + \Sigma^2)^{1/2}$ of the curve connecting the endpoints. This is a standard exercise in the calculus of variations, and carrying it out opens the door to deriving counterparts to the theorems of Euclidean geometry for non-Euclidean spaces.
9. This *hyperbolic* geometry can be embedded in 3-dimensional Minkowski space, and we could have used our understanding of special relativity to explore it by analogy with what is done with the positively curved space. Had we done so, we would have discovered that it is the same as the mass hyperboloid depicted in Fig. 3.2. However, the spirit of Riemannian geometry is not to do this but instead to explore the space through its *inner* properties.

HOMOGENEOUS 3-DIMENSIONAL SPACES

The generalization to 3 dimensions is straightforward. If space is isotropic, then the geometrical statements we made about the subspace ϕ =const must be true about any 2-dimensional subspace; so we arrive at the same three possibilities for Σ in the 3-dimensional metric (28.1). We replace the circumference of a circle by the area of a sphere of radius χ, which is less than (more than) $4\pi\chi^2$ for positive (negative) curvature. The total volume of the positively curved, closed 3-space is $4\pi \int_0^{\pi R} d\chi\, \Sigma^2 = 2\pi^2 R^3$.[10]

ROBERTSON-WALKER METRIC

Relativistic cosmology is transacted in spacetime, not space (Sec. 25.2). How do we generalize our understanding of these homogeneous subspaces to include a time coordinate and allow for the expansion of the universe? The first step is to assume that the 3-spaces in the past were also homogeneous and so had the same geometry as today. The only thing that can change is the radius of curvature, and so we write

scale factor

$$R(t) = a(t) R_0, \tag{28.8}$$

where R_0 is its value today when $t = t_0$ and $a = 1$. We call $a(t)$ the *scale factor.* (We handle a flat space by taking the limit $R_0 \to \infty$.) An FO then moves from one spatial hypersurface to the next carrying the same spherical coordinates $\{\chi, \theta, \phi\}$ (Fig. 28.2). We call these three coordinates *comoving coordinates.*

comoving coordinates

The basis vector $\partial_t \equiv \partial/\partial t$ is tangent to the world line of an FO, which is orthogonal to the spatial hypersurface $t =$ const. Thus, $g_{ti} \equiv \partial_t \cdot \partial_i = 0$.[11] Next consider g_{tt}. We have so far implicitly measured time using the temperature of the CMB, but as we want to be quantitative, we choose as a time coordinate the *proper time* (or *cosmic time* or simply, *time*) Δt that elapses, as measured by a clock carried by an exceedingly patient FO with $u^\alpha = (1, 0, 0, 0)$, from one hypersurface to the next. (We set $c = 1$ in all subsequent equations.) It is implicit in the assumption of homogeneity that this interval of time will be the same for all FOs, and we can add up all the intervals to make our time coordinate t the *total age of the universe* since the big bang. With this choice, $g_{tt} = -1$.[12]

time coordinate

10. Although homogeneity and isotropy force the cosmological model's hypersurfaces to have one of the three metrics we have described, the topologies of those hypersurfaces need not be the obvious ones adopted here. For example, a flat model could have a closed topology with finite volume rather than an open topology with infinite volume. This could come about in much the same way as a flat piece of paper can be rolled into a cylinder. The geometry is still Euclidean, but one of the two coordinates, say x, is identified with $x + C$, where C is the length of the circumference of the cylinder. In 3 dimensions we can make similar identifications for the y and z coordinates. There are no credible observations to suggest that the universe is like this, but the possibility is worth keeping in mind.
11. Such a coordinate system is termed *synchronous.*
12. Other time coordinates, such as *conformal time,* $\int dt/a$, are in common use for a variety of technical reasons, but we shall eschew these.

Our full metric is then

$$ds^2 = -dt^2 + a(t)^2[d\chi^2 + \Sigma^2(d\theta^2 + \sin^2\theta d\phi^2)]. \tag{28.9}$$

This is known as the *Robertson-Walker metric*.[13] Using this metric, we can easily verify that the acceleration $\nabla_{\vec{u}}\vec{u}$ of an FO vanishes, so FOs with fixed χ follow geodesics.

meaning of $a(t)$

The scale factor $a(t)$ is very important. Not only does it measure the size of the universe in the past, it also measures the separation of any two FOs as a fraction of their separation today. Insofar as the universe is well described as homogeneous and isotropic, this single function tells us all we need to know about how the universe expanded and how it will continue to expand. We shall use it—actually, its logarithm—as our independent variable, because physical quantities such as densities and temperatures scale simply with it and because it is closely related to what astronomers commonly measure. Measuring $t(a)$ is the kinematic challenge to observational cosmology; explaining it is the dynamical challenge to general relativity.

EINSTEIN TENSOR

Our final geometrical task is to calculate the Einstein tensor (Sec. 25.8). Although the calculation can be simplified by exploiting symmetry, it is best to use computer algebra. The only nonzero elements are

components of Einstein tensor in orthonormal basis

$$\begin{aligned}
G^{\hat{t}\hat{t}} &= -G_t{}^t = -\frac{2\Sigma''\Sigma + \Sigma'^2 - 3\Sigma^2\dot{a}^2 - 1}{\Sigma^2 a^2}, \\
G^{\hat{\chi}\hat{\chi}} &= G_\chi{}^\chi = \frac{\Sigma'^2 - \Sigma^2(2\ddot{a}a + \dot{a}^2) - 1}{\Sigma^2 a^2}, \\
G^{\hat{\theta}\hat{\theta}} &= G_\theta{}^\theta = G^{\hat{\phi}\hat{\phi}} = G_\phi{}^\phi = \frac{\Sigma'' - \Sigma(2\ddot{a}a + \dot{a}^2)}{\Sigma a^2}.
\end{aligned} \tag{28.10}$$

Here each dot or prime denotes a derivative with respect to t or χ; the hatted components are in the proper reference frame of an FO [with orthonormal basis vectors $\vec{e}_{\hat{t}} = \partial_t, \vec{e}_{\hat{\chi}} = a^{-1}\partial_\chi, \vec{e}_{\hat{\theta}} = (a\Sigma)^{-1}\partial_\theta, \vec{e}_{\hat{\phi}} = (a\Sigma \sin\theta)^{-1}\partial_\phi$; cf. the metric (28.9)]; the unhatted components are those of the coordinate basis. The mixed-coordinate components (one index up, the other down) are the easiest to evaluate mathematically, but the orthonormal components are the best for physical interpretation, since they are what an FO measures. The two sets of components are equal up to a sign, because the metric is diagonal.

13. Historically, the three possible choices for the geometry of a homogeneous, isotropic cosmological model were discovered by the Russian meteorologist Alexander Friedmann (1922). These solutions were found independently by a Belgian priest, Georges Lemaître (1927), who included a cosmological constant and discussed the growth of perturbations. The first proof that these three choices are the only possibilities was due to Robertson (1935, 1936a,b) and Walker (1935).

inclusion of geometry

If we now substitute our geometrical conditions (28.4) and (28.5) with $K = kR_0^{-2}$ [Eqs. (28.7) and (28.8)], we obtain

$$G^{\hat{t}\hat{t}} = 3\frac{\dot{a}^2 + kR_0^{-2}}{a^2}, \quad G^{\hat{\chi}\hat{\chi}} = G^{\hat{\theta}\hat{\theta}} = G^{\hat{\phi}\hat{\phi}} = -\frac{2\ddot{a}a + \dot{a}^2 + kR_0^{-2}}{a^2}. \tag{28.11}$$

Note that the spatial part of the Einstein tensor is proportional to the 3-metric $g^{\hat{i}\hat{j}} = \delta^{ij}$. In other words, it is isotropic, as we expect.

EXERCISES

Exercise 28.1 *Example: Alternative Derivation of the Spatial Metric*
Not surprisingly, there are several other approaches to deriving the possible forms of $\Sigma(\chi)$.[14] Another derivation exploits the symmetries of the Riemann tensor.

(a) The 3-dimensional Riemann curvature tensor of the hypersurface must be homogeneous and isotropic. Explain why it should therefore only involve the metric tensor and not any other tensor, for example, not the Levi-Civita tensor.

(b) Show that the only combination of these quantities that exhibits the full symmetry properties of the Riemann tensor is

$$^{(3)}R_{ijkl} = K(g_{ik}g_{jl} - g_{il}g_{jk}), \tag{28.12}$$

where K is a constant.

(c) Write a computer algebra routine to evaluate the Riemann tensor from the metric tensor (28.1) directly. By equating it to Eq. (28.12), show that two differential equations must be satisfied and these are identical to Eqs. (28.5) and (28.6).

(d) Compute the Ricci tensor, and compare it with the metric tensor. Comment.

Exercise 28.2 *Problem: Area of a Spherical Triangle*
Consider the triangle formed by the three geodesics in Fig. 28.3. In a flat space, the exterior angle ζ must equal $\theta + \psi$. However, if the space is homogeneous and positively curved, then the angle deficit $\Delta \equiv \theta + \psi - \zeta$ will be positive.

(a) By considering the geometry of the 2-dimensional surface of a sphere embedded in 3-dimensional Euclidean space, show that the area of the triangle is Δ/K. [Hint: We know that the area of a *lune* lying between two lines of longitude separated by an angle ϕ is $2\phi/K$.]

(b) Make a conjecture (or, better still, devise a demonstration) as to the formula for the area of a triangle in a negatively curved homogeneous space.

These results are special cases of the famous *Gauss-Bonnet* theorem,[15] which allows for the possibility that the topology of the space might not be simple.

14. For example, there is an elegant, group-theoretic approach (e.g., Ryan and Shepley, 1975).
15. See, e.g., Peacock (1999) and Carroll (2004).

28.2.3 Kinematics

RAYS AND WORLD LINES

We are interested in events which lie at the intersection of our past light cone and the world lines of other FOs. Imagine a single wavelength of light emitted in the χ direction at time t_e. Let there be two FOs at either end of the wavelength, and let their comoving radii be χ_e and $\chi_e + d\chi_e$. Their physical separation at the time of emission is λ_e, the emitted wavelength. Their separation in terms of the comoving coordinate is therefore $d\chi = \lambda_e/a(t_e)$. Let the light be observed by us today. Let the front of the wavelength be observed at time t_0 and the end of the wavelength at $t_0 + P_0$, where P_0 is the observed period. Using the metric (28.9) with $ds = d\theta = d\phi = 0$, we see that $\chi_e = \int_{t_e}^{t_0} dt/a$ and $\chi_e + \lambda_e/a(t_e) = \int_{t_e}^{t_0+P_0} dt/a$, from which we deduce that $\lambda_e/a(t_e) = P_e/a(t_e) = P_0$ or in terms of the frequencies, $\nu_0 = a(t_e)\nu_e$.[16] Put another way, as a photon crosses the universe, its frequency, as measured by the FOs, satisfies $\nu \propto a^{-1}$. So, if we observe a spectral line that was emitted by the stars in a distant galaxy with frequency ν_e and we observe it today with frequency ν_0, then we can deduce immediately that the scale factor a at the time of emission was ν_0/ν_e. The first stars and galaxies to form, and therefore the most distant ones that we can see, emitted their light when $a \sim 0.1$, so that a Lyman α spectral line of hydrogen emitted with a wavelength of $\lambda_e = 122$ nm is observed in the infrared today with $\lambda_0 \sim 1.2\ \mu$m.

cosmological redshift

THERMODYNAMICS

Construct a small, imaginary sphere around us with radius χ and let the sphere's surface be carried by a population of FOs, so it comoves and expands with them. These observers each see isotropic blackbody radiation. Every time a photon leaves the sphere, it will be replaced, on average, by an entering photon. The FOs could therefore construct a spherical mirror, perfectly reflecting on both sides and expanding with the universe. Nothing would change. The radiation inside this sphere will undergo slow adiabatic expansion. If it starts off as thermal blackbody radiation, it will remain so. The blackbody temperature T_γ, photon number density n_γ, pressure P_γ, and energy density ρ_γ will therefore vary as (cf. Secs. 3.2.4 and 3.5.5)

radiation in an expanding universe

$$T_\gamma \propto a^{-1}; \quad n_\gamma \propto a^{-3}; \quad \rho_\gamma, P_\gamma \propto a^{-4} \propto n_\gamma^{4/3}, \tag{28.13}$$

and the specific heat ratio is 4/3. If the radiation is isotropic but not blackbody, its distribution function $\eta_\gamma(p, a)$ will not change along a trajectory in phase space (Sec. 3.6), and so $\eta_\gamma(p', a') = \eta_\gamma(p = p'a'/a, a)$. Here $p = |\mathbf{p}|$ is the magnitude of the photon's momentum, or equally well its energy $\mathcal{E} = p$.

By parallel arguments, the individual momenta p of massive particles vary as $p \propto a^{-1}$—their de Broglie wavelengths expand with the universe just like photon

16. This argument neatly avoids a generally unhelpful separation into the Doppler shift and gravitational redshift. The quantity *redshift*, $z = (\nu_e - \nu_0)/\nu_0 = 1/a - 1$, is in common use by astronomers. However, we shall continue to work with the scale factor $a(t)$, as we are focusing on the kinematics of the expansion and not the observations through which this has been inferred.

wavelengths—so long as their behavior is adiabatic. When nonrelativistic, their kinetic energies and temperatures vary according to $T \propto p^2 \propto a^{-2}$, and their pressure P scales with their number density n as $P \propto nT \propto a^{-5} \propto n^{5/3}$, recovering the familiar specific heat ratio for a monatomic gas.

gas in an expanding universe

EXPANSION, DECELERATION, AND JERK

kinematics of the expansion

The kinematic behavior of the homogeneous universe is fully described by the single function $a(t)$. Not surprisingly, its derivatives turn out to be very useful. Adopting standard (if archaic) conventions, we define the *expansion rate* $H(t)$, the *deceleration function* $q(t)$, and the *jerk function* $j(t)$ by

expansion rate H, deceleration q and jerk j

$$H = \frac{\dot{a}}{a} \equiv \left(\frac{dt}{d\ln a}\right)^{-1}, \quad q = -\frac{\ddot{a}a}{\dot{a}^2}, \quad j = \frac{\dddot{a}a^2}{\dot{a}^3}, \tag{28.14}$$

respectively. Note that $H^{-1} = dt/d\ln a$ converts a derivative with respect to time (denoted with a dot) to a derivative with respect to $\ln a$ (denoted with a prime), which is our preferred independent variable. The expansion rate H is a reciprocal time; q and j are dimensionless. A galaxy at small radius χ emitted its photons when the scale factor was $a \simeq 1 - \chi\dot{a}$, and so the shift in the wavelengths of spectral lines observed on Earth will satisfy to first order $\delta\lambda/\lambda \equiv v = \dot{a}\chi \equiv H_0\chi$,[17] where v is the inferred recession speed, and the *Hubble constant*[18] $H_0 \equiv H(t_0)$ is the contemporary value of H and is one of the key parameters of observational cosmology. It measures the age and size of the universe. We now know that the universe is accelerating $q_0 \equiv q(t_0) < 0$ and $j_0 \equiv j(t_0) > 0$ (Riess et al., 1998; Perlmutter et al., 1999).

Hubble constant H_0

DISTANCE MEASUREMENT

There are two common measures of distance in cosmology. The first is based on observing a small source of known physical size, η, perpendicular to the line of sight at radius χ. The source's angular size measured at Earth will be $\theta = \eta/(a\Sigma)$ (Fig. 28.3, where O at $\chi = 0$ is Earth's location). This motivates us to define the angular diameter distance $d_A \equiv a\Sigma$.

angular diameter distance d_A

The second measure is based on an isotropic source of known luminosity L. If the universe were flat and static, the source's measured flux would be $F = L/(4\pi\chi^2)$. Relativistic cosmology introduces three modifications. First, the area of a sphere centered on the source with comoving radius χ, at the time of observation (today), is $4\pi\Sigma(\chi)^2$. Second, the source emits photons, and their individual energies $h\nu$ as observed at O will be reduced by a factor a from their energies when emitted. Third,

17. The quantity χ must be large enough for the recession speed to exceed the random motion with respect to a local FO.

18. The Hubble "law," that the recession velocity is proportional to the distance, was published in 1929 by Edwin Hubble, following theory by Lemaître and observations by himself, Slipher, and others. It provided the first compelling demonstration that the universe was expanding, although Hubble's inferred constant of proportionality, $H_0 \sim 500$ km s^{-1} Mpc^{-1}, was over seven times larger than the contemporary measurement.

the time it takes to emit a fixed number of photons will be shorter by another factor a than the time it takes these same photons to pass the observer on Earth. The flux will therefore be modified to $F = a^2 L/(4\pi \Sigma^2)$. This motivates us to define a *luminosity distance* $d_L = \Sigma/a = a^{-2} d_A$.

luminosity distance d_L

Making cosmological distance measurements is challenging. There are many astronomical candidates for sources of known size or luminosity for use in measuring d_A or d_L. However, they should all be greeted with suspicion, as it is not just the universe that evolves with time; its contents can likewise change. The most natural distance measures use variable and exploding stars. As we discuss below, statistical measures in the CMB and distribution of galaxies as well as gravitational lenses are also important. Many methods are anchored by local trigonometric surveys. Inevitably, there are systematic errors (*inaccuracy*), which are now more limiting than random error (*imprecision*). Remarkably, the spread in measurements of H_0 has been reduced to roughly 10%. For specificity, we shall use cosmological parameters from Planck Collaboration (2016b), starting with $H_0 = 67.7$ km s^{-1} Mpc$^{-1} \equiv (4.56 \times 10^{17}\ \text{s})^{-1} \equiv (14.4\ \text{Gyr})^{-1} \equiv (1.37 \times 10^{26}\ \text{m})^{-1} \equiv (4.43\ \text{Gpc})^{-1}$.

measurement of H_0

HORIZON

The final kinematic quantity we introduce is the *horizon*[19] radius $\chi_H(t)$ of the event on our past world line at time t. This is the comoving radius of a sphere, centered on a point on our past world line, that contains the region of the universe that can have sent signals to the event since the big bang. We also find it useful to introduce the *acoustic horizon radius* $\chi_A = C/\dot{a}$. This is the comoving distance a sound wave can travel at the local sound speed C during one expansion time scale $a/\dot{a} = H^{-1}$:

horizon radius χ_H

acoustic horizon radius χ_A

$$\chi_H(t) = \int_0^t \frac{dt'}{a(t')} = \int_0^{a(t)} \frac{da'}{a'^2 H(a')}; \ \chi_A = \frac{C}{\dot{a}}. \tag{28.15}$$

When the fluid is ultrarelativistic, C is simply $3^{-1/2}$ and $\chi_A \equiv \chi_R = 3^{-1/2} H^{-1}$.

EXERCISES

Exercise 28.3 *Example: Mildly Relativistic Particles*

Suppose that the universe contained a significant component in the form of isotropic but noninteracting particles with momentum p and rest mass m. Suppose that they were created with a distribution function $f(p, a) \propto p^{-q}$, with $4 < q < 5$ extending from $p \ll m$ to $p \gg m$ (Sec. 3.5.3).

(a) Show that the pressure and the internal energy density (not including rest mass) of these particles are both well defined.

(b) Show that the adiabatic index of this component is $q/3$ and interpret your answer.

(c) Explain how P and ρ for this component should vary with the scale factor a.

19. Or more properly the *particle horizon*, to distinguish it from an *event horizon* (Sec. 26.4.5).

Exercise 28.4 *Problem: Observations of the Luminosity Distance*

Astronomers find it convenient to use the redshift $z = 1/a - 1$ to measure the size of the universe when the light they observed was emitted.

(a) Perform a Taylor expansion in z to show that the luminosity distance of a source is given to quadratic order by

$$d_L = H_0^{-1}[z + 1/2(1 - q_0)z^2 + \cdots].$$

(b) The cubic term in this expansion involves the curvature K_0 and the jerk j_0 today. Calculate it.

28.2.4

28.2.4 Dynamics

FRIEDMANN EQUATIONS

Ultimately, we must understand how the universe's individual constituents—matter, photons, neutrinos—behave and interact. However, for the initial purposes of providing an idealized description of the average contents of the universe that can match our idealized description of its geometry and kinematics, we approximate everything as a homogeneous perfect fluid at rest in the FOs' proper reference frame, with pressure $T^{\hat{\chi}\hat{\chi}} = T^{\hat{\theta}\hat{\theta}} = T^{\hat{\phi}\hat{\phi}} = P$ (Sec. 2.13). This is related to the Einstein tensor through the Einstein field equation (Sec. 25.8). We start with the time-time part and employ geometrized units (Sec. 25.8.1): $G^{\hat{t}\hat{t}} = 8\pi T^{\hat{t}\hat{t}}$, which becomes, with the aid of Eq. (28.11):

equation of motion

$$\dot{a}^2 = \frac{8\pi}{3}\rho a^2 - \frac{k}{R_0^2}. \tag{28.16}$$

This equation of motion is the foundation of relativistic cosmology. It relates the universe's expansion rate $\dot{a}$ to its mean mass-energy density ρ and curvature k/R_0^2.

It is illuminating to write Eq. (28.16) in the form

$$\frac{1}{2}(\dot{a}\chi)^2 - \frac{4\pi\rho(a\chi)^3}{3(a\chi)} = -\frac{k\chi^2}{2R_0^2} = \text{const}, \tag{28.17}$$

for fixed χ. Again imagine that we are at the center of a small, imaginary, expanding sphere of radius $a\chi$ carried by FOs. We can regard the first term as the kinetic energy of a unit mass resting on the surface of the sphere and the second as its Newtonian gravitational potential energy. So this looks just like an equation of Newtonian energy conservation. Could we not have written it down without recourse to general relativity? The answer is "No." Equation (28.16) addresses three crucial issues on which Newtonian cosmology must remain silent. First, it neatly handles the influence of the exterior matter. Second, it relates the total energy of the expansion, the right-hand side of Eq. (28.17), to the spatial curvature. Third, it has taken account of the pressure, which as we saw with neutron stars [(Eqs. 26.38)], can contribute an active gravitational mass. The absence of pressure in Eq. (28.16) is correct but does not

inadequacy of Newtonian treatment

have a simple Newtonian explanation. These three features demonstrate the power of general relativistic cosmology and the inadequacy of a purely Newtonian approach.

Next, consider the spatial part of the Einstein equation $G^{\hat{i}\hat{j}} = 8\pi T^{\hat{i}\hat{j}} = 8\pi P\delta^{\hat{i}\hat{j}}$. With the aid of Eqs. (28.11) and (28.16), this gives a second-order differential equation:

$$\ddot{a} = -\frac{4\pi}{3}(\rho + 3P)a. \tag{28.18}$$

cosmic acceleration

The divergence of the stress-energy tensor must vanish. We have seen many times in this book [Ex. 2.26, Eqs. (13.86), Secs. 13.8.1, 26.3.3] that for any perfect fluid, in the fluid's rest frame (which for cosmology is the same as the FO rest frame), the time component $T^{\hat{t}\hat{\alpha}}{}_{;\hat{\alpha}} = 0$ is energy conservation, or equivalently, the first law of thermodynamics. For a fluid element with unit comoving volume, the physical volume is a^3, so the first law states

$$d(\rho a^3) = -P d(a^3). \tag{28.19}$$

first law of thermodynamics

If the fluid comprises two or more independent components that do not exchange heat and evolve separately, then Eq. (28.19) must apply to each component.

We have also seen many times [Ex. 2.26, Eqs. (13.86), Sec. 26.3.3] that in the fluid's rest frame, the spatial component $T^{\hat{i}\hat{\alpha}}{}_{;\hat{\alpha}} = 0$ is force balance; it equates the fluid's pressure gradient ∇P to its inertial mass per unit volume $(\rho + P)$ times its 4-acceleration. However, homogeneity and isotropy guarantee that in our cosmological situation, the fluid's pressure gradient and 4-acceleration vanish, so $T^{\hat{i}\hat{\alpha}}{}_{;\hat{\alpha}} = 0$ is satisfied identically and automatically. It teaches us nothing new.

Equations (28.16), (28.18), and (28.19) are called the *Friedmann equations* in honor of Alexander Friedmann. They are not independent; the contracted Bianchi identity guarantees that from any two of them, one can deduce the third (Sec. 25.8).

CRITICAL DENSITY

Not only does the Hubble constant define a scale of length and of time, it also defines a critical density, ρ_{cr}—the value of ρ for which the universe's spatial curvature $K = k/R_0^2$ vanishes today [Eqs. (28.16) and (28.14)]:

$$\rho_{cr} = \frac{3H_0^2}{8\pi} = 8.6 \times 10^{-27}\ \text{kg m}^{-3} \equiv 7.7 \times 10^{-10}\ \text{J m}^{-3}. \tag{28.20}$$

It is conventional to express the energy density of constituent i today (e.g., radiation or baryonic matter) as a fraction by introducing

density fraction

$$\Omega_i = \frac{\rho_i}{\rho_{cr}}; \quad \Omega = \frac{\rho}{\rho_{cr}} = \sum_i \Omega_i = 1 - \Omega_k, \tag{28.21}$$

where Ω refers to the total energy density, and

$$\Omega_k = -\frac{k}{H_0^2 R_0^2} \tag{28.22}$$

takes account of curvature.

expansion rate, deceleration function, and jerk function

The kinematic quantities, H, q, j [Eq. (28.14)] are then given by

$$H = H_0 \left(\frac{\rho}{\rho_{cr}} + \frac{\Omega_k}{a^2} \right)^{1/2}; \quad q = \frac{1}{2} \frac{\rho + 3P}{\rho + \frac{\Omega_k \rho_{cr}}{a^2}}; \quad j = \frac{\rho - \frac{3}{2} \frac{dP}{d \ln a}}{\rho + \frac{\Omega_k \rho_{cr}}{a^2}}, \tag{28.23}$$

respectively. Note that if the curvature is negligible, $q = \frac{1}{2}$ when $P = 0$ and $j = 1$ when P is constant.

A FLAT UNIVERSE

At this point we introduce a key simplification. Essentially geometrical arguments, based largely on observations of CMB fluctuations, have shown that the radius of curvature today is $|R_0| \gtrsim 14/H_0$. Equivalently, $\Omega_k \lesssim 0.005$. This is so small, and was even smaller in the past (when we replace R_0 by R) that henceforth we shall set $\Omega_k = 0$ and $\Sigma = \chi$. If significant spatial curvature is measured one day, then small corrections will be required to what follows, while the principles are unaffected. However, none of this absolves us from the obligation to explain *why* the universe is so flat, an issue to which we shall return in Sec. 28.7.1.

SUMMARY

We have described an idealized universe that is homogeneous and isotropic everywhere. We have shown its spatial geometry can take one of three different forms and have invoked observation to restrict attention to the spatially flat case. We have also introduced some useful cosmological measures, most notably the scale factor $a(t)$, to characterize the kinematics of the universe. We have married the universe's geometry to its kinematics to calculate the two independent components of the Einstein tensor, which we then combined with the volume-averaged stress-energy tensor of the contents of the universe to derive the Friedmann equations. We now turn to cataloging the contributions to the stress-energy tensor today.

EXERCISES

Exercise 28.5 *Example: Einstein–de Sitter Universe*

Suppose, as was once thought to be the case, that the universe today is flat and dominated by cold (pressure-free) matter.

(a) Show that $a \propto t^{2/3}$ and evaluate the age of the universe assuming the Hubble constant given in Sec. 28.2.3: $H_0 = 67.7$ km s^{-1} Mpc^{-1}.

(b) Evaluate an expression for the angular diameter distance as a function of a and find its maximum value.

(c) Calculate the comoving volume within the universe back to very early times.

28.3 The Universe Today

28.3

28.3.1 Baryons

28.3.1

Among all the constituents in our universe, baryonic matter is the easiest form to identify, because it is capable of radiating when hot and absorbing when cool. The baryons from the very early universe are mostly in the form of hydrogen, with mass fraction 0.75, and helium, with mass fraction $0.25 \equiv Y_{He}$. Some of these baryons found their way into stars, which produced more helium and created about 1% by mass, on average, of heavier elements. Stars have masses ranging from a tenth to more than 100 times that of the Sun. The most massive stars shine for roughly a million years, while the least massive ones will last more than a trillion years, much longer than the age of the universe.

stars at a glance

Very roughly 10^{-4} of the total mass of a typical galaxy is found in a massive, spinning black hole residing in its nucleus. When this black hole is supplied with gas, roughly 10% of the rest mass energy of this gas is converted into radiation. The galaxy is then said to have an *active galactic nucleus*, and when this outshines the entire galaxy, it is called a *quasar*.

active galactic nuclei and quasars

It is common practice to use the luminosity of a galaxy as a measure of its stellar mass. However, this must be done with care, because the answer depends on the relative proportions of low- and high-mass stars. The latter are far more luminous per unit mass. It also depends on the time that has elapsed since the stars were formed. If the stars are old, the luminous high-mass ones will have consumed all their nuclear fuel and evolved to form neutron stars and black holes. Absorption is also an issue. Despite all these difficulties, astronomers estimate that the average fraction of stellar mass today is $\Omega_* \sim 0.005$.

The baryons that are not contained in stars exist as gas. We can assay this gas fraction by measuring the X-ray emission from clusters of galaxies and find that $\Omega_{gas} \sim 0.05$. More accurate measurements of the total baryon fraction Ω_b are made possible by measuring the tiny fraction of deuterium and other light elements that are created in trace amounts in the early universe (Sec. 28.4.2) and through interpreting the spectrum of CMB fluctuations (Sec. 28.6.1). The best estimate today is $\Omega_b = 0.049$, implying that the mean energy and number densities of baryons in the universe are $\rho_{b0} = 3.8 \times 10^{-11}$ J m^{-3} $= 4.2 \times 10^{-28}$ kg m^{-3} and $n_{b0} = 0.25$ m^{-3}. The equivalent proton, electron, and helium densities (Sec. 28.4.2) are $n_{p0} = 0.19$ m^{-3}, $n_{e0} = 0.22$ m^{-3}, and $n_{\alpha 0} = 0.016$ m^{-3} (Table 28.1). Of course, these densities are substantially higher in galaxies.

measuring Ω_b

EXERCISES

Exercise 28.6 *Example: Stars and Massive Black Holes in Galaxies*

Assume that a fraction ~0.2 of the baryons in the universe is associated with galaxies, split roughly equally between stars and gas. Also assume that a fraction $\sim 10^{-3}$ of the baryons in each galaxy is associated with a massive black hole and that most of the radiation from stars and black holes was radiated when a ~0.3.

TABLE 28.1: The universe today

Geometric	$\Omega_k = 0$		
Kinematic	Hubble constant	Deceleration parameter	Jerk parameter
	$H_0 = 67.7$ km s^{-1} Mpc^{-1}	$q_0 = -0.54$	$j_0 = 1$
Derived	Hubble time	Hubble distance	Critical density
	$t_H = H_0^{-1} = 4.6 \times 10^{17}$ s	$d_H = ct_H = 1.4 \times 10^{26}$ m	$\rho_{cr} = 7.7 \times 10^{-10}$ J m^{-3}
Constituent	Energy fraction	Energy density (J m^{-3})	Number density (m^{-3})
Baryons	$\Omega_b = 0.049$	$\rho_{b0} = 3.8 \times 10^{-11}$	$n_{b0}, n_{e0} = 0.25, 0.22$
Dark matter	$\Omega_D = 0.26$	$\rho_{D0} = 2.0 \times 10^{-10}$	$n_{D0} = 0.0013 m_{D,12}^{-1}$
Cosmological	$\Omega_\Lambda = 0.69$	$\rho_\Lambda = 5.3 \times 10^{-10}$	
Photons	$\Omega_\gamma = 5.4 \times 10^{-5}$	$\rho_{\gamma 0} = 4.2 \times 10^{-14}$	$n_{\gamma 0} = 4.1 \times 10^8$
Neutrinos	$\Omega_\nu = 0.0014 m_{\nu,-1}$	$\rho_{\nu 0} = 1.1 \times 10^{-12} m_{\nu,-1}$	$n_{\nu 0} = 3.4 \times 10^8$

Note: The notation is explained in the text.

(a) The current energy density in light from all the galaxies is $\sim 4\times10^{-16}$ J m^{-3}. Estimate what fraction of stellar hydrogen has been converted by nuclear reactions into helium. (You may assume that heat is created by these reactions at a rate of 6.4 MeV per nucleon.)

(b) The current energy density of light radiated by observed accreting black holes—dominated by quasars—is $\sim 6 \times 10^{-17}$ J m^{-3}. Estimate the actual efficiency with which the rest mass energy of the accreting gas is released.

(c) Estimate the total entropy per baryon associated with the horizons of massive black holes today and compare with the entropy per baryon associated with the microwave background and the intergalactic medium (see Sec. 4.10.2).

28.3.2

28.3.2 Dark Matter

clusters of galaxies

After studying the Coma cluster of galaxies, Fritz Zwicky (1933)[20] argued that most of the gravitational mass in clusters of galaxies is neither in the form of stars nor of gas but in a *dark* form that has only been detectable through its gravity (Fig. 28.1c). In effect, the gas and the galaxies move in giant gravitational potential wells formed by this dark matter. Careful measurements of the motions of gas and stars have led to an estimate of the dark matter density about five times the baryon density. Again, CMB observations improve the accuracy, and today we find that the dark matter fraction is $\Omega_D = 0.26$ so that $\Omega_b + \Omega_D = 0.31$ is the total matter fraction. It is widely suspected that dark

20. Oort (1932) realized that there is invisible matter in the solar neighborhood. Important subsequent evidence came from optical and radio observations of nearby spiral galaxies by Rubin, Roberts, and others; e.g., Rubin and Ford (1970); and Roberts and Whitehurst (1975).

matter mostly comprises new elementary particles. Furthermore, the development of gravitational clustering on small linear scales points to this matter being collisionless and having negligible pressure when the perturbations grow.[21] The inferred energy and number density of putative dark matter particles is $\rho_{D0} = 2.0 \times 10^{-10}$ J m^{-3}, $n_{D0} = 0.0013 m_{D,12}{}^{-1}$ m^{-3}, where $m_{D,12}$ is the mass of the hypothesized particle in units of 1 TeV. Of course, the local dark matter density in our galaxy is much larger than this.

cold dark matter

EXERCISES

Exercise 28.7 *Challenge: Galaxies*

Make a simple (numerical) model of a spherical galaxy in which the dark matter particles moving in the (Newtonian) gravitational field they create behave like collisionless plasma particles moving in an electromagnetic field. Ignore the baryons.

(a) Adopt the fluid approximation, treat the pressure P as isotropic and equal to $K\rho^{4/3}$, and use the equation of hydrostatic equilibrium (Sec. 13.3) to solve for the mass and radius. Comment on the answer you get, and make a suitable approximation to define an effective mass and radius.

(b) Consider a line passing through the galaxy with impact parameter relative to the center given by b. Solve for the dark matter particles' rms velocity along this line as a function of b.

(c) It is observed that the central densities of galaxies scale as the inverse square of the rms velocity for $b = 0$. How does the mass scale with the velocity?

(d) Solve for the distribution function of the dark matter particles assuming that it is just a function of the energy (Sec. 22.2).

28.3.3 Photons

28.3.3

The next contributor to the energy density of the universe is the CMB. With its measured temperature of $T_{\gamma 0} = 2.725$ K, the energy density is $\rho_{\gamma 0} = 4.2 \times 10^{-14}$ J m^{-3} (Sec. 3.5). Equivalently, $\Omega_\gamma = 5.4 \times 10^{-5}$. This is dynamically insignificant today but was very important in the past. We can also compute the number density of photons to be $n_{\gamma 0} = 4.1 \times 10^8$ m$^{-3} = 1.6 \times 10^9 n_{b0}$. An equivalent measure, which we need below, is the photon entropy per baryon (Secs. 4.8 and 4.10):

photon number density

$$\sigma_{\gamma 0} = \frac{\rho_{\gamma 0} + P_{\gamma 0}}{n_{b0} T_{\gamma 0}} = \frac{32\pi^5}{45 n_{b0}} \left(\frac{k_B T_{\gamma 0}}{h}\right)^3 k_B = 5.9 \times 10^9 k_B. \quad (28.24)$$

entropy of the CMB

The photon entropy per baryon—the entropy in a box that contains, on average, one baryon and expands with the universe—is therefore conserved to high accuracy.

21. It is often called *cold dark matter*. However, "cold" is inappropriate at early times, when the particles are probably relativistic, and at late times, when they virialize in dark matter potential wells with thermal speeds ~100–1,000 km s^{-1}.

28.3.4

28.3.4 Neutrinos

There is also a (currently) undetectable background of neutrinos that was in thermal equilibrium with the photons at early epochs and then became thermodynamically isolated with an approximate Fermi-Dirac distribution. As we show in the next section, the current total neutrino density is $n_{\nu 0} = 3.4 \times 10^8\ \mathrm{m}^{-3}$. There are three *flavors* of neutrinos (ν_e, ν_μ, ν_τ) plus their antiparticles, and they are known to have small masses. The contemporary total neutrino energy density and mass fraction are then $\rho_{\nu 0} = 1.8 \times 10^{-12} m_{\nu,-1}\ \mathrm{J\ m}^{-3}$, $\Omega_\nu = 0.0023 m_{\nu,-1}$, respectively, where $m_{\nu,-1} = (m_{\nu_e} + m_{\nu_\mu} + m_{\nu_\tau})/100\ \mathrm{meV}$. This total m_ν is bounded below through neutrino oscillation measurements (e.g., Cahn and Goldhaber, 2009) at 60 meV and above by cosmological observations at 230 meV. For illustration purposes, we shall assume that there is a single, dominant neutrino of mass 100 meV (i.e., 0.1 eV).

neutrino flavors

28.3.5

28.3.5 Cosmological Constant

In 1998, it was discovered that the universe is accelerating. This was first inferred from observations of exploding stars called *supernovae*,[22] which turn out to be surprisingly good distance indicators. As we know the scale factor at the time of the explosion, we can measure $\chi(a)$ and infer contemporary values for the deceleration and jerk functions: $q_0 = -0.54$ and j_0 consistent with 1. Using Eq. (28.23), we infer that the pressure is also negative, $P = -0.69\rho$, where ρ includes the matter density.

As we have mentioned, Einstein anticipated this possibility in 1917 when he noted that the field equations would remain "covariant"—in our language, would continue to obey the Principle of Relativity (be expressible in geometric language)—if an additional term proportional to the metric tensor $\boldsymbol{g}$ were added to the field equation (Sec. 25.8):[23]

$$\boldsymbol{G} + \Lambda \boldsymbol{g} = 8\pi \boldsymbol{T}. \tag{28.25}$$

the cosmological constant Λ

The constant of proportionality Λ is known as the *cosmological constant*. As the Einstein tensor involves second derivatives of the metric tensor, a cosmological constant term that is significant on a cosmological scale should be undetectable on the scale of the solar system or a compact object.

Although we might interpret Eq. (28.25) as a modification to the field equation, we can instead incorporate Λ into the stress-energy tensor $\mathbf{T}$ [move $\Lambda \boldsymbol{g}$ to the right-hand side of Eq. (28.25) and absorb it into $8\pi \boldsymbol{T}$], resulting in a *cosmological energy density* ρ_Λ and *cosmological pressure* P_Λ satisfying

$$\rho_\Lambda = -P_\Lambda = \frac{\Lambda}{8\pi}. \tag{28.26}$$

22. Perlmutter et al. (1999), Riess et al. (1998).
23. Einstein then went on to propose a static universe, made possible by the cosmological constant Λ. But it was proved wrong when observations revealed the Hubble expansion of the universe, and it was also unstable; so he renounced Λ (Einstein, 1931).

In the absence of another major component of the universe, we deduce that $\Omega_\Lambda = 1 - \Omega_b - \Omega_D - \Omega_\gamma - \Omega_\nu = 0.69$ and $\rho_\Lambda = -P_\Lambda = 5.3 \times 10^{-10}\ \mathrm{J\,m^{-3}}$.

Although the large negative pressure may seem strange, it is precedented in classical electromagnetism. Consider a uniform magnetic field B permeating a cylinder plus piston and aligned with their axis. The energy density is $\rho_{\mathrm{mag}} = B^2/(2\mu_0)$, and the total stress acting on the piston—the combination of the isotropic magnetic pressure and the magnetic tension—is $P_{\mathrm{mag}} = -B^2/(2\mu_0) = -\rho_{\mathrm{mag}}$. When we withdraw the piston, ρ_{mag} and P_{mag} will not change, in agreement with the first law of thermodynamics, just like the cosmological constant. What is different about the cosmological constant is that the stress is isotropic.

Λ as density and negative pressure

The cosmological constant contribution to the stress-energy tensor is, by assumption, constant on a hypersurface of simultaneity—it is *ubiquitous.* Applying the first law of thermodynamics [Eq. (28.19)], we see that it was the same in the past and will be the same in the future—it is (past and future) *eternal.* In addition, as the contribution to the stress-energy tensor is directly proportional to the metric tensor, it takes the same form in all frames—it is *invariant.* Combining these features, we see that from a relativistic perspective, it is *universal.*

Λ is universal

28.3.6 Standard Cosmology

This cosmological energy density ρ_Λ and pressure P_Λ are the most conservative explanation for the universe's acceleration. They motivate us to define a *standard cosmology,* in which there are no other major components of the universe besides ρ_Λ, P_Λ, baryonic matter, dark matter, photons, and neutrinos, and in which the spatial geometry is flat.

EXERCISES

Exercise 28.8 *Example: The Pressure of the Rest of the Universe*

Calculate three contributions to the pressure of the contemporary universe.

(a) Baryons. Assume that most of the baryons in the universe outside of stars make up a uniform, hot intergalactic medium with temperature 10^6 K.

(b) Radiation.

(c) Neutrinos. Assume that almost all the neutrino pressure is associated with one flavor with an associated mass of 100 meV.

28.4 Seven Ages of the Universe

Given this description of the present contents of the universe, we are now in a position to describe its history. As our main purpose is to use cosmology to illustrate and apply many features of classical physics, we do not retrace the tortuous, inferential path that led to the standard description of the universe. Instead, we proceed more or less deductively from the earliest time and describe many of the essentially classical processes that must have occurred along the way. To help us do this, we use the device of dividing the history into seven ages, each dominated by different physical processes.

28.4.1

28.4.1 Particle Age

CONSTITUENTS

As explained at the start of this chapter, the universe began expanding from a very hot, dense state; we commence our story at a time when we still have confidence in essentially classical principles. To be specific, this is when $a \sim 10^{-11}$, $T \sim 10$ MeV $\sim 10^{11}$ K, and $t \sim 10$ ms. At earlier times, thermodynamic equilibrium required the presence of a large density of nucleon-antinucleon pairs and pions at near-nuclear density. A description in terms of quarks and gluons is needed (Cahn and Goldhaber, 2009). However, for $a \gtrsim 10^{-11}$, only a tiny but permanent residue of neutrons n and protons p was present, and it was dynamically and thermodynamically irrelevant. We use the comoving density of these baryons as a reference:

baryon density

$$n_b \equiv n_p + n_n = n_{b0} a^{-3} = 2.5 \times 10^{32} a_{-11}^{-3}\ \mathrm{m}^{-3}; \quad a \gtrsim 10^{-11}, \tag{28.27}$$

where $a_{-11} \equiv 10^{11} a$.

There were also muons and τs along with their antiparticles, but like the pions, they disappeared from thermodynamic equilibrium and quickly decayed. However, when $a \sim 10^{-11}$, the much lighter electrons and positrons were present with combined density n_e comparable to that of the photons n_γ. The photons and pairs interacted electromagnetically and remained in thermodynamic equilibrium with each other, as well as with the baryons, at a common temperature $T_\gamma(a)$. Neutrinos, with combined density n_ν, were also present at early times. Each of the three neutrino flavors had an approximate Fermi-Dirac distribution with zero chemical potential (Secs. 4.4.3, 5.5.3); and as they were effectively massless, like the photons, the neutrinos' temperature T_ν decreased $\propto a^{-1}$ until comparatively recently, when their finite masses became significant. However, unlike the photons, they did so in thermodynamic isolation (they were *decoupled* from other particles) as the rate at which they self-interacted or exchanged energy with the other constituents quickly became less than the rate of expansion, as we demonstrate below.

pair density

neutrino decoupling

ENTROPY AND TEMPERATURE

Prior to neutrino decoupling, the photons and neutrinos were in equilibrium at the same temperature. However, when $T_\gamma \lesssim m_e/k_B \sim 10^{10}$ K, the pairs became nonrelativistic; they annihilated, and their number (and entropy) density, relative to that of the photons, quickly declined to a very small value—ultimately just that required to equalize the proton charge density—while the photon entropy [Eq. (28.24)] was augmented. Now, the photon and pair entropies per (conserved) baryon, σ_γ and σ_e, are given by Eq. (28.24) with $T_{\gamma 0}$ replaced by T_γ and n_{b0} by n_b, and by

pair annihilation

$$\sigma_e = \frac{\rho_e + P_e}{n_b T_\gamma} = \frac{16\pi a^3}{n_{b0} h^3 T_\gamma} \int_0^\infty \frac{dp\, p^2 (E + p^2/(3E))}{e^{E/(k_B T_\gamma)} + 1} = \frac{7}{4} \sigma_\gamma f_{\sigma e}(y), \tag{28.28}$$

where $E = (p^2 + m_e^2)^{1/2}$, $y = m_e/(k_B T_\gamma)$, and

$$f_{\sigma e}(y) = \frac{90 y^4}{7\pi^4} \int_0^\infty \frac{\mathrm{d}x\, x^2 (1 + 4x^2/3)}{(1 + x^2)^{1/2} (e^{y(1+x^2)^{1/2}} + 1)} \tag{28.29}$$

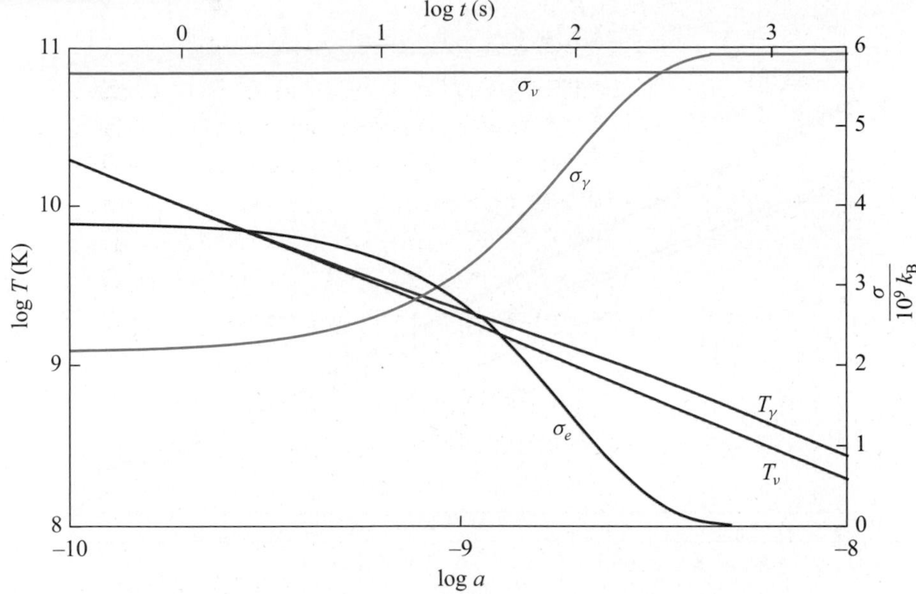

FIGURE 28.4 Temperature T and entropy σ variation during the particle age. At early times t (or equivalently, small scale factors a), the photons, electron-positron pairs, and neutrinos (designated by subscripts γ, e, and ν, respectively) were in equilibrium with a common temperature. However, when the temperature approached m_e/k_B, the pairs began to annihilate, and their contribution to the total entropy was taken up by the photons. (The neutrinos had decoupled by this time, and their entropy per baryon remained constant.) The photon temperature T_γ increased relative to the neutrino temperature T_ν by a factor of 1.4.

varies between 1 when $y \to 0$ ($T_\gamma \to \infty$) and 0 when $y \to \infty$ ($T_\gamma \to 0$) (Secs. 4.8, 4.10.3). Entropy conservation dictates that $\sigma_e + \sigma_\gamma = \sigma_{\gamma 0}$. We can therefore solve for the scale factor $a = (y/y_0)(1 + 7f_{\sigma e}(y)/4)^{-1/3}$, where $y_0 = m_e/(k_B T_{\gamma 0}) = 2.2 \times 10^9$, and hence for the photon temperature as a function of a. At early times and high temperatures, $\sigma_\gamma = (4/11)\sigma_{\gamma 0}$, and so $T_\nu/T_\gamma = (4/11)^{1/3} = 0.71$ at later times.[24] The neutrino and photon densities satisfy $n_\nu = \frac{9}{4}(T_\nu/T_\gamma)^3 n_\gamma \propto n_b$, and so n_ν/n_γ decreased from $\frac{9}{4}$ to $\frac{9}{11}$, whence $n_\nu = 3.4 \times 10^{41} a_{-11}^{-3}$ m^{-3} (Fig. 28.4).

neutrino density

ENERGY DENSITY

We can now sum ρ_γ, ρ_e, ρ_ν, ρ_D, ρ_b, and ρ_Λ to give the total energy density for $T_\gamma \lesssim 10^{11}$ K (Fig. 28.5; Secs. 3.5.4, 3.5.5)

variation of energy density over cosmic time

$$\rho = a_B T_\gamma^4 + \frac{7}{4} a_B T_\gamma^4 f_{\rho e}(y) + \frac{21}{8} a_B T_\nu^4 + \rho_{D0} a^{-3} + \rho_{b0} a^{-3} + \rho_\Lambda, \qquad (28.30)$$

where

$$f_{\rho e} = \frac{120 y^4}{7\pi^4} \int_0^\infty \frac{dx\, x^2 (1+x^2)^{1/2}}{(e^{y(1+x^2)^{1/2}} + 1)}. \qquad (28.31)$$

24. As neutrinos have mass yet do not interact, they do not maintain a thermal distribution function when they eventually become nonrelativistic, and so $T_{\nu 0}$ is meaningless.

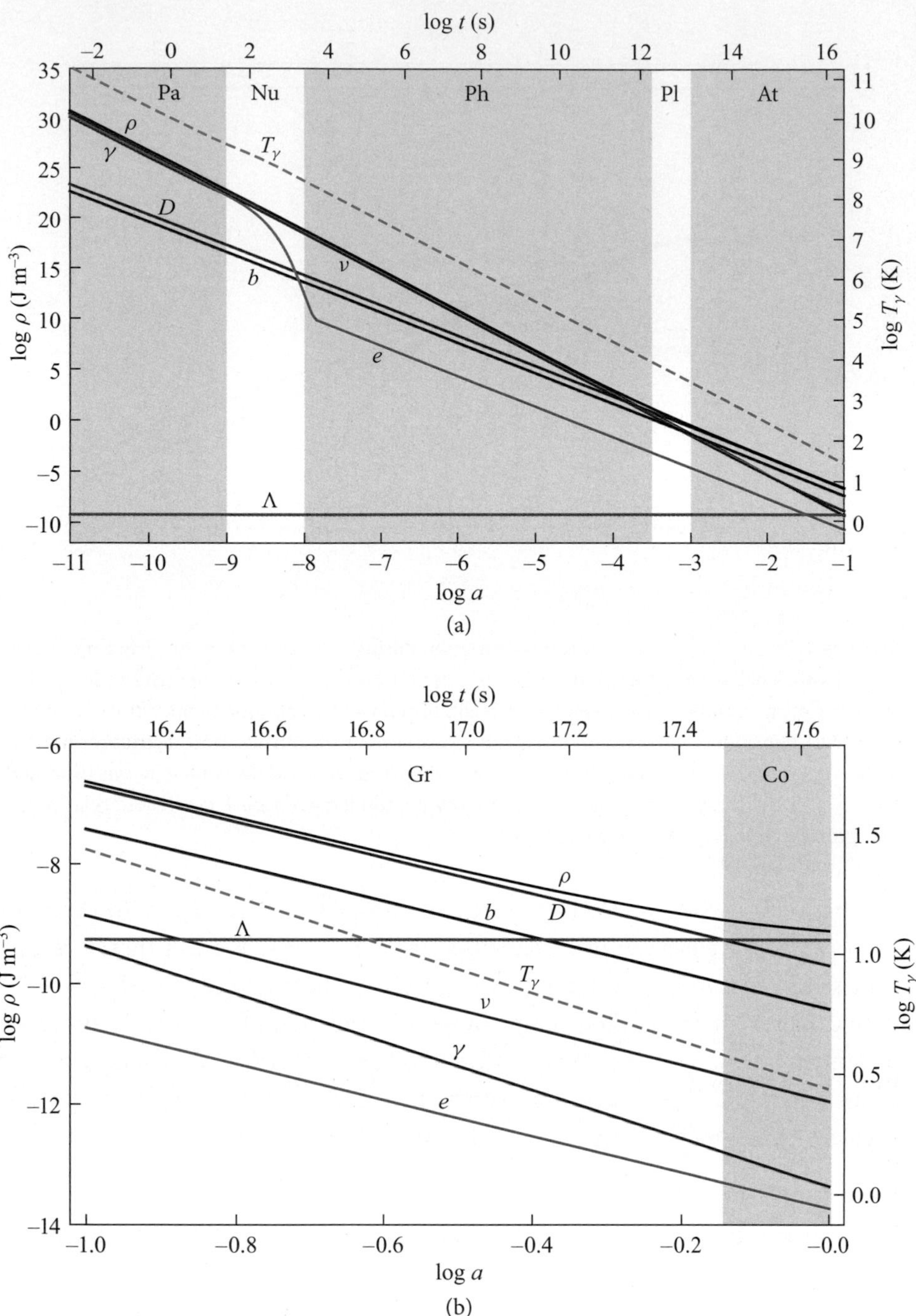

FIGURE 28.5 Energy density in the expanding universe. The energy density ρ (in J m^{-3}) for photons (γ), neutrinos (ν), dark matter (D), baryons (b), and electrons (e) as a function of the scale factor a while the universe evolved. (a) Evolution through the particle (Pa), nuclear (Nu), photon (Ph), plasma (Pl), and atomic (At) ages ending at the *epoch of reionization,* when $a \sim 0.1$. (b) Evolution through the gravitational (Gr) and cosmological (Co) ages. The total energy density is depicted by a thick black line. Also displayed is the cosmic time (proper time) t and the photon temperature T_γ.

Only the first three terms in Eq. (28.30) are significant at early times, but we give the complete expression here.[25] We also make a correction for the (unknown) neutrino rest mass density as discussed in Sec. 28.3.4; it has small, observable effects at late times.

AGE

Given the density, we can use the equation of motion to compute the age of the universe (cosmic time) t as a function of the scale factor a:

time as a function of scale factor

$$t(a) = \left(\frac{3}{8\pi}\right)^{1/2} \int_0^a \frac{da'}{a'} \rho(a')^{-1/2} \tag{28.32}$$

[Eq. (28.16)]. The age today ($a = 1$) is $t_0 = 4.4 \times 10^{17}$ s $= 13.8$ Gyr.

HORIZON

At early times when matter is ultrarelativistic with $T_\gamma \gtrsim 10^{11}$ K, t is proportional to a^2, ignoring a weak dependence on the number of types of particle contributing to the expansion. In this case, using Eq. (28.15), we find for the horizon radius $\chi_H = 2t/a \propto a$. This tells us that the horizon sphere gets smaller and smaller in terms of comoving baryon mass $M_{Hb} = 4\pi \chi_H^3 \rho_{b0}/3 \propto (t/a)^3$ as we go back in time. When $a \sim 10^{-11}$, $\chi_H \sim 3 \times 10^{16}$ m and $M_H \sim 10^{25}$ kg (roughly an Earth mass). As we have emphasized, the universe that we see around us today grew deterministically out of a hot big bang and is believed to be homogeneous as far as we can see. Yet the regions that were able to establish this smoothness through some form of mixing were comparatively tiny at any time when this could have happened (Fig. 28.6). A plausible resolution of this longstanding paradox, *inflation,* will be discussed in Sec. 28.7.1.

horizon problem

28.4.2 Nuclear Age

28.4.2

NEUTRON-PROTON RATIO

During the early particle age, neutrons and protons were able to establish thermodynamic equilibrium primarily through the weak reactions $n + \nu \leftrightarrow p + e^-$ and $n + e^+ \leftrightarrow p + \bar{\nu}$. When 10^{12} K $\gtrsim T_\gamma \gtrsim 10^{11}$ K, $n_n \sim n_p \sim 0.5 n_b$. As the universe cooled, a Boltzmann distribution was established: $n_n/n_p = \exp[-\gamma_{np} m_e/(k_B T)]$, where we write the mass difference $m_n - m_p$ as $\gamma_{np} m_e = 2.53 m_e = 1.29$ MeV. However, when $T_\gamma \lesssim 3 \times 10^{10}$ K, the reaction rate became slower than the expansion rate, and n_n/n_p declined slowly to ~ 0.14. If these were the only possible interactions, then the neutrons would have eventually undergone β-decay, $n \rightarrow p + e^- + \bar{\nu}$, when the universe had expanded further and the age was of order the mean life of a neutron, ~ 15 min. ($\gamma_{np} m_e$ is the maximum electron energy in this decay process.) However, before this happened, strong interactions intervened, and helium was formed.

neutron freeze out

25. Although the free electrons are dynamically insignificant at late times, they are very important for the evolution of the microwave background, and their density evolution is discussed below.

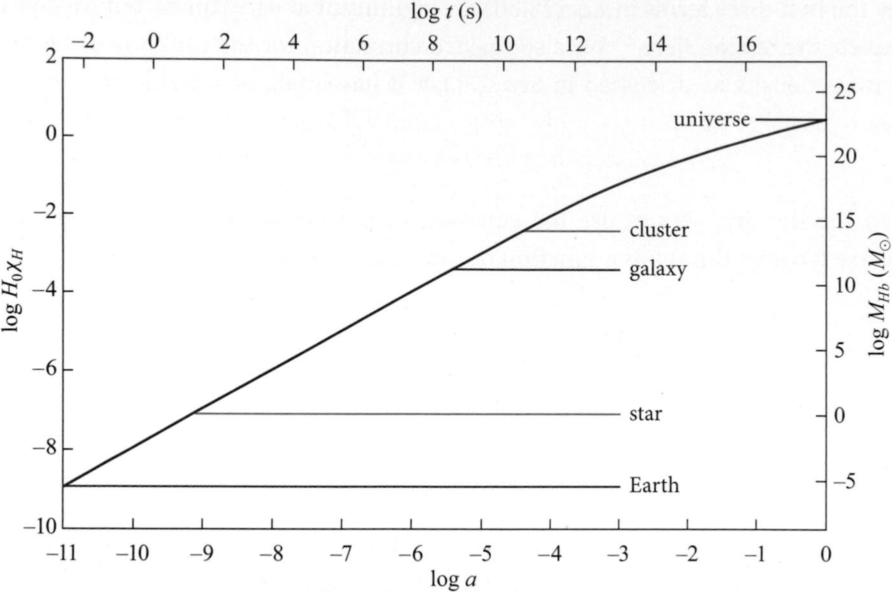

FIGURE 28.6 Comoving horizon radius χ_H as a function of the scale factor a. Also displayed is the associated baryon mass M_{Hb} measured in solar masses (see footnote 1 in this chapter). Note that the mass of baryons in causal contact when $a \sim 10^{-11}$ was equal to only an Earth mass. The mass associated with a cluster of galaxies entered our horizon when $t \sim 300$ yr. The horizon mass when the microwave background was last scattered, $a \sim 10^{-3}$, is equivalent to $\sim 10^4$ galaxy clusters but is only $\sim 10^{-5}$ of the total mass that is observed today, which appears to be impressively homogeneous.

Let us begin analyzing this process by ignoring the expansion of the universe and considering the conversion of n to p through the single reaction $n + \nu \rightarrow p + e^-$ (cf. Sec. 5.5.3). Let us define the *rate constant* $\lambda_{n\nu}$ by the expression

forward reaction

$$\dot{n}_p = -\dot{n}_n = \lambda_{n\nu} n_n, \tag{28.33}$$

where a dot denotes differentiation with respect to t. To compute $\lambda_{n\nu}$, we make some well-justified simplifications. First, we ignore the motion of the nucleons and the mass of the neutrinos, which is justified when $T \sim 10^9$ K. Second, we suppose that all distribution functions are isotropic, which is ensured by frequent scattering. Third, we suppose that the electrons and positrons have a common Fermi-Dirac distribution function $f_e = (e^{\gamma m_e/(k_B T_\gamma)} + 1)^{-1}$ with zero chemical potential and temperature T_γ, which is reasonable, as there were many more electrons than nucleons at this time and they were strongly coupled to the photons through Compton scattering. Likewise, the neutrinos and antineutrinos maintain a similar Fermi-Dirac distribution f_ν with an independent temperature T_ν.

On general grounds [cf. Eqs. (3.28), (23.50), and (23.56)], we expect to be able to write the rate coefficient in the form:

$$\lambda_{n\nu} = \int \left(\frac{d^3 p_\nu}{h^3} f_\nu\right) \left(\frac{h^6 W}{m_e^5} \delta(E_\nu - \gamma m_e + \gamma_{np} m_e)\right) \left(2 \frac{d^3 p_e}{h^3}(1 - f_e)\right). \tag{28.34}$$

The first term in this expression describes the number density of neutrinos. The second is the *rate per state,* defining W and including a delta function to ensure energy conservation, while the third describes the electron density of states, taking account of both spins and including a blocking factor when the states are occupied. The quantity W can be calculated using the theory of electro-weak interactions and can be measured experimentally. It has the value $W = 2.0 \times 10^{-6}\ \mathrm{s}^{-1}$ (e.g., Weinberg, 2008). As everything is isotropic, we can write $d^3p = 4\pi(E^2 - m^2)^{1/2} E dE$ and integrate over the delta function to obtain

integration over phase space

$$\lambda_{n\nu} = 32\pi^2 W \int_{\gamma_{np}}^{\infty} d\gamma \frac{\gamma(\gamma^2 - 1)^{1/2}(\gamma - \gamma_{np})^2}{(e^{(\gamma-\gamma_{np})m_e/(k_B T_\nu)} + 1)(e^{-\gamma m_e/(k_B T_\gamma)} + 1)}. \quad (28.35)$$

The integration is over the range of γ permitted by energy conservation.

Note that if we set $T_\nu = T_\gamma = T$, the neutrons and protons will attain Boltzmann equilibrium (Sec. 4.4), and so the rate constant λ_{pe} for the inverse reaction will be $e^{-\gamma_{np} m_e/(kT)} \lambda_{n\nu}$.[26] This, in turn, implies that W, in the integral corresponding to Eq. (28.34), is the same for the forward and backward reactions. Furthermore, the theory of electro-weak interactions informs us that W is the same for the $n + e^+ \leftrightarrow p + \bar{\nu}$ and $n \rightarrow p + e^- + \bar{\nu}$ reactions. The five relevant rate constants are exhibited in Fig. 28.7(a).

inverse reaction

Including the expansion of the universe is a straightforward matter. We replace the number density of neutrons and protons with the number contained in a fixed volume expanding with the universe. If we set this volume as n_b^{-1}, recalling that baryons are conserved, then we can allow for the expansion by the device of replacing n_p with the proton fraction $X_p \equiv n_p/n_b$, and similarly treating n_n in Eq. (28.33).

including expansion

DEUTERIUM AND HELIUM FORMATION

Neutrons and protons can combine to form deuterons ($^2\mathrm{H} \equiv d$) through the reactions $p + n \leftrightarrow d + \gamma$ (e.g., Cyburt et al., 2016, and Ex. 4.10). If we just consider the forward reaction and, temporarily, ignore the expansion, d would have formed at the rate $\dot{n}_d = n_p n_n \langle \sigma v \rangle_{np}$. Here the product of the cross section σ and the relative speed v of the nucleons, averaged over the velocity distribution at a given temperature and over the evolution of the universe when these reactions are most significant, is $\langle \sigma v \rangle_{np} = 5 \times 10^{-26}\ \mathrm{m}^3\ \mathrm{s}^{-1}$.

deuterium formation

We can now include the inverse reaction by observing that, in equilibrium, we can relate the density n_i of each species i to the chemical potential μ_i through

including inverse reaction

$$n_i = g_i \int \left(\frac{d^3p}{h^3}\right) e^{(\mu_i - E)/(k_B T_\gamma)} = g_i \left(\frac{2\pi m_i k_B T_\gamma}{h^2}\right)^{3/2} e^{\mu_i/(k_B T_\gamma)} \quad (28.36)$$

26. As we could have inferred quantum mechanically.

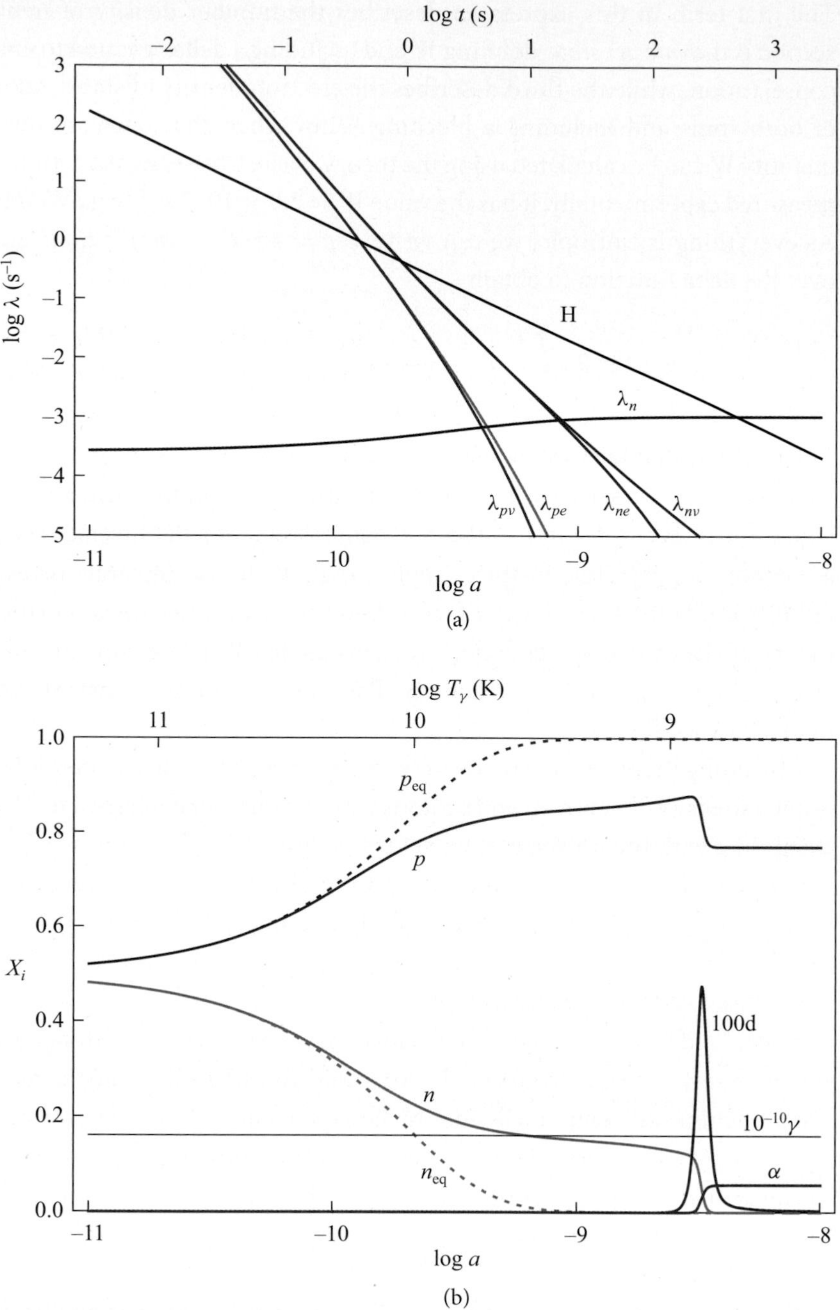

FIGURE 28.7 (a) Rate coefficients for the five reactions that determine the neutron-proton ratio n/p; the subscripts on the rates λ correspond to the left-hand side of the reaction. Note that when time $t \gtrsim 300$ ms, the equilibration rate is slower than the Hubble expansion rate H, and the neutron fraction is stabilized. When $t \gtrsim 600$ s, the few remaining neutrons are able to undergo β-decay before being incorporated in αs. (b) Number fractions (number of particles per baryon) for photons γ, protons p, neutrons n, deuterons d, and alpha particles α. When $T_\gamma \lesssim 2 \times 10^{10}$ K, the nucleons depart from thermodynamic equilibrium. The deuterons are produced when $T_\gamma \sim 10^9$ K before being quickly incorporated into αs. These two panels demonstrate that the current deuterium and helium fractions depend sensitively on the expansion rate and the nuclear physics details; the agreement with observation is a nontrivial validation of standard cosmology.

(Sec. 5.2), where g_i is the degeneracy of species i, and we treat the nucleons as nonrelativistic particles with $E = p^2/(2m_i) + \ldots$. Under equilibrium conditions, the relativistic chemical potentials satisfy $\tilde{\mu}_p + \tilde{\mu}_n = \tilde{\mu}_d + \tilde{\mu}_\gamma = \tilde{\mu}_d$, and so we can eliminate the chemical potentials to obtain the Saha-like (Ex. 5.10) relation:

$$n_p n_n = S(T_\gamma) n_d, \tag{28.37a}$$

where

$$S(T_\gamma) = 2^{1/2} 3^{-1} \left(\frac{2\pi m_p k_B T_\gamma}{h^2} \right)^{3/2} e^{-\gamma_d m_e/(k_B T_\gamma)}, \tag{28.37b}$$

we have used $g_p = g_n = 2$ and $g_d = 3$ and $\gamma_d\, m_e = m_n + m_p - m_d = 4.34\, m_e$ is the deuteron binding energy. The combined forward-reverse reaction rate must then be described by $dn_d/dt = \langle \sigma v \rangle_{np} [n_p n_n - S(T_\gamma) n_d]$, as this is the only way that we can maintain equilibrium at all temperatures. At early times, $S \langle \sigma v \rangle_{np} \gg H$ and equilibrium must have been maintained.

The final step is to explain how the tritons (nuclei of $^3\mathrm{H} \equiv t$), helions (nuclei of $^3\mathrm{He} \equiv h$), and alpha particles (nuclei of $^4\mathrm{He} \equiv \alpha$) were formed. There are two important pathways: $d + d \to t + p$, $t + d \to \alpha + n$; and $d + d \to h + n$, $h + d \to \alpha + p$ (cf. Sec. 19.3.1). The effective, combined forward reaction rate is calculated using $\langle \sigma v \rangle_{dd} = 7 \times 10^{-24}\ \mathrm{m^3\ s^{-1}}$. The reverse reaction rates can be calculated using a simple generalization of the argument leading to Eq. (28.37) for reactions involving four or more nucleons. We can now allow for the expansion as before, introducing the d number fraction $X_d = n_d/n_b$, to obtain three coupled rate equations for the proton, neutron, and deuteron number fractions $X_p = n_p/n_b$, $X_n = n_n/n_b$, and $X_d = n_d/n_b$:

deuterium reactions

rate equations

$$\begin{aligned}
HX'_p &= (\lambda_{n\nu} + \lambda_{ne} + \lambda_n) X_n - (\lambda_{pe} + \lambda_{p\nu}) X_p - \langle \sigma v \rangle_{np} (n_b X_p X_n - S(T_\gamma) X_d), \\
HX'_n &= (\lambda_{pe} + \lambda_{p\nu}) X_p - (\lambda_{n\nu} + \lambda_{ne} + \lambda_n) X_n - \langle \sigma v \rangle_{np} (n_b X_p X_n - S(T_\gamma) X_d), \\
HX'_d &= \langle \sigma v \rangle_{np} (n_b X_p X_n - S(T_\gamma) X_d) - 2 \langle \sigma v \rangle_{dd} n_b X_d^2,
\end{aligned} \tag{28.38}$$

where a prime denotes differentiation with respect to $\ln a$. It is easiest to solve these equations by assuming the neutrons and protons are in equilibrium until $a \sim 10^{-11}$ and the deuterons are in equilibrium until $a \sim 10^{-9}$.

numerical solution

The numerical solution to these equations (Fig. 28.7) shows that the neutrons and protons remained in thermodynamic equilibrium until the temperature fell to $\sim 2 \times 10^{10}$ K, at which time the neutron fraction declined slowly, reaching a value ~ 0.15 by the time the temperature had fallen to $\sim 10^9$ K. At this temperature, the deuteron equilibrium fraction climbed quickly, and the deuteron density became large enough for the $d + d$ reactions to produce α at $T_\gamma = 8 \times 10^8$ K, and then declined. Most neutrons were incorporated into helium, though a small fraction was left to decay freely. The final helium fraction is[27] $X_\alpha = n_\alpha/n_b = \frac{1}{4}(1 - X_p) = 0.058$ (Fig. 28.7), slightly smaller than the (observed) value $X_\alpha = 0.062$ obtained by more

nucleosynthesis

final helium and deuterium fractions

27. Astronomers reserve the symbol X for hydrogen and use the symbol $Y = 4X_\alpha$ for the helium mass fraction and Z for the mass fraction of all other elements which they call "metals"!

detailed calculations that include about ten more reactions, temperature dependence of the reaction rates, and other refinements. The late-time deuterium number fraction computes to be $X_d = 2.3 \times 10^{-5}$, consistent with the detailed calculations and with observations. Yields of h, ^{6}Li, and ^{7}Li are also computed and can be reconciled with observations if one takes account of large astrophysical uncertainty. The t decays.

It is the absence of stable nuclei with $A = 5, 8$ that prevented the build-up of elements beyond helium in the early universe. Instead, the synthesis of these elements had to await the formation and evolution of stars.

28.4.3

28.4.3 Photon Age

Primordial nucleosynthesis was complete by the time the scale factor had increased to $a \sim 10^{-8}$ and the temperature had fallen to $T_\gamma \sim 3 \times 10^8$ K. This marks the beginning of the photon age, during which the energy density of the universe was still dominated by photons (and neutrinos) and $\rho_\gamma \propto a^{-4}$, $a \propto t^{1/2}$. Now is a good time to examine the implicit assumption that we have been making that, with the conspicuous exception of the neutrinos, all major constituents of the universe are maintained in thermal equilibrium during the nuclear and photon ages.

PLASMA EQUILIBRATION

Let us first consider protons and electrons. The plasma frequency $\omega_P = [ne^2/(m\epsilon_0)]^{1/2} \propto a^{-3/2}$ (Sec. 20.3) is very high in comparison with the expansion rate. In fact, $\omega_p H \sim 10^{17} a_{-8}^{1/2}$, where $a_{-8} = a/10^{-8}$. Likewise, the ratio of the Debye length $\lambda_D = [\epsilon_0 k_B T/(ne^2)]^{1/2}$ to the horizon radius is $\lambda_D \chi_H \sim 10^{-18} a_{-8}^{-1}$. Therefore, the use of fluid mechanics in place of plasma physics is amply justified for the whole expansion of the universe.

validity of the fluid approximation

Now let us question our implicit assumption of thermodynamic equilibrium. Using Eqs. (20.23), (28.27), and (28.13), we find that the ratios of the e-e, p-p, and p-e equilibration timescales ($\propto T^{3/2} n^{-1} \propto a^{3/2}$) to the expansion timescale ($\propto a^2$) are given by $H\{t_{ee}, t_{pp}, t_{ep}\} \sim \{1, 40, 1800\} \times 10^{-10} a_{-8}^{-1/2}$ throughout the photon age, ensuring that baryons remained in thermal equilibrium.

equilibration timescales

COMPTON SCATTERING

Next consider the interaction between the plasma and the photons, which is mediated by Compton scattering. If there had been no such interaction, then the (nonrelativistic) plasma would have maintained its Maxwellian distribution with a temperature that would have fallen as $T \propto a^{-2}$ (Sec. 28.2.3). However, each Compton scattering will lead to an energy transfer from the photons to the electrons through a combination of the Doppler effect and Compton recoil. This energy change will then be quickly shared with the protons, so that the plasma is only allowed to become a tiny bit cooler than the radiation. To be quantitative, the mean fractional energy exchange per scattering is $|\Delta \ln E| \sim k_B T/m_e$. Therefore, it will take $\sim m_e/(k_B T)$ scatterings to equilibrate the plasma with the far more numerous photons. The scattering time is $t_{e\gamma} \sim (n_\gamma \sigma_T)^{-1}$, where $\sigma_T = 8\pi r_e^2/3 = 6.65 \times 10^{-29}$ m^{-2} is the *Thomson cross section*, and $r_e = 2.8$ fm is the classical electron radius (Sec. 3.7.1). The ratio of the timescale for the plasma to equilibrate with the radiation, to the expansion timescale, is $Ht_{e\gamma} = Hm_e/(\rho_\gamma \sigma_T) \sim 2 \times 10^{-16} a_{-8}^2$. The electrons therefore exchange energy with the radiation field even faster than they share it among themselves and with protons. This justifies our assumption of a common matter and radiation temperature in the early universe. Insofar as the universe is homogeneous, the far more numerous photons will automatically maintain their initial Planck distribution without energy redistribution by electrons. Their response to small, inhomogeneous perturbations is considered below.

electron-photon coupling

28.4.4 Plasma Age

28.4.4

Eventually, the rest-mass energy density of matter (mostly dark matter) exceeded that of radiation and neutrinos, and the plasma age began. This happened when $a = a_{\rm eq} = 0.00030$, $t = t_{\rm eq} = 52$ kyr, and $T_\gamma = T_{\rm eq} = 9{,}100$ K. Thereafter, according to Eq. (28.30), $\rho \propto a^{-3}$ and $a \propto t^{2/3}$, approximately.[28] The Friedmann equation (28.16) can be integrated in this plasma age to give, more precisely,

onset of matter dominance

$$\frac{t}{t_{\rm eq}} = \left(1 + 2^{-1/2}\right)\left[2 - \left(2 - \frac{a}{a_{\rm eq}}\right)\left(1 + \frac{a}{a_{\rm eq}}\right)^{1/2}\right], \qquad (28.39)$$

which remains valid until the cosmological age.

Next the helium ions became singly ionized through capturing one electron and then became neutral by taking on a second electron, leaving a proton-electron plasma. After a further interval, the hydrogen recombined.[29] Unlike the helium recombination, the details of this process are highly significant.

helium recombination

The total (atomic plus ionized) hydrogen density is $n_{{\rm H}+p} = 1.9 \times 10^8 a_{-3}{}^{-3}\ {\rm m}^{-3}$, where $a_{-3} = a/10^{-3}$. In equilibrium, the atomic fraction would satisfy the Saha equation (5.68). However, just as happened with nucleosynthesis, the universe expanded too fast for the reactions to keep up. The basic problem is that when an electron and a proton recombine, they emit one or more photons that have a short mean free path and are mostly reabsorbed by neighboring atoms, leading to no net change in ionization. This is especially true of the Lyman α photons emitted with frequency $\nu_\alpha = 2.47 \times 10^{15}$ Hz when a hydrogen atom transitions from its first excited state, designated by quantum number $n = 2$, to its ground state with $n = 1$. A good approximation is to treat the $n = 2$ level as the effective ground state, changing the effective ionization potential from $I_1 = 13.6$ eV to $I_2 = 13.6/4 = 3.4$ eV and modifying the degeneracy in the Saha equation, and then to allow for the slow permanent population of the true, $n = 1$, ground state.

hydrogen recombination

28. The thermal energy density of the plasma was only 6×10^{-10} times the radiation energy density at this time.
29. The use of the term *recombination* is conventional but misleading, because it is the first occurrence of this process.

If we denote by X_1 and X_2 the fraction of the hydrogen in the $n = 1$ and $n = 2$ states, respectively, and ignore higher energy levels,[30] then the *ionization fraction* is $x = n_e/n_{H+p} = 1 - X_1 - X_2$. The rate equations analogous to Eqs. (28.38) are

rate equations

$$HX_2{}' = \alpha_2 \left(n_{H+p}x^2 - \frac{1}{4}X_2 \left(\frac{2\pi m k_B T_\gamma}{h^2}\right)^{3/2} e^{-I_2/(k_B T_\gamma)} \right) - \lambda_{21}X_2,$$

$$HX_1{}' = \lambda_{21}X_2. \tag{28.40}$$

The use of fractional densities once again takes account of the expansion of the universe. The quantity $\alpha_2 = 2.8 \times 10^{-19} T_4^{-1/2}$ m^3 s^{-1} is the *recombination coefficient* into the $n = 2$ level, which is computed by summing over all pathways, excluding transitions to $n = 1$. The $\frac{1}{4}$ takes into account the 2-fold degeneracy of the $n = 1$ level and 8-fold degeneracy of the $n = 2$ level, and the "Saha" factor ensures equilibrium in the absence of transitions to $n = 1$, just like the factor S in Eq. (28.37). The rate constant λ_{21} describes the permanent transitions to the ground state, corrected for reverse transitions.

allowing for inverse reactions

We compute λ_{21} as follows. The sublevels of the $n = 2$ level are well mixed by collisions, so one-quarter of the excited atoms will be in the 2s sublevel, while three-quarters will be in the 2p sublevel. The 2s atoms can permanently deexcite by emitting two photons, neither of which is reabsorbed. This process has a *forbidden* spontaneous rate $A_{2s} = 8.2$ s^{-1}, and the inverse process can be ignored. The 2p atoms create Lyman α photons with a *permitted* rate $A_{2p} = 4.7 \times 10^8$ s^{-1}. The spectral line will have a small, combined natural and Doppler width. The cross section for absorbing a Lyman α photon in the low-frequency wing of the line decreases with decreasing frequency and eventually becomes small enough that the expansion of the universe allows the photon to avoid absorption altogether and therefore leads to the formation of a hydrogen atom in its ground state. Let us define by P_{esc} the probability that one of these photons avoids absorption in this manner. Next let us describe the line profile by $P_\nu(\nu)$, the cumulative probability that an emitted Lyman α photon has frequency less than ν. (See Ex. 28.9 for this and some other details of the analysis.)

two-photon deexcitation

escape probability

Kirchhoff's law of radiation (e.g., Sec. 10.2.1) ensures that the net absorption cross section has the same frequency dependence as the emissivity. Using the Einstein coefficients, we can show that the net absorption cross section associated with the same atoms as those that are emitting the Lyman α photons is $\sigma_\alpha = [A_{2p}/(8\pi\nu_\alpha^2)]dP_\nu/d\nu$.[31] The probability that a Lyman α photon with any frequency will escape due to expansion is then given by

30. The fractional occupancy of the $n = 2$ level never exceeds $\sim 10^{-13}$, and so this is a good approximation.

31. Another way of expressing this is $\int d\nu\sigma_\alpha = \pi r_e f_\alpha$, where $f_\alpha = 0.42$ is the *oscillator strength* (Cohen-Tannoudji, Diu, and Laloë, 1977).

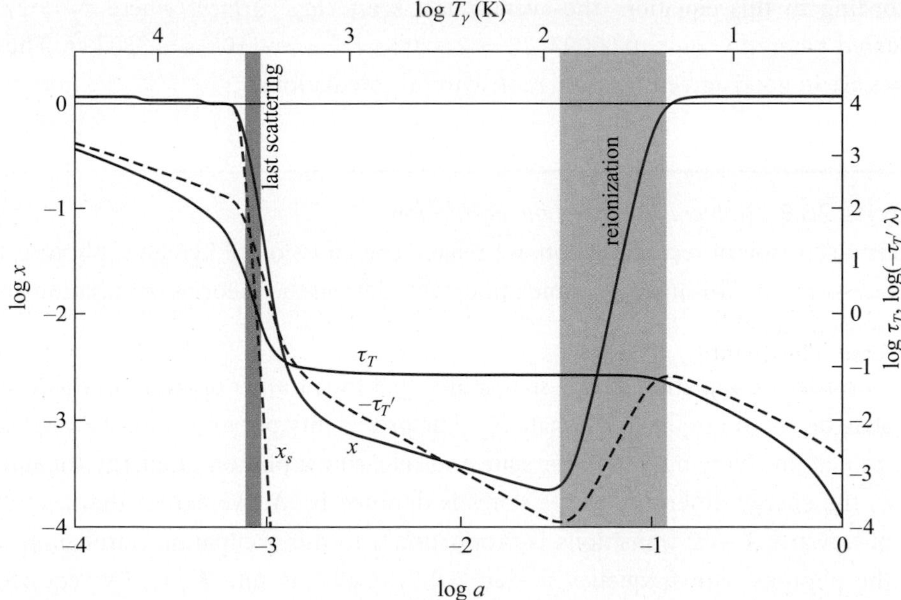

FIGURE 28.8 Ionization fraction of the universe, x, plotted as a function of the scale factor a. The blue dashed curve shows the result from assuming Saha equilibrium x_S. The solid blue curve shows the actual ionization fraction, x, deduced from Eqs. (28.40) after including in λ_{21} the atomic processes that delay recombination of the electrons. The solid red curve shows the Thomson optical depth τ_T from now back to scale factor a, while the dashed red curve is its derivative with respect to $-\ln a$: $-\tau_T' = n_e\sigma_T H^{-1}$. The shaded region to the left delineates the short interval when τ_T fell from $\sim$3 to $\sim$1. The shaded region to the right delineates the poorly understood epoch of reionization (Sec. 28.4.5), when newly formed stars and black holes are thought to have created sufficient ultraviolet photons to change most hydrogen in the universe back to a plasma. The ionization fraction adopted for $0.05 < a < 1$ is $x = 1.16[1 + 1.5(a/0.1)^{-8.7}]^{-1}$, which allows for the presence of helium.

$$P_{\rm esc} = \int_0^1 dP_\nu e^{-n_1\int d\tau\sigma_\alpha} = \int_0^1 dP_\nu e^{-\frac{n_1}{H\nu_\alpha}\int d\nu\sigma_\alpha} = \int_0^1 dP_\nu e^{-\frac{n_1 A_{2p} P_\nu}{8\pi\nu_\alpha^3 H}} \simeq \frac{8\pi\nu_\alpha^3 H}{n_1 A_{2p}}, \tag{28.41}$$

where n_1 is the number density of H atoms in the $n = 1$ state, which is effectively constant in the short time it takes the universe to Doppler shift the frequency of the photon by enough to escape absorption. Therefore $\lambda_{21} = \frac{1}{4}A_{2s} + \frac{3}{4}A_{2p}P_{\rm esc}$, independent of A_{2p}.

To solve Eqs. (28.40), note that $X_2 \ll X_1$, and so $X_2{}'$ can be set to zero. The resulting ionization x (solid blue curve in Fig. 28.8) follows the full Saha evolution as long as $T \gtrsim 4{,}000$ K (Fig. 28.8). Thereafter the ionization fraction is significantly larger. The universe is half-ionized when $T_\gamma \sim 3{,}500$ as opposed to $\sim$3,700 K, according to the Saha equation.

We will need the Thomson optical depth τ_I on our past light cone:

$$\tau_T(\ln a) = \int_{\ln a}^0 (d\ln a) n_e\sigma_T/H. \tag{28.42}$$

last scattering surface

According to this equation, the average last scattering surface, where $\tau_T = 2/3$, occurred when $a = a_{ls} = 0.00093$, $T_\gamma = 2{,}920$ K, $t = 1.2 \times 10^{13}$ s $= 380$ kyr. These values are in good agreement with more careful calculations.

EXERCISES

Exercise 28.9 *Problem: Spectral Line Formation*

In our discussion of recombination, we related the emission of Lyman α photons to their absorption. This involves some important ideas in the theories of radiation and thermodynamics.

(a) Consider a population of two-state atoms. Let the number of atoms in the lower state be N_1 and in the upper state N_2. The probability per unit time of an upper-state atom changing to a lower state and releasing a photon of energy $h\nu$ equal to the energy difference of the states is denoted by A. We expect that the rate of upward, $1 \rightarrow 2$ transitions is proportional to the occupation number η_γ of the photons with frequency ν (Sec. 3.2.5). Call this rate $K_u \eta_\gamma$. By requiring that the atoms should be able to remain in Boltzmann equilibrium with the Planckian radiation field of the same temperature, show that there must also be downward, *stimulated emission* at a rate per state-2 atom of $K_d \eta_\gamma$, and show that $K_u = K_d = A$.

(b) The absorption cross section σ for an atom at rest can be written in the *Lorentz* or *Breit-Wigner* form as:

$$\sigma = \frac{\pi A^2}{(\nu - \nu_0)^2 + A^2}.$$

Either make a classical model of an atom as an electron oscillator with natural frequency ν_0, or use time-dependent perturbation theory in quantum mechanics to justify the form of this formula.

(c) Identify the frequency probability function P_ν introduced in Sec. 28.4.4, and plot the natural line profile for an emission line.

(d) Atoms also have thermal motions, which Doppler shift the photon frequencies. Modify the line profile by numerically convolving the natural profile with a 1-dimensional Gaussian velocity distribution and replot it, drawing attention to its behavior when A is much more than the thermal Doppler shift.

(e) We have restricted our attention to a two-state system. It is usually the case that we are dealing with energy levels containing several distinct states, and the formalism we have described has to be modified to include the *degeneracies* g_i of these levels. Make the necessary corrections and recover the formulas used in the text.

(f) A second complication is polarization (Sec. 7.7). Discuss how to include this.

28.4.5 Atomic Age

28.4.5

The next age is the atomic age.[32] If atomic hydrogen were completely decoupled from the radiation field, then it would cool with temperature $T_H \propto a^{-2}$ and eventually become cryogenic! However, this is not what happened, as the electrons that remained were still able to keep in thermal contact with the radiation and with the protons and atomic hydrogen. The plasma was maintained at roughly the electron temperature. Eventually, when $a \sim 0.03$, the temperature had fallen to $T \sim 100$ K, and molecular hydrogen appeared. However, this was also about the time when the very first self-luminous stars and black holes formed and emitted ultraviolet radiation, which caused the molecular hydrogen to dissociate and the atoms to ionize. This is known as the *epoch of reionization* and must have continued until $a \sim 0.14$, because neutral hydrogen is actually detected at this time through Lyman α absorption of quasar light. Characterizing the epoch of reionization is a major goal of modern research and involves many considerations that lie beyond the scope of this book. The evolution adopted in Fig. 28.8 and Sec. 28.5 is consistent with current observations but is not yet well constrained.

electron temperature

first stars and black holes

EXERCISES

Exercise 28.10 *Problem: Reionization of the Universe*

(a) Estimate the minimum fraction of the rest mass energy of the hydrogen that must have undergone nuclear reactions inside stars to have ionized the remaining gas when $a \sim 0.1$.

(b) Suppose that these stars radiated 30 times this minimum energy at optical frequencies. Estimate the energy density and frequency of this stellar radiation background today.

You may find Exercises 4.11 and 28.6 helpful.

28.4.6 Gravitational Age

28.4.6

SCALE FACTOR

As we discuss further in Sec. 28.5, after recombination small inhomogeneities in the early universe grew under the influence of gravity to form galaxies and larger-scale structures. The influence of stars was supplemented by that of accreting massive black holes that formed and grew in the nuclei of galaxies. After reionization this radiative onslaught kept most baryons in the universe in a multiphase, high-temperature state.

However, this is also the time when the influence of the cosmological constant started to become significant. If we ignore photons, neutrinos, and spatial curvature, then Eq. (28.16) describing the expansion of the universe in the gravitational age has

influence of cosmological constant

32. Sometimes called the *dark age*.

a simple analytical solution derived and discussed by Bondi (1952):

$$t = \frac{2}{3H_0(1-\Omega_M)^{1/2}} \sinh^{-1}\left[(\Omega_M^{-1}-1)^{1/2} a^{3/2}\right], \tag{28.43}$$

where Ω_M is the current matter density in units of the critical density. We see that the influence of the cosmological constant is negligible at early enough times, $t \ll t_0$, and the energy density of the matter dominates the expansion. It also dominates any curvature that there might be today. Such a universe is usually called an *Einstein–de Sitter universe* and has $a = (3H_0t/2)^{2/3}$. Similarly, at late times, $t \gg t_0$, we can ignore the matter and find that $a \propto \exp(Ht)$, where $H = (1-\Omega_M)^{1/2}H_0$. This is called a *de Sitter universe*. The kinematic properties of our standard universe relative to an Einstein–de Sitter universe and an open ($k=-1$) one are best expressed using the deceleration parameter q [Eqs. (28.14) and (28.23)], which is exhibited in Fig. 28.9. Note that the jerk parameter is $j = 1$ throughout the gravitational and cosmological ages.

Einstein-de Sitter universe

DISTANCE AND VOLUME

Excepting the CMB, most cosmological measurements are made when $a \gtrsim 0.1$. Therefore, now is a good time to calculate distance and volume. The comoving distance to a source whose light was emitted when the expansion parameter was a, $\Sigma(a)$ (equal to the radius $\chi(a)$ in our flat universe), can be computed from the Friedmann equations and is exhibited in Fig. 28.10. We can also compute the angular diameter distance $d_A = a\chi(a)$, which is seen to reach a maximum value of $0.41/H_0 = 5.6 \times 10^{25}$ m at $a = 0.39$. More distant sources of fixed physical length will actually appear progressively larger. By contrast, the luminosity distance $d_L = a^{-1}\chi(a)$ increases rapidly with distance, and individual sources become unobservably faint. The total distance to the early universe is $\chi_H(t_0) = 3.2/H_0 = 14$ Gpc $= 4.4 \times 10^{26}$ m and the associated comoving volume is $V_H(t_0) = 4\pi\chi_H(t_0)^3/3 = 140H_0^{-3} = 1.2 \times 10^4$ Gpc^3 $= 3.6 \times 10^{80}$ m^3. The universe is a big place!

angular diameter and luminosity distances

volume of observable universe

EXERCISES

Exercise 28.11 *Problem: Type 1a Supernovae and the Accelerating Universe*

Rather surprisingly, it turns out that a certain type of supernova explosion (called "Type 1a" and associated with detonating white-dwarf stars) has a peak luminosity L that can be determined by studying the way its brightness subsequently declines. Astronomers can measure the peak fluxes F for a population of supernovae at a range of distances.

Calculate the flux measured at Earth for a given L as a function of the scale factor $0.3 < a < 1$ at the time of emission for the following.

(a) An Einstein–de Sitter universe.

(b) A nonaccelerating universe with $a \propto t$.

(c) Our standard model universe.

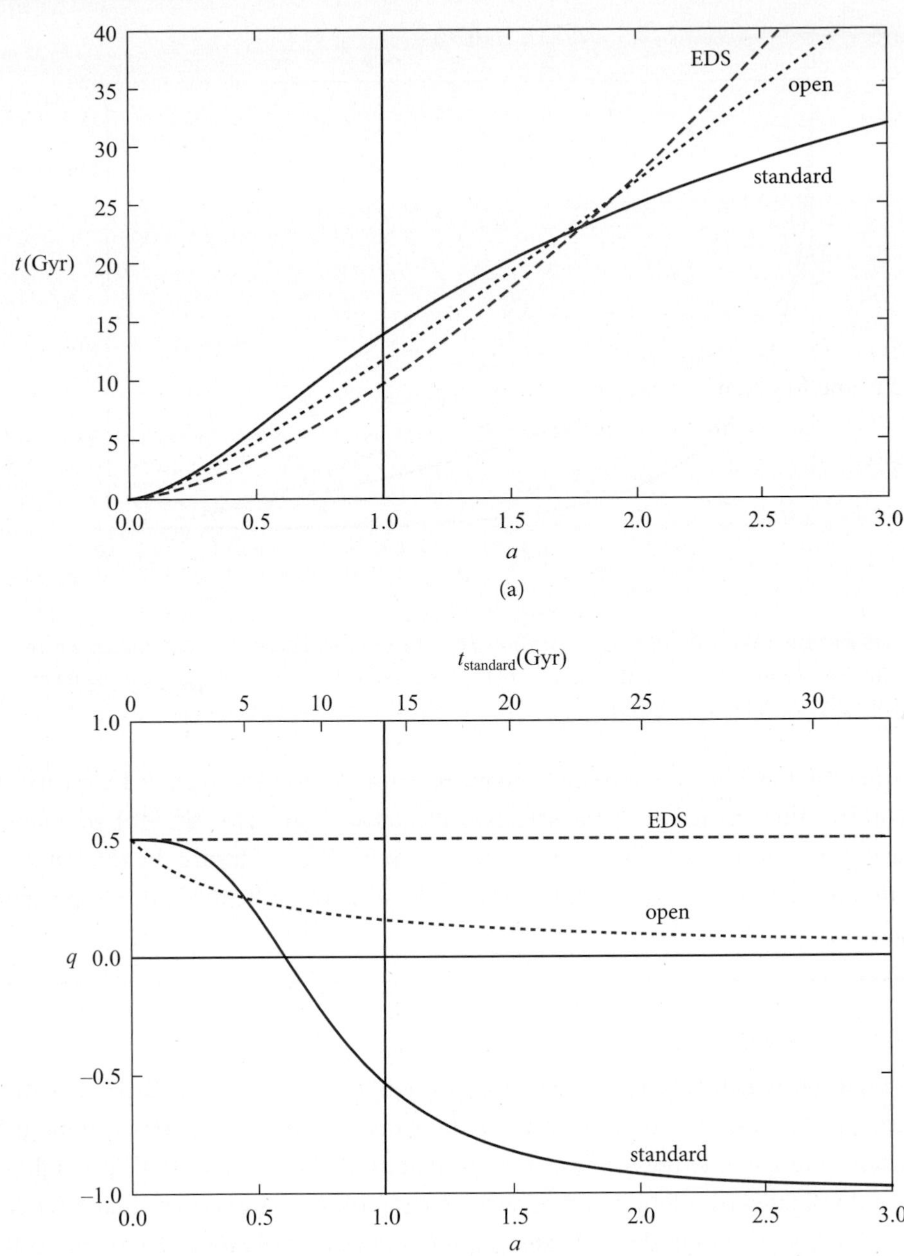

FIGURE 28.9 Expansion of the universe during the gravitational and cosmological ages. (a) The variation of age t with scale factor a is shown for standard cosmology (solid curve). Note that the expansion of the universe initially decelerated under the pull of gravity, but then began to accelerate at age ~6 Gyr under the influence of the cosmological constant. If this continues for the next ~15 Gyr, then the universe will embark on an exponential growth. (b) This exponential growth is brought out in a plot of the associated deceleration parameter q versus a. Also shown in both panels are the solutions for a (dashed) Einstein–de Sitter (EDS) model, which is flat and matter dominated, with an identical Hubble constant as standard cosmology, and a negatively curved (dotted) open model with the same contemporary density parameter as used in standard cosmology. Neither of these models exhibits acceleration; the former decelerates forever, the latter eventually expands with constant speed.

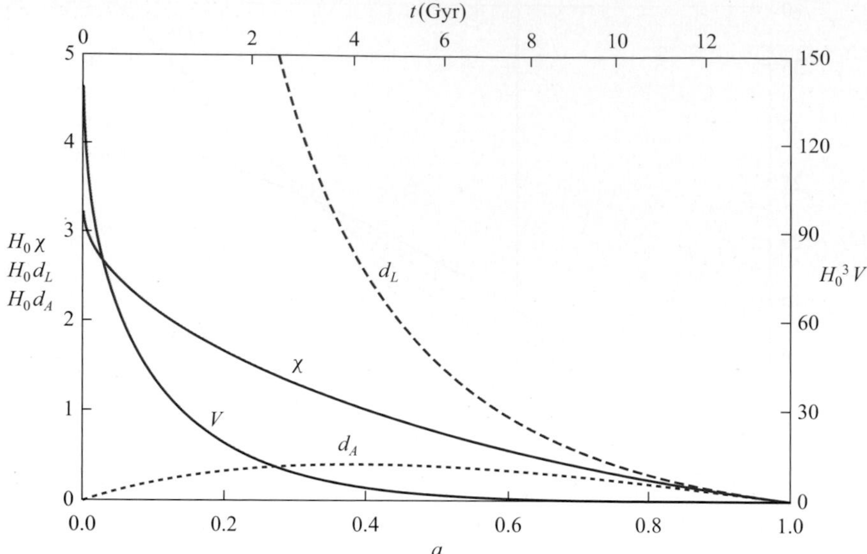

FIGURE 28.10 Comoving distance χ (in units H_0^{-1}) to a source whose light was emitted when the scale factor was $a < 1$. Also shown are the luminosity distance d_L, the angular diameter distance d_A, and the comoving volume V associated with a sphere of radius χ (red).

Assume the same Hubble constant today, H_0. How accurately must F be measured to confirm the prediction of the standard model with an error of $\sim$0.1 in a single measurement at $a \sim 0.7$? Astronomers do not, in practice, measure the total flux but the flux in a specific spectral band, but this adjustment can be made if the spectrum is known.

28.4.7

28.4.7 Cosmological Age

eschatology!

The final age, which began about 5 Gyr ago, when $a \sim a_\Lambda \sim 0.7$, is called the cosmological age because the energy density is thereafter dominated by the cosmological constant. We are entering a phase of exponential, de Sitter expansion, presaging a future dominated by dilution and decay—an agoraphobic's worst nightmare! Operationally, the acceleration slows the development of large-scale structure in the distribution of galaxies, which provides one way to measure the value of the cosmological constant. Of course, as a pure cosmological constant is still a weakly constrained fit to the observations, the future could be more subtle, as we discuss in Sec. 28.7.3.

EXERCISES

Exercise 28.12 *Problem: Future Evolution of the Universe*

Assume that the universe will continue to expand according to Eq. (28.43).

(a) Calculate the behavior of the angular diameter distance and the associated volume as a function of the scale factor for the next 20 billion years.

(b) Interpret your answer physically.

(c) Explain qualitatively what will happen if the universe accelerates even faster than this.

We return to this topic in Sec. 28.7.3.

28.5 Galaxy Formation T2

28.5

The universe we have described so far is homogeneous and isotropic and completely ignores the large density fluctuations on small scales, observed today as clustered galaxies. Our task now is to set up a formalism to describe the growth under gravity of the perturbations that produce this structure as the universe ages. Much of this problem can be handled using Newtonian physics, but as the most interesting questions are intrinsically relativistic and as we have already developed the necessary formalism, we shall dive right into a fully relativistic analysis (Peebles and Yu, 1970; Sunyaev and Zel'dovich, 1970).

28.5.1 Linear Perturbations T2

28.5.1

METRIC

relativistic perturbation theory

We generalize the Robertson-Walker metric [Eq. (28.9)] to include linear perturbations in a manner inspired by our discussions of weak fields (Sec. 25.9) and the Schwarzschild spacetime (Sec. 26.2):

$$ds^2 = -(1 + 2\Phi)dt^2 + a^2(1 - 2\Psi)\delta_{ij}d\chi^i d\chi^i. \tag{28.44}$$

Here a is the same function of t as in the unperturbed model, and FOs (by definition) continue to move with fixed comoving coordinate $\boldsymbol{\chi}$. The changes to the spacetime geometry are all contained in the *curvature perturbation* Ψ and in the *potential perturbation* Φ, which agrees with its Newtonian counterpart when it is small in magnitude (relative to unity) and inhomogeneity scale[33] (relative to the horizon).[34]

curvature and potential perturbations

KINEMATICS

An FO no longer follows a timelike geodesic. We use $\vec{u} \cdot \vec{u} = -1$ to evaluate the components of its 4-velocity and 4-acceleration $a_\alpha = u^\beta u_{\alpha;\beta}$ to first order in the perturbation:

$$u^t = 1 - \Phi, \quad u^j = 0; \quad a^t = 0, \quad a_i = \partial_i \Phi. \tag{28.45}$$

33. Note that Φ does not include the Newtonian potential difference $2\pi\rho r^2/3$ that we might be tempted to associate with two points separated by r in the background medium if we had not appreciated the way that general relativity neatly resolves this ambiguity.

34. The coordinate choice is sometimes called a *gauge* [cf. Eqs. (25.87), (25.88), and (27.18)] by analogy with classical electromagnetism and particle physics and is often motivated by considerations of symmetry. Physical observables should not (and do not) depend on the coordinate/gauge choice. Our choice is known as the *Newtonian gauge* and is useful for perturbations that can be expressed as scalar quantities. (We will encounter tensor perturbations in Sec. 28.7.1.) This gauge choice is appropriate because, when Φ and Ψ are small, it becomes the weak-gravitational-field limit of general relativity (Sec. 25.9), which we need to interpret the actual cosmological observations we make today and which can be used to discuss the initial conditions, as we shall see in Sec. 28.7.1.)

T2

acceleration with respect to FO

These expressions exhibit gravitational time dilation—$dt/d\tau = 1 - \Phi$ (Sec. 27.2.1)—and Einstein's equivalence principle (Sec. 25.4). A particle that moves with small 3-velocity $\mathbf{v}$, as measured in an FO's local Lorentz frame, has a 4-velocity $u^t = 1 - \Phi$, $u^i = v_i/a$ and a 4-acceleration $a_t = 0$, $a_i = d(av_i)/dt + \partial_i \Phi$. In the limit $\Phi = 0$, $\mathbf{v} \propto a^{-1}$ if the particle is freely moving (Sec. 28.2.3).

FOURIER MODES

We are interested in the linear evolution of perturbations, and it is convenient to work with the spatial Fourier transform of the perturbed quantities (Sec. 8.3). For the potential, which is our primary concern, we write

potential oscillations

$$\Phi(t, \boldsymbol{\chi}) = \int \frac{d^3k}{(2\pi)^3} e^{i\mathbf{k}\cdot\boldsymbol{\chi}} \tilde{\Phi}(t, \mathbf{k}), \tag{28.46}$$

where the $\sim$ denotes spatial Fourier transform, the comoving wave vector $\mathbf{k}$ does not change with time, and we treat t and a as interchangeable coordinates. As Φ is real, $\tilde{\Phi}(t, -\mathbf{k}) = \tilde{\Phi}^*(t, \mathbf{k})$. Implicit in this Fourier expansion is a box, which we presume is much larger than the current horizon but which will not feature in our development. The actual modes can be considered as traveling waves moving in antiparallel directions or as standing modes in quadrature, which is a better way to think about their nonlinear development. In what follows, we shall consider the temporal development of linear perturbations, which can be thought of as either individual wave modes or as continuous Fourier transforms.

PERTURBED EINSTEIN EQUATION

Now focus on a single Fourier oscillation and use computer algebra to evaluate the nonzero, linear perturbations to the Einstein field equations in a local orthonormal basis:

perturbed field equations

$$\tilde{G}^{\hat{t}\hat{t}} = -2[3H(\dot{\tilde{\Psi}} + H\tilde{\Phi}) + (k/a)^2\tilde{\Psi}] = 8\pi \tilde{T}^{\hat{t}\hat{t}} = 8\pi\tilde{\rho}, \tag{28.47a}$$

$$\tilde{G}^{\hat{t}\hat{\parallel}} = -\frac{2ik}{a}(\dot{\tilde{\Psi}} + H\tilde{\Phi}) = 8\pi \tilde{T}^{\hat{t}\hat{\parallel}} = 8\pi(\rho + P)\tilde{v}, \tag{28.47b}$$

$$\tilde{G}^{\hat{\parallel}\hat{\parallel}} = 2[\ddot{\tilde{\Psi}} + H(\dot{\tilde{\Phi}} + 3\dot{\tilde{\Psi}}) + (1-2q)H^2\tilde{\Phi}] = 8\pi\tilde{T}^{\hat{\parallel}\hat{\parallel}}, \tag{28.47c}$$

$$\tilde{G}^{\hat{\perp}\hat{\perp}} = \tilde{G}^{\hat{\parallel}\hat{\parallel}} + \frac{k^2(\tilde{\Psi} - \tilde{\Phi})}{a^2} = 8\pi\tilde{T}^{\hat{\perp}\hat{\perp}}, \tag{28.47d}$$

where $\parallel$ and $\perp$ are components parallel and perpendicular to $\mathbf{k}$. Equation (28.47b) defines the mean velocity perturbation, which is purely parallel, $\tilde{v} = \tilde{v}^{\parallel}$. Note that if the cosmological fluid is perfect, there is no shear stress ($\tilde{T}^{\hat{\parallel}\hat{\parallel}} = \tilde{T}^{\hat{\perp}\hat{\perp}} = \tilde{P}$) and, consequently, $\tilde{\Psi} = \tilde{\Phi}$. This is a major simplification, echoing our treatment of Schwarzschild spacetime (Sec. 26.2). However, when neutrinos or photons have decoupled from matter and free stream through primordial (dark-matter) density perturbations, their stresses become sufficiently anisotropic to produce a measurable distinction between $\tilde{\Phi}$ and $\tilde{\Psi}$.

T2

Because these equations only involve the component of mean velocity parallel to **k**, they describe longitudinal waves, generalizations of the sound waves discussed in Sec. 16.5. This mean velocity is irrotational and can therefore be written as a carefully chosen function of time, multiplied by the gradient of a velocity potential $\tilde{\psi}$ (Sec. 13.5.4):[35]

velocity potential

$$\tilde{\mathbf{v}} = i\boldsymbol{\beta}\tilde{\psi}. \tag{28.48a}$$

Here we introduce a *scaled wave vector*

$$\boldsymbol{\beta} = \frac{\mathbf{k}}{3^{1/2}\dot{a}}, \tag{28.48b}$$

scaled wave vector

which can be regarded as a function of either t or a. (Note that $q \equiv (\partial \ln \beta / \partial \ln a)_{\mathbf{k}}$, which is a useful relation.)

We also define the *total relative density perturbation:*

$$\tilde{\delta} \equiv \tilde{\rho}/\rho = (\rho_b\tilde{\delta}_b + \rho_D\tilde{\delta}_D + \rho_\gamma\tilde{\delta}_\gamma + \rho_\nu\tilde{\delta}_\nu)/\rho, \quad \text{with} \quad \tilde{\delta}_{b,D,\gamma,\nu} \equiv \tilde{\rho}_{b,D,\gamma,\nu}/\rho_{b,D,\gamma,\nu}, \tag{28.49}$$

relative density perturbation

where the subscripts b, D, γ, and ν continue to refer to baryons, dark matter, photons, and neutrinos. Equation (28.47a) then becomes

$$\tilde{\Psi}' + \beta^2\tilde{\Psi} + \tilde{\Phi} = -\frac{1}{2}\tilde{\delta}, \tag{28.50}$$

potential evolution equation

where the prime denotes a derivative with respect to $\ln a$.

EARLY EVOLUTION

Before neutrino decoupling, when $a \lesssim 10^{-11}$, the stress-energy tensor $\boldsymbol{T}$ is dominated by a well-coupled, relativistic fluid consisting of radiation, neutrinos, and elementary particles with $P = \rho/3$ and $a \propto t^{1/2}$, so that $q = 1$, $\tilde{\Psi} = \tilde{\Phi}$, and $\tilde{T}^{\hat{l}\hat{l}} = \tilde{\rho}/3$. This remains a pretty good approximation until dark matter dominates the density during the plasma era when $a \sim a_{eq}$, and it can be used to bring out some important features of the general evolution. We can identify β as the ratio of the (relativistic) acoustic horizon χ_R [Eq. (28.15) and subsequent line] to the size of the perturbation, measured by $1/k$, both in comoving coordinates. Further simplification results from changing the independent variable to $\beta \propto a$ and combining Eqs. (28.47b) and (28.47d) to obtain a single second-order, homogeneous differential equation:[36]

$$\frac{d^2\tilde{\Phi}}{d\beta^2} + \frac{4}{\beta}\frac{d\tilde{\Phi}}{d\beta} + \tilde{\Phi} = 0. \tag{28.51}$$

evolution under radiation dominance

35. Actually there are modes with vorticity (embodied in the perpendicular part of the velocity $\tilde{\mathbf{v}}^{\perp}$), and they evolve according to the relativistic generalization of the equations discussed in Sec. 14.2.1. In principle, they could have been created by some sort of primordial turbulence. However, in practice they decay quickly as the universe expands and so will be ignored.

36. This equation, like many other equations describing the evolution of perturbations, has the form of a damped simple harmonic oscillator equation. The first derivative term is then often called a "friction" term. However, this is only a mathematical analogy. Physically, it represents the loss of energy in work done on the expanding medium, not a true dissipation.

T2

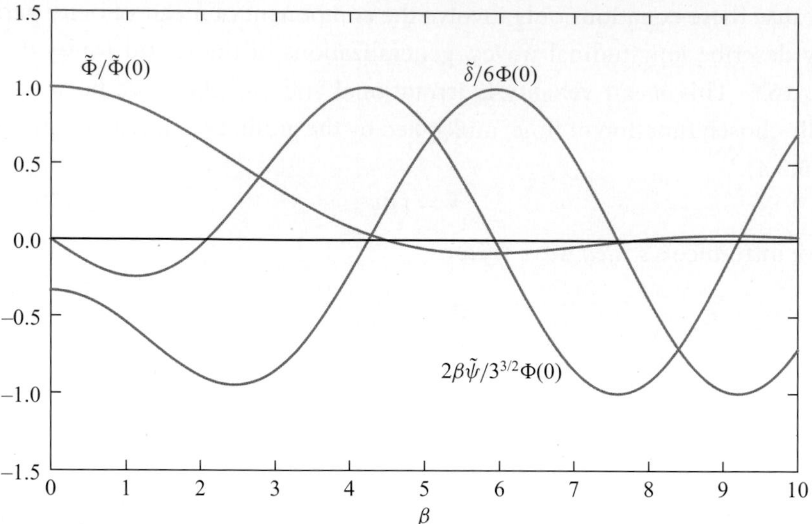

FIGURE 28.11 Early growth of a single spatial Fourier component of the perturbations. The amplitude of the potential $\tilde{\Phi}$, relative density $\tilde{\delta}$, and velocity potential $\tilde{\psi}$ perturbations are shown as functions of $\beta = k\chi_R$ for a single spatial Fourier component. The perturbations are frozen until they "enter the horizon" when $\beta \sim 1$. The perturbations then convert oscillations in which the amplitude of $\tilde{\delta}$ is constant while $\tilde{\Phi}$ is in antiphase and $\tilde{\psi}$ is in quadrature.

This has a unique solution, nonsingular as $\beta, a \to 0$ and valid for all scales:

$$\tilde{\Phi}(t, \mathbf{k}) = \frac{3\tilde{\Phi}(0, \mathbf{k})}{\beta^2}\left(\frac{\sin\beta - \beta\cos\beta}{\beta}\right) = \tilde{\Phi}(0, \mathbf{k})(1 - \beta^2/10 + \ldots). \quad (28.52)$$

The mode does not evolve significantly until it is contained by the acoustic horizon.[37]

Using Eq. (28.50), the relative density and velocity potential perturbations are then given by

$$\tilde{\delta} = -6\tilde{\Phi}(0)\left(\frac{2(\beta^2 - 1)\sin\beta - \beta(\beta^2 - 2)\cos\beta}{\beta^3}\right) = -2\tilde{\Phi}(0)(1 + 7\beta^2/10 + \ldots),$$

$$\tilde{\psi} = \frac{-3^{3/2}\tilde{\Phi}(0)}{2}\left(\frac{(\beta^2 - 2)\sin\beta + 2\beta\cos\beta}{\beta^3}\right) = \frac{-3^{1/2}\tilde{\Phi}(0)}{2}(1 - 3\beta^2/10 + \ldots) \quad (28.53)$$

evolution after entering horizon

(see Fig. 28.11). When $\beta \gg 1$, the wavelength is smaller than the acoustic horizon and the amplitude of the velocity perturbation is $3^{1/2}\delta/4$, in agreement with the expectation for a sound wave in a stationary, relativistic fluid. The angular frequency of the

37. What this really means is that we have chosen a coordinate system in which the expectation is clearly expressed that a physical perturbation not change significantly until a signal can cross it. Even in the absence of a genuine physical perturbation, we could have created the illusion of one simply by changing to a "wrinkled" set of coordinates.

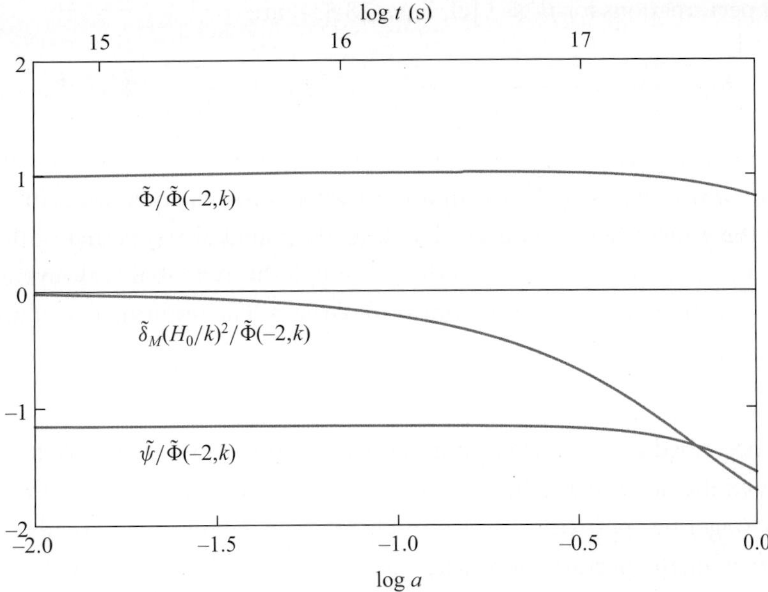

FIGURE 28.12 Evolution of the potential $\tilde{\Phi}$, the matter density perturbation $\tilde{\delta}_M$, and the velocity potential $\tilde{\psi}$ for $-2.0 \le \log a \le 0.0$. Note that $\tilde{\Phi}$ and $\tilde{\psi}$ only change slowly during the cosmological era when the cosmological constant is significant. By contrast, the density perturbation $\tilde{\delta}_M$ is $\propto k^2$ and grows rapidly to create the structure we observe today.

wave is $\omega = d\beta/dt = k/(3^{1/2}a)$, just what is expected for a sound wave (Sec. 16.5) in a relativistic fluid. The amplitude of the potential Φ decays as $\sim 3\tilde{\Phi}_0/\beta^2$, in accord with Poisson's equation.

The constancy of the wave amplitude $\tilde{\delta}$ turns out to be a nice illustration of adiabatic invariance (cf. Ex. 7.4). The locally measured wave energy in a single wavelength is $\propto \tilde{\delta}^2 \rho a^3$. This should scale with the wave frequency, implying that $\tilde{\delta}$ is constant (since $\omega \propto a^{-1}$).

LATE EVOLUTION

We can also describe the evolution during the gravitational and cosmological ages, when photons, neutrinos, and pressure can be ignored.[38] The space-space part of the Einstein tensor [Eq. (28.47d)] gives

$$\tilde{\Phi}'' + (3-q)\tilde{\Phi}' + (1-2q)\tilde{\Phi} = 0, \tag{28.54}$$

where $q = \frac{1}{2} - 3\rho_\Lambda/2\rho$ is initially $\sim\frac{1}{2}$. This says that the Fourier transform of the potential $\tilde{\Phi}$ is almost constant until the cosmological constant becomes important; then it decreases to $\sim$0.80 times its starting value (Fig. 28.12). The density and velocity

slow potential evolution

38. Photons and neutrinos contribute small corrections right after recombination, and there are transients associated with the sudden decrease of the coupling of baryons to photons. These are preserved in the full solution below but can be ignored in this approximate treatment.

T2

growth of density perturbations

potential perturbations for $\beta \gg 1$ [cf. Eqs. (28.53)] are

$$\tilde{\delta}_M \equiv \frac{\tilde{\rho}_M}{\rho_M} = \frac{\tilde{\rho}_D + \tilde{\rho}_b}{\rho_D + \rho_b} \sim -\frac{3\beta^2\tilde{\Phi}}{1+q}; \quad \tilde{\psi}_M \sim -\frac{3^{1/2}}{1+q}(\tilde{\Phi}' + \tilde{\Phi}) \tag{28.55}$$

for short wavelengths $\beta \gg 1$, where we have introduced the relative matter perturbation $\tilde{\delta}_M$, which grows $\propto \beta^2 \propto a$ in accord with Poisson's equation until ρ_Λ takes over and the growth rate is reduced. The velocity potential $\tilde{\psi}_M$ is that of the matter (baryons and dark matter). Note also that, although this potential is always small, the density fluctuation $\tilde{\delta}_M$ becomes nonlinear for large β. The resulting corrections must be computed numerically.

SUMMARY

We have described the early evolution of linear perturbations that were frozen until they entered the horizon and became sound waves with constant $\tilde{\delta}$ and the universe could no longer be approximated as a single relativistic fluid. We have also outlined the growth of matter perturbations when galaxies are visible—specifically when $0.1 \lesssim a < 1$. To connect these two limits, we must examine the behavior of the separate perturbations to dark matter, neutrinos, photons, and baryons.

28.5.2

28.5.2 Individual Constituents T2

DARK MATTER

dark matter density perturbation and velocity potential

Because the nongravitational interactions of dark matter, neutrinos, baryons, and photons are negligible during the epoch of galaxy formation, we handle the evolution of the different constituent perturbations by equating the 4-divergence of their individual stress-energy tensors to zero in the given spacetime.[39] Dark matter has no pressure, so the nonzero, mixed, orthonormal stress-energy tensor components are $\tilde{T}_D^{\hat{t}\hat{t}} = \rho_D\tilde{\delta}_D$, $\tilde{T}_D^{\hat{t}\hat{\parallel}} = \rho_D\tilde{v}_D = i\beta\rho_D\tilde{\psi}_D$. Setting its divergence to zero leads to two independent equations:

$$\tilde{\delta}_D' - 3\tilde{\Psi}' - 3^{1/2}\beta^2\tilde{\psi}_D = 0; \quad \tilde{\psi}_D' + (1+q)\tilde{\psi}_D + 3^{1/2}\tilde{\Phi} = 0, \tag{28.56}$$

where we have used the conservation law $\rho_D' + 3\rho_D = 0$ [cf. Eq. (28.19)].

Importantly for what follows, we can derive the first of Eqs. (28.56) from the flux of dark matter particles, $(\rho_D/m_D)[1+\tilde{\delta}_D, \tilde{\mathbf{v}}_D]$ in orthonormal coordinates, setting its divergence to zero.

What is the initial dark matter density perturbation? It could have been quite independent of the perturbation to the photons, neutrinos, pairs, etc. However, the simplest assumption to make is that it just depended on local physics and that equilibrium was established on a timescale short compared with the expansion time.[40]

39. The derivations in this section are only sketched. Confirming them will take some work.

40. It is not necessary that the initial conditions be established simultaneously—for example, during inflation—only that they be fixed before the neutrinos start to decouple (when $a \sim 10^{-11}$) and before the pairs annihilate ($a \sim 10^{-9}$).

Equivalently, the number of photons per dark matter particle $\propto \rho_\gamma^{3/4}/\rho_D$ was a fixed number and so, using Eqs. (28.53), $\tilde{\delta}_D(0) = \frac{3}{4}\tilde{\delta}_\gamma(0) = -\frac{3}{2}\tilde{\Phi}(0)$. Using a similar argument we can deduce that $\tilde{\delta}_b(0) = \tilde{\delta}_D(0)$, $\tilde{\delta}_\nu(0) = \tilde{\delta}_\gamma(0)$. This type of perturbation is called *adiabatic* and is found to describe the observations very accurately, vindicating our trust in basic principles. The initial velocity potential perturbation, common to all constituents, is $\tilde{\psi}_D(0) = -(3^{1/2}/2)\tilde{\Phi}$, from Eqs. (28.53).

T2

adiabatic perturbations

NEUTRINOS

Neutrinos add pressure but are effectively massless and travel at the speed of light until recent epochs, and are collisionless after decoupling. Let us throw caution to the winds and follow our treatment of a warm plasma (Sec. 22.3.5) to develop a fluid model. We set

$$\tilde{T}_\nu^{\hat{t}\hat{t}} = \rho_\nu \tilde{\delta}_\nu, \quad \tilde{T}_\nu^{\hat{t}\hat{\parallel}} = 4\rho_\nu \tilde{v}_\nu/3, \quad \tilde{T}_\nu^{\hat{\parallel}\hat{\parallel}} = \tilde{T}_\nu^{\hat{\perp}\hat{\perp}} = \rho_\nu \tilde{\delta}_\nu/3. \tag{28.57}$$

fluid approximation

When we set the divergence of the neutrino stress-energy tensor to zero, we obtain two equations for the perturbations. These equations are the same as those that we would have gotten if we had treated neutrinos as collisional and would lead to oscillations after the mode entered the horizon. However, the neutrinos can free stream through the mode to damp δ_ν and ψ_ν. If we ignore the potential $\tilde{\Phi}$ and imagine starting with a simple sine wave, we can use Jeans' theorem to solve approximately for the time evolution of the distribution function. We find that the wave will decay in a time $\sim 2a/k$. A neutrino wave initialized in this fashion and with no other perturbations would then decay according to $\delta_\nu' \sim -\beta\delta_\nu$, $\delta\psi_\nu' \sim -\beta\delta\psi_\nu$. We therefore add these terms to the perturbation equations to account for free-streaming. The final results are:

$$\tilde{\delta}_\nu' - 4\tilde{\Psi}' - (4/3^{1/2})\beta^2\tilde{\psi}_\nu = -\beta\tilde{\delta}_\nu; \quad \tilde{\psi}_\nu' + q\tilde{\psi}_\nu + 3^{1/2}\tilde{\Phi} + (3^{1/2}/4)\tilde{\delta}_\nu = -\beta\tilde{\psi}_\nu. \tag{28.58}$$

BARYONS AND PHOTONS

Prior to recombination at the end of the plasma age, the photons and baryons were tightly coupled by Thomson scattering and behaved as a single fluid with the photons dominating the density. The baryons were therefore prevented from falling into dark-matter gravitational potential wells. However, around the time of recombination, the baryon density grew larger than the photon density, and the photon mean free paths lengthened, causing the photon-baryon fluctuations to damp through heat conduction and viscosity (Sec. 18.2). This effect is known as Silk damping (Silk, 1968). If there were no cold dark matter, structure would have been erased on small scales and we would have had to find some other explanation for galaxy formation. Instead, as baryons released themselves from photons, they fell into potential wells formed by the dark matter. It is this complex evolution that we must now try to address.

baryon-photon coupling

Silk damping

The baryons are relatively easy. They can be treated as a cold fluid, just like the dark matter (because their pressure is never significant in the linear regime), with a single fluid velocity $\tilde{\mathbf{v}}_b = i\boldsymbol{\beta}\tilde{\psi}_b$ parallel to $\mathbf{k}$ (because the ions and electrons must have

T2

net zero charge density on all scales larger than the Debye length; Sec. 20.3). Their conservation law can be obtained by setting the divergence of the flux of baryons to zero just like we did for dark matter, Eq. (28.56):

baryon density evolution

$$\tilde{\delta}_b' - 3\tilde{\Psi}' - 3^{1/2}\beta^2\tilde{\psi}_b = 0. \tag{28.59}$$

Now turn to the photons. Just as with the neutrinos, the photons contribute $\rho_\gamma\tilde{\delta}_\gamma$ to the energy density and $\frac{1}{3}\rho_\gamma\tilde{\delta}_\gamma$ to the pressure. Initially, they shared the baryon velocity and so also contributed a term $\frac{4}{3}\rho_\gamma\tilde{v}_b$ to the momentum density/energy flux. Under the diffusive approximation, their heat flux in the baryon rest frame is $-ik\rho_\gamma\tilde{\delta}_\gamma/(3n_e\sigma_T a)$, where we do not have to worry about frequency shifts of the photons. We now define a photon velocity potential, $\tilde{\psi}_\gamma$, in the frame of the FOs, by equating the photon heat flux to $(4i/3)\boldsymbol{\beta}\rho_\gamma(\tilde{\psi}_\gamma - \tilde{\psi}_b)$ [cf. Eq. (28.48a)]. However, this relation breaks down when the photon mean free path approaches the wavelength of the perturbation. The heat flux will then be limited by $\sim\rho_\gamma\tilde{\delta}_\gamma/3$, and we simply modify the photon velocity potential by adding a flux-limiter:

heat flux

$$\tilde{\psi}_\gamma = \tilde{\psi}_b + 3^{1/2}\tilde{\delta}_\gamma/4(\tau_T' - 3^{1/2}\beta), \tag{28.60}$$

substituting the Thomson optical depth from Eq. (28.42). The combined baryon-photon energy flux in the FO frame is then $\tilde{T}_{b\gamma}^{\hat{t}\Vert} = i\beta[\rho_b\tilde{\psi}_b + 4\rho_\gamma(\tilde{\psi}_b + \tilde{\psi}_\gamma)/3]$.

We can now take the divergence of the stress-energy tensor to obtain

$$\begin{aligned}
&\tilde{\delta}_\gamma' - 4\tilde{\psi}' - 4\beta^2\tilde{\psi}_\gamma/3^{1/2} = 0,\\
&\quad \rho_b[\tilde{\psi}_b' + (1+q)\tilde{\psi}_b + 3^{1/2}\tilde{\Phi}] + (4/3)\rho_\gamma[\tilde{\psi}_\gamma' + q\tilde{\psi}_\gamma + 3^{1/2}\tilde{\Phi} + 3^{1/2}\tilde{\delta}_\gamma/4] = 0.
\end{aligned} \tag{28.61}$$

need for kinetic treatment

As with the neutrinos, accurate calculation mandates a kinetic treatment (Ex. 28.12; Sec. 28.6.1), but this simplified treatment captures most of the kinetic results.

SUMMARY

We have now derived a complete set of linear equations describing the evolution of a single mode with wave vector $\mathbf{k}$. These equations can be used over the observable range, $10^{-4} \lesssim k/H_0 \lesssim 0.3$, and for the whole range of evolution for $10^{-11} < a < 1$, although different terms are significant during different epochs, as we have described. We next turn to the solution of these equations.

EXERCISES

Exercise 28.13 *Challenge: Kinetic Treatment of Neutrino Perturbations* T2

The fluid treatment of the neutrino component would only be adequate if the neutrinos were self-collisional, which they are not.[41] The phenomenon of Landau damping (Sec. 22.3) alerts us to the need for a kinetic approach. We develop this in stages.

41. If we were to introduce shear stress, then viscous damping should also be included (cf. the discussion of stars in Sec. 3.7.1).

T2

Following the discussion in Sec. 3.2.5, we introduce the neutrino distribution function $\eta_\nu(t, x^i, p_j)$, where x^i is the (contravariant) comoving (spatial) coordinate, and p_j is the (covariant) conjugate 3-momentum (Sec. 3.6, Box 3.2, and Ex. 4.1). The function η_ν satisfies the collisionless Boltzmann equation [Eq. (3.65)]:

$$\frac{\partial \eta_\nu}{\partial t} + \frac{dx^i}{dt}\frac{\partial \eta_\nu}{\partial x^i} + \frac{dp_j}{dt}\frac{\partial \eta_\nu}{\partial p_j} = 0. \tag{28.62}$$

We work with this equation to linear order.

(a) Show that the neutrino equation of motion in phase space can be written as

$$\frac{d\chi^i}{dt} = \frac{p_i}{a(t)(p_k p_k)^{1/2}}(1 + \Phi + \Psi); \quad \frac{dp_j}{dt} = -\frac{(p_k p_k)^{1/2}}{a(t)}\frac{\partial(\Phi + \Psi)}{\partial x^j}, \tag{28.63}$$

and explain why it is necessary to express the right-hand sides in terms of t, x^i, and p_i.

(b) Interpret the momentum equation in terms of the expansion of the universe (Sec. 28.2.3) and gravitational lensing (Sec. 7.6.1).

(c) Introduce locally orthonormal coordinates in the rest frame of the FOs, and define $p^{\hat{\alpha}} = \{\mathcal{E}, p^{\hat{1}}, p^{\hat{2}}, p^{\hat{3}}\}$. Carefully interpret the density, velocity, and pressure in this frame, remembering that it is only necessary to work to linear order in Φ.

(d) Multiply the Boltzmann equation successively by 1, $\mathcal{E}$, and $p^{\hat{i}}$, and integrate over the momentum space volume element $dp^{\hat{1}}dp^{\hat{2}}dp^{\hat{3}}$ to show that

$$\tilde{\delta}'_{n_\nu} - 3\tilde{\Psi}' + 3^{1/2} i \boldsymbol{\beta} \cdot \tilde{\mathbf{S}}_\nu = 0; \quad \tilde{\delta}'_\nu - 4\tilde{\Psi}' + \frac{4i\beta}{3^{1/2}}\tilde{v}_\nu = 0;$$

$$i\tilde{\mathbf{v}}'_\nu - 3^{1/2}\boldsymbol{\beta}\left(\frac{\tilde{\delta}_\nu}{4} + \tilde{\Phi} + \tilde{\Psi}\right), \tag{28.64}$$

where $\tilde{\delta}_{n_\nu}$ is the fractional fluctuation in neutrino number density, and $\tilde{\mathbf{S}}_\nu$ is the associated number flux. In deriving these equations, it is necessary to impose the same closure relation (cf. Sec. 22.2.2) as was used to derive the fluid equations (28.57).[42] These kinetic equations have the same form as the fluid equations, although the coefficients are different and would change again if we changed the closure relation. This demonstrates that fluid equations can only be approximate, even when derived using the Boltzmann equation.

(e) One standard way to handle the neutrino perturbations accurately is to expand the distribution function in spherical harmonics. Outline how you would carry this out in practice, and how you would then use the more accurate neutrino distribution to improve the evolution equation for the dark-matter, baryon, and photon components.

42. Thus the trace of the stress-energy tensor vanishes in all Lorentz frames.

T2

Exercise 28.14 *Problem: Neutrino Mass* T2

Neutrinos have mass, which becomes measurable at late times through its influence on the growth of structure.

(a) Explain how the expansion of the universe is changed if there is a single dominant neutrino species of mass 100 meV.

(b) Modify the equations for neutrino phase-space trajectories described in the preceding problem to allow for neutrino rest mass, and outline how this will affect the growth of perturbations.

(c) Describe how you could, in principle, measure the individual neutrino masses using cosmological observations. (In practice, this would be extremely difficult.)

28.5.3

28.5.3 Solution of the Perturbation Equations T2

GROWTH VECTOR

If we ignore shear stress and equate Ψ to Φ [as discussed after Eq. (28.47d)], we have eight first-order, coupled, linear differential equations [Eqs. (28.50), (28.56), (28.58), (28.59), and (28.61)] plus an algebraic equation (28.60), describing the evolution of small perturbations in the presence of a single wave mode. A simple reorganization gives

linear growth of perturbations

$$\tilde{\boldsymbol{\Upsilon}}' = \mathcal{M}\tilde{\boldsymbol{\Upsilon}}, \tag{28.65}$$

where $\tilde{\boldsymbol{\Upsilon}} = \{\tilde{\Phi}, \tilde{\delta}_D, \tilde{\psi}_D, \tilde{\delta}_\nu, \tilde{\psi}_\nu, \tilde{\delta}_b, \tilde{\Psi}_b, \tilde{\delta}_\gamma, \tilde{\psi}_\gamma\}$ is dimensionless, and the elements of the matrix $\mathcal{M}$ are all real functions of a that we have already calculated and that describe the unperturbed universe. The solution of Eq. (28.65) can be written as $\tilde{\boldsymbol{\Upsilon}} = \boldsymbol{\Gamma}(a, \mathbf{k})\tilde{\Phi}(0, \mathbf{k})$, where $\boldsymbol{\Gamma}$ is the *growth vector* for adiabatic perturbations, initialized by

$$\boldsymbol{\Gamma}(0, k) = \left\{1, -\frac{3}{2}, -\frac{3^{1/2}}{2}, -2, -\frac{3^{1/2}}{2}, -\frac{3}{2}, -\frac{3^{1/2}}{2}, -2, -\frac{3^{1/2}}{2}\right\}. \tag{28.66}$$

[Note that we here switch from the scaled wave vector $\boldsymbol{\beta}$, Eq. (28.48b), to its unscaled, comoving form $\mathbf{k}$, as that is what we will need in using the growth vector.]

In the above equations and analysis, we have neither allowed for uncertainty in the governing parameters nor the addition of speculative physical processes (Ex. 28.26). We have also ignored nonlinear corrections (cf. Sec. 23.2, Ex. 28.15). Nonetheless, in their domain of applicability, these equations allow us to exhibit much that is observed and measured (Fig. 28.13).

POTENTIAL POWER SPECTRUM

What determined the initial amplitudes $\tilde{\Phi}(0, \mathbf{k})$? It turns out that a simple early conjecture[43] accounts very well for a large number of independent cosmological mea-

43. First proposed by Harrison (1970), who considered the mode amplitudes when they entered the horizon, not when they were initialized much earlier, as we shall do. The idea was put on a more general footing by Zel'dovich (1972); see also Peebles and Yu (1970).

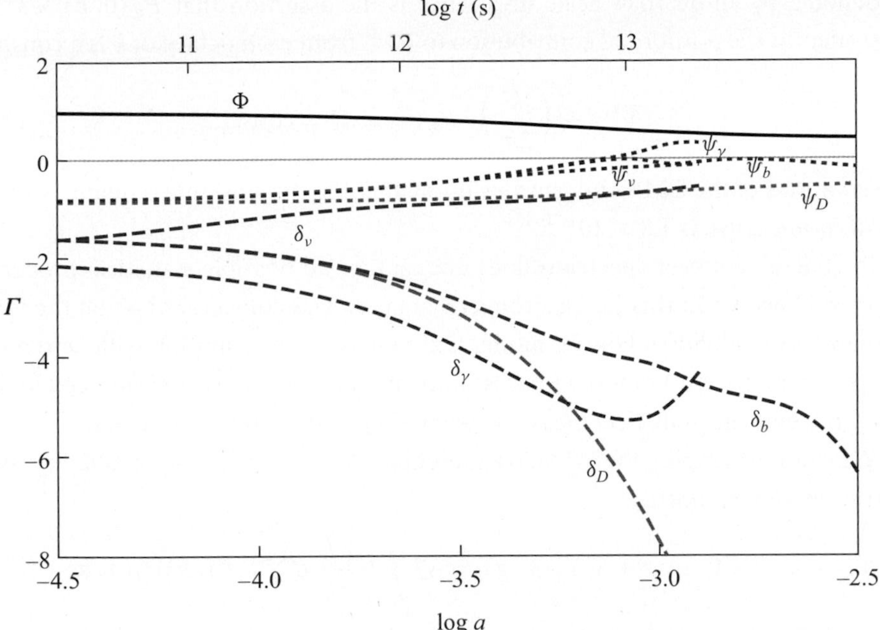

FIGURE 28.13 Variation of the growth vector $\boldsymbol{\Gamma}$ for one choice of wave number $k_{\rm eq} = 78H_0$, which enters the horizon, $\beta = 1$, when the universe becomes matter dominated ($a = a_{\rm eq} = 0.00030$; $\log a_{\rm eq} = -3.52$). The solid black line is the potential perturbation; the dashed lines are the density perturbations, δ; the dotted lines are the velocity potential perturbations. The potential decreases slowly as the universe expands through the radiation, plasma, and atomic ages. The dark matter density perturbation starts to grow $\propto a$ after it enters the horizon. By contrast, the neutrino and photon perturbations, being hot, do not fall into the dark matter potential wells. The baryon perturbations are initially coupled to the photon perturbations, as can be seen by their common velocity potentials, but after recombination, they are released to fall into the dark matter potential wells. The model for the neutrino and photon perturbations is decreasingly realistic after recombination but irrelevant for our purpose here, and so is not shown.

surements. The conjecture is that the initial metric perturbations were *scale invariant and isotropic*. To explain what this means, adopt the formalism developed in Sec. 6.4, and imagine an ensemble of universes[44] defining a dimensionless *power spectrum* of potential fluctuations $P_\Phi(a, k)$. Specifically, we define [cf. Eq. (6.31)]

$$\langle \tilde{\Phi}(a, \mathbf{k})\tilde{\Phi}^*(a, \mathbf{k}')\rangle = (2\pi)^3 P_\Phi(a, k)\delta(\mathbf{k} - \mathbf{k}') = (2\pi)^3 \Gamma_1(a, k)^2 P_\Phi(0, k)\delta(\mathbf{k} - \mathbf{k}'). \tag{28.67}$$

Henceforth, we use $\langle\cdot\rangle$ to denote an ensemble average; all other averages that can be computed by integrating known functions over position, angle, frequency, etc. will

ensemble average

44. Of course, we only have one universe to observe, but when we study many small regions, the average properties are well defined. By contrast, when we examine large regions, the *cosmic variance* is also large, and no matter how precisely we make our measurements, there is a limit to how much we can learn about the statistical properties of the ensemble.

T2

be denoted by an overbar. Scale invariance is the assertion that $P_\Phi(0, k) \propto k^{-3}$.[45] Equivalently, the primordial contribution to $\langle\Phi^2\rangle$ from each octave of k is a constant:

$$\langle\Phi^2\rangle_k \equiv \left(\frac{k}{2\pi}\right)^3 P_\Phi(k) \equiv \mathcal{Q} = \text{const.} \tag{28.68}$$

initial cosmic noise

Observations of the CMB and galaxies imply that the dimensionless quantity $\mathcal{Q}$, the *initial cosmic noise,* is 1.8×10^{-10}.[46]

Gaussian fluctuation spectrum

In general, a power spectrum does not capture all possible statistical properties of noise. However, in this case a stronger statement was conjectured—that the initial amplitudes of individual Fourier modes had a Gaussian distribution with zero mean, constant variance, and random phases (implying no covariance and no need for any other independent statistical measures beyond $\mathcal{Q}$; cf. Secs 6.3.2 and 6.3.3).

We can now employ the Wiener-Khintchine theorem (Sec. 6.4) to obtain a symmetric *correlation matrix:*

correlation matrix

$$C_{ij}(\mathbf{s}, a) \equiv \langle \Upsilon_i(a, \boldsymbol{\chi} + \mathbf{s})\Upsilon_j(a, \boldsymbol{\chi})\rangle = \mathcal{Q}\int \frac{d^3k}{4\pi k^3} e^{i\mathbf{k}\cdot\mathbf{s}}\Gamma_i(a, \mathbf{k})\Gamma_j(a, \mathbf{k}) \tag{28.69}$$

(cf. Ex. 9.8). Many entries in this matrix have been verified observationally, thereby validating the remarkably simple physical model that we have outlined. The birth of the universe was accompanied by a hum, not a fanfare!

SUMMARY

We have set up a general formalism for describing, approximately and linearly, the growth of perturbations under general relativity in the expanding universe all the way from neutrino decoupling to the present day. The principal output is the evolution of the potential functions Ψ, Φ and the accompanying density perturbations through recombination and after reionization. This provides a basis for a more careful treatment of the photons, to which we turn in Sec. 28.6, and the observed clustering of galaxies, which we now address.

28.5.4

28.5.4 Galaxies T2

SURVEYS

observations of galaxies

Much of what we have learned about cosmology has come from systematic surveys of distant galaxies over large areas of sky. As we have emphasized (cf. Sec. 28.2.1), galaxies are not well standardized; they are more like people than elementary particles! However, it is possible to average over this diversity to study their clustering. To date,

45. This spectrum may be thought of as the (3-dimensional) spatial generalization of flicker or "$1/f$" noise that is commonly measured in time series, such as music (Sec. 6.6.1). The common property is that there are no characteristic spectral features, and the power diverges logarithmically at both small and large scales. The measured spectrum has slightly more power at small k, as we discuss in Sec. 28.7.1.

46. For reasons that make perfect sense to astronomers, the conventional normalization is expressed as the rms relative density fluctuation in a sphere of radius 11 Mpc, assuming linear evolution of the perturbations. This quantity, known as σ_8, has the value $\sim$0.82.

roughly a billion galaxies (out of the roughly one trillion that are observable) have had their positions, shapes, and fluxes measured in a few spectral bands. Over a million of these have had their spectra taken, so their distances can be determined using the Hubble law.

T2

GALAXY POWER SPECTRUM

The power spectrum of density perturbations in the gravitational and cosmological ages, $P_M(a, k)$, defined by $\langle\tilde{\delta}_M(a, \mathbf{k})\tilde{\delta}^*_M(a, \mathbf{k}')\rangle = (2\pi)^3 P_M(a, k)\delta(\mathbf{k} - \mathbf{k}')$, can be simply related to the power spectrum for the gravitational perturbations $P_\Phi(a, k)$ [the evolved Eq. (28.68)] using Eqs. (28.55). If we also assume that, despite the different evolution of baryons and dark matter, the space density of galaxies is directly proportional to their combined density,[47] then we can use galaxy counts to measure the total matter power spectrum. The computed power spectrum is exhibited in Fig. 28.14, and its high-k region is explored in Ex. 6.7. Note that after recombination, the small-scale (large k) structure was relatively more important than the larger-scale structure in comparison with today. In other words, smaller *groups* of protogalaxies merged to create larger groups as the universe expanded. This principle is also apparent before reionization, when smaller, pregalactic dark matter *halos* merged to form larger halos, which eventually made observable galaxies when the gas was able to cool and form luminous stars.

merging of galaxies and groups

To describe merging requires that we handle the perturbations nonlinearly. Mild nonlinearity can be handled by a variety of analytical techniques, but the best approach is to assume that the dark matter is collisionless and to perform N-body numerical simulations, which can now (2016) follow over a trillion test particles. These calculations can then be supplemented with prescriptions for handling the non-gravitational behavior of the baryons on the smallest length scales as they cool to form stars and massive black holes. This happened most vigorously when $a \sim 0.3$. The current incidence of small structure appears to be significantly less than expected. This may be a result of nongravitational effects, or it could signify a high-k cutoff in the initial potential power spectrum [Eq. (28.68)].

nonlinearity of evolution

BARYON ACOUSTIC OSCILLATIONS

The perturbations at recombination are basically sound waves whose amplitude depends on the phase of the oscillation measured since the time when the wave mode entered the horizon. As a result, at recombination (380 kyr), the amplitude oscillates with the (comoving) wavelength. These oscillations are observed directly in angular fluctuations of the CMB (see Fig. 28.15), where they are known as *acoustic peaks.* Baryons were released from the grip of photons during recombination and fell into dark matter potential wells. There should thus be preferred scales imprinted on the

47. This turns out to be a better approximation than might be imagined, but it fails on small scales, where cooling and stellar activity become more important than gravity in determining what we see. These effects are addressed by attempting to compute *bias* factors.

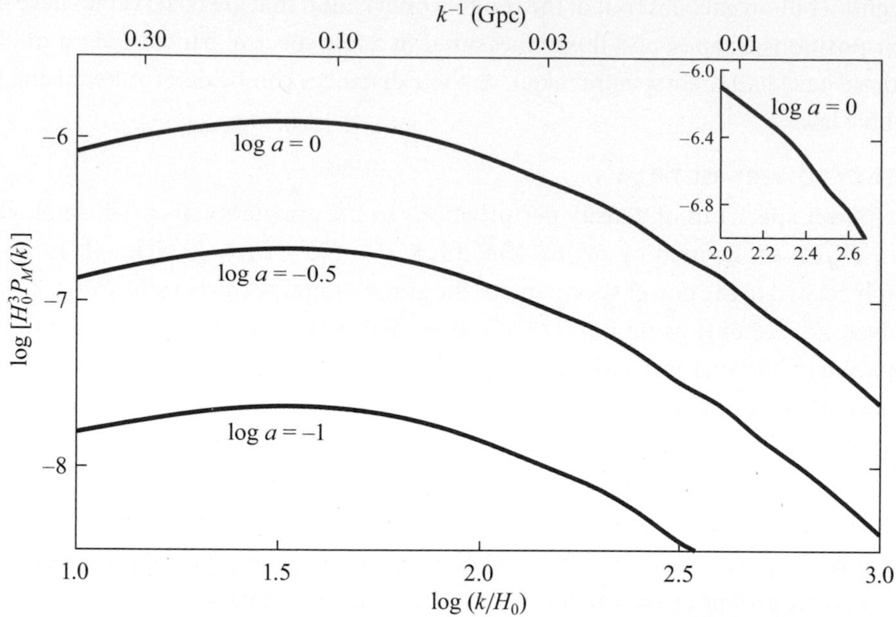

FIGURE 28.14 Matter fluctuation power spectrum $P_M(k)$ for log $a = -1, -0.5$, and 0, corresponding to recombination, reionization, the epoch of maximal galaxy and star formation, and today, respectively. The spectrum demonstrates clearly that structure grows first on small scales, and it provides a fair description of the results of large-scale galaxy surveys. Note the wiggles in the power spectrum for $2 < \log(k/H_0) < 2.7$, shown more clearly in the inset for log $a = 0$. These are baryon acoustic oscillations—echoes of the acoustic oscillations observed using CMB measurements at recombination.

distribution of galaxies we see around us today. These *baryon acoustic oscillations* are very important, because they allow astronomers to follow the expansion of a comoving ruler over time—in other words, to measure $a(t)$. As we shall see, the calibration length at recombination is very well determined by observations of the radiation (Sec. 28.6.1). Baryon acoustic oscillations can be seen in the angular correlation functions measured in large galaxy surveys and also in studies of the radial velocities of these galaxies.

EXERCISES

Exercise 28.15 *Example: Nonlinearity* T2

Explore nonlinear effects in the growth of perturbations in the gravitational age—when radiation and the cosmological constant can be ignored—by considering the evolution of a sphere in which the matter density is uniform and exceeds the external density by a small quantity.

(a) Use the Friedmann equations (28.16), (28.18), and (28.19) to show that the sphere behaves like a universe with density greater than the critical density and stops expanding when its density exceeds the external value by a factor of $9\pi^2/16$.

T2

(b) Assume that the perturbation remains strictly spherical, and determine by what additional scale factor the external universe will have expanded when the perturbation collapses to a point.

(c) Argue that realistic perturbations behave quite differently, that nonspherical perturbations grow during the collapse, and that the infall kinetic energy effectively randomizes during the collapse. Show that the collapse stops when the radius of the sphere is roughly half its maximum value and that this occurs when the average density exceeds that in the still-expanding external universe by a factor of $\sim$150.

28.6 Cosmological Optics T2

28.6

28.6.1 Cosmic Microwave Background T2

28.6.1

OVERVIEW

So far, we have emphasized the dynamical effects that govern the evolution of small perturbations in the expanding universe and have shown how these lead to a statistical description of the potential, density, and velocity perturbations. We now consider the effect of these perturbations on extragalactic observations where the radiation propagates passively through them. We start with the CMB. As we have explained in Sec. 28.4.4, most of the action happens at recombination—over an interval of time short compared with the age of the universe—when free electrons are rapidly captured and retained by protons and the rate of Thomson scattering plummets. We need to describe photons as they transition from belonging to a perfect fluid to uninterrupted, free propagation along null geodesics. We then discuss the statistical properties of the relative temperature and polarization fluctuations.

radiative transfer

MONTE CARLO RADIATIVE TRANSFER

The standard way to compute the radiative transfer is to generalize the moments of the Boltzmann equation (28.64) to include baryon motion and the potentials $\tilde{\Phi}$ and $\tilde{\Psi}$. Hundreds of spherical harmonics are necessary to achieve the requisite accuracy. As we have already discussed many calculations of this general character in the preceding chapters, we shall elucidate the underlying physical processes by using a Monte Carlo description (cf. Sec. 5.8.4). Monte Carlo methods are often used for problems that are too complex for a Boltzmann approach.

To do this, we first ignore the perturbations and follow backward in time a photon observed by us, today (at time t_0), with initial direction $\mathbf{n}$ and polarization (electric field unit vector) $\hat{\mathbf{E}}$. (The backward transition probabilities are just the same as for the actual, forward path.) We then assign the Thomson optical depth to the first scattering (going backward) according to $\tau_T = -\ln R_1$, where R_1 is a random number distributed uniformly in [0, 1]. Using the discussion in Sec. 28.4.4, we associate this optical depth with the location of the scatterer in spacetime, which is

Thomson optical depth

T2

scattering probability

polarization

specified by t_1 and $\boldsymbol{\chi}_1 = \mathbf{n}\int_{t_1}^{t_0} dt/a$. The differential cross section for electron scattering into direction $\mathbf{n}_1$ is $d\sigma/d\Omega = r_e^2(1-\mu_s^2)$, where r_e is the classical electron radius, and $\mu_s = \hat{\mathbf{E}}\cdot\mathbf{n}_1$ (e.g., Jackson, 1999).[48] The cumulative probability distribution for μ_s is $(3\mu_s/4 - \mu_s^3/4 + 1/2)$, and so we equate this to another uniformly distributed random number R_2 and solve for μ_s. The scattered photon's azimuth is likewise assigned as $2\pi R_3$, with R_3 a third random number uniform on [0, 1], and the new polarization vector $\hat{\mathbf{E}}_1$ is along the direction of $(\hat{\mathbf{E}}\times\mathbf{n}_1)\times\mathbf{n}_1$. We iterate and trace the photon path backward until the scatterings were so frequent that the evolution of the radiation can be treated as adiabatic. This happened at time $t_{\rm ad}$ and location $\chi_{\rm ad}$.

Having determined the path, we consider a photon traveling forward along it. According to Eq. (28.63) with $\Phi = \Psi = 0$, the covariant momentum p_i is constant between scatterings, and so the photon's frequency is $\nu \propto a^{-1}$. This frequency is unchanged by scattering, as the electrons are assumed to be at rest with respect to the FOs. When we repeat this exercise many times, we find no net polarization to statistical accuracy, as we must.

starting frequency shift

starting time shift

We now switch on a single perturbation with wave vector $\mathbf{k}$. [It is helpful to express the perturbation's Ψ and Φ as standing waves rather than running waves; cf. the passage following Eq. (28.46).] And we make the approximation that $\Psi = \Phi$ [i.e., we neglect gravitational effects of anisotropies in the free-streaming photon and neutrino stresses; see passage following Eq. (28.47d).] We need to calculate the linear relative frequency shift $\delta_\nu \equiv \delta\nu/\nu$ induced by the perturbation for a photon propagating forward in time along the above path, superposing four effects. The first is that the perturbation mode changes the *initial* frequency through the radiation density perturbation, the Doppler shift associated with the baryon velocity, and the gravitational frequency shift (Sec. 25.9). Specifically, $\delta_\nu = \frac{1}{4}\tilde{\delta}_\gamma(t_{\rm ad}) - \mathbf{n}_{\rm ad}\cdot\tilde{\mathbf{v}}_b(t_{\rm ad}) + \tilde{\Phi}(t_{\rm ad})$. The second effect is more subtle. The proper time (age of the universe) at the start of the path differs from that of an FO in a homogeneous universe by $\delta\tilde{t} = \int_0^{t_{\rm ad}} dt\,\tilde{\Phi}(t)$ (cf. Eq. 28.44). Since the universe is expanding according to local laws, the local scale factor (as measured, for example, by the local temperature) is modified by $\delta\tilde{a}/a = H\delta\tilde{t}$; and correspondingly, the frequency of the photon moving along the above Monte Carlo path is modified by $\delta_\nu = -H(t_{\rm ad})\int_0^{t_{\rm ad}} dt\,\tilde{\Phi}(t)$.[49]

The third effect is related to the second one. The scattering rate is $n_e\sigma(1-\mathbf{n}_i\cdot\tilde{\mathbf{v}}_b)$, and its perturbation leads to a change in the propagation time and consequently to

48. The Compton recoil is ignorable, and so the Thomson cross section suffices.

49. This is known as the Sachs-Wolfe effect (Sachs and Wolfe, 1967). The largest influence on the photon frequency comes from long-wavelength (low-k) gravitational perturbations, for which $\tilde{\Phi}$—which arises from dark matter—is nearly time independent during $0 < t < t_{\rm ad}$. Combining this with the value $H(t_{\rm ad}) = 2/(3t_{\rm ad})$ in the plasma age, we obtain for the second effect $\delta_\nu = -\frac{2}{3}\tilde{\Phi}(t_{\rm ad})$. And adding this to our first effect's gravitational frequency shift $\delta_\nu = +\tilde{\Phi}(t_{\rm ad})$, we obtain a combined direct gravitational frequency shift $\delta_\nu = +\frac{1}{3}\tilde{\Phi}(t_{\rm ad})$.

the time and frequency at the start of the path. The baryon density perturbations $\tilde{\delta}_b$ induce relative electron density perturbations $\tilde{\delta}_e$. They can be estimated using

T2

$$\tilde{\delta}_e = \left(\frac{\partial \ln(n_p x)}{\partial \ln n_b}\right)\tilde{\delta}_b + \frac{1}{4}\left(\frac{\partial \ln(n_p x)}{\partial \ln T_\gamma}\right)\tilde{\delta}_\gamma, \tag{28.70}$$

ionization shift

where x is the ionization fraction (Sec. 28.4.4), and the partial derivatives are computed at the recombination surface using the formalism of Sec. 28.4.4. (A more careful treatment includes many more atomic processes.) When the scattering rate increased and the total duration of the path decreased, the universe became colder, which contributes a negative frequency shift $\delta_\nu = -\int_{t_{\rm ad}}^{t_0} dt\, H\tau_T'[\tilde{\delta}_e + \mathbf{n}(t)\cdot\tilde{\mathbf{v}}_b]/\tau_T'(t_{\rm ad})$. The fourth and final effect is the Doppler shift applied at each scattering: $\delta_\nu = \sum_i (\mathbf{n}_i - \mathbf{n}_{i-1})\cdot\tilde{\mathbf{v}}(t_i)$.

Doppler shift

Now, the point of the Boltzmann equation is that the photon distribution function $\eta_\gamma = \{\exp[h\nu/(k_B T_\gamma)] - 1\}^{-1}$ is conserved along a trajectory in phase space (Sec. 3.6). Furthermore, it is not changed by scattering in the electron rest frame or by Lorentz transformation into and out of this frame. Therefore, the relative temperature fluctuation, which is what is actually measured, satisfies $\delta_{T_\gamma} \equiv \delta_T = \langle\delta_\nu\rangle$, averaging over the sum of the four contributions.

temperature fluctuation

We can also consider the effect on the polarization. The natural basis for the electric vector is $\mathbf{e}_a = \mathbf{k}\times\mathbf{n}/|\mathbf{k}\times\mathbf{n}|$ and $\mathbf{e}_b = \mathbf{n}\times\mathbf{e}_a$, and we expect any measured polarization to be perpendicular or parallel to the projection of $\mathbf{k}$ on the sky. (See Ex. 28.16.) However, $\mathbf{v}_b$ is along $\mathbf{k}$, and transforming into and out of the electron rest frame does not rotate the polarization vector in the $\{\mathbf{e}_a, \mathbf{e}_b\}$ basis. And the influence of gravitational deflections on the polarization is also negligible.

polarization

SPHERICAL HARMONIC EXPANSION

The Monte-Carlo calculation that we have just outlined allows us to compute the expected temperature fluctuation and write it in the form

$$\tilde{\delta}_T(\mathbf{n},\mathbf{k}) = \mathcal{T}_k(\mathbf{n}\cdot\hat{\mathbf{k}})\tilde{\Phi}(0,\mathbf{k}) \tag{28.71}$$

where $\mathcal{T}_k$ is the *initial potential-temperature transfer function*. As $\mathcal{T}_k$ is defined on a sphere, it is natural to expand it in Legendre polynomials, the functional equivalent of a Fourier series:

$$\mathcal{T}_k(\mathbf{n}\cdot\hat{\mathbf{k}}) = \sum_{l=0}^{\infty}\mathcal{T}_{kl}P_l(\mathbf{n}\cdot\hat{\mathbf{k}}), \quad \text{where} \quad \mathcal{T}_{kl} = \frac{(2l+1)}{2}\int_{-1}^{1} d(\mathbf{n}\cdot\hat{\mathbf{k}})\mathcal{T}_k P_l(\mathbf{n}\cdot\hat{\mathbf{k}}). \tag{28.72}$$

We are interested in the cross correlation of the temperature fluctuations $\langle\delta_T(\mathbf{n})\delta_T(\mathbf{n}')\rangle$, where the average is over all directions $\mathbf{n}$, $\mathbf{n}'$ separated by a fixed angle and we take the ensemble average over perturbations using Eqs. (28.67), (28.68). This

T2

must depend only on that angle and can therefore be expanded as another sum over Legendre polynomials:[50]

expansion in Legendre polynomials

$$\langle \delta_T(\mathbf{n})\delta_T(\mathbf{n}')\rangle = \int \frac{d^3k}{(2\pi)^3} P_\Phi(0,k) \sum_{l,\,l'} \mathcal{T}_{kl}\mathcal{T}^*_{kl'} \langle P_l(\hat{\mathbf{k}}\cdot\mathbf{n}) P_{l'}(\hat{\mathbf{k}}\cdot\mathbf{n}')\rangle$$

$$= \int \frac{d^3k}{(2\pi)^3} P_\Phi(0,k) \sum_{l=0}^{\infty} |\mathcal{T}_{kl}|^2 \frac{P_l(\mathbf{n}\cdot\mathbf{n}')}{2l+1}. \tag{28.73}$$

This is the functional equivalent of the Wiener-Khintchine theorem (Sec. 6.4.4). It is conventional to express $\langle \delta_T(\mathbf{n})\delta_T(\mathbf{n}')\rangle$ in terms of the total *multipole coefficient* C_l:

$$\langle \delta_T(\mathbf{n})\delta_T(\mathbf{n}')\rangle = \sum_{l=0}^{\infty} \frac{2l+1}{4\pi\, T_{\gamma o}^2} C_l\, P_l(\mathbf{n}\cdot\mathbf{n}') \;; \tag{28.74a}$$

$$C_l = \frac{4\pi T_{\gamma 0}^2}{(2l+1)^2} \int \frac{d^3k}{(2\pi)^3} P_\Phi(0,k) |\mathcal{T}_{lk}|^2$$

$$= \frac{16\pi^2\, \mathcal{Q}\, T_{\gamma 0}^2}{(2l+1)^2} \int_0^\infty d\ln k\, |\mathcal{T}_{lk}|^2 \;, \tag{28.74b}$$

where we have used the scale-invariant form (28.68) of the initial perturbation spectrum $P_\Phi(0,k)$.

REIONIZATION SCATTERING

influence of intervening electrons

The rapid increase in electron density and Thomson scattering following reionization at the end of the atomic age (see Fig. 28.8) can actually be detected in the observations of the CMB and is described by essentially the same equations that we used for recombination. The Thomson optical depth, backward in time from us through reionization (as defined in Sec. 28.4.4) averages to $\tau_T \sim 0.066$.

INTEGRATED SACHS-WOLFE EFFECT

influence of intervening structure

Another late-time effect is better described in configuration space. Consider a photon crossing a large negative gravitational potential well, associated with an excess of matter during the cosmological age. The cosmological constant causes the potential perturbation to decrease [cf. Eq. (28.54)] while the photon crosses it and so the photon loses less energy climbing out of it than it gained by falling into it, so there is a net positive temperature fluctuation. This is most apparent at long wavelengths and is most easily detected by cross-correlating the matter distribution with the temperature fluctuations.

GRAVITATIONAL LENSING

As we describe in more detail in Sec. 28.6.2, the gravitational deflection of rays crossing the universe leads to the distortion of the images of background sources. Of

50. To verify this identity, consider the special case $\mathbf{n} = \mathbf{n}'$.

T2

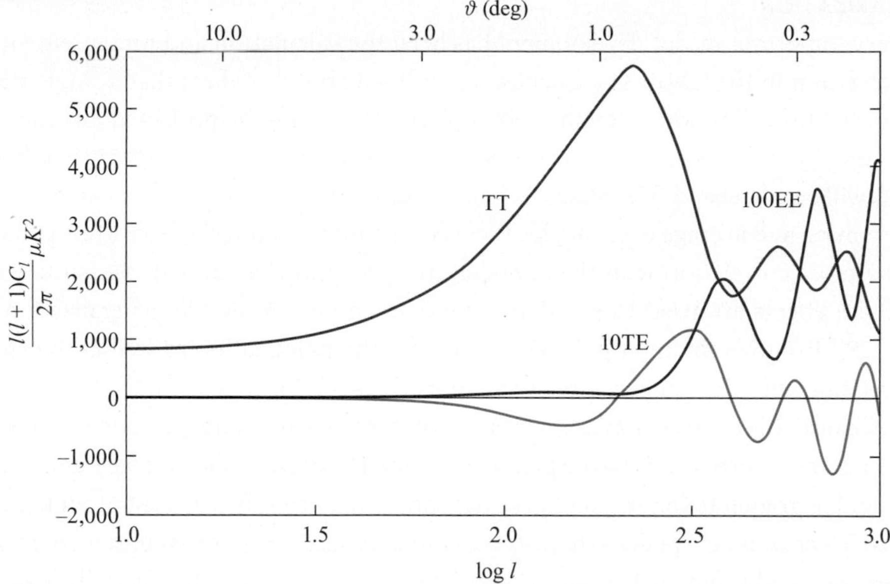

FIGURE 28.15 Theoretical spectra for anisotropy of the microwave background fluctuations as measured by the coefficients C_l, where l is the spherical harmonic quantum number as defined in Eqs. (28.74). The curve labeled TT shows the C_ls for the temperature fluctuations [Eqs. (28.74); multiplied by $l(l+1)/(2\pi)$]; that labeled EE shows the C_ls for the E-mode polarization fluctuations; and that labeled TE shows the C_ls for the temperature-polarization cross correlation. These curves are adapted from theoretical calculations by Planck Collaboration (2016b) for parameter values that best fit the observations. The low-l portion of the TT curve can be reproduced with the formalism presented in this section. The fluctuations' angular scale on the top axis is $\vartheta = 180^o/l$. Note the prominence of the first "acoustic peak" at $l \sim 200$, which corresponds to waves that have reached maximum amplitude at recombination. The large-angle fluctuations with $l \lesssim 70$ are basically gravitational redshifts associated with perturbations that have not yet entered the horizon at the time of recombination. Modes with $l \gtrsim 300$ are dominated by density changes. Velocity effects contribute heavily to intermediate l harmonics. Ten of the predicted acoustic peaks have been measured in the TT spectrum out to $l \sim 3{,}000$. The observations agree extremely well with these predictions after correcting for some additional effects listed in the text.

course this makes no difference to the appearance of a uniform background radiation. However, it will change the fluctuation spectrum. The formalism used is an adaption of that developed below. This has turned out to be a powerful probe of intervening structure and, consequently, an important consistency check on the standard model.

TEMPERATURE FLUCTUATION SPECTRUM

the spectrum of temperature fluctuations

The theoretical temperature fluctuation spectrum computed using more detailed calculations than ours (Planck Collaboration, 2016b) is shown in Fig. 28.15. It fits the observational data extremely well with only six adjustable parameters. Our Monte Carlo results (not shown) roughly recover this spectrum for low l. The detailed fit of the observations to the theory is responsible for many of the features of the standard model described earlier in this chapter.

T2

POLARIZATION

A very important recent development has been the calculation and measurement of polarization in the CMB. The calculation outlined above predicts that a single wave perturbation will produce roughly 10% polarization along the projected direction of **k** on the sky and also predicts that the polarization observed along neighboring directions will be correlated (Ex. 28.22). When we sum over all modes, expand in spherical harmonics, and average over the sky, there is a net polarization signal of a few percent and a cross correlation with the temperature spectrum (Fig. 28.15). Measurements of these effects are used to refine the standard model. A simple generalization of Eq. (28.74b) gives the multipole coefficients for the polarization as well as the cross correlation with the temperature fluctuations (Fig. 28.15).

Because this polarization arises from photon scattering in the presence of density fluctuations, it turns out to have a pattern described by tensor spherical harmonics that are double gradients (on the sky) of scalar spherical harmonics; these are sometimes called "electric-type" spherical harmonics, and polarization patterns constructed from them are called *E-modes*. Primordial gravitational waves, interacting with the plasma during recombination, can catalyze a second type of polarization pattern called *B-modes*, whose "magnetic-type" spherical harmonics are constructed by operating on a scalar spherical harmonic with one gradient $\boldsymbol{\nabla}$ and one angular momentum operator $\mathbf{L} = \mathbf{e}_{\hat{r}} \times \boldsymbol{\nabla}$. For some details of E-modes, B-modes, and other aspects of the predicted polarization, see Ex. 28.22.

RADIATION STRESS-ENERGY TENSOR

The least satisfactory aspect of our treatment of the growth of perturbations is the approximation of the radiation as a perfect fluid. Our treatment of radiative transfer allows one to refine this approximation by including in Eq. (28.61) an estimate of the anisotropic part of the photon stress-energy tensor.

EXERCISES

Exercise 28.16 *Example: Stokes' Parameters* T2

There are many ways to represent the polarization of electromagnetic radiation (Sec. 7.7). A convenient one that is used in the description of CMB fluctuations was introduced by Stokes.

(a) Consider a monochromatic wave propagating along $\mathbf{e}_z$ with electric vector $\mathbf{E} = \{E_x, E_y\}e^{i\omega t}$, where the components are complex numbers. Explain why this wave is completely polarized, and introduce the *Stokes' parameters* (following Jackson, 1999) $I = E_x E_x^* + E_y E_y^*$, $Q = E_x E_x^* - E_y E_y^*$, $U = 2\Re(E_x E_y^*)$, and $V = -2\Im(E_x E_y^*)$. Sketch the behavior of the electric vector as the complex ratio $r = E_y/E_x$ is varied, and hence associate Q, U, and V with different states of polarization.

(b) Derive the transformation laws for the Stokes' parameters if we rotate the $\mathbf{e}_x$, $\mathbf{e}_y$ directions about $\mathbf{e}_z$ through an angle ψ. T2

(c) Show that $Q^2 + U^2 + V^2 = I^2$ and that the polarization of the wave may be represented as a point on a sphere, which you should identify.

(d) Now suppose that the wave is polychromatic and partially polarized, and replace the definitions of I, Q, U, and V with the time averages $\langle E_x E_x^* \rangle$, and so forth. Show that $Q^2 + U^2 + V^2 < I^2$, and give expressions for the degree of polarization and the associated position angle in terms of I, Q, U, and V.

Exercise 28.17 *Problem: Cosmic Variance* T2

The precision with which the low-l spherical harmonic power spectrum can be determined observationally is limited because of the low number of independent measurements that can be averaged over. Give an approximate expression for the *cosmic variance* that should be associated with the CMB fluctuation spectrum.

Exercise 28.18 *Problem: Acoustic Peaks* T2

We have explained how the peaks in the CMB temperature fluctuation spectrum arise because the sound waves all began at the same time and are all effectively observed at the same time, while they entered the horizon at different times. Suppose that the universe had been radiation dominated up to recombination, so that Eq. (28.51) is valid and oscillatory waves of constant amplitude were created with different values of k. Calculate the total relative density perturbation $\tilde{\delta}(k)$ at recombination. Describe the main changes in standard cosmology that are introduced.

Exercise 28.19 *Example: Cosmic Dawn* T2

The cosmic dawn that preceded the epoch of reionization can be probed by low-frequency CMB observations using a special radio hyperfine line emitted and absorbed by hydrogen atoms. This line is associated with a flip in direction of the magnetic dipole associated with the central proton relative to the magnetic field created by the orbiting electron. The line's frequency and strength can be calculated using quantum mechanics. For our purposes, all that we need to know is that the frequency associated with the transition is $\nu_H = 1.42$ GHz, the degeneracy of the ground/upper state is 1/3, and the rate of spontaneous transition is $A = 3 \times 10^{-15}\ \text{s}^{-1}$ (Ex. 28.9).

(a) Explain why these hyperfine transitions should produce a change in the measured CMB spectrum over a range of frequencies $\sim$20–150 MHz, where the lower limit is due to the practicality of making the measurement.

(b) Consider hydrogen atoms with total number density n_1 in the ground state and $n_2 = 3n_1 \exp(-h\nu_H / k_B T_S)$ in the excited state, where T_S defines the *spin temperature,* and we measure frequencies and rates locally. Show that the net creation rate

T2

of photons per unit volume and frequency can be written as $3n_1 A\delta(\nu - \nu_H)(1 - T_\gamma/T_S)$. [Hint: $T_S, T_\gamma \gg h\nu_H/k_B$.]

(c) Hence, show that the CMB temperature fluctuation at frequency $\nu_0 = \nu_H/a$, produced when the expansion factor was a, is

$$\delta_T = \left(\frac{3}{32\pi}\right)\left(\frac{n_H}{\nu_H^3}\right)\left(\frac{A}{H(a)}\right)[1 - T_{\gamma 0}/T_s(a)],$$

where n_H is the atomic hydrogen density, $H(a)$ is the expansion rate, and $T_{\gamma 0}$ is the CMB temperature today while T_γ is the CMB temperature at the point of emission.

(d) It is predicted that $T_S \sim 10$ K when $a \sim 0.05$. Estimate the associated temperature perturbation, δ_T.

The spin temperature T_s will follow the nonrelativistic gas and fall faster than the radiation temperature T_γ (Sec. 28.2.3), creating absorption above ~10 MHz. The first stars will heat the gas and create Lyman alpha photons (Sec. 28.4.4), which end up populating the upper hyperfine state (cf. Sec. 10.2.1), increasing the spin temperature and reducing the absorption above ~50 MHz. Black holes are expected to create highly penetrating X-rays, which may make $T_s > T_\gamma$ and lead to emission above 100 MHz. Eventually the gas will be fully ionized, so that the spectrum above ~200 MHz should be unaffected.

At the lowest frequencies observable today, the radiation from our galaxy is ~300 times brighter than the CMB and has to be carefully removed, along with the influence of the ionosphere (cf. Sec. 21.5.4). Most attention is now (2016) focused on measuring a signal associated with the growing density perturbations from the time just before reionization. If the measurements are successful, we will have another powerful probe of the growth of matter perturbations (Sec. 28.5.3).

28.6.2

28.6.2 Weak Gravitational Lensing T2

NULL GEODESIC CONGRUENCE

We introduced strong gravitational lensing in Sec. 7.6. Such lensing is important for rare lines of sight where the galaxy-induced gravitational deflections of light rays are strong enough to image background sources more than once. There is a complementary effect called *weak gravitational lensing,* which is a consequence of the growth of perturbations in the universe and is present for all images (e.g., Schneider, Ehlers, and Falco, 1992). Basically, the tidal actions of gravitational perturbations distort galaxy images, inducing a correlated ellipticity that we can measure if we assume that the galaxies' intrinsic shapes are randomly oriented on the sky. To quantify this effect, we need to consider the propagation of neighboring rays through the inhomogeneous universe, under the geometrical optics approximation.

weak lensing

T2

What we actually do is a little more subtle and much more powerful. We consider one *fiducial* ray and a *congruence* of rays that encircle it—a generalization of the paraxial optics developed in Sec. 7.4. We imagine this congruence as propagating backward in time from us, now (in a scholastically correct manner!), toward a distant galaxy. We label rays that belong to the congruence by the vectorial angle $\boldsymbol{\psi}$ they make with the fiducial ray here and now—what an astronomer observes. The fiducial ray will follow a crooked path, but we are concerned with the proper transverse separations of neighboring rays $\boldsymbol{\xi}(\chi; \boldsymbol{\psi})$. This is a job for the equation of geodesic deviation [Eq. (25.31)]. (For a more detailed analysis along the lines of the following, see Blandford et al., 1991.)

ray congruence

As we have discussed in Sec. 25.4, we parameterize distance along a null geodesic using an affine parameter ζ, which must satisfy $dt/d\zeta \propto p^0$ [cf. Eq. (2.14)]. Now, p^0 is the energy of a photon measured by an FO; in the homogeneous universe, this will vary $\propto a^{-1}$. A convenient choice for ζ is therefore

affine parameter

$$\zeta(a) = \int_{t(a)}^{t_0} dt' a(t') = \int_a^1 \frac{da'}{H(a')} = \int_0^{\chi(a)} d\chi' a(\chi')^2. \tag{28.75}$$

Note that we use the scale factor appropriate to the unperturbed universe in defining ζ, because the overall expansion of the universe is dictated by the behavior of the stress-energy tensor $\boldsymbol{T}$ on the largest scales where it is, by assumption, homogeneous. The associated tangent vector to use in the equation of geodesic deviation is $dx^\alpha/d\zeta = \{a^{-1}, 0, 0, a^{-2}\}$.

The Riemann tensor for the perturbed metric (28.44), like the Einstein tensor (28.47), is easily computed to linear order in the perturbations and then inserted into the equation of geodesic deviation (25.31). In contrast to our treatment of the CMB, for weak lensing we explicitly assume that the relevant perturbations are of short wavelength and are effectively static when crossed by the photons we see today, allowing us to use local Lorentz coordinates parallel-propagated along the ray with $\hat{\mathbf{e}}_3$ aligned along the ray. Assuming that $\Phi = \Psi$ as dictated by the Einstein equations (28.47d) for a perfect fluid, and just retaining lowest order terms, the equation of geodesic deviation becomes

$$a^2\frac{d^2\xi^i}{d\zeta^2} - \dot{H}\xi^i = -\left(\Phi_{,3}{}^3\xi^i + 2\Phi_{,j}{}^i\xi^j\right) = -\left(4\pi\delta\rho_M\xi^i + 2\bar{\Phi}_{,j}{}^i\xi^j\right),$$
$$\text{for } i, j = 1, 2, \tag{28.76}$$

equation of geodesic deviation

where we have used Poisson's equation $\Phi_{,k}{}^k + \Phi_{,3}{}^3 = 4\pi\delta\rho_M$ and have introduced the trace-free *tidal tensor* $\bar{\Phi}_{,j}{}^i \equiv \Phi_{,j}{}^i - \frac{1}{2}\Phi_{,k}{}^k\delta_j{}^i$; and where spatial indices are raised and lowered with the flat metric, and all indices following a comma represent partial derivatives.

T2

CONVERGENCE AND SHEAR

It is instructive and helpful to express this equation in terms of comoving coordinates. Substituting $\boldsymbol{\xi} \to a\boldsymbol{\eta}$ and $d\zeta \to a^2 d\chi$, we obtain[51]

$$\frac{d^2\eta^i}{d\chi^2} = -\left(4\pi\,\delta\rho_M \delta_j{}^i + 2\bar{\Phi}_{,j}{}^{i}\right) a^2\eta^j. \qquad (28.77)$$

The right-hand side vanishes in the absence of perturbations, which is what we expect, as the 3-space associated with the homogeneous universe is flat (Sec. 28.2.2). Equation (28.77), made linear by inserting the unperturbed η^i on the right-hand side, admits a Green function solution:

$$\boldsymbol{\eta} = \frac{\boldsymbol{\xi}}{a} = \chi[(1-\kappa)\mathbf{I} - \boldsymbol{\gamma}]\cdot\boldsymbol{\psi};$$

$$\{\kappa, \gamma_j{}^i\} = \int_0^{\chi} d\chi' \left(\chi' - \frac{\chi'^2}{\chi}\right) a(\chi')^2 \left\{4\pi\,\delta\rho_M(\chi'), 2\bar{\Phi}_{,j}{}^{,i}(\chi')\right\}. \qquad (28.78)$$

Here $\boldsymbol{\psi} = (d\boldsymbol{\xi}/d\zeta)_0 = (d\boldsymbol{\xi}/d\chi)_0$ is the observed angular displacement from the fiducial ray, and $\mathbf{I}$ is the 2-dimensional unit tensor (metric). Also, we introduce the convergence κ; it is the analog of the expansion that appears in the theory of elastostatics (Sec. 11.2), and it is produced by matter density perturbations inside the congruence. The quantity $\boldsymbol{\gamma}$ is the trace-free *cosmic shear tensor* produced by matter distributed anisotropically outside the congruence.

convergence

cosmic shear

If we replace χ with d_A/a, Eq. (28.78) appears to be a simple linear generalization of the formalism introduced in our discussion of strong gravitational lensing (Sec. 7.6). However, the equation of geodesic deviation is necessary to justify the neglect of the cosmological constant, to handle the potential derivatives along the ray, and to include large density perturbations.

Now we jump into **k**-space, write $\bar{\Phi}$ in terms of the Fourier transform $\tilde{\Phi}$, and express the two components of the distortion as:

distortion

$$\{\kappa, \gamma_j{}^i\} = -\int_0^{\chi} d\chi' \left(\chi' - \frac{\chi'^2}{\chi}\right) \int \frac{d^3k}{(2\pi)^3} \left\{k^2, k_j k^i - \frac{1}{2}k_l k^l \delta_j{}^i\right\} \tilde{\Phi} e^{i\mathbf{k}\cdot\boldsymbol{\chi}'}. \qquad (28.79)$$

The first component is a scalar; the second is a tensor.[52]

CORRELATION FUNCTIONS

The ensemble average distortion vanishes, because the wave phases are random. When we consider the convergence and shear along two congruences labeled 1, 2,

51. The cosmological constant does not contribute to the focusing, but we should include radiation and neutrinos when considering weak lensing of the CMB.

52. Formally, κ should be a tensor with an antisymmetric part describing image rotation. In the case of elastostatics, this term is dropped, because it induces no stress. Weak lensing rotation vanishes in the linear, scalar approximation, but there is a tiny nonlinear contribution. In principle, an antisymmetric part of κ could be created by a hypothetical cosmic torsion field, which has been sought (unsuccessfully!) using polarization observations.

T2

and separated by an angle ϑ, the contribution from the products of different waves likewise vanishes. Even the contribution from a single wave is small except when it is directed almost perpendicular to the line of sight and we are integrating along its crest or trough. We formalize these expectations by using the potential power spectrum [Eq. (28.67)] and keeping faith with the magic of delta functions. Let us assume that $\vartheta \ll 1$, so the sky can be treated as flat, and introduce local basis vectors $\mathbf{e}_{||}$ parallel to the separation of the congruences and $\mathbf{e}_{\perp}$ perpendicular to the separation. We then define two independent components of shear by $\gamma_+ \equiv \gamma_{||}{}^{||} - \gamma_{\perp}{}^{\perp}$ and $\gamma_\times \equiv 2\gamma_{\perp}{}^{||}$. The convergence correlation function is $C_{\kappa\kappa} = \langle \kappa_1 \kappa_2 \rangle$, where the average is over all pairs of rays on the sky separated by an angle ϑ. We can likewise define correlation functions involving the shear. Then the values of the nonzero correlation functions are

$$\begin{pmatrix} C_{\kappa\kappa} \\ C_{\kappa+} \\ C_{++} \\ C_{\times\times} \end{pmatrix} = \int_0^\chi d\chi' \int_0^\chi d\chi'' \left(\chi' - \frac{\chi'^2}{\chi}\right)\left(\chi'' - \frac{\chi''^2}{\chi}\right) \times$$

$$\int \frac{d^3k}{(2\pi)^3} e^{i(k_{||}\vartheta\chi' + k_3(\chi' - \chi''))} P_\Phi(a, k) \begin{pmatrix} k^4 \\ k^2(k_{||}^2 - k_\perp^2) \\ (k_{||}^2 - k_\perp^2)^2 \\ 4k_{||}^2 k_\perp^2 \end{pmatrix}. \tag{28.80}$$

(The vanishing of the two remaining averages provides a check on the accuracy of the measurements.) As $k\chi \gg 1$, we approximate the integral over χ'' by an infinite integral over $\chi'' - \chi'$ to produce a delta function, which can then be integrated over to obtain

$$\begin{pmatrix} C_{\kappa\kappa} \\ C_{\kappa+} \\ C_{++} \\ C_{\times\times} \end{pmatrix} = \int_0^\chi d\chi' \left(\chi' - \frac{\chi'^2}{\chi}\right)^2 \int \frac{dk}{2\pi} k^5 P_\Phi(a, k) \int_0^{2\pi} \frac{d\phi_k}{2\pi} e^{ik\vartheta\chi' \cos\phi_k} \begin{pmatrix} 1 \\ \cos 2\phi_k \\ \cos^2 2\phi_k \\ \sin^2 2\phi_k \end{pmatrix}$$

$$= (2\pi)^2 Q \int_0^\chi d\chi' \left(\chi' - \frac{\chi'^2}{\chi}\right)^2 \int dk\, k^2 \Gamma_1^2(a, k) \begin{pmatrix} J_0(k\chi\vartheta) \\ -J_2(k\chi\vartheta) \\ \frac{1}{2}[J_0(k\chi\vartheta) + J_4(k\chi\vartheta)] \\ \frac{1}{2}[J_0(k\chi\vartheta) - J_4(k\chi\vartheta)] \end{pmatrix}, \tag{28.81}$$

where we have substituted the initial cosmic noise [Eq. (28.68)] for P_Φ, $\Gamma_1(a, k)$ is the first component of the growth vector Eq. (28.66), and J_0, J_2, J_4 are Bessel functions. The remaining integrals must be performed numerically; see Fig. 28.16. Note that

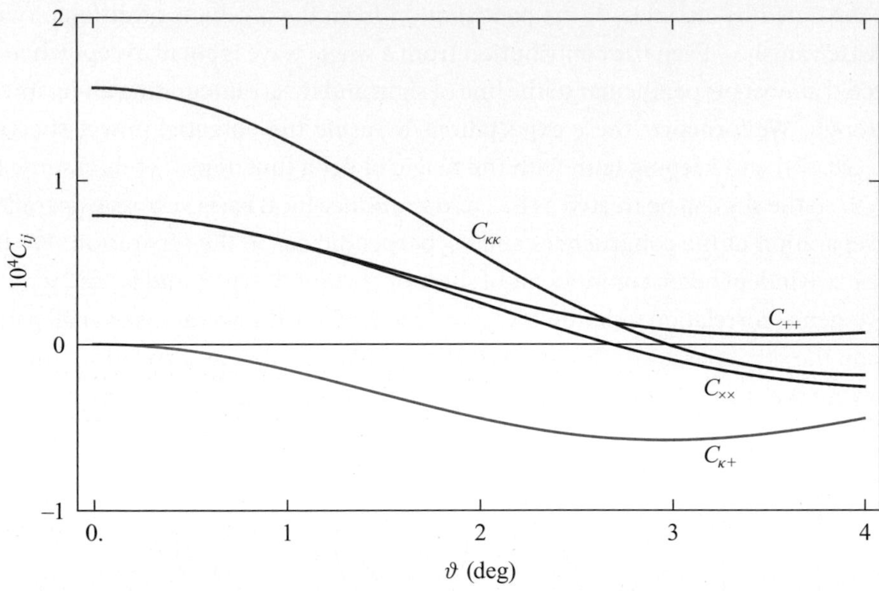

FIGURE 28.16 Two-point correlation functions for weak gravitational lensing for sources located at $\log a = -0.5$, calculated from Eq. (28.81) and adopting the growth factor $\Gamma_1(a, k)$ calculated numerically from Eq. (28.65). The cross correlation of magnification fluctuations is $C_{\kappa\kappa}$. The correlation function for parallel stretching of galaxy images is C_{++}. The correlation function for stretching along a direction inclined at 45° to the separation vector is $C_{\times\times}$. The magnification-shear cross correlation is $C_{\kappa+}$.

shear correlation

$C_{++} + C_{\times\times}$, the total shear cross correlation, equals $C_{\kappa\kappa}$, as can be seen directly from Eq. (28.80) if we recall that $k_3 \sim 0$. Note also that $C_{\times\times}(\vartheta)$ can change sign. This effect has been observed. The calculation is performed here in the linear approximation, which breaks down for small scales, and so numerical simulations must be used in practice.

tomography

These four correlation functions refer to galaxies at a fixed distance. In practice, a given survey averages over a range of distances and, by varying this range, can make a *tomographic* examination of the growth of the potential power spectrum P_Φ.

STATISTICAL MEASUREMENT

flux perturbation

How do we actually use observations of galaxies to measure convergence and shear? Weak lensing does not change the intensity of radiation. Therefore a source's flux F is determined by the solid angle it subtends. Using Eq. (28.78) to linear order, we see that $\Delta F/F = 2\kappa$. Now, we do not know the intrinsic galaxy luminosities. However, their distribution should be the same in different directions at the same distance if the universe is homogeneous, and the predicted magnifications have been measured. Unfortunately, absorption in our galaxy and other problems limit the utility of this approach. Conversely, $C_{\kappa\kappa}(0)$ is a measure of the variance of a flux measurement and a limit on how precisely a measurement need be made in a single case when attempting to use "standard candles" (sources of known luminosity) to measure the universe.

T2

Statistical shear measurements are more useful. We expect the orientations of galaxies to be uncorrelated,[53] and so, if we associate the fiducial ray with the center of an observed galaxy, the tensor $\langle \boldsymbol{\xi} \otimes \boldsymbol{\xi} \rangle$ should be proportional to the unit tensor. Inverting the first part of Eq. (28.78) then provides a linear estimator of the shear:

intrinsic galaxy alignment

shear estimator

$$\{\gamma_+, \gamma_\times\} = \frac{\langle \{\psi_\parallel^2 - \psi_\perp^2, 2\psi_\parallel \psi_\perp\} \rangle}{\langle \psi_\parallel^2 + \psi_\perp^2 \rangle}, \tag{28.82}$$

where the averages are over the photons associated with observed images of individual galaxies and then over galaxies adjacent on the sky but at differing distances. This estimator is quite general, and different types of source[54] and weighting can be employed in evaluating it observationally.

There are many observational challenges. For example, the dominance of Fourier modes with $\mathbf{k}$ almost perpendicular to the line of sight means that occasional structures elongated along the line of sight can hinder the measurement of an unbiased estimate of the ensemble-averaged correlations. Despite this, cosmic shear measurements are consistent with the predicted perturbation spectrum (28.68) and its evolution and show great promise for future surveys that will image 20 billion galaxies.

observing cosmic shear

EXERCISES

Exercise 28.20 *Example: Weak Lensing in an Empty Universe* T2

Consider an extended congruence propagating through an otherwise homogeneous universe, from which all matter has been removed. Show that the affine distance functions as an effective angular diameter distance in this congruence. Now reinstate the matter as compact galaxies and modify Eq. (28.79) for the convergence and shear. Bookkeep the average density of matter as purely positive density perturbations. Explain qualitatively how the convergence and shear are changed when the ray passes through one or more galaxies, when it misses all of them, and on average.

Exercise 28.21 *Problem: Mean Deviation* T2

Consider a single light ray propagating across the universe from a source at $\log a = -0.5$ to us. The cumulative effect of all the deflections caused by large-scale inhomogeneities makes the observed direction of this ray deviate by a small angle from the direction it would have had in the absence of the inhomogeneities. Estimate this deviation angle. Is it large enough to be observable?

53. In fact, neighboring galaxies are systematically aligned, but provided we average over large enough volumes, this is ignorable. In addition, systematic biases associated with telescope and atmospheric distortion must be removed carefully.

54. For example, the CMB fluctuations and the galaxy correlation function at the smallest angular scales can be used.

T2

Exercise 28.22 *Challenge: CMB Polarization* T2

Polarization observations of the CMB provide an extremely important probe of fluctuations in the early universe.

(a) By invoking the electromagnetic features of Thomson scattering by free electrons, give a heuristic demonstration of why a net linear polarization signal is expected.

(b) Using the Monte Carlo formalism sketched in Sec. 28.6.1, calculate the polarization expected from a single fluctuational mode $\tilde{\Phi}e^{i\kappa\cdot\chi}$ of given amplitude.

(c) The description of polarization has many similarities with the formalism we have outlined to describe cosmic shear. Linear polarization is unchanged by rotation through π just like a shear deformation. In addition, the polarization pattern that should be seen from a single inhomogeneity mode should have an electric vector that alternates between parallel and perpendicular to the projection of the mode's wave vector **k** on the sky, just like the elongation of the images of background galaxies in weak gravitational lensing. We do not expect to produce a signal in either case along a direction at 45° to the projection of **k**. These predicted polarization/shear patterns are commonly called "E-modes." However, as we discuss in Sec. 28.7.1, primordial gravitational wave modes may also be present. Explain qualitatively how a single gravitational wave mode can produce a "B-mode" polarization pattern with electric vectors inclined at ±45° to the direction of the projection of **k** on the sky.

(d) When one sums over inhomogeneity modes, and over gravitational-wave modes, the resulting polarization E-modes and B-modes have distinctive patterns that differ from each other. In what ways do they differ? Read, explain, and elaborate the discussion of E-modes and B-modes near the end of Sec. 28.6.1.

(e) Outline how our perturbed metric, Eq. (28.44), would have to be modified to accommodate the presence of primordial gravitational waves.

28.6.3

28.6.3 Sunyaev-Zel'dovich Effect T2

When we discussed CMB radiative transfer in Sec. 28.6.1, we assumed that the plasma was cold. This was appropriate when the temperature was $T_e \sim 3{,}000$ K. However,

clusters of galaxies

the gas that settles in rich clusters of galaxies (see Sec. 28.2.1 and Fig. 28.1) has a temperature of $\sim 10^8$ K, and thermal effects are very important. To understand these important effects in general, consider a homogeneous and isotropic radiation field with distribution function $\eta_\gamma(\nu)$. Every time a photon is scattered by an electron, its energy changes through small increments by Doppler shifting and Compton recoil. This problem is ideally suited for a Fokker-Planck treatment (Sec. 6.9.1). If we ignore emission and absorption, photons are conserved, and the Fokker-Planck equation must have the form

Fokker-Planck formalism

$$\frac{\partial \eta_\gamma}{\partial t} + \frac{1}{\nu^2}\frac{\partial}{\partial \nu}\left(\nu^2 \mathcal{F}_\gamma\right) = 0, \tag{28.83}$$

T2

where the flux $\mathcal{F}_\gamma$ in frequency space depends on η_γ and its frequency derivative. When the electrons are very cold, each scattering produces a Compton recoil with average frequency change $\langle \Delta\nu \rangle = -h\nu^2/m_e$, and so $\mathcal{F}_\gamma = -(n_e\sigma_T h\nu^2/m_e)\eta_\gamma$. Now, our experience with the Fokker-Planck equation suggests that we add a term proportional to $\partial\eta_\gamma/\partial\nu$ to account for the heating of the photons by hot electrons with temperature T_e. However, $\mathcal{F}_\gamma$ must vanish when η_γ is Planckian with temperature T_e. A quick inspection shows that to deal with this, we must also add a term quadratic in η_γ:

Kompaneets equation

$$\mathcal{F}_\gamma = -\frac{n_e\sigma_T h\nu^2}{m_e}\left(\eta_\gamma + \eta_\gamma^2 + \frac{k_B T_e}{h}\frac{\partial \eta_\gamma}{\partial \nu}\right). \tag{28.84}$$

This $\mathcal{F}_\gamma$ vanishes when η_γ has the Bose-Einstein form $\{\exp[(h\nu - \mu)/(k_B T_e)] - 1\}^{-1}$, with μ the chemical potential. This is entirely reasonable, as the total photon number density is fixed in pure electron scattering and the equilibrium photon distribution function under these conditions has the Bose-Einstein form, not the Planck form (Sec. 4.4). The kinetic equation (28.83), adopting the flux $\mathcal{F}_\gamma$ of Eq. (28.84), is known as the *Kompaneets* equation (Kompaneets, 1957). The third term in parentheses in Eq. (28.84) describes the diffusion of photons in frequency space due to the Doppler shift $\Delta\nu \sim \nu(k_B T_e/m_e)^{1/2}$, leading to the familiar second-order frequency increase. The diffusion coefficient in frequency space, $D_{\nu\nu} = n_e\sigma_T \nu^2 k_B T_e/m_e$, can be calculated explicitly by averaging over all electron velocities, assuming a Maxwell-Boltzmann distribution function.

induced Compton scattering

The quadratic term in parentheses in Eq. (28.84) describes the *induced Compton effect,* in which the scattering rate between two photon states (from unprimed to primed) is $\propto \eta_\gamma(1+\eta'_\gamma)$. To lowest order, the product term is canceled by inverse scattering. However, when we make allowance for the electron recoil, there is a finite effect, which is important at low frequency. This term is also derivable from classical electrodynamics without recourse to quantum mechanics.[55]

frequency shift

Now return to what happens in a galaxy cluster. The radiation is still effectively isotropic, and time should be interpreted as path length through the cluster. The derivative term in $\mathcal{F}_\nu$ dominates at high T_e, and so the observed CMB relative temperature change measured at a fixed frequency is given by

$$\begin{aligned}\delta_{T\,SZ} &= \left(\frac{\partial \eta_\gamma}{\partial \ln T_\gamma}\right)_\nu^{-1}\left(\frac{k_B T_e \tau_T}{m_e \nu^2}\right)\left[\frac{\partial}{\partial\nu}\left(\nu^4\frac{\partial\eta_\gamma}{\partial\nu}\right)\right]_{T_\gamma} \\ &= \left(\frac{k_B T_e \tau_T}{m_e}\right)\left[\left(\frac{h\nu}{k_B T_\gamma}\right)\coth\left(\frac{h\nu}{2k_B T_\gamma}\right) - 4\right]. \end{aligned} \tag{28.85}$$

55. The first realization that *induced* Compton scattering could be important was by Kapitsa and Dirac (1933).

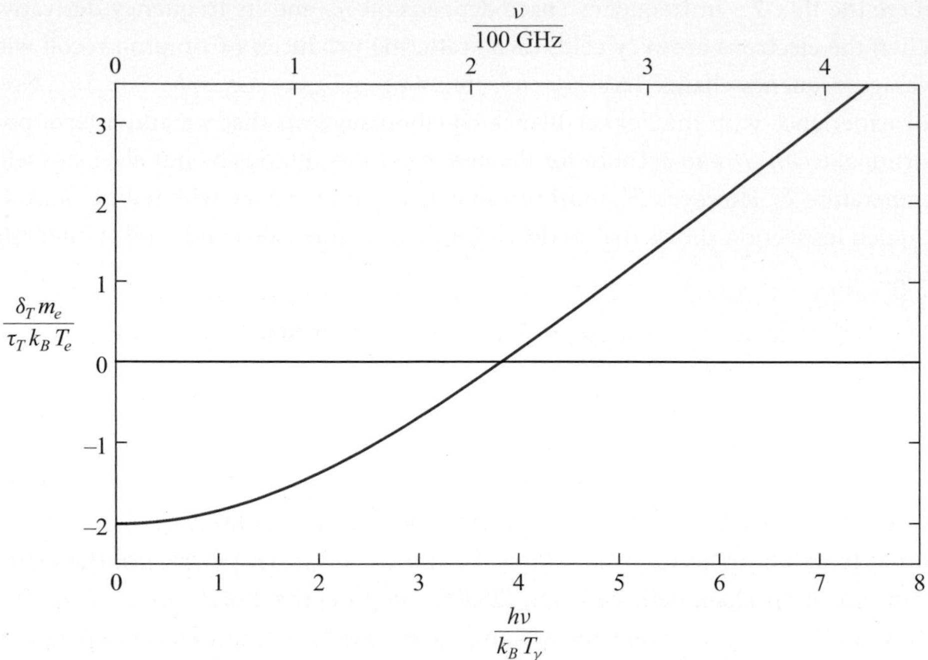

FIGURE 28.17 Sunyaev-Zel'dovich effect: Distortion of the CMB blackbody spectrum by passage of radiation through a cluster of galaxies, where the hot electrons Doppler boost (on average) the frequencies of individual photons. At frequencies well below the peak in the spectrum, the intensity decreases, while at high frequencies, it increases.

This is known as the *Sunyaev-Zel'dovich* (1970) *effect* (Fig. 28.17). In a typical cluster, the optical depth is $\tau_T \sim 0.1$ and $\delta_{T\ SZ} \sim 10^{-3}$. The photon temperature decreases by a fractional amount $\delta_T = -2k_B T_e \tau_T / m_e$ in the Rayleigh-Jeans part of the spectrum and increases in the Wien part. The crossover occurs at $\nu = 3.83 k_B T_\gamma / h = 217$ GHz, as observed.

EXERCISES

Exercise 28.23 *Example and Challenge: The Music of the Sphere* T2

We have hitherto focused on the statistical properties of the cosmological perturbations as probed by a variety of observations. However, we on Earth occupy a unique location in a specific realization of wave modes that we have argued are drawn from a specific set of waves with particular amplitudes and phases, despite these supposedly being drawn from a statistical distribution (in much the same way that different pieces of music are distinct despite having similar power spectra). It should not be too long before we can produce a 3-dimensional "map" of our universe out to recombination based on a suite of observations, presuming that our standard cosmology and theory for the evolution of perturbations is correct.

(a) Calculate the comoving radii of recombination, reionization, and the most distant galaxies and quasars.

(b) Suppose that we have noise-free CMB fluctuation maps in temperature and polarization up to spherical harmonic quantum number $l = 100$. How many numbers would be contributed by these measurements?

(c) How many numbers would we need to measure to describe the potential and the associated density perturbations out to the radius of recombination with comparable resolution?

(d) Would these modes still be in the linear regime today?

(e) *Challenge.* Explore some of the practical challenges of carrying out this program, paying attention to the investigations we have described in Sec. 28.6 and assuming them to be carried out over the whole sky.

28.7 Three Mysteries T2

The cosmology that we have described depends on the application of basic physical laws, mostly classical—including (among others) general relativity with a cosmological constant—in locales where the laws are not independently tested; and it also depends on the "simplest" assumptions, including flatness and an initial, scale-free spectrum of adiabatic fluctuations. In this cosmology's most elementary version, on which we have focused, what we observe follows from just four dimensionless parameters, which we can choose to be Ω_b, Ω_D, Ω_γ, and $\mathcal{Q}$, plus a single length/timescale which we can choose to be t_0; and we learn these parameters by fitting observations. In a more comprehensive version of this cosmology, one must also fit additional astrophysical parameters associated with reionization and galaxy formation.

We conclude this chapter with a brief summary of contemporary views on the more fundamental processes that presumably determine this cosmology's assumptions and parameters.

28.7.1 Inflation and the Origin of the Universe T2

KINEMATICS

Early in our discussion of cosmology, we introduced the horizon and emphasized that it was smaller and smaller when the universe was younger and younger, so less and less of the universe was in causal contact at early times (Fig. 28.6). However, the universe we observe today is nearly homogeneous and isotropic. In particular, large-scale, spatially coherent fluctuations in the CMB are seen at recombination, when the horizon was less than 1% of their size. We have argued that the universe began in thermodynamic equilibrium at a very high temperature and that key properties (such as the net number of baryons and dark matter particles per photon and the amplitude of the potential perturbations) were determined by specific, though unknown, local physical laws. If this is so, how can there be spacelike slices of simultaneity that are homogeneous on the largest scale? It is also surprising that the universe is as flat as

horizon problem

T2

it appears to be.[56] As was realized surprisingly early in the development of modern cosmology, these mysteries may have a common explanation called *inflation*.[57]

causal behavior

Under the inflationary scenario, it is proposed that the material of the universe we see today was initially in causal contact and was well mixed and therefore homogeneous. There followed an epoch of runaway expansion—inflation—when parts of the universe on different world lines from our own exited our horizon and lost contact with us. After this phase ended, these parts independently followed the evolution we described in Sec. 28.4 and, primed with the same features as us, their world lines then reentered our horizon, the last to leave being the first to return. "Hello, goodbye, hello!"

We have not demonstrated that the universe had such an origin, but all recent observations are consistent with the simplest version of inflation, and nothing that we have learned is in conflict with it. At the very least, the theory of inflation illustrates some fascinating features of general relativity, bringing out clearly the challenge that will have to be met by any rational description of the very early universe.[58]

We do not understand the physics that underlies inflation, but let us make a guess, inspired by observation. We have explained that the inferred initial amplitude of a potential perturbation is almost independent of its comoving length scale (scale invariance). We presume that each perturbation was laid down by local physics just prior to that scale exiting the horizon and argue that the physical conditions were therefore constant in time. The one homogeneous cosmology that has this character is a de Sitter expansion with $\rho = \text{const}$ and $a \propto e^{H_i t}$ for $t_i < t < t_h$, with a constant *inflation Hubble constant* H_i and where the subscript h denotes the end of inflation.

de Sitter expansion

It is commonly supposed that inflation began during the epoch of "grand unification" of the electroweak and strong forces, when $t \sim t_i \sim H_i^{-1} \sim 10^{-36}$ s. If we also assume that the universe was homogeneous on the scale of the horizon $\sim t_i$, and inflation went on for long enough to encompass our horizon, then $a_i \sim t_i/\chi_{0H} \sim 3 \times 10^{-55}$. However, we also know that $a_h \sim a_i \exp(H_i t_h)$; and if we denote the start of the particle era by $a_p \sim 10^{-11}$ when $t_p \sim 3$ ms, then we also have that $a_h \sim a_p (t_h/t_p)^{1/2}$. These two relations and the above numbers imply that $t_h \sim 64 t_i$, $a_h \sim 2 \times 10^{-27}$. In this simple example, 64 e-foldings of inflationary expansion suffice to explain the homogeneity of our universe today.

Furthermore, any significant curvature that may have been present when the universe started to inflate would have made a fractional contribution to the Friedmann equation (28.16) $\sim(a_i/a_h)^2 \sim 10^{-56}$ at t_h. This fractional contribution would then

56. There are additional quantum field theory puzzles, especially the apparent scarcity of topological defects like monopoles, strings, and domain walls that are addressed by this theory.
57. Pioneers of these ideas included Kazanas (1980), Starobinsky (1980), Guth (1981), Sato (1981), Albrecht and Steinhardt (1982), and Linde (1982).
58. We exclude "Just So" stories—for example, flood geology—that assert that the world began at a specific time in the relatively recent past with just the right initial conditions necessary to evolve to the world of today.

grow by a factor $\sim(a_{\rm eq}/a_h)^2 \sim 10^{46}$ by the end of the radiation era and by a further factor $\sim a_\Lambda/a_{\rm eq} \sim 3{,}000$ by the start of the cosmological era. In this example it is only $\sim 10^{-56} \times 10^{46} \times 3{,}000 \sim 3 \times 10^{-7}$ in the early cosmological age (today). However, with other plausible assumptions, it may just be detectable. Therefore, the observed flatness, which otherwise requires very careful fine tuning of the initial conditions, also has a natural explanation.

T2

CLASSICAL ELECTROMAGNETIC FIELD THEORY

The constant energy density that we have argued is needed to drive inflation is commonly associated with a classical scalar field, sometimes called the *inflaton.* To describe its properties necessitates a short digression into classical field theory (cf. Ex. 7.4).

inflaton field

Lagrangian methods were devised to solve problems in celestial mechanics and turned out to be useful for a larger class of classical problems (Goldstein, Poole, and Safko, 2002). To summarize the approach (with which the reader is presumed to be quite familiar), the coordinates $\mathbf{x}$ of all particles are replaced by a sufficient number of generalized coordinates $\mathbf{q}(\mathbf{x}, t)$—for example, three Euler angles for a spinning top—to describe the system. A scalar *lagrangian* $L(\mathbf{q}, \dot{\mathbf{q}}, t)$ is introduced as the difference of the kinetic and potential energies, and the system evolves so as to make the *action* $\int dt\, L$ stationary.[59] This implies the Euler-Lagrange equations $\partial_{\mathbf{q}} L = d(\partial_{\dot{\mathbf{q}}} L)/dt$, where $\partial_{\mathbf{q}}$ is shorthand notation for $\partial/\partial q^i$. A *hamiltonian,* $\partial_{\dot{\mathbf{q}}} L \cdot \dot{\mathbf{q}} - L$, is introduced, which equals the conserved energy if the system does not interact with its environment and evolve explicitly with time.

Lagrangian dynamics

This Lagrangian approach was generalized to describe the classical electromagnetic field[60] (where it is not very much used in practice; Jackson, 1999).[61] For this, three changes need to be made to the particle lagrangian. First, as we are dealing with a relativistic theory, we work in spacetime coordinates. Second, we use the lagrangian density $\mathcal{L}$, so that the action is $\int dt dV \mathcal{L}$; $\mathcal{L}$ must be a Lorentz-invariant scalar (cf. Sec. 2.12 and Ex. 7.4). Third, we treat the electromagnetic field itself as a generalized coordinate. However, instead of being an N-dimensional vector, we take the limit $N \to \infty$ and treat it as a continuous variable.

The natural choice of field coordinate is the 4-vector potential A^α constrained by the Lorenz gauge condition $\partial_\alpha A^\alpha = 0$. (We use indices here to avoid notational confusion, we use ∂_α for partial derivatives, $\partial_\alpha A^\alpha \equiv \partial A^\alpha/\partial x^\alpha$, and for simplicity we assume spacetime is flat.) For the free electromagnetic field, the only choice for the

59. In a phrase that captures the philosophical, political, theological, and literary context in which this revolutionary approach to physics was created, we live "in the best of all possible worlds."

60. Faraday first conceptualized a field description of electromagnetism and gravity in the 1820s, contrasting it with the Newtonian "action at a distance" and gradually developed this idea (Faraday, 1846). Although Maxwell and others sought a variational description of electromagnetism, it was the astronomer Karl Schwarzschild (1903) who first got it right, 2 years before the advent of special relativity.

61. It is, of course, indispensible for understanding quantum mechanics, quantum electrodynamics, and quantum field theory and is extremely useful for probing the fundamental character of general relativity.

T2

lagrangian density for electromagnetic field

lagrangian density (except for an additive or multiplicative constant) consistent with these requirements is $\mathcal{L}_{\rm EM} = \frac{1}{4\pi}\partial^{[\alpha}A^{\beta]}\partial_{[\beta}A_{\alpha]}$, in Gaussian units. Comparing with the particle lagrangian, we recognize this as kinetic energy–like with no potential energy–like contribution. Varying the action leads to the Euler-Lagrange equations:

$$\frac{\delta\mathcal{L}_{\rm EM}}{\delta A_\alpha} \equiv -\partial_\beta\left(\frac{\partial\mathcal{L}_{\rm EM}}{\partial(\partial_\beta A_\alpha)}\right) = 0 \Rightarrow F^{\alpha\beta}{}_{,\beta} = 0 \quad \text{where } F_{\alpha\beta} = 2\partial_{[\beta}A_{\alpha]}. \tag{28.86}$$

These are the free-field Maxwell equations.[62]

The electromagnetic stress-energy tensor is a natural generalization of the hamiltonian for particle dynamics:[63]

$$T_\beta{}^\alpha = \mathcal{L}_{\rm EM}\delta_\beta{}^\alpha - \left(\frac{\partial\mathcal{L}_{\rm EM}}{\partial(\partial_\alpha A_\gamma)}\right)\partial_\beta A_\gamma = \frac{1}{4\pi}\left(\partial^{[\gamma}A^{\delta]}\partial_{[\delta}A_{\gamma]}\delta^\alpha_\beta - 4\partial^{[\alpha}A^{\gamma]}\partial_{[\gamma}A_{\beta]}\right). \tag{28.87}$$

This agrees with the standard form, Eq. (2.75).

SCALAR FIELD THEORY

The observed isotropy of the universe suggests that the fundamental inflaton field we seek—henceforth designated as φ—is a real scalar and not a vector field (as in electromagnetism) or a tensor field (as in general relativity) or a complex or spinorial quantum field (like the Higgs field). In this section we set $G = c = 1$, so that φ is dimensionless and we deal with the real universe where spacetime is curved. The simplest form for the lagrangian density, by analogy with $\mathcal{L}_{\rm EM}$, is $\mathcal{L} = -\frac{1}{2}\vec{\nabla}\varphi\cdot\vec{\nabla}\varphi - V(\varphi)$ (where we no longer need to use indices, as the field is a scalar). The first term is the only invariant choice we have for the kinetic energy–like part (except that the $-\frac{1}{2}$ is a convention); the second term is the simplest potential energy–like part, which is absent for classical electromagnetism but necessary here. Continuing the analogy, the stress-energy tensor is given by

lagrangian density for scalar field

stress-energy tensor

$$\boldsymbol{T} = \mathcal{L}\boldsymbol{g} - \frac{\partial\mathcal{L}}{\partial(\vec{\nabla}\varphi)}\otimes\vec{\nabla}\varphi = \mathcal{L}\boldsymbol{g} + \vec{\nabla}\varphi\otimes\vec{\nabla}\varphi. \tag{28.88}$$

potential minimum

Now consider a harmonic potential $V = \frac{1}{2}\omega_h^2(\varphi - \varphi_h)^2$, in which φ_h is the (vacuum) value of the field, where the potential vanishes. The (Euler-Lagrange) field equation for our lagrangian density is $\nabla_\alpha\nabla^\alpha\varphi = dV/d\varphi$. If we seek a wave solution in a local Lorentz frame, then we recover the dispersion relation: $\omega^2 - k^2 = \omega_h^2$, where ω_h is Lorentz invariant. This describes a longitudinal scalar wave propagating in vacuo but has a similar dispersion relation to a transverse vector wave propagating in an unmagnetized plasma [cf. Eq. (21.24)]. Note that we can Lorentz transform into a frame

62. Including a current source requires the addition of an interaction lagrangian, which is straightforward but need not concern us here.

63. The formal justification of this heuristic argument hinges on the celebrated theorem (Noether, 1918) that relates conserved quantities to symmetry.

in which $\omega = \omega_h$ and $k = 0$, and the wave becomes a pure oscillation with no spatial gradients. In this frame, the field is directly analogous to a classical particle moving in a stationary potential well. In a general frame, the stress tensor for an individual wave is anisotropic and oscillatory.

T2

potential maximum

Next, consider a potential maximum $V = V_i - \frac{1}{2}\omega_i^2\varphi^2$, and transform into the frame where there are no spatial gradients. The equation of motion for the field is $\ddot{\varphi} - \omega_i^2\varphi = 0$. It is simplest to imagine the field as starting with $\varphi(0) = 0$ and a small positive velocity $\dot{\varphi}(0)$, so that $\dot{\varphi}(t) = \dot{\varphi}(0)\cosh\omega_i t$, which soon increases exponentially with time.

SLOW-ROLL INFLATION

We have now introduced all the ideas we need to design a potential that allows the universe to *slow roll* and inflate for ~64 *e*-foldings before transitioning classically to a decelerating expansion. Qualitatively, we require a potential maximum with small enough curvature for sufficient inflation to take place, joined to a potential minimum into which the field can settle and allow fundamental particles to take over the dynamics.[64]

model potential

A convenient and illustrative choice for the potential is $V = V_i[1 - (\varphi/\varphi_h)^2]^2$ (Fig. 28.18a). The two parameters V_i and φ_h measure the height and the width of the potential and ought to be derivable from basic physics. Let us work in the frame in which $\varphi = \varphi(t)$ with no spatial gradients, which will define and evolve into the sequence of homogeneous spatial hypersurfaces that we have been using. Using Eq. (28.88) and comparing with the perfect fluid stress-energy tensor, we can identify the energy density ρ and the pressure P:

density and pressure for field

$$\rho_\varphi = V + \frac{1}{2}\dot{\varphi}^2, \quad P_\varphi = -V + \frac{1}{2}\dot{\varphi}^2. \tag{28.89}$$

evolution of the field

If we ignore all other possible contributions and insert these expressions into the first law of thermodynamics [Eq. (28.19)], we obtain the cosmological scalar field equation:

$$\ddot{\varphi} + 3H\dot{\varphi} + dV/d\varphi = 0, \tag{28.90}$$

where $H^2 = 8\pi\rho_\varphi/3$. We set $V_i = 3H_i^2/8\pi = 1.6 \times 10^{98}\ \mathrm{J\,m^{-3}}$ and we choose $\varphi_h = 1.5$ and the value of the scalar field $\dot{\varphi}(0)$ so as to prolong inflation for 64 *e*-foldings and to match the expansion during the particle age (Fig. 28.18).

particle production

Now, this model thus far is seriously incomplete, because its evolution asymptotes to an empty, static universe with $H = 0$ and $a = \text{const}$. We have completely ignored the relativistic, fundamental particles that drive the post-inflationary expansion. These cannot be primordial, because their contribution would have inflated away; instead, they must have been generated at the end of inflation, specifically, as the field oscillates in the potential well. From a quantum mechanical perspective, this is quite reasonable,

64. In the original theory, this transition—called the *graceful exit*—was attributed, unsuccessfully, to quantum mechanical tunneling.

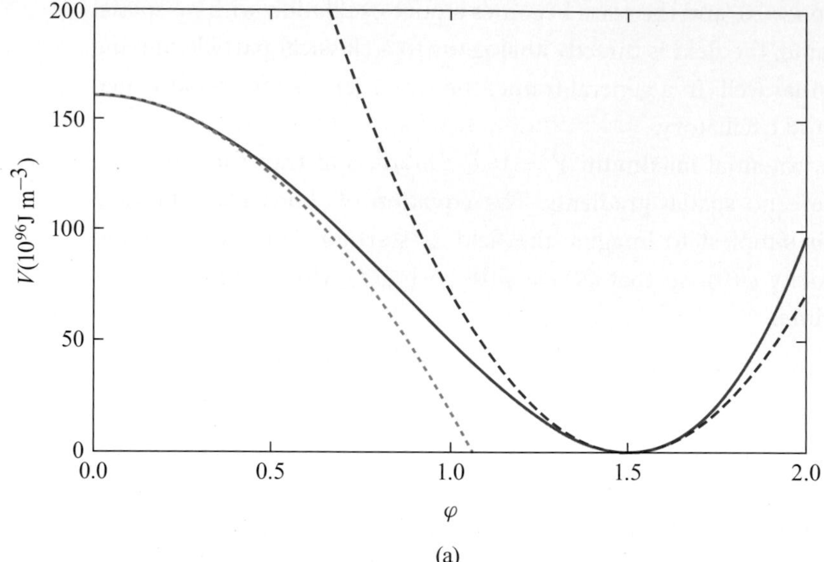

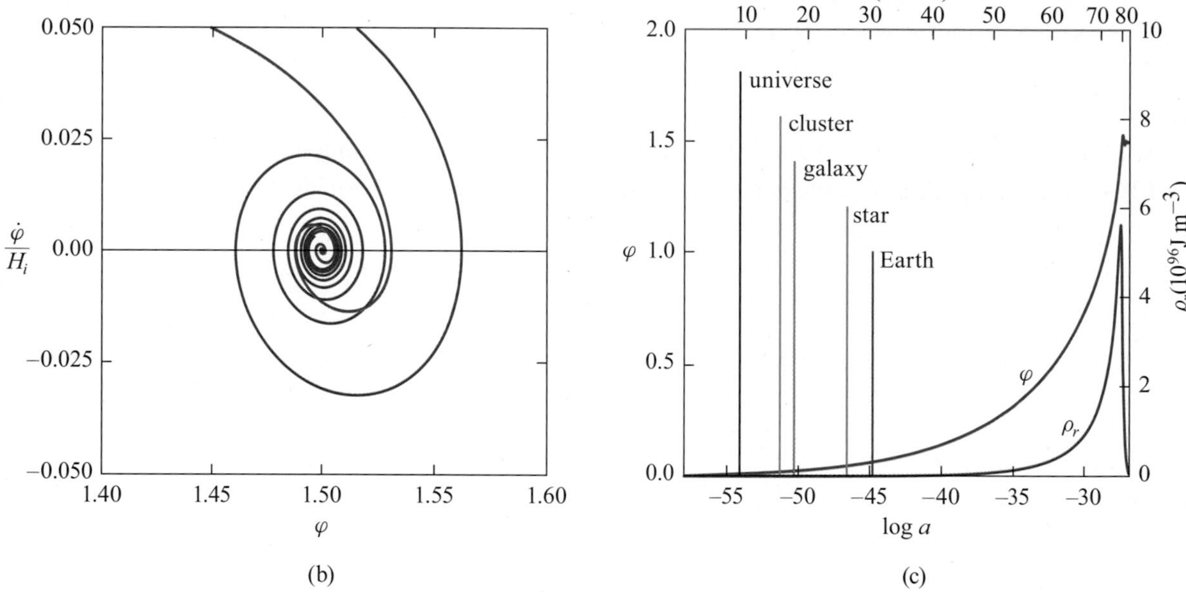

FIGURE 28.18 (a) The three types of potential considered in the text. The dashed line is a (shifted) potential minimum, the dotted line is a potential maximum, and the solid line is the simple model inflaton potential that joins these two solutions. (b) Variation of φ as it oscillates about the potential minimum exhibited in the φ–$\dot{\varphi}$ plane. The red curve is the free field variation; the blue curve includes the ad hoc particle production. (c) Variation of the field φ, the scale factor a, and the relativistic particle energy density ρ_r as a function of a. The rapid decline in ρ_r at late time is due to expansion. The times at which perturbations, on the comoving scale of various cosmic structures, leave the horizon are indicated by vertical lines.

T2

neutrino species. Using the current dark matter and neutrino densities from Table 28.1, we find that today $Z_D = 4 \times 10^{-12} m_{D,12}{}^{-1}$ (Table 28.1).

So, most of the dark matter particles must have vanished. The simplest explanation is that during binary collisions they annihilated to form less massive, relativistic particles when they became mildly relativistic, like electrons and positrons annihilated, consistent with thermodynamic equilibrium (cf. Sec 28.4.1). Eventually the annihilation rate fell below the Hubble expansion rate, and a small relict density was left over, out of equilibrium, constituting the dark matter we see today (Lee and Weinberg, 1977). Using the principles we established in our discussion of nucleosynthesis (Sec. 28.4.2), we can write down immediately a kinetic equation for the dark particles that balances annihilation with the reverse reactions:

relict density

$$\dot{Z}_D = n_\nu \langle \sigma_{DD} v_D \rangle \left(\frac{n_{\mathrm{Dth}}^2}{n_\nu^2} - Z_D^2 \right), \tag{28.92a}$$

$$\text{where} \quad n_{\mathrm{Dth}}(T, m_D) = 2 \int_0^\infty \frac{4\pi p^2 dp}{h^3} \frac{1}{e^{(m_D^2 + p^2)^{1/2}/(k_B T)} + 1} \tag{28.92b}$$

[cf. Eqs. (28.38)], and where $\langle \sigma_{DD} v_D \rangle$ is the annihilation rate, $n_{\mathrm{Dth}}(m_D, T)$ is the density the dark particles would have in thermal equilibrium, and the neutrino density is $n_\nu(T) = 3 n_{\mathrm{Dth}}(0, T)$. (The particle speeds v_D are nonrelativistic when they stop annihilating and typically, $\sigma_{DD} \propto v_D^{-1}$.) We assume that the dark particles are kept at the same temperature as the rest of the universe and that the expansion obeyed the same $t \propto a^2$ law as at the start of the photon age. Eq. (28.92a) is now easily solved numerically for different masses to derive the relation $\sigma_{DD}(m_D)$ needed to produce the contemporary value of Z_D. More simply, though less accurately, we note that $\langle \sigma_{DD} v_D \rangle \sim H/n_D \propto m_D a_D$ when $a \sim a_D$ and annihilation ceased. Guided by pair annihilation, we estimate that the associated temperature was $T_D \sim 0.1\, m_D/k_B \propto a_D^{-1}$. In order for this to be the origin of dark matter, we require that $\langle \sigma_{DD} v_D \rangle \sim 3 \times 10^{-32}\ \mathrm{m^3\ s^{-1}}$, which is also the result from the more careful calculation and is pretty insensitive to the more detailed assumptions and entirely consistent with the expectations from particle physics.

"WIMP miracle"

We do not understand the properties of the dark particle, and the choices we have just made look like they have been delivered by a committee! (Indeed, every single one of them can be negated and still lead to a viable explanation.) However, these choices do describe the most widely supported explanation for dark matter, namely, that it is the lightest supersymmetric particle.[72] *Supersymmetry* is a promising extension of the standard model of particle physics that postulates the existence of fermionic partners to the bosons of the standard model and vice versa. A major experimental program

supersymmetry

72. Alternatives that have been seriously considered and sought include axions (Peccei and Quinn, 1977) and sterile neutrinos (Pontecorvo, 1968).

T2

dark matter searches

under way at the Large Hadron Collider seeks evidence for it. Many *direct* searches are also being conducted in deep mines (to filter out the cosmic ray background) for very rare collisions between dark particles and atomic nuclei. Finally, the small rate of annihilations still going on today might lead to γ-rays and positrons that can be seen, indirectly, by astronomers. All three searches—below, on, and above ground—are under way or are being undertaken in 2016. What is interesting and encouraging is that the sensitivity attainable, in each case, is roughly compatible with the value of $\langle\sigma_{DD}v\rangle$ inferred on the basis of cosmology observations and theoretical calculations.

To date, despite exquisite experiments, no convincing evidence for dark particles has been found. Instead, significant constraints on their properties are being measured by all three techniques, and improvements in sensitivity should be forthcoming over the next several years. Only Nature knows whether we are now on the threshold of identifying most of the matter in the universe and exploring a second standard model of particle physics or if we must look elsewhere for an explanation, but the hunt is on.

BARYOGENESIS

The puzzle over baryons is why there are any of them at all! Protons have antiparticles with which they can annihilate with very large cross sections, so symmetry would seem to suggest that none should have been left over after the temperature fell far below $\sim m_p/k_B \sim 10^{13}$ K. The only way residual baryons could have been created is if some imbalance were created between baryons and antibaryons.[73] The protons (and neutrons, which decay into protons) derived from quarks and primordial asymmetry can be preserved through the quark phase of the universe by a conserved quantum number—the *baryon number*. Essentially, what we seek is some process occurring during the very early universe and involving a significant constituent that creates such an asymmetry. Any serious discussion of this topic involves highly speculative high-energy-physics considerations that we cannot go into here. However, there are essentially classical considerations that should underlie any proposed mechanism, and these we now discuss.

baryon number

Let us suppose that a particle—one with a distinct antiparticle—undergoes a decay that creates a net baryon number. Such a possibility is permitted—and indeed, suggested—by the existence of finite neutrino masses, though it has never been observed. The problem is that on quite general grounds, an antiparticle will create the opposite baryon number and so we have to explain why there is a difference between the amounts of matter and antimatter. Now, the laws of classical physics are tacitly assumed to be the same when the spatial coordinates are inverted: $\mathbf{x} \rightarrow -\mathbf{x}$. In the language of particle physics, we say that they are invariant under a *parity* (P) transformation. Likewise, the laws of classical electromagnetism are unchanged if we change

73. Equally effective would be a process that created and maintained an asymmetry between electrons and positrons, so that a baryon asymmetry would be needed to preserve charge neutrality. However, baryogenesis is thought to be more likely than *leptogenesis*.

T2

the signs of all the charges. This is invariance under a *charge* (C) transformation. Finally, the microscopic classical laws (e.g., those describing collisions of particles or the evolution of an electromagnetic field) are invariant under time reversal—a *time* (T) transformation. This last observation does not imply that all physical phenomena are reversible. As discussed in Sec. 4.7, the act of averaging (or "coarse graining") introduces the arrow of time, embodied in the second law of thermodynamics.

This need not have been the case and is not true in particle physics. Specifically, it was shown in the 1950s that β decay—a weak interaction—produces left-handed (not right-handed) neutrinos, which violates P symmetry. This was a surprise to many physicists. Likewise, if we were to change the signs of all charges in this decay, we would expect to produce a left-handed antineutrino, and this is not seen either, so C symmetry is violated as well. However, combined C and P transformations lead to the preservation of CP symmetry in β decay. This is important, because a fundamental theorem states that CPT symmetry must be respected, and so a violation of CP would imply a violation of T.

It therefore came as a second, even greater, shock when in the 1960s, experimental measurements of neutral K mesons showed that their decay into two channels that were CP equivalent occurred at different rates. CP symmetry is violated, and Nature can distinguish matter from antimatter and forward from backward in time. For the universe to have actually made this distinction also requires a departure from thermodynamic equilibrium in the past. If this did not happen, the particles and antiparticles would have had identical distribution functions with zero chemical potentials, and no net baryons would have been made. Now, thermodynamic equilibrium would not have been maintained had the expansion been too fast for the particle reactions (cf. Secs. 28.4.1 and 28.4.2). An alternative possibility is that the particles did not scatter, so that their momenta evolved according to $p \propto a^{-1}$ (cf. Sec. 28.2.3). This automatically leads to a nonthermal distribution function[74] and allows baryon asymmetry to proceed. We are probably a long way from understanding the detailed particle physics of baryogenesis, but determining that its general requirements could plausibly have been satisfied is a good first step.

EXERCISES

Exercise 28.25 *Challenge: Including More Details* T2

We have made many simplifying assumptions in this chapter to demonstrate the strong connection to the principles and techniques developed in the preceding 27 chapters. It is possible to improve on our standard cosmological model by being more careful without introducing anything fundamentally new. (The research frontier, of course, is advancing fast and contains much we have not attempted to explain.) Consider how to implement the following corrections to standard cosmology that

74. That all three violations—baryon number, C/CP symmetry, and thermal equilibrium—were necessary was first recognized by Sakharov (1965).

T2

are mandated by observation, and then repeat the calculations with these changes, comparing with the research literature.

(a) Changing the slope of the initial cosmic noise spectrum (Sec. 28.5.3).

(b) Including more spherical harmonics in the description of the radiation field and using the Boltzmann equation instead of a Monte Carlo simulation (Sec. 28.6.1).

(c) Including tensor perturbations in the Robertson-Walker metric (Secs. 28.5.1, 28.5.3, and 28.7.1).

(d) Accounting for neutrino mass (Sec. 28.3.4).

Exercise 28.26 *Challenge: Testing Standard Cosmology* T2

There are many elaborations of standard cosmology either involving new features following from known physics or involving new physics. While no convincing evidence exists for any of them as of this writing, they are all being actively sought. Explain how to generalize standard cosmology to accommodate these possibilities and to test for them, repeating calculations, where possible.

(a) Space curvature (Sec. 28.2.2).

(b) Dark energy. Adopt an empirical equation of state with the parameter w being constant and slightly greater than -1 (Sec. 28.7.3).

(c) Additional neutrino flavors, including *sterile* neutrinos (Secs. 28.4.1 and 28.4.2).

(d) Nonadiabatic and non-Gaussian initial fluctuations (Sec. 28.5.3). [Hint: Consulting an advanced textbook is recommended.]

28.7.3

28.7.3 The Cosmological Constant and the Fate of the Universe T2

Einstein's cosmological constant, Λ

We have explained (Sec. 28.3.5) that the observed acceleration of the universe requires an effective negative pressure in the stress-energy tensor, and how such a possibility was anticipated by Einstein (1917). Furthermore, we have explained that Einstein's cosmological constant is consistent with the data. This is important, because it is a simple prescription whose consequences can be computed without too many extraneous assumptions and is therefore readily falsifiable. However, there are also theoretical reasons for questioning this interpretation; they, too, lie at the interface between quantum and classical physics.

The earliest view of the cosmological constant (implicitly, that of Einstein) was that its presence in the field equations is an expression of the true law of gravitation necessary on a large scale, in an analogous fashion to Newton proposing the inverse square law for planets and expanding our understanding of gravitational force beyond what is needed on the surface of Earth. The ultimate connection between the general relativity of, say, stellar-mass black holes and that of the universe at large has yet to be divined, but it should exist independent of the messy details of our cosmic environment. Many interesting ideas have been proposed, in particular those involving

T2

extra spatial dimensions and oscillating universes. There have even been proposals to dispense with dark matter particles altogether and to interpret the observed motions of stars and galaxies in terms of modified Newtonian gravity or entanglement.

Today it is more common to view the field equations (Sec. 25.8) as providing a complete framework for describing gravitation and the cosmological constant as being just one of several contributions to the stress-energy tensor. This is the approach we followed in Sec. 28.3.5. Its most natural identification is with the quantum mechanical vacuum. Having $\boldsymbol{T}_\Lambda \propto \boldsymbol{g}$ ensures that the vacuum looks the same in all local Lorentz frames, consistent with the principle of relativity. However, attempts to develop this relationship quantitatively have mostly foundered on the tiny size of Λ relative to the Planck scale of quantum gravity[75]—$G^2\hbar\rho_\Lambda \sim 10^{-122}$. Either this represents an unprecedented degree of fine tuning, or the cosmological constant has even less to do with quantum mechanics than, say, oceanography, and it is instead some ungrasped expression of the "fabric" of spacetime on a supraclassical scale.[76] If so, it throws down a challenge to the *reductionist* view of physics, under which physics at essentially all scales (especially classical physics) is viewed as being derivative of physics at the smallest lengthscales and highest energies. The properties of materials depend on the behavior of atoms and molecules, which depend on electrons and nuclei, which depend on quantum electrodynamics and the interactions of quarks, and so on. This does not preclude the existence, interest, or importance of *emergent* phenomena (e.g., ferromagnetism, shock fronts, or astrophysical black holes) that require appealing to the properties of matter in bulk, but it is a statement of faith that ultimately, the governing principles of these phenomena are reducible to the physics of the smallest scales, even if this is not useful in describing what we observe. In this sense Λ might be like ferromagnetism.

Λ as a contribution to the stress-energy tensor

Λ as the vacuum

reductionist view of physics

Λ as emergent phenomena

However, the square peg of quantum field theory need not be forced into the round hole of Riemannian geometry. The very existence of something like a cosmological constant should also cause us to inquire whether some physics is derivative of the largest scales and lowest energies instead of the smallest scales and the highest energies. Consider, allegorically, a bug in a still pond, living its low-Reynolds-number life. A fish swims by, and the bug finds itself in the fish's turbulent wake. It observes that its food is moving more rapidly with an average speed that increases as the cube root of its distance [Eq. (15.23)] and, if it is very sensitive, it might find the water

75. A more intriguing possibility is that the natural, quantum mechanical mass scale for Λ is $\sim(\hbar^3\rho_\Lambda/c^3)^{1/4} \sim 10$ meV, similar to the neutrino mass scale. However, there is no known good reason for this.

76. It is interesting that the cosmological constant was taken very seriously before the 1980s (e.g., Lemaître, 1934; Tolman, 1934; Bondi, 1952; Zel'dovich, 1968) because it allowed the universe to be older than its contents and because it provided a scale distinct from those associated with atoms, nuclei, and elementary particles (e.g., Eddington, 1933). However, the challenge of reconciling it with quantum field theory then led to its near abandonment. The subsequent discovery of cosmic acceleration therefore came as a surprise to many physicists.

T2

heating up a little. While it is true that the Navier-Stokes equations can be derived by suitable averaging of a kinetic theory, if the bug desires to reconcile its observations with their causes, it is to the outer scale of a turbulent spectrum that it should turn, not the properties of water molecules. The speculations involving the anthropic principle and the multiverse contain some of this spirit. Perhaps Λ is "situational"—our first glimpse of physics on the large scale—just like blackbody radiation and the photoelectric effect opened the door to the physics of small scales and quantum mechanics more than a century ago. Of course, this physics, dependent on conditions at and beyond our current horizon, might reflect what happened immediately prior to our observable universe leaving the horizon during inflation; alternatively, it might reflect the accidental properties of our contemporary cosmological environment.

Λ as a "situational" phenomenon

A more pragmatic approach is to consider generalizations of the cosmological constant. These include *dark energy* (or *quintessence*; Perlmutter et al., 1999). In particular, the behavior of the stress-energy tensor has been parameterized by introducing an equation of state $P_\Lambda = w\rho_\Lambda$, where w is negative, although there is no compelling theoretical reason for doing this. Several observational studies have concluded that $|w+1| \lesssim 0.08$, and more accuracy is in the works. Another common approach is to invoke classical field theory (just as we did for inflation) and to use the same formalism to develop a description of contemporary inflation. It is tempting to suppose that there have been many instances of inflation over the history of the universe, and that Nature has managed to find graceful exits from every one of these expansions in the past and, in the fullness of time, it will do so again. However, there is no requirement that this will ever happen. Another way to connect to inflation is to suppose that the bottom of the inflaton potential well is associated with some sort of quantum mechanical zero-point energy or a classical offset, and that this is what the universe is experiencing now.

dark energy

A particularly interesting outcome would be if future observations demonstrated that $-P_\Lambda > \rho_\Lambda > 0$; for example, if $w < -1$.[77] This condition corresponds to negative cosmological enthalpy and negative kinetic energy for a scalar field (cf. Sec. 28.7.1). Not only would it exclude a simple scalar field; it would also, if taken literally, predict an unusual fate for the universe. If $w =$ const, then the first law of thermodynamics implies that eventually $\rho \propto a^{-3(w+1)}$. The energy density increases as the universe expands if $w < -1$; the universe reaches infinite size in finite time, while the horizon shrinks and closer and closer neighbors disappear, a behavior dubbed "the big rip." Despite its eschatological fascination, many cosmologists reject this outcome on grounds of physics inconsistency and regard $w \geq -1$ as a prediction.

big rip

As with the experimental searches for dark matter, the prospects for learning more about the expansion and fate of the universe from astronomical studies over the next decade are bright, and the observations will presumably continue to corral speculative

future measurements

77. This is sometimes known as a violation of the *weak energy condition*.

interpretation. At present, progress in cosmology seems to be following that in the standard model of particle physics, where a relatively simple physical model suffices to account for essentially all data pertaining to questions addressed by the model, while leaving other questions to be confronted by new physics and future experiments. Three familiar constituents, baryons, neutrinos, and photons—supplemented by dark matter, zero curvature, and a cosmological constant and imprinted with an almost scale-free spectrum of adiabatic, Gaussian potential perturbations—suffice to account for most of what we measure. It will be fascinating to see whether this apparent simplicity is maintained through the next phase of cosmological exploration, when the true nature of inflation, dark matter, and the cosmological constant will likely be a focus of attention.

T2

Bibliographic Note

The literature on cosmology is enormous, and it is not hard to find excellent texts and research papers to elaborate on the many topics we have touched on in this chapter. Arguably, the most useful advanced text is Weinberg (2008). Other books that are helpful on the physics but are rather out of date observationally are Padmanabhan (1993); Peebles (1993); Kolb and Turner (1994); Peacock (1999); Dodelson (2003); and Hobson, Efstathiou, and Lasenby (2006). Texts that emphasize the early universe include Liddle and Lyth (2000) and Mukhanov (2005). An especially lucid and up-to-date textbook accessible to undergraduates that emphasizes observational cosmology is Schneider (2015). Excellent discussions of cosmology from a more elementary standpoint include Ryden (2002) and Hartle (2003). Among the most important contemporary observations are the CMB results from the Planck (Planck Collaboration, 2016a, 2016b) and Wilkinson Microwave Anisotropy Probe (WMAP) (Komatsu et al., 2011) satellites, one of several careful analyses of weak-lensing observations (Heymans et al., 2012), and the preliminary results on galaxy clustering from the Baryon Oscillation Spectroscopic Survey (BOSS) galaxy survey (Anderson et al., 2014).

Special Relativity: Geometric Viewpoint T2

(Chapter 2 of *Modern Classical Physics*)

Henceforth space by itself, and time by itself, are doomed to fade away into mere shadows, and only a kind of union of the two will preserve an independent reality.

HERMANN MINKOWSKI, 1908

2.1 Overview

2.1

This chapter is a fairly complete introduction to special relativity at an intermediate level. We extend the geometric viewpoint, developed in Chap. 1 for Newtonian physics, to the domain of special relativity; and we extend the tools of differential geometry, developed in Chap. 1 for the arena of Newtonian physics (3-dimensional Euclidean space) to that of special relativity (4-dimensional Minkowski spacetime).

We begin in Sec. 2.2 by defining inertial (Lorentz) reference frames and then introducing fundamental, geometric, reference-frame-independent concepts: events, 4-vectors, and the invariant interval between events. Then in Sec. 2.3, we develop the basic concepts of tensor algebra in Minkowski spacetime (tensors, the metric tensor, the inner product and tensor product, and contraction), patterning our development on the corresponding concepts in Euclidean space. In Sec. 2.4, we illustrate our tensor-algebra tools by using them to describe—without any coordinate system or reference frame—the kinematics (world lines, 4-velocities, 4-momenta) of point particles that move through Minkowski spacetime. The particles are allowed to collide with one another and be accelerated by an electromagnetic field. In Sec. 2.5, we introduce components of vectors and tensors in an inertial reference frame and rewrite our frame-independent equations in slot-naming index notation; then in Sec. 2.6, we use these extended tensorial tools to restudy the motions, collisions, and electromagnetic accelerations of particles. In Sec. 2.7, we discuss Lorentz transformations in Minkowski spacetime, and in Sec. 2.8, we develop spacetime diagrams and use them to study length contraction, time dilation, and simultaneity breakdown. In Sec. 2.9, we illustrate the tools we have developed by asking whether the laws of physics permit a highly advanced civilization to build time machines for traveling backward in time as well as forward. In Sec. 2.10, we introduce directional derivatives, gradients, and the Levi-Civita tensor in Minkowski spacetime, and in Sec. 2.11, we use these tools to discuss Maxwell's equations and the geometric nature of electric and magnetic fields.

T2

BOX 2.1. READERS' GUIDE

- Parts II (Statistical Physics), III (Optics), IV (Elasticity), V (Fluid Dynamics), and VI (Plasma Physics) of this book deal almost entirely with Newtonian physics; only a few sections and exercises are relativistic. Readers who are inclined to skip those relativistic items (which are all labeled Track Two) can skip this chapter and then return to it just before embarking on Part VII (General Relativity). Accordingly, this chapter is Track Two for readers of Parts II–VI and Track One for readers of Part VII—and in this spirit we label it Track Two.
- More specifically, this chapter is a prerequisite for the following: sections on relativistic kinetic theory in Chap. 3, Sec. 13.8 on relativistic fluid dynamics, Ex. 17.9 on relativistic shocks in fluids, many comments in Parts II–VI about relativistic effects and connections between Newtonian physics and relativistic physics, and all of Part VII (General Relativity).
- We recommend that those readers for whom relativity is relevant—and who already have a strong understanding of special relativity—not skip this chapter entirely. Instead, we suggest they browse it, especially Secs. 2.2–2.4, 2.8, and 2.11–2.13, to make sure they understand this book's geometric viewpoint and to ensure their familiarity with such concepts as the stress-energy tensor that they might not have met previously.

In Sec. 2.12, we develop our final set of geometric tools: volume elements and the integration of tensors over spacetime; finally, in Sec. 2.13, we use these tools to define the stress-energy tensor and to formulate very general versions of the conservation of 4-momentum.

2.2

2.2 Foundational Concepts

2.2.1

2.2.1 Inertial Frames, Inertial Coordinates, Events, Vectors, and Spacetime Diagrams

Because the nature and geometry of Minkowski spacetime are far less obvious intuitively than those of Euclidean 3-space, we need a crutch in our development of the geometric viewpoint for physics in spacetime. That crutch will be inertial reference frames.

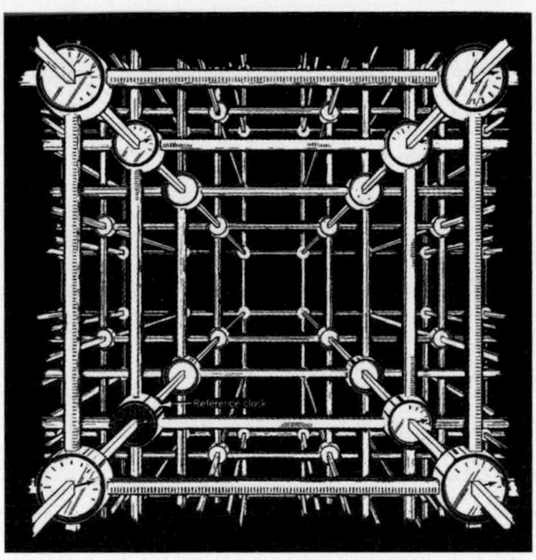

FIGURE 2.1 An inertial reference frame. From Taylor and Wheeler (1966). Used with permission of E. F. Taylor and the estate of J. A. Wheeler.

inertial reference frame

An inertial reference frame is a 3-dimensional latticework of measuring rods and clocks (Fig. 2.1) with the following properties:

- The latticework is purely conceptual and has arbitrarily small mass, so it does not gravitate.
- The latticework moves freely through spacetime (i.e., no forces act on it) and is attached to gyroscopes, so it is inertially nonrotating.
- The measuring rods form an orthogonal lattice, and the length intervals marked on them are uniform when compared to, for example, the wavelength of light emitted by some standard type of atom or molecule. Therefore, the rods form an orthonormal Cartesian coordinate system with the coordinate x measured along one axis, y along another, and z along the third.
- The clocks are densely packed throughout the latticework so that, ideally, there is a separate clock at every lattice point.
- The clocks tick uniformly when compared to the period of the light emitted by some standard type of atom or molecule (i.e., they are *ideal clocks*).
- The clocks are synchronized by the Einstein synchronization process: if a pulse of light, emitted by one of the clocks, bounces off a mirror attached to another and then returns, the time of bounce t_b, as measured by the clock that does the bouncing, is the average of the times of emission and reception, as measured by the emitting and receiving clock: $t_b = \frac{1}{2}(t_e + t_r)$.[1]

ideal clocks and their synchronization

1. For a deeper discussion of the nature of ideal clocks and ideal measuring rods see, for example, Misner, Thorne, and Wheeler (1973, pp. 23–29 and 395–399).

T2

(That inertial frames with these properties can exist, when gravity is unimportant, is an empirical fact; it tells us that, in the absence of gravity, spacetime is truly Minkowski.)

event

Our first fundamental, frame-independent relativistic concept is the *event*. An event is a precise location in space at a precise moment of time—a precise location (or *point*) in 4-dimensional spacetime. We sometimes denote events by capital script letters, such as $\mathcal{P}$ and $\mathcal{Q}$—the same notation used for points in Euclidean 3-space.

4-vector

A *4-vector* (also often referred to as a *vector in spacetime* or just a *vector*) is a straight[2] arrow $\Delta\vec{x}$ reaching from one event $\mathcal{P}$ to another, $\mathcal{Q}$. We often deal with 4-vectors and ordinary (3-space) vectors simultaneously, so we shall use different notations for them: boldface Roman font for 3-vectors, $\Delta\mathbf{x}$, and arrowed italic font for 4-vectors, $\Delta\vec{x}$. Sometimes we identify an event $\mathcal{P}$ in spacetime by its vectorial separation $\vec{x}_{\mathcal{P}}$ from some arbitrarily chosen event in spacetime, the origin $\mathcal{O}$.

An inertial reference frame provides us with a coordinate system for spacetime. The coordinates $(x^0, x^1, x^2, x^3) = (t, x, y, z)$ that it associates with an event $\mathcal{P}$ are $\mathcal{P}$'s location (x, y, z) in the frame's latticework of measuring rods and the time t of $\mathcal{P}$ *as measured by the clock that sits in the lattice at the event's location*. (Many apparent paradoxes in special relativity result from failing to remember that the time t of an event is always measured by a clock that resides at the event—never by clocks that reside elsewhere in spacetime.)

spacetime diagrams

It is useful to depict events on *spacetime diagrams*, in which the time coordinate $t = x^0$ of some inertial frame is plotted upward; two of the frame's three spatial coordinates, $x = x^1$ and $y = x^2$, are plotted horizontally; and the third coordinate $z = x^3$ is omitted. Figure 2.2 is an example. Two events $\mathcal{P}$ and $\mathcal{Q}$ are shown there, along with their vectorial separations $\vec{x}_{\mathcal{P}}$ and $\vec{x}_{\mathcal{Q}}$ from the origin and the vector $\Delta\vec{x} = \vec{x}_{\mathcal{Q}} - \vec{x}_{\mathcal{P}}$ that separates them from each other. The coordinates of $\mathcal{P}$ and $\mathcal{Q}$, which are the same as the components of $\vec{x}_{\mathcal{P}}$ and $\vec{x}_{\mathcal{Q}}$ in this coordinate system, are $(t_{\mathcal{P}}, x_{\mathcal{P}}, y_{\mathcal{P}}, z_{\mathcal{P}})$ and $(t_{\mathcal{Q}}, x_{\mathcal{Q}}, y_{\mathcal{Q}}, z_{\mathcal{Q}})$. Correspondingly, the components of $\Delta\vec{x}$ are

$$\Delta x^0 = \Delta t = t_{\mathcal{Q}} - t_{\mathcal{P}}, \qquad \Delta x^1 = \Delta x = x_{\mathcal{Q}} - x_{\mathcal{P}},$$
$$\Delta x^2 = \Delta y = y_{\mathcal{Q}} - y_{\mathcal{P}}, \qquad \Delta x^3 = \Delta z = z_{\mathcal{Q}} - z_{\mathcal{P}}. \tag{2.1}$$

We denote these components of $\Delta\vec{x}$ more compactly by Δx^α, where the index α and all other lowercased Greek indices range from 0 (for t) to 3 (for z).

When the physics or geometry of a situation being studied suggests some preferred inertial frame (e.g., the frame in which some piece of experimental apparatus is at rest), then we typically use as axes for our spacetime diagrams the coordinates of that preferred frame. By contrast, when our situation provides no preferred inertial frame, or when we wish to emphasize a frame-independent viewpoint, we use as axes

2. By "straight" we mean that in any inertial reference frame, the coordinates along $\Delta\vec{x}$ are linear functions of one another.

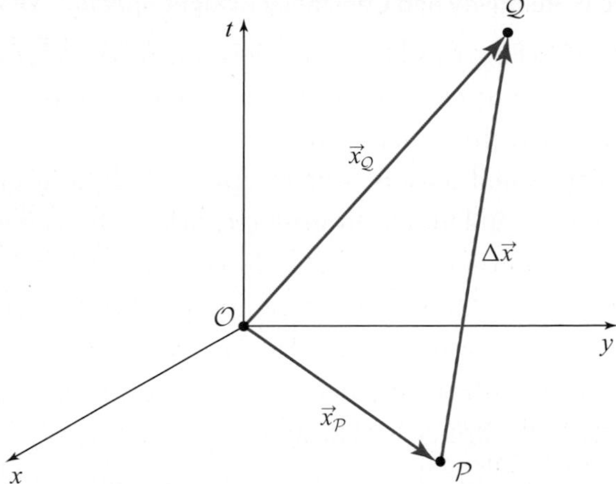

FIGURE 2.2 A spacetime diagram depicting two events $\mathcal{P}$ and $\mathcal{Q}$, their vectorial separations $\vec{x}_{\mathcal{P}}$ and $\vec{x}_{\mathcal{Q}}$ from an (arbitrarily chosen) origin $\mathcal{O}$, and the vector $\Delta\vec{x} = \vec{x}_{\mathcal{Q}} - \vec{x}_{\mathcal{P}}$ connecting them. The laws of physics cannot involve the arbitrary origin; we introduce it only as a conceptual aid.

the coordinates of a completely arbitrary inertial frame and think of the diagram as depicting spacetime in a coordinate-independent, frame-independent way.

inertial coordinates (Lorentz coordinates)

We use the terms *inertial coordinate system* and *Lorentz coordinate system* interchangeably[3] to mean the coordinate system (t, x, y, z) provided by an inertial frame; we also use the term *Lorentz frame* interchangeably with *inertial frame*. A physicist or other intelligent being who resides in a Lorentz frame and makes measurements using its latticework of rods and clocks will be called an *observer.*

observer

Although events are often described by their coordinates in a Lorentz reference frame, and 4-vectors by their components (coordinate differences), it should be obvious that the concepts of an event and a 4-vector need not rely on any coordinate system whatsoever for their definitions. For example, the event $\mathcal{P}$ of the birth of Isaac Newton and the event $\mathcal{Q}$ of the birth of Albert Einstein are readily identified without coordinates. They can be regarded as points in spacetime, and their separation vector is the straight arrow reaching through spacetime from $\mathcal{P}$ to $\mathcal{Q}$. Different observers in different inertial frames will attribute different coordinates to each birth and different components to the births' vectorial separation, but all observers can agree that they are talking about the same events $\mathcal{P}$ and $\mathcal{Q}$ in spacetime and the same separation vector $\Delta\vec{x}$. In this sense, $\mathcal{P}$, $\mathcal{Q}$, and $\Delta\vec{x}$ are *frame-independent, geometric objects* (points and arrows) that reside in spacetime.

3. It was Lorentz (1904) who first wrote down the relationship of one such coordinate system to another: the Lorentz transformation.

T2 2.2.2

2.2.2 The Principle of Relativity and Constancy of Light Speed

Principle of Relativity

Einstein's Principle of Relativity, stated in modern form, says that *Every (special relativistic) law of physics must be expressible as a geometric, frame-independent relationship among geometric, frame-independent objects* (i.e., such objects as points in spacetime and 4-vectors and tensors, which represent physical quantities, such as events, particle momenta, and the electromagnetic field). This is nothing but our Geometric Principle for physical laws (Chap. 1), lifted from the Euclidean-space arena of Newtonian physics to the Minkowski-spacetime arena of special relativity.

Since the laws are all geometric (i.e., unrelated to any reference frame or coordinate system), they can't distinguish one inertial reference frame from any other. This leads to an alternative form of the Principle of Relativity (one commonly used in elementary textbooks and equivalent to the above): *All the (special relativistic) laws of physics are the same in every inertial reference frame everywhere in spacetime.* This, in fact, is Einstein's own version of his Principle of Relativity; only in the sixty years since his death have we physicists reexpressed it in geometric language.

Because inertial reference frames are related to one another by Lorentz transformations (Sec. 2.7), we can restate Einstein's version of this Principle as *All the (special relativistic) laws of physics are Lorentz invariant.*

A more operational version of this Principle is: Give identical instructions for a specific physics experiment to two different observers in two different inertial reference frames at the same or different locations in Minkowski (i.e., gravity-free) spacetime. The experiment must be self-contained; that is, it must not involve observations of the external universe's properties (the "environment"). For example, an unacceptable experiment would be a measurement of the anisotropy of the universe's cosmic microwave radiation and a computation therefrom of the observer's velocity relative to the radiation's mean rest frame; such an experiment studies the universal environment, not the fundamental laws of physics. An acceptable experiment would be a measurement of the speed of light using the rods and clocks of the observer's own frame, or a measurement of cross sections for elementary particle reactions using particles moving in the reference frame's laboratory. The Principle of Relativity says that in these or any other similarly self-contained experiments, the two observers in their two different inertial frames must obtain identical experimental results—to within the accuracy of their experimental techniques. Since the experimental results are governed by the (nongravitational) laws of physics, this is equivalent to the statement that all physical laws are the same in the two inertial frames.

constancy of light speed

Perhaps the most central of special relativistic laws is the one stating that *The speed of light c in vacuum is frame independent;* that is, it is a constant, independent of the inertial reference frame in which it is measured. In other words, there is no "aether" that supports light's vibrations and in the process influences its speed—a remarkable fact that came as a great experimental surprise to physicists at the end of the nineteenth century.

The constancy of the speed of light, in fact, is built into Maxwell's equations. For these equations to be frame independent, the speed of light, which appears in them, must be frame independent. In this sense, the constancy of the speed of light follows from the Principle of Relativity; it is not an independent postulate. This is illustrated in Box 2.2.

T2

BOX 2.2. MEASURING THE SPEED OF LIGHT WITHOUT LIGHT

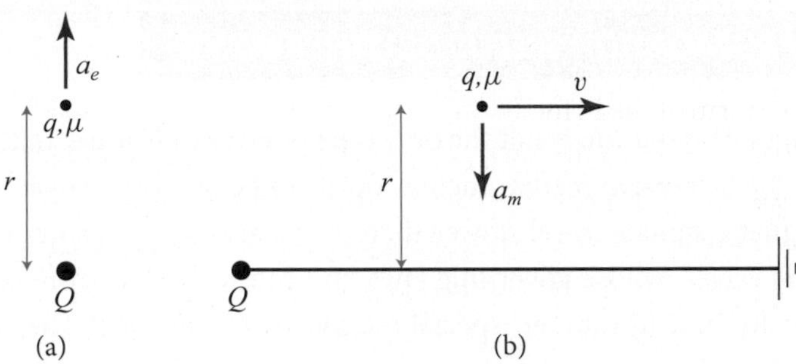

In some inertial reference frame, we perform two thought experiments using two particles, one with a large charge Q; the other, a test particle, with a much smaller charge q and mass μ. In the first experiment, we place the two particles at rest, separated by a distance $|\Delta x| \equiv r$, and measure the electrical repulsive acceleration a_e of q (panel a in the diagram). In Gaussian units (where the speed of light shows up explicitly instead of via $\epsilon_o \mu_o = 1/c^2$), the acceleration is $a_e = qQ/r^2\mu$. In the second experiment, we connect Q to ground by a long wire, and we place q at the distance $|\Delta x| = r$ from the wire and set it moving at speed v parallel to the wire. The charge Q flows down the wire with an e-folding time τ, so the current is $I = dQ/d\tau = (Q/\tau)e^{-t/\tau}$. At early times $0 < t \ll \tau$, this current $I = Q/\tau$ produces a solenoidal magnetic field at q with field strength $B = (2/cr)(Q/\tau)$, and this field exerts a magnetic force on q, giving it an acceleration $a_m = q(v/c)B/\mu = 2vqQ/c^2\tau r\mu$. The ratio of the electric acceleration in the first experiment to the magnetic acceleration in the second experiment is $a_e/a_m = c^2\tau/2rv$. Therefore, we can measure the speed of light c in our chosen inertial frame by performing this pair of experiments; carefully measuring the separation r, speed v, current Q/τ, and accelerations; and then simply computing $c = \sqrt{(2rv/\tau)(a_e/a_m)}$. The Principle of Relativity insists that the result of this pair of experiments should be independent of the inertial frame in which they are performed. Therefore, the speed of light c that appears in Maxwell's equations must be frame independent. In this sense, the constancy of the speed of light follows from the Principle of Relativity as applied to Maxwell's equations.

T2

What makes light so special? What about the propagation speeds of other types of waves? Are they or should they be the same as light's speed? For a digression on this topic, see Box 2.3.

The constancy of the speed of light underlies our ability to use the geometrized units introduced in Sec. 1.10. Any reader who has not studied that section should do so now. We use geometrized units throughout this chapter (and also throughout this book) when working with relativistic physics.

BOX 2.3. PROPAGATION SPEEDS OF OTHER WAVES

Electromagnetic radiation is not the only type of wave in Nature. In this book, we encounter dispersive media, such as optical fibers and plasmas, where electromagnetic signals travel slower than c. We also analyze sound waves and seismic waves, whose governing laws do not involve electromagnetism at all. How do these fit into our special relativistic framework? The answer is simple. Each of these waves involves an underlying medium that is at rest in one particular frame (not necessarily inertial), and the velocity at which the wave's information propagates (the group velocity) is most simply calculated in this frame *from the wave's and medium's fundamental laws.* We can then use the kinematic rules of Lorentz transformations to compute the velocity in another frame. However, if we had chosen to compute the wave speed in the second frame directly, using the same fundamental laws, we would have gotten the same answer, albeit perhaps with greater effort. All waves are in full compliance with the Principle of Relativity. What is special about vacuum electromagnetic waves and, by extension, photons, is that no medium (or "aether," as it used to be called) is needed for them to propagate. Their speed is therefore the same in all frames. (Although some physicists regard the cosmological constant, discussed in Chap. 28, as a modern aether, we must emphasize that, unlike its nineteenth-century antecedent, its presence does not alter the propagation of photons through Lorentz frames.)

This raises an interesting question. What about other waves that do not require an underlying medium? What about electron de Broglie waves? Here the fundamental wave equation, Schrödinger's or Dirac's, is mathematically different from Maxwell's and contains an important parameter, the electron rest mass. This rest mass allows the fundamental laws of relativistic quantum mechanics to be written in a form that is the same in all inertial reference frames and at the same time allows an electron, considered as either a wave or a particle, to travel at a different speed when measured in a different frame.

(continued)

BOX 2.3. (continued)

Some particles that have been postulated (such as gravitons, the quanta of gravitational waves; Chap. 27) are believed to exist without a rest mass (or an aether!), just like photons. Must these travel at the same speed as photons? The answer, according to the Principle of Relativity, is "yes." Why? Suppose there were two such waves or particles whose governing laws led to different speeds, c and $c' < c$, with each speed claimed to be the same in all reference frames. Such a claim produces insurmountable conundrums. For example, if we move with speed c' in the direction of propagation of the second wave, we will bring it to rest, in conflict with our hypothesis that its speed is frame independent. Therefore, all signals whose governing laws require them to travel with a speed that has no governing parameters (no rest mass and no underlying physical medium) must travel with a unique speed, which we call c. The speed of light is more fundamental to relativity than light itself!

2.2.3 The Interval and Its Invariance

2.2.3

the interval

Next we turn to another fundamental concept, the *interval* $(\Delta s)^2$ between the two events $\mathcal{P}$ and $\mathcal{Q}$ whose separation vector is $\Delta\vec{x}$. In a specific but arbitrary inertial reference frame and in geometrized units, $(\Delta s)^2$ is given by

$$(\Delta s)^2 \equiv -(\Delta t)^2 + (\Delta x)^2 + (\Delta y)^2 + (\Delta z)^2 = -(\Delta t)^2 + \sum_{i,j} \delta_{ij} \Delta x^i \Delta x^j; \tag{2.2a}$$

spacelike, timelike, and null

cf. Eq. (2.1). If $(\Delta s)^2 > 0$, the events $\mathcal{P}$ and $\mathcal{Q}$ are said to have a *spacelike* separation; if $(\Delta s)^2 = 0$, their separation is *null* or *lightlike;* and if $(\Delta s)^2 < 0$, their separation is *timelike.* For timelike separations, $(\Delta s)^2 < 0$ implies that Δs is imaginary; to avoid dealing with imaginary numbers, we describe timelike intervals by

$$(\Delta\tau)^2 \equiv -(\Delta s)^2, \tag{2.2b}$$

whose square root $\Delta\tau$ is real.

The coordinate separation between $\mathcal{P}$ and $\mathcal{Q}$ depends on one's reference frame: if $\Delta x^{\alpha'}$ and Δx^{α} are the coordinate separations in two different frames, then $\Delta x^{\alpha'} \neq$

T2

Δx^α. Despite this frame dependence, the Principle of Relativity forces the interval $(\Delta s)^2$ to be the same in all frames:

$$(\Delta s)^2 = -(\Delta t)^2 + (\Delta x)^2 + (\Delta y)^2 + (\Delta z)^2$$
$$= -(\Delta t')^2 + (\Delta x')^2 + (\Delta y')^2 + (\Delta z')^2. \quad (2.3)$$

In Box 2.4, we sketch a proof for the case of two events $\mathcal{P}$ and $\mathcal{Q}$ whose separation is timelike.

Because of its frame invariance, the interval $(\Delta s)^2$ can be regarded as a geometric property of the vector $\Delta\vec{x}$ that reaches from $\mathcal{P}$ to $\mathcal{Q}$; we call it the *squared length* $(\Delta\vec{x})^2$ of $\Delta\vec{x}$:

$$(\Delta\vec{x})^2 \equiv (\Delta s)^2. \quad (2.4)$$

BOX 2.4. PROOF OF INVARIANCE OF THE INTERVAL FOR A TIMELIKE SEPARATION

A simple demonstration that the interval is invariant is provided by a thought experiment in which a photon is emitted at event $\mathcal{P}$, reflects off a mirror, and is then detected at event $\mathcal{Q}$. We consider the interval between these events in two reference frames, primed and unprimed, that move with respect to each other. Choose the spatial coordinate systems of the two frames in such a way that (i) their relative motion (with speed β, which will not enter into our analysis) is along the x and x' directions, (ii) event $\mathcal{P}$ lies on the x and x' axes, and (iii) event $\mathcal{Q}$ lies in the x-y and x'-y' planes, as depicted below. Then evaluate the interval between $\mathcal{P}$ and $\mathcal{Q}$ in the unprimed frame by the following construction: Place the mirror parallel to the x-z plane at precisely the height h that permits a photon, emitted from $\mathcal{P}$, to travel along the dashed line to the mirror, then reflect off the mirror and continue along the dashed path, arriving at event $\mathcal{Q}$. If the mirror were placed lower, the photon would arrive at the spatial location of $\mathcal{Q}$ sooner than the time of $\mathcal{Q}$; if placed higher, it would arrive later. Then the distance the photon travels (the length of the two-segment dashed line) is equal to $c\Delta t = \Delta t$, where Δt is the time between events $\mathcal{P}$ and $\mathcal{Q}$ as measured in the unprimed frame. If the mirror had not been present, the photon would have arrived at event $\mathcal{R}$ after time Δt, so $c\Delta t$ is the distance between $\mathcal{P}$ and $\mathcal{R}$. From the diagram, it is easy to see that the height of $\mathcal{R}$ above the x-axis is $2h - \Delta y$, and the Pythagorean theorem then implies that

$$(\Delta s)^2 = -(\Delta t)^2 + (\Delta x)^2 + (\Delta y)^2 = -(2h - \Delta y)^2 + (\Delta y)^2. \quad \text{(1a)}$$

The same construction in the primed frame must give the same formula, but with primes:

$$(\Delta s')^2 = -(\Delta t')^2 + (\Delta x')^2 + (\Delta y')^2 = -(2h' - \Delta y')^2 + (\Delta y')^2. \quad \text{(1b)}$$

(continued)

T2

BOX 2.4. (continued)

The proof that $(\Delta s')^2 = (\Delta s)^2$ then reduces to showing that the Principle of Relativity requires that distances perpendicular to the direction of relative motion of two frames be the same as measured in the two frames: $h' = h$, $\Delta y' = \Delta y$. We leave it to the reader to develop a careful argument for this (Ex. 2.2).

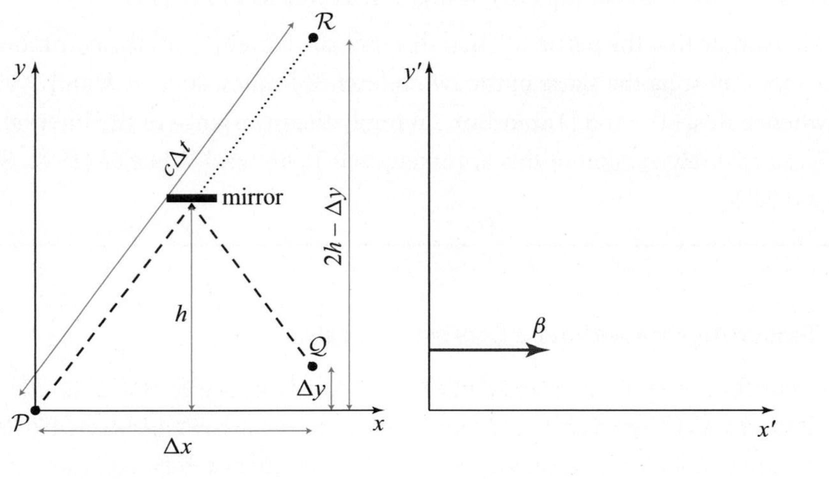

Note that this squared length, despite its name, can be negative (for timelike $\Delta\vec{x}$) or zero (for null $\Delta\vec{x}$) as well as positive (for spacelike $\Delta\vec{x}$).

The invariant interval $(\Delta s)^2$ between two events is as fundamental to Minkowski spacetime as the Euclidean distance between two points is to flat 3-space. Just as the Euclidean distance gives rise to the geometry of 3-space (as embodied, e.g., in Euclid's axioms), so the interval gives rise to the geometry of spacetime, which we shall be exploring. If this spacetime geometry were as intuitively obvious to humans as is Euclidean geometry, we would not need the crutch of inertial reference frames to arrive at it. Nature (presumably) has no need for such a crutch. To Nature (it seems evident), the geometry of Minkowski spacetime, as embodied in the invariant interval, is among the most fundamental aspects of physical law.

EXERCISES

Exercise 2.1 *Practice: Geometrized Units*
Do Ex. 1.15 in Chap. 1.

Exercise 2.2 *Derivation and Example: Invariance of the Interval*
Complete the derivation of the invariance of the interval given in Box 2.4, using the

T2

Principle of Relativity in the form that the laws of physics must be the same in the primed and unprimed frames. Hints (if you need them):

(a) Having carried out the construction in the unprimed frame, depicted at the bottom left of Box 2.4, use the same mirror and photons for the analogous construction in the primed frame. Argue that, independently of the frame in which the mirror is at rest (unprimed or primed), the fact that the reflected photon has (angle of reflection) = (angle of incidence) in its rest frame implies that this is also true for the same photon in the other frame. Thereby conclude that the construction leads to Eq. (1b) in Box 2.4, as well as to Eq. (1a).

(b) Then argue that the perpendicular distance of an event from the common x- and x'-axes must be the same in the two reference frames, so $h' = h$ and $\Delta y' = \Delta y$; whence Eqs. (1b) and (1a) in Box 2.4 imply the invariance of the interval. [Note: For a leisurely version of this argument, see Taylor and Wheeler (1992, Secs. 3.6 and 3.7).]

2.3

2.3 Tensor Algebra without a Coordinate System

Having introduced points in spacetime (interpreted physically as events), the invariant interval $(\Delta s)^2$ between two events, 4-vectors (as arrows between two events), and the squared length of a vector (as the invariant interval between the vector's tail and tip), we can now introduce the remaining tools of tensor algebra for Minkowski spacetime in precisely the same way as we did for the Euclidean 3-space of Newtonian physics (Sec. 1.3), with the invariant interval between events playing the same role as the squared length between Euclidean points.

tensor

In particular: a *tensor* $\boldsymbol{T}(\_\_, \_\_, \_\_)$ is a real-valued linear function of vectors in Minkowski spacetime. (We use slanted letters $\boldsymbol{T}$ for tensors in spacetime and unslanted letters $\mathbf{T}$ in Euclidean space.) A tensor's *rank* is equal to its number of slots. The *inner product* (also called the dot product) of two 4-vectors is

inner product

$$\vec{A} \cdot \vec{B} \equiv \frac{1}{4}\left[(\vec{A} + \vec{B})^2 - (\vec{A} - \vec{B})^2\right], \tag{2.5}$$

where $(\vec{A} + \vec{B})^2$ is the squared length of this vector (i.e., the invariant interval between its tail and its tip). The *metric tensor* of spacetime is that linear function of 4-vectors whose value is the inner product of the vectors:

metric tensor

$$\boldsymbol{g}(\vec{A}, \vec{B}) \equiv \vec{A} \cdot \vec{B}. \tag{2.6}$$

Using the inner product, we can regard any vector $\vec{A}$ as a rank-1 tensor: $\vec{A}(\vec{C}) \equiv \vec{A} \cdot \vec{C}$.

tensor product

Similarly, the *tensor product* $\otimes$ is defined precisely as in the Euclidean domain, Eqs. (1.5), as is the *contraction* of two slots of a tensor against each other, Eqs. (1.6), which lowers the tensor's rank by two.

contraction

2.4 Particle Kinetics and Lorentz Force without a Reference Frame

2.4

T2

2.4.1 Relativistic Particle Kinetics: World Lines, 4-Velocity, 4-Momentum and Its Conservation, 4-Force

2.4.1

In this section, we illustrate our geometric viewpoint by formulating the special relativistic laws of motion for particles.

ideal clock

An accelerated particle moving through spacetime carries an *ideal clock.* By "ideal" we mean that the clock is unaffected by accelerations: it ticks at a uniform rate when compared to unaccelerated atomic oscillators that are momentarily at rest beside the clock and are well protected from their environments. The builders of inertial guidance systems for airplanes and missiles try to make their clocks as ideal as possible in just this sense. We denote by τ the time ticked by the particle's ideal clock, and we call it the particle's *proper time.*

proper time

world line

The particle moves through spacetime along a curve, called its *world line,* which we can denote equally well by $\mathcal{P}(\tau)$ (the particle's spacetime location $\mathcal{P}$ at proper time τ), or by $\vec{x}(\tau)$ (the particle's vector separation from some arbitrarily chosen origin at proper time τ).[4]

momentary rest frame

We refer to the inertial frame in which the particle is momentarily at rest as its *momentarily comoving inertial frame* or *momentary rest frame.* Now, the particle's clock (which measures τ) is ideal, and so are the inertial frame's clocks (which measure coordinate time t). Therefore, a tiny interval $\Delta\tau$ of the particle's proper time is equal to the lapse of coordinate time in the particle's momentary rest frame $\Delta\tau = \Delta t$. Moreover, since the two events $\vec{x}(\tau)$ and $\vec{x}(\tau + \Delta\tau)$ on the clock's world line occur at the same spatial location in its momentary rest frame ($\Delta x^i = 0$, where $i = 1, 2, 3$) to first order in $\Delta\tau$, the invariant interval between those events is $(\Delta s)^2 = -(\Delta t)^2 + \sum_{i,j} \Delta x^i \Delta x^j \delta_{ij} = -(\Delta t)^2 = -(\Delta\tau)^2$. Thus, *the particle's proper time τ is equal to the square root of the negative of the invariant interval, $\tau = \sqrt{-s^2}$, along its world line.*

Figure 2.3 shows the world line of the accelerated particle in a spacetime diagram where the axes are coordinates of an arbitrary Lorentz frame. This diagram is intended to emphasize the world line as a frame-independent, geometric object. Also shown in the figure is the particle's *4-velocity* $\vec{u}$, which (by analogy with velocity in 3-space) is the time derivative of its position

4-velocity

$$\boxed{\vec{u} \equiv d\mathcal{P}/d\tau = d\vec{x}/d\tau} \tag{2.7}$$

and is the tangent vector to the world line. The derivative is defined by the usual limiting process

$$\frac{d\mathcal{P}}{d\tau} = \frac{d\vec{x}}{d\tau} \equiv \lim_{\Delta\tau\to 0} \frac{\mathcal{P}(\tau + \Delta\tau) - \mathcal{P}(\tau)}{\Delta\tau} = \lim_{\Delta\tau\to 0} \frac{\vec{x}(\tau + \Delta\tau) - \vec{x}(\tau)}{\Delta\tau}. \tag{2.8}$$

4. One of the basic ideas in string theory is that an elementary particle is described as a 1-dimensional loop in space rather than a 0-dimensional point. This means that it becomes a cylinder-like surface in spacetime—a world tube.

T2

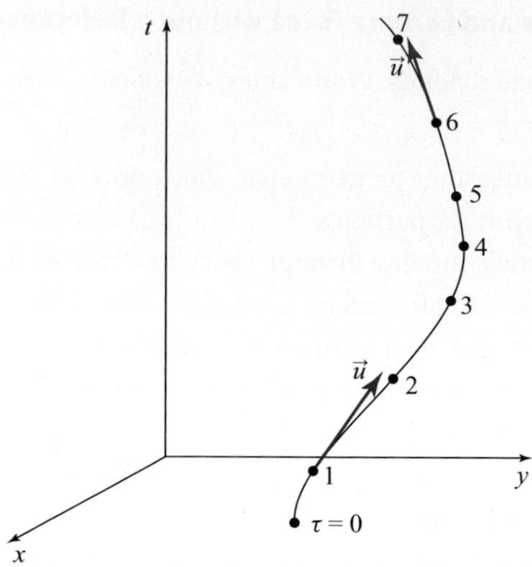

FIGURE 2.3 Spacetime diagram showing the world line $\vec{x}(\tau)$ and 4-velocity $\vec{u}$ of an accelerated particle. Note that the 4-velocity is tangent to the world line.

Here $\mathcal{P}(\tau + \Delta\tau) - \mathcal{P}(\tau)$ and $\vec{x}(\tau + \Delta\tau) - \vec{x}(\tau)$ are just two different ways to denote the same vector that reaches from one point on the world line to another.

The squared length of the particle's 4-velocity is easily seen to be -1:

$$\vec{u}^2 \equiv \mathbf{g}(\vec{u}, \vec{u}) = \frac{d\vec{x}}{d\tau} \cdot \frac{d\vec{x}}{d\tau} = \frac{d\vec{x} \cdot d\vec{x}}{(d\tau)^2} = -1. \tag{2.9}$$

The last equality follows from the fact that $d\vec{x} \cdot d\vec{x}$ is the squared length of $d\vec{x}$, which equals the invariant interval $(\Delta s)^2$ along it, and $(d\tau)^2$ is the negative of that invariant interval.

4-momentum

The particle's *4-momentum* is the product of its 4-velocity and rest mass:

$$\boxed{\vec{p} \equiv m\vec{u} = m d\vec{x}/d\tau \equiv d\vec{x}/d\zeta.} \tag{2.10}$$

Here the parameter ζ is a renormalized version of proper time,

$$\zeta \equiv \tau/m. \tag{2.11}$$

affine parameter

This ζ and any other renormalized version of proper time with a position-independent renormalization factor are called *affine parameters* for the particle's world line. Expression (2.10), together with $\vec{u}^2 = -1$, implies that the squared length of the 4-momentum is

$$\boxed{\vec{p}^2 = -m^2.} \tag{2.12}$$

In quantum theory, a particle is described by a relativistic wave function, which, in the geometric optics limit (Chap. 7), has a wave vector $\vec{k}$ that is related to the classical particle's 4-momentum by

$$\vec{k} = \vec{p}/\hbar. \tag{2.13}$$

T2

The above formalism is valid only for particles with nonzero rest mass, $m \neq 0$. The corresponding formalism for a *particle with zero rest mass* (e.g., a photon or a graviton) can be obtained from the above by taking the limit as $m \to 0$ and $d\tau \to 0$ with the quotient $d\zeta = d\tau/m$ held finite. More specifically, the 4-momentum of a zero-rest-mass particle is well defined (and participates in the conservation law to be discussed below), and it is expressible in terms of the particle's affine parameter ζ by Eq. (2.10):

$$\vec{p} = d\vec{x}/d\zeta. \tag{2.14}$$

By contrast, the particle's 4-velocity $\vec{u} = \vec{p}/m$ is infinite and thus undefined, and proper time $\tau = m\zeta$ ticks vanishingly slowly along its world line and thus is undefined. Because proper time is the square root of the invariant interval along the world line, the interval between two neighboring points on the world line vanishes. Therefore, *the world line of a zero-rest-mass particle is null.* (By contrast, since $d\tau^2 > 0$ and $ds^2 < 0$ along the world line of a particle with finite rest mass, *the world line of a finite-rest-mass particle is timelike.*)

conservation of 4-momentum

The 4-momenta of particles are important because of the *law of conservation of 4-momentum* (which, as we shall see in Sec. 2.6, is equivalent to the conservation laws for energy and ordinary momentum): If a number of "initial" particles, named $A = 1, 2, 3, \ldots$, enter a restricted region of spacetime $\mathcal{V}$ and there interact strongly to produce a new set of "final" particles, named $\bar{A} = \bar{1}, \bar{2}, \bar{3}, \ldots$ (Fig. 2.4), then the total 4-momentum of the final particles must be the same as the total 4-momentum of the initial ones:

$$\boxed{\sum_{\bar{A}} \vec{p}_{\bar{A}} = \sum_{A} \vec{p}_{A}.} \tag{2.15}$$

Note that this law of 4-momentum conservation is expressed in frame-independent, geometric language—in accord with Einstein's insistence that all the laws of physics should be so expressible. As we shall see in Part VII, 4-momentum conservation is a consequence of the translation symmetry of flat, 4-dimensional spacetime. In general relativity's curved spacetime, where that translation symmetry is lost, we lose 4-momentum conservation except under special circumstances; see Eq. (25.56) and associated discussion.

If a particle moves freely (no external forces and no collisions with other particles), then its 4-momentum $\vec{p}$ will be conserved along its world line, $d\vec{p}/d\zeta = 0$. Since $\vec{p}$ is tangent to the world line, this conservation means that the direction of the world line in spacetime never changes: the free particle moves along a straight line through spacetime. To change the particle's 4-momentum, one must act on it with a *4-force* $\vec{F}$,

4-force

$$d\vec{p}/d\tau = \vec{F}. \tag{2.16}$$

T2

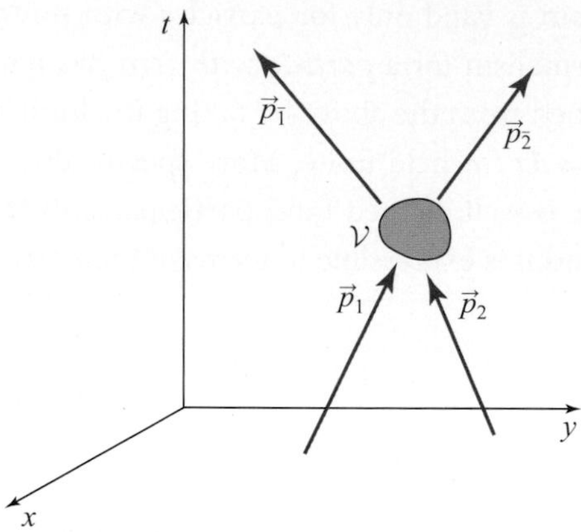

FIGURE 2.4 Spacetime diagram depicting the law of 4-momentum conservation for a situation where two particles, numbered 1 and 2, enter an interaction region $\mathcal{V}$ in spacetime, and there interact strongly and produce two new particles, numbered $\bar{1}$ and $\bar{2}$. The sum of the final 4-momenta, $\vec{p}_{\bar{1}} + \vec{p}_{\bar{2}}$, must be equal to the sum of the initial 4-momenta, $\vec{p}_1 + \vec{p}_2$.

If the particle is a fundamental one (e.g., photon, electron, proton), then the 4-force must leave its rest mass unchanged,

$$0 = dm^2/d\tau = -d\vec{p}^2/d\tau = -2\vec{p}\cdot d\vec{p}/d\tau = -2\vec{p}\cdot\vec{F}; \tag{2.17}$$

that is, the 4-force must be orthogonalxpage4-force|orthogonal to 4–velocity to the 4-momentum in the 4-dimensional sense that their inner product vanishes.

2.4.2

2.4.2 Geometric Derivation of the Lorentz Force Law

As an illustration of these physical concepts and mathematical tools, we use them to deduce the relativistic version of the Lorentz force law. From the outset, in accord with the Principle of Relativity, we insist that the law we seek be expressible in geometric, frame-independent language, that is, in terms of vectors and tensors.

electromagnetic field tensor

Consider a particle with charge q and rest mass $m \neq 0$ interacting with an electromagnetic field. It experiences an electromagnetic 4-force whose mathematical form we seek. The Newtonian version of the electromagnetic force $\mathbf{F} = q(\mathbf{E} + \mathbf{v}\times\mathbf{B})$ is proportional to q and contains one piece (electric) that is independent of velocity $\mathbf{v}$ and a second piece (magnetic) that is linear in $\mathbf{v}$. It is reasonable to expect that, to produce this Newtonian limit, the relativistic 4-force $\vec{F}$ will be proportional to q and will be linear in the 4-velocity $\vec{u}$. Linearity means there must exist some second-rank tensor $\boldsymbol{F}(\_\_, \_\_)$, the *electromagnetic field tensor,* such that

$$d\vec{p}/d\tau = \vec{F}(\_\_) = q\boldsymbol{F}(\_\_, \vec{u}). \tag{2.18}$$

T2

Because the 4-force $\vec{F}$ must be orthogonal to the particle's 4-momentum and thence also to its 4-velocity, $\vec{F} \cdot \vec{u} \equiv \vec{F}(\vec{u}) = 0$, expression (2.18) must vanish when $\vec{u}$ is inserted into its empty slot. In other words, for all timelike unit-length vectors $\vec{u}$,

$$\mathbf{F}(\vec{u}, \vec{u}) = 0. \tag{2.19}$$

It is an instructive exercise (Ex. 2.3) to show that this is possible only if $\mathbf{F}$ is *anti-symmetric*, so the electromagnetic 4-force is

electromagnetic 4-force

$$\boxed{d\vec{p}/d\tau = q\mathbf{F}(\_\_, \vec{u}), \quad \text{where} \quad \mathbf{F}(\vec{A}, \vec{B}) = -\mathbf{F}(\vec{B}, \vec{A}) \quad \text{for all } \vec{A} \text{ and } \vec{B}.} \tag{2.20}$$

Equation (2.20) must be the relativistic form of the Lorentz force law. In Sec. 2.11 we deduce the relationship of the electromagnetic field tensor $\mathbf{F}$ to the more familiar electric and magnetic fields, and the relationship of this relativistic Lorentz force to its Newtonian form (1.7c).

The discussion of particle kinematics and the electromagnetic force in this Sec. 2.4 is elegant but perhaps unfamiliar. In Secs. 2.6 and 2.11, we shall see that it is equivalent to the more elementary (but more complex) formalism based on components of vectors in Euclidean 3-space.

EXERCISES

Exercise 2.3 *Derivation and Example: Antisymmetry of Electromagnetic Field Tensor*
Show that Eq. (2.19) can be true for all timelike, unit-length vectors $\vec{u}$ if and only if $\mathbf{F}$ is antisymmetric. [Hints: (i) Show that the most general second-rank tensor $\mathbf{F}$ can be written as the sum of a symmetric tensor $\mathbf{S}$ and an antisymmetric tensor $\mathbf{A}$, and that the antisymmetric piece contributes nothing to Eq. (2.19), so $\mathbf{S}(\vec{u}, \vec{u})$ must vanish for every timelike $\vec{u}$. (ii) Let $\vec{a}$ be a timelike vector, let $\vec{b}$ be an artibrary vector (timelike, null, or spacelike), and let ϵ be a number small enough that $\vec{A}_\pm \equiv \vec{a} \pm \epsilon\vec{b}$ are both timelike. From the fact that $\mathbf{S}(\vec{A}_+, \vec{A}_+)$, $\mathbf{S}(\vec{A}_-, \vec{A}_-)$, and $\mathbf{S}(\vec{a}, \vec{a})$ all vanish, deduce that $\mathbf{S}(\vec{b}, \vec{b}) = 0$ for the arbitrary vector $\vec{b}$. (iii) From this, deduce that $\mathbf{S}$ vanishes (i.e., it gives zero when any two vectors are inserted into its slots).]

Exercise 2.4 *Problem: Relativistic Gravitational Force Law*
In Newtonian theory, the gravitational potential Φ exerts a force $\mathbf{F} = d\mathbf{p}/dt = -m\nabla\Phi$ on a particle with mass m and momentum $\mathbf{p}$. Before Einstein formulated general relativity, some physicists constructed relativistic theories of gravity in which a Newtonian-like scalar gravitational field Φ exerted a 4-force $\vec{F} = d\vec{p}/d\tau$ on any particle with rest mass m, 4-velocity $\vec{u}$, and 4-momentum $\vec{p} = m\vec{u}$. What must that force law have been for it to (i) obey the Principle of Relativity, (ii) reduce to Newton's law in the nonrelativistic limit, and (iii) preserve the particle's rest mass as time passes?

T2 2.5

2.5 Component Representation of Tensor Algebra

2.5.1

2.5.1 Lorentz Coordinates

In Minkowski spacetime, associated with any inertial reference frame (Fig. 2.1 and Sec. 2.2.1), there is a Lorentz coordinate system $\{t, x, y, z\} = \{x^0, x^1, x^2, x^3\}$ generated by the frame's rods and clocks. (Note the use of superscripts.) And associated with these coordinates is a set of *Lorentz basis vectors* $\{\vec{e}_t, \vec{e}_x, \vec{e}_y, \vec{e}_z\} = \{\vec{e}_0, \vec{e}_1, \vec{e}_2, \vec{e}_3\}$. (Note the use of subscripts. The reason for this convention will become clear below.) The basis vector $\vec{e}_\alpha$ points along the x^α coordinate direction, which is orthogonal to all the other coordinate directions, and it has squared length -1 for $\alpha = 0$ (vector pointing in a timelike direction) and $+1$ for $\alpha = 1, 2, 3$ (spacelike):

Lorentz (orthonormal) basis

$$\vec{e}_\alpha \cdot \vec{e}_\beta = \eta_{\alpha\beta}. \tag{2.21}$$

Here $\eta_{\alpha\beta}$ (a spacetime analog of the Kronecker delta) are defined by

$$\eta_{00} \equiv -1, \qquad \eta_{11} \equiv \eta_{22} \equiv \eta_{33} \equiv 1, \qquad \eta_{\alpha\beta} \equiv 0 \text{ if } \alpha \neq \beta. \tag{2.22}$$

Any basis in which $\vec{e}_\alpha \cdot \vec{e}_\beta = \eta_{\alpha\beta}$ is said to be *orthonormal* (by analogy with the Euclidean notion of orthonormality, $\mathbf{e}_j \cdot \mathbf{e}_k = \delta_{jk}$).

Because $\vec{e}_\alpha \cdot \vec{e}_\beta \neq \delta_{\alpha\beta}$, many of the Euclidean-space component-manipulation formulas (1.9b)–(1.9h) do not hold in Minkowski spacetime. There are two approaches to recovering these formulas. One approach, used in many older textbooks (including the first and second editions of Goldstein's *Classical Mechanics* and Jackson's *Classical Electrodynamics*), is to set $x^0 = it$, where $i = \sqrt{-1}$ and correspondingly make the time basis vector be imaginary, so that $\vec{e}_\alpha \cdot \vec{e}_\beta = \delta_{\alpha\beta}$. When this approach is adopted, the resulting formalism does not depend on whether indices are placed up or down; one can place them wherever one's stomach or liver dictates without asking one's brain. However, this $x^0 = it$ approach has severe disadvantages: (i) it hides the true physical geometry of Minkowski spacetime, (ii) it cannot be extended in any reasonable manner to nonorthonormal bases in flat spacetime, and (iii) it cannot be extended in any reasonable manner to the curvilinear coordinates that must be used in general relativity. For these reasons, most modern texts (including the third editions of Goldstein and Jackson) take an alternative approach, one always used in general relativity. This alternative, which we shall adopt, requires introducing two different types of components for vectors (and analogously for tensors): *contravariant components* denoted by superscripts (e.g., $T^{\alpha\beta\gamma}$) and *covariant components* denoted by subscripts (e.g., $T_{\alpha\beta\gamma}$). In Parts I–VI of this book, we introduce these components only for orthonormal bases; in Part VII, we develop a more sophisticated version of them, valid for nonorthonormal bases.

2.5.2

2.5.2 Index Gymnastics

contravariant components

A vector's or tensor's contravariant components are defined as its expansion coefficients in the chosen basis [analogs of Eqs. (1.9b) and (1.9d) in Euclidean 3-space]:

$$\vec{A} \equiv A^\alpha \vec{e}_\alpha, \qquad \boldsymbol{T} \equiv T^{\alpha\beta\gamma} \vec{e}_\alpha \otimes \vec{e}_\beta \otimes \vec{e}_\gamma. \tag{2.23a}$$

T2

Here and throughout this book, *Greek (spacetime) indices are to be summed when they are repeated with one up and the other down;* this is called the Einstein summation convention.

covariant components

The covariant components are defined as the numbers produced by evaluating the vector or tensor on its basis vectors [analog of Eq. (1.9e) in Euclidean 3-space]:

$$A_\alpha \equiv \vec{A}(\vec{e}_\alpha) = \vec{A} \cdot \vec{e}_\alpha, \qquad T_{\alpha\beta\gamma} \equiv \boldsymbol{T}(\vec{e}_\alpha, \vec{e}_\beta, \vec{e}_\gamma). \tag{2.23b}$$

(Just as there are contravariant and covariant components A^α and A_α, so also there is a second set of basis vectors $\vec{e}^\alpha$ dual to the set $\vec{e}_\alpha$. However, for economy of notation we delay introducing them until Part VII.)

These definitions have a number of important consequences. We derive them one after another and then summarize them succinctly with equation numbers:

(i) The covariant components of the metric tensor are $g_{\alpha\beta} = \boldsymbol{g}(\vec{e}_\alpha, \vec{e}_\beta) = \vec{e}_\alpha \cdot \vec{e}_\beta = \eta_{\alpha\beta}$. Here the first equality is the definition (2.23b) of the covariant components, the second equality is the definition (2.6) of the metric tensor, and the third equality is the orthonormality relation (2.21) for the basis vectors.

(ii) The covariant components of any tensor can be computed from the contravariant components by

$$\begin{aligned} T_{\lambda\mu\nu} &= \boldsymbol{T}(\vec{e}_\lambda, \vec{e}_\mu, \vec{e}_\nu) = T^{\alpha\beta\gamma} \vec{e}_\alpha \otimes \vec{e}_\beta \otimes \vec{e}_\gamma (\vec{e}_\lambda, \vec{e}_\mu, \vec{e}_\nu) \\ &= T^{\alpha\beta\gamma} (\vec{e}_\alpha \cdot \vec{e}_\lambda)(\vec{e}_\beta \cdot \vec{e}_\mu)(\vec{e}_\gamma \cdot \vec{e}_\nu) = T^{\alpha\beta\gamma} g_{\alpha\lambda} g_{\beta\mu} g_{\gamma\nu}. \end{aligned}$$

The first equality is the definition (2.23b) of the covariant components, the second is the expansion (2.23a) of $\boldsymbol{T}$ on the chosen basis, the third is the definition (1.5a) of the tensor product, and the fourth is one version of our result (i) for the covariant components of the metric.

raising and lowering indices: sign flip if temporal

(iii) This result, $T_{\lambda\mu\nu} = T^{\alpha\beta\gamma} g_{\alpha\lambda} g_{\beta\mu} g_{\gamma\nu}$, together with the numerical values (i) of $g_{\alpha\beta}$, implies that when one lowers a spatial index there is no change in the numerical value of a component, and when one lowers a temporal index, the sign changes: $T_{ijk} = T^{ijk}$, $T_{0jk} = -T^{0jk}$, $T_{0j0} = +T^{0j0}$, $T_{000} = -T^{000}$. We call this the "sign-flip-if-temporal" rule. As a special case, $-1 = g_{00} = g^{00}$, $0 = g_{0j} = -g^{0j}$, $\delta_{jk} = g_{jk} = g^{jk}$—that is, the metric's covariant and contravariant components are numerically identical; they are both equal to the orthonormality values $\eta_{\alpha\beta}$.

components of metric tensor

(iv) It is easy to see that this sign-flip-if-temporal rule for lowering indices implies the same sign-flip-if-temporal rule for raising them, which in turn can be written in terms of metric components as $T^{\alpha\beta\gamma} = T_{\lambda\mu\nu} g^{\lambda\alpha} g^{\mu\beta} g^{\nu\gamma}$.

mixed components

(v) It is convenient to define *mixed components* of a tensor, components with some indices up and others down, as having numerical values obtained

T2

by raising or lowering some but not all of its indices using the metric, for example, $T^{\alpha}{}_{\mu\nu} = T^{\alpha\beta\gamma} g_{\beta\mu} g_{\gamma\nu} = T_{\lambda\mu\nu} g^{\lambda\alpha}$. Numerically, this continues to follow the sign-flip-if-temporal rule: $T^{0}{}_{0k} = -T^{00k}$, $T^{0}{}_{jk} = T^{0jk}$, and it implies, in particular, that the mixed components of the metric are $g^{\alpha}{}_{\beta} = \delta_{\alpha\beta}$ (the Kronecker-delta values; $+1$ if $\alpha = \beta$ and 0 otherwise).

summary of index gymnastics

These important results can be summarized as follows. *The numerical values of the components of the metric in Minkowski spacetime are expressed in terms of the matrices* $[\delta_{\alpha\beta}]$ *and* $[\eta_{\alpha\beta}]$ *as*

$$g_{\alpha\beta} = \eta_{\alpha\beta}, \quad g^{\alpha}{}_{\beta} = \delta_{\alpha\beta}, \quad g_{\alpha}{}^{\beta} = \delta_{\alpha\beta}, \quad g^{\alpha\beta} = \eta_{\alpha\beta}; \tag{2.23c}$$

indices on all vectors and tensors can be raised and lowered using these components of the metric:

$$A_{\alpha} = g_{\alpha\beta} A^{\beta}, \quad A^{\alpha} = g^{\alpha\beta} A_{\beta}, \quad T^{\alpha}{}_{\mu\nu} \equiv g_{\mu\beta} g_{\nu\gamma} T^{\alpha\beta\gamma}, \quad T^{\alpha\beta\gamma} \equiv g^{\beta\mu} g^{\gamma\nu} T^{\alpha}{}_{\mu\nu}, \tag{2.23d}$$

which is equivalent to the sign-flip-if-temporal rule.

This index notation gives rise to formulas for tensor products, inner products, values of tensors on vectors, and tensor contractions that are obvious analogs of those in Euclidean space:

$$[\text{Contravariant components of } \boldsymbol{T}(\_,\_,\_) \otimes \boldsymbol{S}(\_,\_)] = T^{\alpha\beta\gamma} S^{\delta\epsilon}, \tag{2.23e}$$

$$\vec{A} \cdot \vec{B} = A^{\alpha} B_{\alpha} = A_{\alpha} B^{\alpha}, \quad \boldsymbol{T}(\mathbf{A}, \mathbf{B}, \mathbf{C}) = T_{\alpha\beta\gamma} A^{\alpha} B^{\beta} C^{\gamma} = T^{\alpha\beta\gamma} A_{\alpha} B_{\beta} C_{\gamma}, \tag{2.23f}$$

$$\text{Covariant components of } [1\&3\text{contraction of } \boldsymbol{R}] = R^{\mu}{}_{\alpha\mu\beta},$$

$$\text{Contravariant components of } [1\&3\text{contraction of } \boldsymbol{R}] = R^{\mu\alpha}{}_{\mu}{}^{\beta}. \tag{2.23g}$$

Notice the very simple pattern in Eqs. (2.23b) and (2.23d), which universally permeates the rules of index gymnastics, a pattern that permits one to reconstruct the rules without any memorization: *Free indices (indices not summed over) must agree in position (up versus down) on the two sides of each equation.* In keeping with this pattern, one can regard the two indices in a pair that is summed as "annihilating each other by contraction," and one speaks of "lining up the indices" on the two sides of an equation to get them to agree. These rules provide helpful checks when performing calculations.

In Part VII, when we use nonorthonormal bases, all these index-notation equations (2.23) will remain valid except for the numerical values [Eq. (2.23c)] of the metric components and the sign-flip-if-temporal rule.

2.5.3

2.5.3 Slot-Naming Notation

In Minkowski spacetime, as in Euclidean space, we can (and often do) use slot-naming index notation to represent frame-independent geometric objects and equations and physical laws. (Readers who have not studied Sec. 1.5.1 on slot-naming index notation should do so now.) T2

For example, we often write the frame-independent Lorentz force law $d\vec{p}/d\tau = q\boldsymbol{F}(\_, \vec{u})$ [Eq. (2.20)] as $dp_\mu/d\tau = qF_{\mu\nu}u^\nu$.

Notice that, because the components of the metric in any Lorentz basis are $g_{\alpha\beta} = \eta_{\alpha\beta}$, we can write the invariant interval between two events x^α and $x^\alpha + dx^\alpha$ as

$$ds^2 = g_{\alpha\beta}dx^\alpha dx^\beta = -dt^2 + dx^2 + dy^2 + dz^2. \tag{2.24}$$

This is called the special relativistic *line element*. line element

EXERCISES

Exercise 2.5 *Derivation: Component Manipulation Rules*
Derive the relativistic component manipulation rules (2.23e)–(2.23g).

Exercise 2.6 *Practice: Numerics of Component Manipulations*
In some inertial reference frame, the vector $\vec{A}$ and second-rank tensor $\boldsymbol{T}$ have as their only nonzero components $A^0 = 1$, $A^1 = 2$; $T^{00} = 3$, $T^{01} = T^{10} = 2$, $T^{11} = -1$. Evaluate $\boldsymbol{T}(\vec{A}, \vec{A})$ and the components of $\boldsymbol{T}(\vec{A}, \_)$ and $\vec{A} \otimes \boldsymbol{T}$.

Exercise 2.7 *Practice: Meaning of Slot-Naming Index Notation*

(a) Convert the following expressions and equations into geometric, index-free notation: $A^\alpha B_{\gamma\delta}$; $A_\alpha B_\gamma{}^\delta$; $S_\alpha{}^{\beta\gamma} = S^{\gamma\beta}{}_\alpha$; $A^\alpha B_\alpha = A_\alpha B^\beta g^\alpha{}_\beta$.

(b) Convert $\boldsymbol{T}(\_, \boldsymbol{S}(\boldsymbol{R}(\vec{C}, \_), \_), \_)$ into slot-naming index notation.

Exercise 2.8 *Practice: Index Gymnastics*

(a) Simplify the following expression so the metric does not appear in it:

$$A^{\alpha\beta\gamma} g_{\beta\rho} S_{\gamma\lambda} g^{\rho\delta} g^\lambda{}_\alpha.$$

(b) The quantity $g_{\alpha\beta}g^{\alpha\beta}$ is a scalar since it has no free indices. What is its numerical value?

(c) What is wrong with the following expression and equation?

$$A_\alpha{}^{\beta\gamma} S_{\alpha\gamma}; \qquad A_\alpha{}^{\beta\gamma} S_\beta T_\gamma = R_{\alpha\beta\delta} S^\beta.$$

2.6 Particle Kinetics in Index Notation and in a Lorentz Frame 2.6

As an illustration of the component representation of tensor algebra, let us return to the relativistic, accelerated particle of Fig. 2.3 and, from the frame-independent equations for the particle's 4-velocity $\vec{u}$ and 4-momentum $\vec{p}$ (Sec. 2.4), derive the component description given in elementary textbooks.

We introduce a specific inertial reference frame and associated Lorentz coordinates x^α and basis vectors $\{\vec{e}_\alpha\}$. In this Lorentz frame, the particle's world line

T2

$\vec{x}(\tau)$ is represented by its coordinate location $x^\alpha(\tau)$ as a function of its proper time τ. The contravariant components of the separation vector $d\vec{x}$ between two neighboring events along the particle's world line are the events' coordinate separations dx^α [Eq. (2.1)]; and correspondingly, the components of the particle's 4-velocity $\vec{u} = d\vec{x}/d\tau$ are

$$u^\alpha = dx^\alpha/d\tau \tag{2.25a}$$

(the time derivatives of the particle's spacetime coordinates). Note that Eq. (2.25a) implies

$$v^j \equiv \frac{dx^j}{dt} = \frac{dx^j/d\tau}{dt/d\tau} = \frac{u^j}{u^0}. \tag{2.25b}$$

This relation, together with $-1 = \vec{u}^2 = g_{\alpha\beta}u^\alpha u^\beta = -(u^0)^2 + \delta_{ij}u^i u^j = -(u^0)^2(1 - \delta_{ij}v^i v^j)$, implies that the components of the 4-velocity have the forms familiar from elementary textbooks:

$$u^0 = \gamma, \quad u^j = \gamma v^j, \quad \text{where } \gamma = \frac{1}{(1 - \delta_{ij}v^i v^j)^{\frac{1}{2}}}. \tag{2.25c}$$

ordinary velocity

slice of simultaneity

It is useful to think of v^j as the components of a 3-dimensional vector $\mathbf{v}$, the *ordinary velocity*, that lives in the 3-dimensional Euclidean space $t =$ const of the chosen Lorentz frame (the green plane in Fig. 2.5). This 3-space is sometimes called the frame's *slice of simultaneity* or *3-space of simultaneity*, because all events lying in it are simultaneous, as measured by the frame's observers. This 3-space is not well defined until a Lorentz frame has been chosen, and correspondingly, $\mathbf{v}$ relies for its existence on a specific choice of frame. However, once the frame has been chosen, $\mathbf{v}$ can be regarded as a coordinate-independent, basis-independent 3-vector lying in the frame's slice of simultaneity. Similarly, the spatial part of the 4-velocity $\vec{u}$ (the part with components u^j in our chosen frame) can be regarded as a 3-vector $\mathbf{u}$ lying in the frame's 3-space; and Eqs. (2.25c) become the component versions of the coordinate-independent, basis-independent 3-space relations

$$\mathbf{u} = \gamma \mathbf{v}, \quad \gamma = \frac{1}{\sqrt{1 - \mathbf{v}^2}}, \tag{2.25d}$$

where $\mathbf{v}^2 = \mathbf{v} \cdot \mathbf{v}$. This γ is called the "Lorentz factor."

The components of the particle's 4-momentum $\vec{p}$ in our chosen Lorentz frame have special names and special physical significances: The time component of the 4-momentum is the particle's (relativistic) *energy* $\mathcal{E}$ as measured in that frame:

relativistic energy

$$\begin{aligned} \mathcal{E} \equiv p^0 = mu^0 = m\gamma = \frac{m}{\sqrt{1-\mathbf{v}^2}} &= \text{(the particle's energy)} \\ \simeq m + \frac{1}{2}m\mathbf{v}^2 \quad &\text{for } |\mathbf{v}| \ll 1. \end{aligned} \tag{2.26a}$$

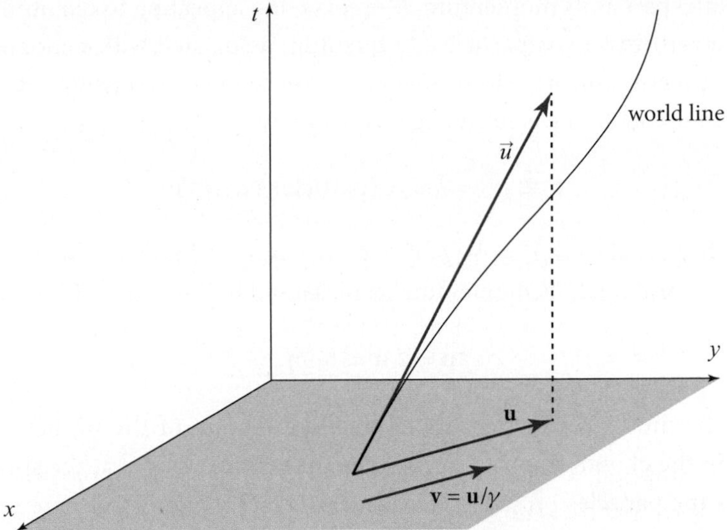

FIGURE 2.5 Spacetime diagram in a specific Lorentz frame, showing the frame's 3-space $t = 0$ (green region), the world line of a particle, the 4-velocity $\vec{u}$ of the particle as it passes through the 3-space, and two 3-dimensional vectors that lie in the 3-space: the spatial part $\mathbf{u}$ of the particle's 4-velocity and the particle's ordinary velocity $\mathbf{v}$.

Note that this energy is the sum of the particle's *rest mass-energy* $m = mc^2$ and its *kinetic energy*

rest mass-energy

$$E \equiv \mathcal{E} - m = m\left(\frac{1}{\sqrt{1-\mathbf{v}^2}} - 1\right)$$

kinetic energy

$$\simeq \frac{1}{2} m\mathbf{v}^2 \quad \text{for } |\mathbf{v}| \ll 1. \tag{2.26b}$$

The spatial components of the 4-momentum, when regarded from the viewpoint of 3-dimensional physics, are the same as the components of the *momentum*, a 3-vector residing in the chosen Lorentz frame's 3-space:

momentum

$$p^j = mu^j = m\gamma v^j = \frac{mv^j}{\sqrt{1-\mathbf{v}^2}} = \mathcal{E}v_j$$

$$= (j \text{ component of particle's momentum}); \tag{2.26c}$$

or, in basis-independent, 3-dimensional vector notation,

$$\mathbf{p} = m\mathbf{u} = m\gamma\mathbf{v} = \frac{m\mathbf{v}}{\sqrt{1-\mathbf{v}^2}} = \mathcal{E}\mathbf{v} = (\text{particle's momentum}). \tag{2.26d}$$

For a zero-rest-mass particle, as for one with finite rest mass, we identify the time component of the 4-momentum, in a chosen Lorentz frame, as the particle's energy,

T2

and the spatial part as its momentum. Moreover, if—appealing to quantum theory—we regard a zero-rest-mass particle as a quantum associated with a monochromatic wave, then quantum theory tells us that the wave's angular frequency ω as measured in a chosen Lorentz frame is related to its energy by

$$\mathcal{E} \equiv p^0 = \hbar\omega = \text{(particle's energy)}; \tag{2.27a}$$

and, since the particle has $\vec{p}^2 = -(p^0)^2 + \mathbf{p}^2 = -m^2 = 0$ (in accord with the lightlike nature of its world line), its momentum as measured in the chosen Lorentz frame is

$$\mathbf{p} = \mathcal{E}\mathbf{n} = \hbar\omega\mathbf{n}. \tag{2.27b}$$

Here $\mathbf{n}$ is the unit 3-vector that points in the direction of the particle's travel, as measured in the chosen frame; that is (since the particle moves at the speed of light $v = 1$), $\mathbf{n}$ is the particle's ordinary velocity. Eqs. (2.27a) and (2.27b) are respectively the temporal and spatial components of the geometric, frame-independent relation $\vec{p} = \hbar\vec{k}$ [Eq. (2.13), which is valid for zero-rest-mass particles as well as finite-mass ones].

3+1 split of spacetime into space plus time

The introduction of a specific Lorentz frame into spacetime can be said to produce a 3+1 split of every 4-vector into a 3-dimensional vector plus a scalar (a real number). The 3+1 split of a particle's 4-momentum $\vec{p}$ produces its momentum $\mathbf{p}$ plus its energy $\mathcal{E} = p^0$. Correspondingly, the 3+1 split of the law of 4-momentum conservation (2.15) produces a law of conservation of momentum plus a law of conservation of energy:

$$\sum_{\bar{A}} \mathbf{p}_{\bar{A}} = \sum_{A} \mathbf{p}_{A}, \qquad \sum_{\bar{A}} \mathcal{E}_{\bar{A}} = \sum_{A} \mathcal{E}_{A}. \tag{2.28}$$

The unbarred quantities in Eqs. (2.28) are momenta and energies of the particles entering the interaction region, and the barred quantities are those of the particles leaving (see Fig. 2.4).

Because the concept of energy does not even exist until one has chosen a Lorentz frame—and neither does that of momentum—the laws of energy conservation and momentum conservation separately are frame-dependent laws. In this sense, they are far less fundamental than their combination, the frame-independent law of 4-momentum conservation.

By learning to think about the 3+1 split in a geometric, frame-independent way, one can gain conceptual and computational power. As an example, consider a particle with 4-momentum $\vec{p}$, being studied by an observer with 4-velocity $\vec{U}$. In the observer's own Lorentz reference frame, her 4-velocity has components $U^0 = 1$ and $U^j = 0$, and therefore, her 4-velocity is $\vec{U} = U^\alpha \vec{e}_\alpha = \vec{e}_0$; that is, it is identically equal to the time basis vector of her Lorentz frame. Thus the particle energy that she measures is $\mathcal{E} = p^0 = -p_0 = -\vec{p} \cdot \vec{e}_0 = -\vec{p} \cdot \vec{U}$. This equation, derived in the observer's Lorentz frame, is actually a geometric, frame-independent relation: the inner product of two 4-vectors. It says that *when an observer with 4-velocity $\vec{U}$ measures the energy of a* *particle with 4-momentum $\vec{p}$, the result she gets (the time part of the 3+1 split of $\vec{p}$ as seen by her) is*

$$\boxed{\mathcal{E} = -\vec{p} \cdot \vec{U}.} \tag{2.29}$$

We shall use this equation in later chapters. In Exs. 2.9 and 2.10, the reader can gain experience deriving and interpreting other frame-independent equations for 3+1 splits. Exercise 2.11 exhibits the power of this geometric way of thinking by using it to derive the Doppler shift of a photon.

EXERCISES

Exercise 2.9 ***Practice: Frame-Independent Expressions for Energy, Momentum, and Velocity*

An observer with 4-velocity $\vec{U}$ measures the properties of a particle with 4-momentum $\vec{p}$. The energy she measures is $\mathcal{E} = -\vec{p} \cdot \vec{U}$ [Eq. (2.29)].

(a) Show that the particle's rest mass can be expressed in terms of $\vec{p}$ as

$$m^2 = -\vec{p}^2. \tag{2.30a}$$

(b) Show that the momentum the observer measures has the magnitude

$$|\mathbf{p}| = [(\vec{p} \cdot \vec{U})^2 + \vec{p} \cdot \vec{p}]^{\frac{1}{2}}. \tag{2.30b}$$

(c) Show that the ordinary velocity the observer measures has the magnitude

$$|\mathbf{v}| = \frac{|\mathbf{p}|}{\mathcal{E}}, \tag{2.30c}$$

where $|\mathbf{p}|$ and $\mathcal{E}$ are given by the above frame-independent expressions.

(d) Show that the ordinary velocity $\mathbf{v}$, thought of as a 4-vector that happens to lie in the observer's slice of simultaneity, is given by

$$\vec{v} = \frac{\vec{p} + (\vec{p} \cdot \vec{U})\vec{U}}{-\vec{p} \cdot \vec{U}}. \tag{2.30d}$$

Exercise 2.10 ***Example: 3-Metric as a Projection Tensor*

Consider, as in Ex. 2.9, an observer with 4-velocity $\vec{U}$ who measures the properties of a particle with 4-momentum $\vec{p}$.

(a) Show that the Euclidean metric of the observer's 3-space, when thought of as a tensor in 4-dimensional spacetime, has the form

$$\boxed{\mathbf{P} \equiv \boldsymbol{g} + \vec{U} \otimes \vec{U}.} \tag{2.31a}$$

Show, further, that if $\vec{A}$ is an arbitrary vector in spacetime, then $-\vec{A} \cdot \vec{U}$ is the component of $\vec{A}$ along the observer's 4-velocity $\vec{U}$, and

$$\mathbf{P}(\_\_, \vec{A}) = \vec{A} + (\vec{A} \cdot \vec{U})\vec{U} \tag{2.31b}$$

T2

is the projection of $\vec{A}$ into the observer's 3-space (i.e., it is the spatial part of $\vec{A}$ as seen by the observer). For this reason, $\boldsymbol{P}$ is called a *projection tensor.* In quantum mechanics, the concept of a *projection operator* $\hat{P}$ is introduced as one that satisfies the equation $\hat{P}^2 = \hat{P}$. Show that the projection tensor $\boldsymbol{P}$ is a projection operator in the same sense:

$$P_{\alpha\mu} P^{\mu}{}_{\beta} = P_{\alpha\beta}. \tag{2.31c}$$

(b) Show that Eq. (2.30d) for the particle's ordinary velocity, thought of as a 4-vector, can be rewritten as

$$\vec{v} = \frac{\boldsymbol{P}(\_, \vec{p})}{-\vec{p} \cdot \vec{U}}. \tag{2.32}$$

Exercise 2.11 ***Example: Doppler Shift Derived without Lorentz Transformations*

(a) An observer at rest in some inertial frame receives a photon that was emitted in direction $\mathbf{n}$ by an atom moving with ordinary velocity $\mathbf{v}$ (Fig. 2.6). The photon frequency and energy as measured by the emitting atom are $\nu_{\rm em}$ and $\mathcal{E}_{\rm em}$; those measured by the receiving observer are $\nu_{\rm rec}$ and $\mathcal{E}_{\rm rec}$. By a calculation carried out solely in the receiver's inertial frame (the frame of Fig. 2.6), and without the aid of any Lorentz transformation, derive the standard formula for the photon's Doppler shift:

$$\frac{\nu_{\rm rec}}{\nu_{\rm em}} = \frac{\sqrt{1 - v^2}}{1 - \mathbf{v} \cdot \mathbf{n}}. \tag{2.33}$$

[Hint: Use Eq. (2.29) to evaluate $\mathcal{E}_{\rm em}$ using receiver-frame expressions for the emitting atom's 4-velocity $\vec{U}$ and the photon's 4-momentum $\vec{p}$.]

(b) Suppose that instead of emitting a photon, the emitter produces a particle with finite rest mass m. Using the same method as in part (a), derive an expression for the ratio of received energy to emitted energy, $\mathcal{E}_{\rm rec}/\mathcal{E}_{\rm em}$, expressed in terms of the emitter's ordinary velocity $\mathbf{v}$ and the particle's ordinary velocity $\mathbf{V}$ (both as measured in the receiver's frame).

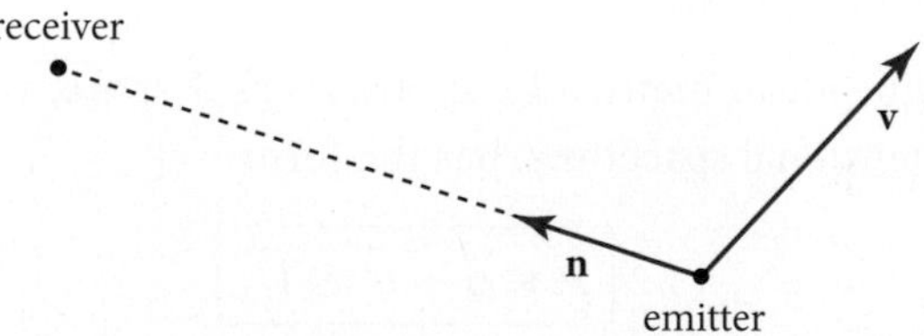

FIGURE 2.6 Geometry for Doppler shift, drawn in a slice of simultaneity of the receiver's inertial frame.

is unchanged. This is called *Lorentz contraction.* As one consequence, heavy ions moving at high speeds in a particle accelerator appear to act like pancakes, squashed along their directions of motion.

T2

Exercise 2.15 *Problem: Allowed and Forbidden Electron-Photon Reactions*
Show, using spacetime diagrams and also using frame-independent calculations, that the law of conservation of 4-momentum forbids a photon to be absorbed by an electron, $e + \gamma \rightarrow e$, and also forbids an electron and a positron to annihilate and produce a single photon, $e^+ + e^- \rightarrow \gamma$ (in the absence of any other particles to take up some of the 4-momentum); but the annihilation to form two photons, $e^+ + e^- \rightarrow 2\gamma$, is permitted.

2.9 Time Travel

2.9

2.9.1 Measurement of Time; Twins Paradox

2.9.1

Time dilation is one facet of a more general phenomenon: time, as measured by ideal clocks, is a personal thing, different for different observers who move through spacetime on different world lines. This is well illustrated by the infamous "twins paradox," in which one twin, Methuselah, remains forever at rest in an inertial frame and the other, Florence, makes a spacecraft journey at high speed and then returns to rest beside Methuselah.

twins paradox

The twins' world lines are depicted in Fig. 2.8a, a spacetime diagram whose axes are those of Methuselah's inertial frame. The time measured by an ideal clock that Methuselah carries is the coordinate time t of his inertial frame; and its total time lapse, from Florence's departure to her return, is $t_{\text{return}} - t_{\text{departure}} \equiv T_{\text{Methuselah}}$. By contrast, the time measured by an ideal clock that Florence carries is her proper time τ (i.e., the square root of the invariant interval (2.4) along her world line). Thus her total time lapse from departure to return is

$$T_{\text{Florence}} = \int d\tau = \int \sqrt{dt^2 - \delta_{ij} dx^i dx^j} = \int_0^{T_{\text{Methuselah}}} \sqrt{1 - v^2}\, dt. \tag{2.40}$$

Here (t, x^i) are the time and space coordinates of Methuselah's inertial frame, and v is Florence's ordinary speed, $v = \sqrt{\delta_{ij}(dx^i/dt)(dx^j/dt)}$, as measured in Methuselah's frame. Obviously, Eq. (2.40) predicts that T_{Florence} is less than $T_{\text{Methuselah}}$. In fact (Ex. 2.16), even if Florence's acceleration is kept no larger than one Earth gravity throughout her trip, and her trip lasts only $T_{\text{Florence}} =$ (a few tens of years), $T_{\text{Methuselah}}$ can be hundreds or thousands or millions or billions of years.

Does this mean that Methuselah actually "experiences" a far longer time lapse, and actually ages far more than Florence? Yes! The time experienced by humans and the aging of the human body are governed by chemical processes, which in turn are governed by the natural oscillation rates of molecules, rates that are constant to high accuracy when measured in terms of ideal time (or, equivalently, proper time τ).

T2

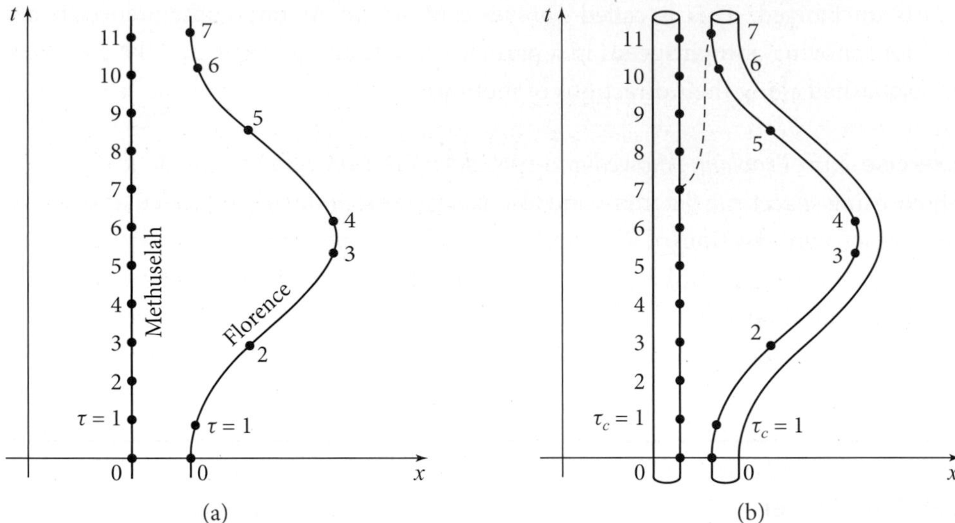

FIGURE 2.8 (a) Spacetime diagram depicting the so-called "twins paradox." Marked along the two world lines are intervals of proper time as measured by the two twins. (b) Spacetime diagram depicting the motions of the two mouths of a wormhole. Marked along the mouths' world tubes are intervals of proper time τ_c as measured by the single clock that sits in the common mouths.

Therefore, a human's experiential time and aging time are the same as the human's proper time—so long as the human is not subjected to such high accelerations as to damage her body.

In effect, then, Florence's spacecraft has functioned as a time machine to carry her far into Methuselah's future, with only a modest lapse of her own proper time (i.e., ideal, experiential, or aging time). This may be a "paradox" in the sense that it is surprising. However, it is in no sense a contradiction. This type of time dilation is routinely measured in high-energy physics storage rings.

2.9.2

2.9.2 Wormholes

Is it also possible, at least in principle, for Florence to construct a time machine that carries her into Methuselah's past—and also her own past? At first sight, the answer would seem to be "yes." Figure 2.8b shows one possible method, using a *wormhole.* [See Frolov and Novikov (1990), Friedman and Higuchi (2006), Everett and Roman (2011) for other approaches.]

wormhole

Wormholes are hypothetical handles in the topology of space. A simple model of a wormhole can be obtained by taking a flat 3-dimensional space, removing from it the interiors of two identical spheres, and identifying the spheres' surfaces so that if one enters the surface of one of the spheres, one immediately finds oneself exiting through the surface of the other. When this is done, there is a bit of strongly localized spatial curvature at the spheres' common surface, so to analyze such a wormhole properly, one must use general relativity rather than special relativity. In particular, it is the laws of general relativity, combined with the laws of quantum field theory, that tell

T2

one how to construct such a wormhole and what kinds of materials are required to hold it open, so things can pass through it. Unfortunately, despite considerable effort, theoretical physicists have not yet deduced definitively whether those laws permit such wormholes to exist and stay open, though indications are pessimistic (Everett and Roman, 2011; Friedman and Higuchi, 2006). However, assuming such wormholes *can* exist, the following special relativistic analysis (Morris et al., 1988) shows how one might be used to construct a machine for backward time travel.

2.9.3 Wormhole as Time Machine

2.9.3

The two identified spherical surfaces are called the wormhole's mouths. Ask Methuselah to keep one mouth with him, forever at rest in his inertial frame, and ask Florence to take the other mouth with her on her high-speed journey. The two mouths' *world tubes* (analogs of world lines for a 3-dimensional object) then have the forms shown in Fig. 2.8b. Suppose that a single ideal clock sits in the wormhole's identified mouths, so that from the external universe one sees it both in Methuselah's wormhole mouth and in Florence's. As seen in Methuselah's mouth, the clock measures his proper time, which is equal to the coordinate time t (see tick marks along the left world tube in Fig. 2.8b). As seen in Florence's mouth, the clock measures her proper time, Eq. (2.40) (see tick marks along the right world tube in Fig. 2.8b). The result should be obvious, if surprising: When Florence returns to rest beside Methuselah, the wormhole has become a time machine. If she travels through the wormhole when the clock reads $\tau_c = 7$, she goes backward in time as seen in Methuselah's (or anyone else's) inertial frame; and then, in fact, traveling along the everywhere timelike world line (dashed in Fig. 2.8b), she is able to meet her younger self before she entered the wormhole.

world tube

This scenario is profoundly disturbing to most physicists because of the dangers of science-fiction-type paradoxes (e.g., the older Florence might kill her younger self, thereby preventing herself from making the trip through the wormhole and killing herself). Fortunately perhaps, it seems likely (though far from certain) that vacuum fluctuations of quantum fields will destroy the wormhole at the moment its mouths' motion first makes backward time travel possible. It may be that this mechanism will always prevent the construction of backward-travel time machines, no matter what tools one uses for their construction (Kay et al., 1997; Kim and Thorne, 1991); but see also contrary indications in research reviewed by Everett and Roman (2011) and Friedman and Higuchi (2006). Whether this is so we likely will not know until the laws of quantum gravity have been mastered.

chronology protection

EXERCISES

Exercise 2.16 *Example: Twins Paradox*

(a) The 4-acceleration of a particle or other object is defined by $\vec{a} \equiv d\vec{u}/d\tau$, where $\vec{u}$ is its 4-velocity and τ is proper time along its world line. Show that, if an observer carries an accelerometer, the magnitude $|\mathbf{a}|$ of the 3-dimensional acceleration $\mathbf{a}$ measured by the accelerometer will always be equal to the magnitude of the observer's 4-acceleration, $|\mathbf{a}| = |\vec{a}| \equiv \sqrt{\vec{a} \cdot \vec{a}}$.

T2

(b) In the twins paradox of Fig. 2.8a, suppose that Florence begins at rest beside Methuselah, then accelerates in Methuselah's x-direction with an acceleration a equal to one Earth gravity, g, for a time $T_{\rm Florence}/4$ as measured by her, then accelerates in the $-x$-direction at g for a time $T_{\rm Florence}/2$, thereby reversing her motion; then she accelerates in the $+x$-direction at g for a time $T_{\rm Florence}/4$, thereby returning to rest beside Methuselah. (This is the type of motion shown in the figure.) Show that the total time lapse as measured by Methuselah is

$$T_{\rm Methuselah} = \frac{4}{g} \sinh\left(\frac{g T_{\rm Florence}}{4}\right). \tag{2.41}$$

(c) Show that in the geometrized units used here, Florence's acceleration (equal to acceleration of gravity at the surface of Earth) is $g = 1.033/\text{yr}$. Plot $T_{\rm Methuselah}$ as a function of $T_{\rm Florence}$, and from your plot estimate $T_{\rm Florence}$ if $T_{\rm Methuselah}$ is the age of the Universe, 14 billion years.

Exercise 2.17 *Challenge: Around the World on TWA*
In a long-ago era when an airline named Trans World Airlines (TWA) flew around the world, Josef Hafele and Richard Keating (1972a) carried out a real live twins paradox experiment: They synchronized two atomic clocks and then flew one around the world eastward on TWA, and on a separate trip, around the world westward, while the other clock remained at home at the Naval Research Laboratory near Washington, D.C. When the clocks were compared after each trip, they were found to have aged differently. Making reasonable estimates for the airplane routing and speeds, compute the difference in aging, and compare your result with the experimental data in Hafele and Keating (1972b). [Note: The rotation of Earth is important, as is the general relativistic gravitational redshift associated with the clocks' altitudes; but the gravitational redshift drops out of the difference in aging, if the time spent at high altitude is the same eastward as westward.]

2.10

2.10 Directional Derivatives, Gradients, and the Levi-Civita Tensor

Derivatives of vectors and tensors in Minkowski spacetime are defined in precisely the same way as in Euclidean space; see Sec. 1.7. Any reader who has not studied that section should do so now. In particular (in extreme brevity, as the explanations and justifications are the same as in Euclidean space):

directional derivative

gradient

The directional derivative of a tensor $\boldsymbol{T}$ along a vector $\vec{A}$ is $\nabla_{\vec{A}}\boldsymbol{T} \equiv \lim_{\epsilon\to 0}(1/\epsilon)[\boldsymbol{T}(\vec{x}_{\mathcal{P}} + \epsilon\vec{A}) - \boldsymbol{T}(\vec{x}_{\mathcal{P}})]$; the gradient $\vec{\nabla}\boldsymbol{T}$ is the tensor that produces the directional derivative when one inserts $\vec{A}$ into its last slot: $\nabla_{\vec{A}}\boldsymbol{T} = \vec{\nabla}\boldsymbol{T}(\_, \_, \_, \vec{A})$. In slot-naming index notation (or in components on a basis), the gradient is denoted $T_{\alpha\beta\gamma;\mu}$. In a Lorentz basis (the basis vectors associated with an inertial reference frame), the components of the gradient are simply the partial derivatives of the tensor,

$T_{\alpha\beta\gamma;\mu} = \partial T_{\alpha\beta\gamma}/\partial x^\mu \equiv T_{\alpha\beta\gamma,\mu}$. (The comma means partial derivative in a Lorentz basis, as in a Cartesian basis.) T2

The gradient and the directional derivative obey all the familiar rules for differentiation of products, for example, $\nabla_{\vec{A}}(\boldsymbol{S} \otimes \boldsymbol{T}) = (\nabla_{\vec{A}}\boldsymbol{S}) \otimes \boldsymbol{T} + \boldsymbol{S} \otimes \nabla_{\vec{A}}\boldsymbol{T}$. The gradient of the metric vanishes, $g_{\alpha\beta;\mu} = 0$. The divergence of a vector is the contraction of its gradient, $\vec{\nabla} \cdot \vec{A} = A_{\alpha;\beta}g^{\alpha\beta} = A^\alpha{}_{;\alpha}$.

Recall that the divergence of the gradient of a tensor in Euclidean space is the Laplacian: $T_{abc;jk}g_{jk} = T_{abc,jk}\delta_{jk} = \partial^2 T_{abc}\partial x^j \partial x^j$. By contrast, in Minkowski spacetime, because $g^{00} = -1$ and $g^{jk} = \delta^{jk}$ in a Lorentz frame, the divergence of the gradient is the wave operator (also called the d'Alembertian):

d'Alembertian

$$T_{\alpha\beta\gamma;\mu\nu}g^{\mu\nu} = T_{\alpha\beta\gamma,\mu\nu}g^{\mu\nu} = -\frac{\partial^2 T_{\alpha\beta\gamma}}{\partial t^2} + \frac{\partial^2 T_{\alpha\beta\gamma}}{\partial x^j \partial x^k}\delta^{jk} = \Box T_{\alpha\beta\gamma}. \tag{2.42}$$

When one sets this to zero, one gets the wave equation.

As in Euclidean space, so also in Minkowski spacetime, there are two tensors that embody the space's geometry: the metric tensor $\boldsymbol{g}$ and the Levi-Civita tensor $\boldsymbol{\epsilon}$. The Levi-Civita tensor in Minkowski spacetime is the tensor that is completely antisymmetric in all its slots and has value $\boldsymbol{\epsilon}(\vec{A}, \vec{B}, \vec{C}, \vec{D}) = +1$ when evaluated on any *right-handed set of orthonormal 4-vectors*—that is, by definition, any orthonormal set for which $\vec{A}$ is timelike and future directed, and $\{\vec{B}, \vec{C}, \vec{D}\}$ are spatial and right-handed. This means that in any right-handed Lorentz basis, the only nonzero components of $\boldsymbol{\epsilon}$ are

Levi-Civita tensor

$$\begin{aligned} \epsilon_{\alpha\beta\gamma\delta} = &+1 \text{ if } \alpha, \beta, \gamma, \delta \text{ is an even permutation of } 0, 1, 2, 3; \\ &-1 \text{ if } \alpha, \beta, \gamma, \delta \text{ is an odd permutation of } 0, 1, 2, 3; \\ &\ 0 \text{ if } \alpha, \beta, \gamma, \delta \text{ are not all different.} \end{aligned} \tag{2.43}$$

By the sign-flip-if-temporal rule, $\epsilon_{0123} = +1$ implies that $\epsilon^{0123} = -1$.

2.11 Nature of Electric and Magnetic Fields; Maxwell's Equations

2.11

Now that we have introduced the gradient and the Levi-Civita tensor, we can study the relationship of the relativistic version of electrodynamics to the nonrelativistic (Newtonian) version. In doing so, we use Gaussian units (with the speed of light set to 1), as is conventional among relativity theorists, and as does Jackson (1999) in his classic textbook, switching from SI to Gaussian when he moves into the relativistic domain.

Consider a particle with charge q, rest mass m, and 4-velocity $\vec{u}$ interacting with an electromagnetic field $\boldsymbol{F}(\_\_, \_\_)$. In index notation, the electromagnetic 4-force acting on the particle [Eq. (2.20)] is

$$dp^\alpha/d\tau = qF^{\alpha\beta}u_\beta. \tag{2.44}$$

T2

Let us examine this 4-force in some arbitrary inertial reference frame in which the particle's ordinary-velocity components are $v^j = v_j$ and its 4-velocity components are $u^0 = \gamma$, $u^j = \gamma v^j$ [Eqs. (2.25c)]. Anticipating the connection with the nonrelativistic viewpoint, we introduce the following notation for the contravariant components of the antisymmetric electromagnetic field tensor:

$$F^{0j} = -F^{j0} = +F_{j0} = -F_{0j} = E_j, \qquad F^{ij} = F_{ij} = \epsilon_{ijk} B_k. \tag{2.45}$$

Inserting these components of $\boldsymbol{F}$ and $\vec{u}$ into Eq. (2.44) and using the relationship $dt/d\tau = u^0 = \gamma$ between t and τ derivatives, we obtain for the components of the 4-force $dp_j/d\tau = \gamma dp_j/dt = q(F_{j0}u^0 + F_{jk}u^k) = qu^0(F_{j0} + F_{jk}v^k) = q\gamma(E_j + \epsilon_{jki}v_k B_i)$ and $dp^0/d\tau = \gamma dp^0/dt = qF^{0j}u_j = q\gamma E_j v_j$. Dividing by γ, converting into 3-space index notation, and denoting the particle's energy by $\mathcal{E} = p^0$, we bring these into the familiar Lorentz-force form

Lorentz force

$$d\mathbf{p}/dt = q(\mathbf{E} + \mathbf{v} \times \mathbf{B}), \qquad d\mathcal{E}/dt = q\mathbf{v} \cdot \mathbf{E}. \tag{2.46}$$

Evidently, **E** is the electric field and **B** the magnetic field as measured in our chosen Lorentz frame.

This may be familiar from standard electrodynamics textbooks (e.g., Jackson, 1999). Not so familiar, but very important, is the following geometric interpretation of **E** and **B**.

The electric and magnetic fields **E** and **B** are spatial vectors as measured in the chosen inertial frame. We can also regard them as 4-vectors that lie in a 3-surface of simultaneity $t =$ const of the chosen frame, i.e., that are orthogonal to the 4-velocity (denote it $\vec{w}$) of the frame's observers (cf. Figs. 2.5 and 2.9). We shall denote this 4-vector version of **E** and **B** by $\vec{E}_{\vec{w}}$ and $\vec{B}_{\vec{w}}$, where the subscript $\vec{w}$ identifies the 4-velocity of the observer who measures these fields. These fields are depicted in Fig. 2.9.

In the rest frame of the observer $\vec{w}$, the components of $\vec{E}_{\vec{w}}$ are $E^0_{\vec{w}} = 0$, $E^j_{\vec{w}} = E_j = F_{j0}$ [the E_j appearing in Eqs. (2.45)], and similarly for $\vec{B}_{\vec{w}}$; the components of $\vec{w}$ are $w^0 = 1$, $w^j = 0$. Therefore, in this frame Eqs. (2.45) can be rewritten as

electric and magnetic fields measured by an observer

$$\boxed{E^\alpha_{\vec{w}} = F^{\alpha\beta} w_\beta, \qquad B^\beta_{\vec{w}} = \frac{1}{2}\epsilon^{\alpha\beta\gamma\delta} F_{\gamma\delta} w_\alpha.} \tag{2.47a}$$

[To verify this, insert the above components of $\boldsymbol{F}$ and $\vec{w}$ into these equations and, after some algebra, recover Eqs. (2.45) along with $E^0_{\vec{w}} = B^0_{\vec{w}} = 0$.] Equations (2.47a) say that in one special reference frame, that of the observer $\vec{w}$, the components of the 4-vectors on the left and on the right are equal. This implies that in every Lorentz frame the components of these 4-vectors will be equal; that is, Eqs. (2.47a) are true when one regards them as geometric, frame-independent equations written in slot-naming index notation. *These equations enable one to compute the electric and magnetic fields measured by an observer (viewed as 4-vectors in the observer's 3-surface of simultaneity)*

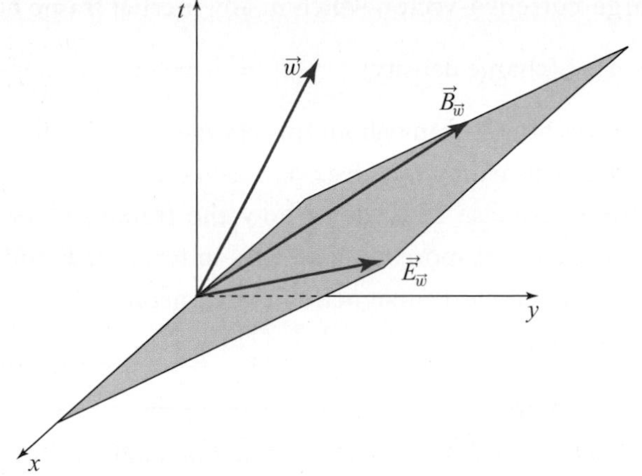

FIGURE 2.9 The electric and magnetic fields measured by an observer with 4-velocity $\vec{w}$, shown as 4-vectors $\vec{E}_{\vec{w}}$ and $\vec{B}_{\vec{w}}$ that lie in the observer's 3-surface of simultaneity (green 3-surface orthogonal to $\vec{w}$).

from the observer's 4-velocity and the electromagnetic field tensor, without the aid of any basis or reference frame.

Equations (2.47a) embody explicitly the following important fact. Although the electromagnetic field tensor $\boldsymbol{F}$ is a geometric, frame-independent quantity, the electric and magnetic fields $\vec{E}_{\vec{w}}$ and $\vec{B}_{\vec{w}}$ individually depend for their existence on a specific choice of observer (with 4-velocity $\vec{w}$), that is, a specific choice of inertial reference frame, or in other words, a specific choice of the split of spacetime into a 3-space (the 3-surface of simultaneity orthogonal to the observer's 4-velocity $\vec{w}$) and corresponding time (the Lorentz time of the observer's reference frame). *Only after making such an observer-dependent 3+1 split of spacetime into space plus time do the electric field and the magnetic field come into existence as separate entities.* Different observers with different 4-velocities $\vec{w}$ make this spacetime split in different ways, thereby resolving the frame-independent $\boldsymbol{F}$ into different electric and magnetic fields $\vec{E}_{\vec{w}}$ and $\vec{B}_{\vec{w}}$.

By the same procedure as we used to derive Eqs. (2.47a), one can derive the inverse relationship, the following expression for the electromagnetic field tensor in terms of the (4-vector) electric and magnetic fields measured by some observer:

$$F^{\alpha\beta} = w^\alpha E^\beta_{\vec{w}} - E^\alpha_{\vec{w}} w^\beta + \epsilon^{\alpha\beta}{}_{\gamma\delta} w^\gamma B^\delta_{\vec{w}}. \tag{2.47b}$$

Maxwell's equations in geometric, frame-independent form are (in Gaussian units)[5]

Maxwell's equations

$$F^{\alpha\beta}{}_{;\beta} = 4\pi J^\alpha,$$

$$\epsilon^{\alpha\beta\gamma\delta} F_{\gamma\delta;\beta} = 0; \quad \text{i.e.,} \quad F_{\alpha\beta;\gamma} + F_{\beta\gamma;\alpha} + F_{\gamma\alpha;\beta} = 0. \tag{2.48}$$

5. In SI units the 4π gets replaced by $\mu_0 = 1/\epsilon_0$, corresponding to the different units for the charge-current 4-vector.

T2

Here $\vec{J}$ is the charge-current 4-vector, which in any inertial frame has components

$$J^0 = \rho_e = (\text{charge density}), \qquad J^i = j_i = (\text{current density}). \tag{2.49}$$

Exercise 2.19 describes how to think about this charge density and current density as geometric objects determined by the observer's 4-velocity or 3+1 split of spacetime into space plus time. Exercise 2.20 shows how the frame-independent Maxwell's equations (2.48) reduce to the more familiar ones in terms of **E** and **B**. Exercise 2.21 explores potentials for the electromagnetic field in geometric, frame-independent language and the 3+1 split.

EXERCISES

Exercise 2.18 *Derivation and Practice: Reconstruction of* **F**

Derive Eq. (2.47b) by the same method as was used to derive Eq. (2.47a). Then show, by a geometric, frame-independent calculation, that Eq. (2.47b) implies Eq. (2.47a).

Exercise 2.19 *Problem: 3+1 Split of Charge-Current 4-Vector*

Just as the electric and magnetic fields measured by some observer can be regarded as 4-vectors $\vec{E}_{\vec{w}}$ and $\vec{B}_{\vec{w}}$ that live in the observer's 3-space of simultaneity, so also the charge density and current density that the observer measures can be regarded as a scalar $\rho_{\vec{w}}$ and 4-vector $\vec{j}_{\vec{w}}$ that live in the 3-space of simultaneity. Derive geometric, frame-independent equations for $\rho_{\vec{w}}$ and $\vec{j}_{\vec{w}}$ in terms of the charge-current 4-vector $\vec{J}$ and the observer's 4-velocity $\vec{w}$, and derive a geometric expression for $\vec{J}$ in terms of $\rho_{\vec{w}}$, $\vec{j}_{\vec{w}}$, and $\vec{w}$.

Exercise 2.20 *Problem: Frame-Dependent Version of Maxwell's Equations*

By performing a 3+1 split on the geometric version of Maxwell's equations (2.48), derive the elementary, frame-dependent version

3+1 split of Maxwell's equations

$$\nabla \cdot \mathbf{E} = 4\pi \rho_e, \qquad \nabla \times \mathbf{B} - \frac{\partial \mathbf{E}}{\partial t} = 4\pi \mathbf{j},$$

$$\nabla \cdot \mathbf{B} = 0, \qquad \nabla \times \mathbf{E} + \frac{\partial \mathbf{B}}{\partial t} = 0. \tag{2.50}$$

Exercise 2.21 *Problem: Potentials for the Electromagnetic Field*

(a) Express the electromagnetic field tensor as an antisymmetrized gradient of a 4-vector potential: in slot-naming index notation

$$F_{\alpha\beta} = A_{\beta;\alpha} - A_{\alpha;\beta}. \tag{2.51a}$$

Show that, whatever may be the 4-vector potential $\vec{A}$, the second of Maxwell's equations (2.48) is automatically satisfied. Show further that the electromagnetic field tensor is unaffected by a gauge change of the form

$$\vec{A}_{\text{new}} = \vec{A}_{\text{old}} + \vec{\nabla}\psi, \tag{2.51b}$$

where ψ is a scalar field (the generator of the gauge change). Show, finally, that it is possible to find a gauge-change generator that enforces *Lorenz gauge*

$$\vec{\nabla} \cdot \vec{A} = 0 \tag{2.51c}$$

on the new 4-vector potential, and show that in this gauge, the first of Maxwell's equations (2.48) becomes (in Gaussian units)

$$\Box \vec{A} = -4\pi \vec{J}; \quad \text{i.e.,} \quad A^{\alpha;\mu}{}_{\mu} = -4\pi J^{\alpha}. \tag{2.51d}$$

(b) Introduce an inertial reference frame, and in that frame split $\boldsymbol{F}$ into the electric and magnetic fields $\mathbf{E}$ and $\mathbf{B}$, split $\vec{J}$ into the charge and current densities ρ_e and $\mathbf{j}$, and split the vector potential into a scalar potential and a 3-vector potential

$$\phi = -A_0, \qquad \mathbf{A} = \text{spatial part of } \vec{A}. \tag{2.51e}$$

Deduce the 3+1 splits of Eqs. (2.51a)–(2.51d), and show that they take the form given in standard textbooks on electromagnetism.

2.12 Volumes, Integration, and Conservation Laws

2.12

2.12.1 Spacetime Volumes and Integration

2.12.1

In Minkowski spacetime as in Euclidean 3-space (Sec. 1.8), the Levi-Civita tensor is the tool by which one constructs volumes. The 4-dimensional parallelepiped whose legs are the four vectors $\vec{A}$, $\vec{B}$, $\vec{C}$, $\vec{D}$ has a 4-dimensional volume given by the analog of Eqs. (1.25) and (1.26):

4-volume

$$\text{4-volume} = \epsilon_{\alpha\beta\gamma\delta} A^{\alpha} B^{\beta} C^{\gamma} D^{\delta} = \boldsymbol{\epsilon}(\vec{A}, \vec{B}, \vec{C}, \vec{D}) = \det \begin{bmatrix} A^0 & B^0 & C^0 & D^0 \\ A^1 & B^1 & C^1 & D^1 \\ A^2 & B^2 & C^2 & D^2 \\ A^3 & B^3 & C^3 & D^3 \end{bmatrix}. \tag{2.52}$$

Note that this 4-volume is positive if the set of vectors $\{\vec{A}, \vec{B}, \vec{C}, \vec{D}\}$ is right-handed and negative if left-handed [cf. Eq. (2.43)].

Equation (2.52) provides us a way to perform volume integrals over 4-dimensional Minkowski spacetime. To integrate a smooth tensor field $\boldsymbol{T}$ over some 4-dimensional region $\mathcal{V}$ of spacetime, we need only divide $\mathcal{V}$ up into tiny parallelepipeds, multiply the 4-volume $d\Sigma$ of each parallelepiped by the value of $\boldsymbol{T}$ at its center, add, and take the limit. In any right-handed Lorentz coordinate system, the 4-volume of a tiny parallelepiped whose edges are dx^{α} along the four orthogonal coordinate axes is $d\Sigma = \boldsymbol{\epsilon}(dt\, \vec{e}_0, dx\, \vec{e}_x, dy\, \vec{e}_y, dz\, \vec{e}_z) = \epsilon_{0123}\, dt\, dx\, dy\, dz = dt\, dx\, dy\, dz$ (the analog of $dV = dx\, dy\, dz$). Correspondingly, the integral of $\boldsymbol{T}$ over $\mathcal{V}$ can be expressed as

$$\int_{\mathcal{V}} T^{\alpha\beta\gamma} d\Sigma = \int_{\mathcal{V}} T^{\alpha\beta\gamma} dt\, dx\, dy\, dz. \tag{2.53}$$

T2

By analogy with the vectorial area (1.27) of a parallelogram in 3-space, any 3-dimensional parallelepiped in spacetime with legs $\vec{A}$, $\vec{B}$, $\vec{C}$ has a vectorial 3-volume $\vec{\Sigma}$ (not to be confused with the scalar 4-volume Σ) defined by

vectorial 3-volume

$$\vec{\Sigma}(\_\_) = \boldsymbol{\epsilon}(\_\_, \vec{A}, \vec{B}, \vec{C}); \qquad \Sigma_\mu = \epsilon_{\mu\alpha\beta\gamma} A^\alpha B^\beta C^\gamma. \tag{2.54}$$

Here we have written the 3-volume vector both in abstract notation and in slot-naming index notation. This 3-volume vector has one empty slot, ready and waiting for a fourth vector ("leg") to be inserted, so as to compute the 4-volume Σ of a 4-dimensional parallelepiped.

Notice that the 3-volume vector $\vec{\Sigma}$ is orthogonal to each of its three legs (because of the antisymmetry of $\boldsymbol{\epsilon}$), and thus (unless it is null) it can be written as $\vec{\Sigma} = V\vec{n}$, where V is the magnitude of the 3-volume, and $\vec{n}$ is the unit normal to the three legs.

Interchanging any two legs of the parallelepiped reverses the 3-volume's sign. Consequently, the 3-volume is characterized not only by its legs but also by the order of its legs, or equally well, in two other ways: (i) by the direction of the vector $\vec{\Sigma}$ (reverse the order of the legs, and the direction of $\vec{\Sigma}$ will reverse); and (ii) by the *sense* of the 3-volume, defined as follows. Just as a 2-volume (i.e., a segment of a plane) in 3-dimensional space has two sides, so a 3-volume in 4-dimensional spacetime has two sides (Fig. 2.10). Every vector $\vec{D}$ for which $\vec{\Sigma} \cdot \vec{D} > 0$ points out the *positive side* of the 3-volume $\vec{\Sigma}$. Vectors $\vec{D}$ with $\vec{\Sigma} \cdot \vec{D} < 0$ point out its *negative side.* When something moves through, reaches through, or points through the 3-volume from its negative side to its positive side, we say that this thing is moving or reaching or pointing in the "positive sense;" similarly for "negative sense." The examples shown in Fig. 2.10 should make this more clear.

positive and negative sides, and sense of 3-volume

Figure 2.10a shows two of the three legs of the volume vector $\vec{\Sigma} = \boldsymbol{\epsilon}(\_\_, \Delta x\vec{e}_x, \Delta y\vec{e}_y, \Delta z\vec{e}_z)$, where $\{t, x, y, z\}$ are the coordinates, and $\{\vec{e}_\alpha\}$ is the corresponding right-handed basis of a specific Lorentz frame. It is easy to show that this $\vec{\Sigma}$ can also be written as $\vec{\Sigma} = -\Delta V\vec{e}_0$, where ΔV is the ordinary volume of the parallelepiped as measured by an observer in the chosen Lorentz frame, $\Delta V = \Delta x \Delta y \Delta z$. Thus, the direction of the vector $\vec{\Sigma}$ is toward the past (direction of decreasing Lorentz time t). From this, and the fact that timelike vectors have negative squared length, it is easy to infer that $\vec{\Sigma} \cdot \vec{D} > 0$ if and only if the vector $\vec{D}$ points out of the "future" side of the 3-volume (the side of increasing Lorentz time t); therefore, the positive side of $\vec{\Sigma}$ is the future side. It follows that the vector $\vec{\Sigma}$ points in the negative sense of its own 3-volume.

Figure 2.10b shows two of the three legs of the volume vector $\vec{\Sigma} = \boldsymbol{\epsilon}(\_\_, \Delta t\vec{e}_t, \Delta y\vec{e}_y, \Delta z\vec{e}_z) = -\Delta t \Delta A\vec{e}_x$ (with $\Delta A = \Delta y \Delta z$). In this case, $\vec{\Sigma}$ points in its own positive sense.

This peculiar behavior is completely general. When the normal to a 3-volume is timelike, its volume vector $\vec{\Sigma}$ points in the negative sense; when the normal is spacelike, $\vec{\Sigma}$ points in the positive sense. And as it turns out, when the normal is null, $\vec{\Sigma}$

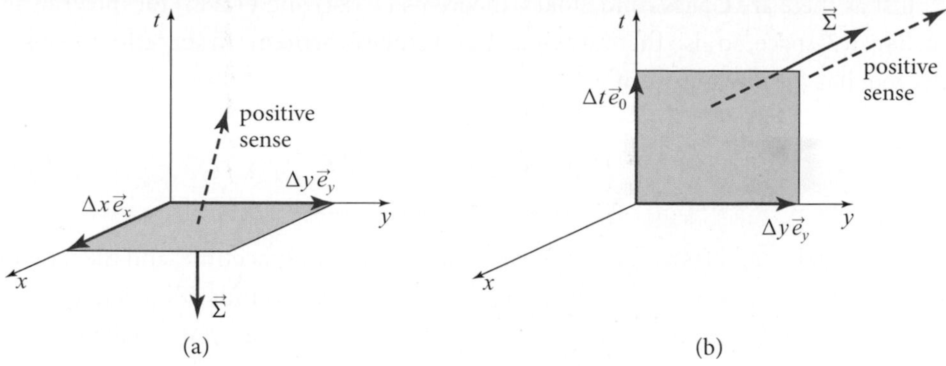

FIGURE 2.10 Spacetime diagrams depicting 3-volumes in 4-dimensional spacetime, with one spatial dimension (that along the z direction) suppressed.

lies in the 3-volume (parallel to its one null leg) and thus points neither in the positive sense nor the negative.[6]

Note the physical interpretations of the 3-volumes of Fig. 2.10. Figure 2.10a shows an instantaneous snapshot of an ordinary, spatial parallelepiped, whereas Fig. 2.10b shows the 3-dimensional region in spacetime swept out during time Δt by the parallelogram with legs $\Delta y\vec{e}_y$, $\Delta z\vec{e}_z$ and with area $\Delta A = \Delta y \Delta z$.

Vectorial 3-volume elements can be used to construct integrals over 3-dimensional volumes (also called 3-dimensional surfaces) in spacetime, for example, $\int_{\mathcal{V}_3} \vec{A} \cdot d\vec{\Sigma}$. More specifically, let (a, b, c) be (possibly curvilinear) coordinates in the 3-surface (3-volume) $\mathcal{V}_3$, and denote by $\vec{x}(a, b, c)$ the spacetime point $\mathcal{P}$ on $\mathcal{V}_3$ whose coordinate values are (a, b, c). Then $(\partial\vec{x}/\partial a)da$, $(\partial\vec{x}/\partial b)db$, $(\partial\vec{x}/\partial c)dc$ are the vectorial legs of the elementary parallelepiped whose corners are at (a, b, c), $(a + da, b, c)$, $(a, b + db, c)$, and so forth; and the spacetime components of these vectorial legs are $(\partial x^\alpha/\partial a)da$, $(\partial x^\alpha/\partial b)db$, $(\partial x^\alpha/\partial c)dc$. The 3-volume of this elementary parallelepiped is $d\vec{\Sigma} = \boldsymbol{\epsilon}\left(\_\_, (\partial\vec{x}/\partial a)da, (\partial\vec{x}/\partial b)db, (\partial\vec{x}/\partial c)dc\right)$, which has spacetime components

$$d\Sigma_\mu = \epsilon_{\mu\alpha\beta\gamma} \frac{\partial x^\alpha}{\partial a} \frac{\partial x^\beta}{\partial b} \frac{\partial x^\gamma}{\partial c} da db dc. \tag{2.55}$$

This is the integration element to be used when evaluating

$$\int_{\mathcal{V}_3} \vec{A} \cdot d\vec{\Sigma} = \int_{\mathcal{V}_3} A^\mu d\Sigma_\mu. \tag{2.56}$$

See Ex. 2.22 for an example.

6. This peculiar behavior gets replaced by a simpler description if one uses one-forms rather than vectors to describe 3-volumes; see, for example, Misner, Thorne, and Wheeler (1973, Box 5.2).

T2

Just as there are Gauss's and Stokes' theorems (1.28a) and (1.28b) for integrals in Euclidean 3-space, so also there are Gauss's and Stokes' theorems in spacetime. Gauss's theorem has the obvious form

Gauss's theorem

$$\boxed{\int_{\mathcal{V}_4} (\vec{\nabla} \cdot \vec{A}) d\Sigma = \int_{\partial \mathcal{V}_4} \vec{A} \cdot d\vec{\Sigma},} \tag{2.57}$$

where the first integral is over a 4-dimensional region $\mathcal{V}_4$ in spacetime, and the second is over the 3-dimensional boundary $\partial\mathcal{V}_4$ of $\mathcal{V}_4$, with the boundary's positive sense pointing outward, away from $\mathcal{V}_4$ (just as in the 3-dimensional case). We shall not write down the 4-dimensional Stokes' theorem, because it is complicated to formulate with the tools we have developed thus far; easy formulation requires *differential forms* (e.g., Flanders, 1989), which we shall not introduce in this book.

2.12.2

2.12.2 Conservation of Charge in Spacetime

In this section, we use integration over a 3-dimensional region in 4-dimensional spacetime to construct an elegant, frame-independent formulation of the law of conservation of electric charge.

We begin by examining the geometric meaning of the charge-current 4-vector $\vec{J}$. We defined $\vec{J}$ in Eq. (2.49) in terms of its components. The spatial component $J^x = J_x = J(\vec{e}_x)$ is equal to the x component of current density j_x: it is the amount Q of charge that flows across a unit surface area lying in the y-z plane in a unit time (i.e., the charge that flows across the unit 3-surface $\vec{\Sigma} = \vec{e}_x$). In other words, $\vec{J}(\vec{\Sigma}) = \vec{J}(\vec{e}_x)$ *is the total charge Q that flows across $\vec{\Sigma} = \vec{e}_x$ in $\vec{\Sigma}$'s positive sense* and similarly for the other spatial directions. The temporal component $J^0 = -J_0 = \vec{J}(-\vec{e}_0)$ is the charge density ρ_e: it is the total charge Q in a unit spatial volume. This charge is carried by particles that are traveling through spacetime from past to future and pass through the unit 3-surface (3-volume) $\vec{\Sigma} = -\vec{e}_0$. Therefore, $\vec{J}(\vec{\Sigma}) = \vec{J}(-\vec{e}_0)$ *is the total charge Q that flows through $\vec{\Sigma} = -\vec{e}_0$ in its positive sense.* This interpretation is the same one we deduced for the spatial components of $\vec{J}$.

This makes it plausible, and indeed one can show, that *for any small 3-surface $\vec{\Sigma}$,*

charge-current 4-vector

$\vec{J}(\vec{\Sigma}) \equiv J^\alpha \Sigma_\alpha$ is the total charge Q that flows across $\vec{\Sigma}$ in its positive sense.

This property of the charge-current 4-vector is the foundation for our frame-independent formulation of the law of charge conservation. Let $\mathcal{V}$ be a compact 4-dimensional region of spacetime and denote by $\partial\mathcal{V}$ its boundary, a closed 3-surface in 4-dimensional spacetime (Fig. 2.11). The charged media (fluids, solids, particles, etc.) present in spacetime carry electric charge through $\mathcal{V}$, from the past toward the future. The law of charge conservation says that all the charge that enters $\mathcal{V}$ through the past part of its boundary $\partial\mathcal{V}$ must exit through the future part of its boundary. If we choose the positive sense of the boundary's 3-volume element $d\vec{\Sigma}$ to point out of $\mathcal{V}$ (toward the past on the bottom boundary and toward the future on the top), then this *global law of charge conservation* can be expressed as

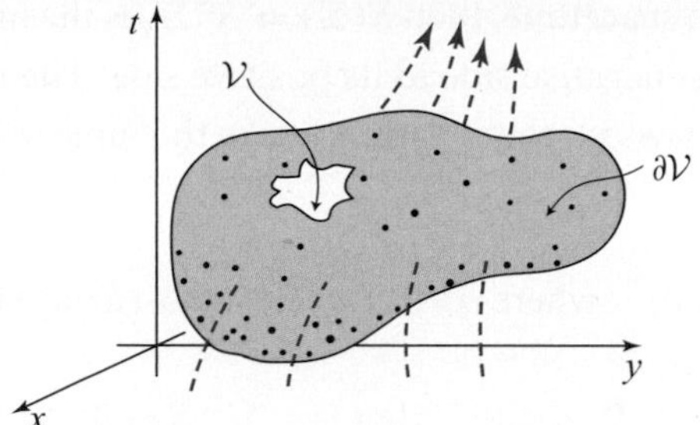

FIGURE 2.11 The 4-dimensional region $\mathcal{V}$ in spacetime and its closed 3-boundary $\partial\mathcal{V}$ (green surface), used in formulating the law of charge conservation. The dashed lines symbolize, heuristically, the flow of charge from the past toward the future.

$$\boxed{\int_{\partial\mathcal{V}} J^{\alpha} d\Sigma_{\alpha} = 0.} \tag{2.58}$$

global law of charge conservation

When each tiny charge q enters $\mathcal{V}$ through its past boundary, it contributes negatively to the integral, since it travels through $\partial\mathcal{V}$ in the negative sense (from positive side of $\partial\mathcal{V}$ toward negative side); and when that same charge exits $\mathcal{V}$ through its future boundary, it contributes positively. Therefore, its net contribution is zero, and similarly for all other charges.

In Ex. 2.23 you will show that, when this global law of charge conservation (2.58) is subjected to a 3+1 split of spacetime into space plus time, it becomes the nonrelativistic integral law of charge conservation (1.29).

This global conservation law can be converted into a *local conservation law* with the help of the 4-dimensional Gauss's theorem (2.57), $\int_{\partial\mathcal{V}} J^{\alpha} d\Sigma_{\alpha} = \int_{\mathcal{V}} J^{\alpha}{}_{;\alpha} d\Sigma$. Since the left-hand side vanishes, so must the right-hand side; and for this 4-volume integral to vanish for every choice of $\mathcal{V}$, it is necessary that the integrand vanish everywhere in spacetime:

$$\boxed{J^{\alpha}{}_{;\alpha} = 0; \quad \text{that is,} \quad \vec{\nabla} \cdot \vec{J} = 0.} \tag{2.59}$$

local law of charge conservation

In a specific but arbitrary Lorentz frame (i.e., in a 3+1 split of spacetime into space plus time), Eq. (2.59) becomes the standard differential law of charge conservation (1.30).

2.12.3 Conservation of Particles, Baryon Number, and Rest Mass

2.12.3

Any conserved scalar quantity obeys conservation laws of the same form as those for electric charge. For example, if the number of particles of some species (e.g., electrons, protons, or photons) is conserved, then we can introduce for that species a *number-flux 4-vector* $\vec{S}$ (analog of charge-current 4-vector $\vec{J}$): in any Lorentz frame, S^0 is the number density of particles, also designated n, and S^j is the particle flux. If $\vec{\Sigma}$ is a small

number-flux 4-vector

T2

3-volume (3-surface) in spacetime, then $\vec{S}(\vec{\Sigma}) = S^\alpha \Sigma_\alpha$ is the number of particles that pass through Σ from its negative side to its positive side. The frame-invariant global and local conservation laws for these particles take the same form as those for electric charge:

laws of particle conservation

$$\int_{\partial\mathcal{V}} S^\alpha d\Sigma_\alpha = 0, \quad \text{where } \partial\mathcal{V} \text{ is any closed 3-surface in spacetime,} \tag{2.60a}$$

$$S^\alpha{}_{;\alpha} = 0; \quad \text{that is,} \quad \vec{\nabla}\cdot\vec{S} = 0. \tag{2.60b}$$

When fundamental particles (e.g., protons and antiprotons) are created and destroyed by quantum processes, the total baryon number (number of baryons minus number of antibaryons) is still conserved—or at least this is so to the accuracy of all experiments performed thus far. We shall assume it so in this book. This law of baryon-number conservation takes the forms of Eqs. (2.60), with $\vec{S}$ the number-flux 4-vector for baryons (with antibaryons counted negatively).

It is useful to express this baryon-number conservation law in Newtonian-like language by introducing a universally agreed-on mean rest mass per baryon $\bar{m}_B$. This $\bar{m}_B$ is often taken to be 1/56 the mass of an ^{56}Fe (iron-56) atomic nucleus, since ^{56}Fe is the nucleus with the tightest nuclear binding (i.e., the endpoint of thermonuclear evolution in stars). We multiply the baryon number-flux 4-vector $\vec{S}$ by this mean rest mass per baryon to obtain a rest-mass-flux 4-vector

rest-mass-flux 4-vector

$$\vec{S}_{\rm rm} = \bar{m}_B \vec{S}, \tag{2.61}$$

which (since $\bar{m}_B$ is, by definition, a constant) satisfies the same conservation laws (2.60) as baryon number.

For such media as fluids and solids, in which the particles travel only short distances between collisions or strong interactions, it is often useful to resolve the particle number-flux 4-vector and the rest-mass-flux 4-vector into a 4-velocity of the medium $\vec{u}$ (i.e., the 4-velocity of the frame in which there is a vanishing net spatial flux of particles), and the particle number density n_o or rest mass density ρ_o as measured in the medium's rest frame:

$$\vec{S} = n_o \vec{u}, \qquad \vec{S}_{\rm rm} = \rho_o \vec{u}. \tag{2.62}$$

See Ex. 2.24.

rest-mass conservation

We make use of the conservation laws $\vec{\nabla}\cdot\vec{S} = 0$ and $\vec{\nabla}\cdot\vec{S}_{\rm rm} = 0$ for particles and rest mass later in this book (e.g., when studying relativistic fluids); and we shall find the expressions (2.62) for the number-flux 4-vector and rest-mass-flux 4-vector quite useful. See, for example, the discussion of relativistic shock waves in Ex. 17.9.

EXERCISES

Exercise 2.22 *Practice and Example: Evaluation of 3-Surface Integral in Spacetime*
In Minkowski spacetime, the set of all events separated from the origin by a timelike interval a^2 is a 3-surface, the hyperboloid $t^2 - x^2 - y^2 - z^2 = a^2$, where $\{t, x, y, z\}$

T2

are Lorentz coordinates of some inertial reference frame. On this hyperboloid, introduce coordinates $\{\chi, \theta, \phi\}$ such that

$$t = a \cosh \chi, \quad x = a \sinh \chi \sin \theta \cos \phi,$$
$$y = a \sinh \chi \sin \theta \sin \phi, \quad z = a \sinh \chi \cos \theta. \tag{2.63}$$

Note that χ is a radial coordinate and (θ, ϕ) are spherical polar coordinates. Denote by $\mathcal{V}_3$ the portion of the hyperboloid with radius $\chi \le b$.

(a) Verify that for all values of (χ, θ, ϕ), the points defined by Eqs. (2.63) do lie on the hyperboloid.

(b) On a spacetime diagram, draw a picture of $\mathcal{V}_3$, the $\{\chi, \theta, \phi\}$ coordinates, and the elementary volume element (vector field) $d\vec{\Sigma}$ [Eq. (2.55)].

(c) Set $\vec{A} \equiv \vec{e}_0$ (the temporal basis vector), and express $\int_{\mathcal{V}_3} \vec{A} \cdot d\vec{\Sigma}$ as an integral over $\{\chi, \theta, \phi\}$. Evaluate the integral.

(d) Consider a closed 3-surface consisting of the segment $\mathcal{V}_3$ of the hyperboloid as its top, the hypercylinder $\{x^2 + y^2 + z^2 = a^2 \sinh^2 b,\ 0 < t < a \cosh b\}$ as its sides, and the sphere $\{x^2 + y^2 + z^2 \le a^2 \sinh^2 b,\ t = 0\}$ as its bottom. Draw a picture of this closed 3-surface on a spacetime diagram. Use Gauss's theorem, applied to this 3-surface, to show that $\int_{\mathcal{V}_3} \vec{A} \cdot d\vec{\Sigma}$ is equal to the 3-volume of its spherical base.

Exercise 2.23 *Derivation and Example: Global Law of Charge Conservation in an Inertial Frame*

Consider the global law of charge conservation $\int_{\partial\mathcal{V}} J^\alpha d\Sigma_\alpha = 0$ for a special choice of the closed 3-surface $\partial\mathcal{V}$: The bottom of $\partial\mathcal{V}$ is the ball $\{t = 0, x^2 + y^2 + z^2 \le a^2\}$, where $\{t, x, y, z\}$ are the Lorentz coordinates of some inertial frame. The sides are the spherical world tube $\{0 \le t \le T, x^2 + y^2 + z^2 = a^2\}$. The top is the ball $\{t = T, x^2 + y^2 + z^2 \le a^2\}$.

(a) Draw this 3-surface in a spacetime diagram.

(b) Show that for this $\partial\mathcal{V}$, $\int_{\partial\mathcal{V}} J^\alpha d\Sigma_\alpha = 0$ is a time integral of the nonrelativistic integral conservation law (1.29) for charge.

Exercise 2.24 *Example: Rest-Mass-Flux 4-Vector, Lorentz Contraction of Rest-Mass Density, and Rest-Mass Conservation for a Fluid*

Consider a fluid with 4-velocity $\vec{u}$ and rest-mass density ρ_o as measured in the fluid's rest frame.

(a) From the physical meanings of $\vec{u}$, ρ_o, and the rest-mass-flux 4-vector $\vec{S}_{\rm rm}$, deduce Eqs. (2.62).

(b) Examine the components of $\vec{S}_{\rm rm}$ in a reference frame where the fluid moves with ordinary velocity $\mathbf{v}$. Show that $S^0 = \rho_o \gamma$, and $S^j = \rho_o \gamma v^j$, where $\gamma = 1/\sqrt{1 - \mathbf{v}^2}$. Explain the physical interpretation of these formulas in terms of Lorentz contraction.

T2

(c) Show that the law of conservation of rest mass $\vec{\nabla} \cdot \vec{S}_{\rm rm} = 0$ takes the form

$$\frac{d\rho_o}{d\tau} = -\rho_o \vec{\nabla} \cdot \vec{u}, \tag{2.64}$$

where $d/d\tau$ is derivative with respect to proper time moving with the fluid.

(d) Consider a small 3-dimensional volume V of the fluid, whose walls move with the fluid (so if the fluid expands, V increases). Explain why the law of rest-mass conservation must take the form $d(\rho_o V)/d\tau = 0$. Thereby deduce that

$$\vec{\nabla} \cdot \vec{u} = (1/V)(dV/d\tau). \tag{2.65}$$

2.13

2.13 Stress-Energy Tensor and Conservation of 4-Momentum

2.13.1

2.13.1 Stress-Energy Tensor

GEOMETRIC DEFINITION

We conclude this chapter by formulating the law of 4-momentum conservation in ways analogous to our laws of conservation of charge, particles, baryon number, and rest mass. This task is not trivial, since 4-momentum is a vector in spacetime, while charge, particle number, baryon number, and rest mass are scalar quantities. Correspondingly, the density-flux of 4-momentum must have one more slot than the density-fluxes of charge, baryon number, and rest mass, $\vec{J}$, $\vec{S}$ and $\vec{S}_{\rm rm}$, respectively;

stress-energy tensor

it must be a second-rank tensor. We call it the *stress-energy tensor* and denote it $\boldsymbol{T}(\_, \_)$. It is a generalization of the Newtonian stress tensor to 4-dimensional spacetime.

Consider a medium or field flowing through 4-dimensional spacetime. As it crosses a tiny 3-surface $\vec{\Sigma}$, it transports a net electric charge $\vec{J}(\vec{\Sigma})$ from the negative side of $\vec{\Sigma}$ to the positive side, and net baryon number $\vec{S}(\vec{\Sigma})$ and net rest mass $\vec{S}_{\rm rm}(\vec{\Sigma})$. Similarly, it transports a net 4-momentum $\boldsymbol{T}(\_, \vec{\Sigma})$ from the negative side to the positive side:

$$\boldsymbol{T}(\_, \vec{\Sigma}) \equiv (\text{total 4-momentum } \vec{P} \text{ that flows through } \vec{\Sigma}); \quad \text{or } T^{\alpha\beta}\Sigma_\beta = P^\alpha. \tag{2.66}$$

COMPONENTS

From this definition of the stress-energy tensor we can read off the physical meanings of its components on a specific, but arbitrary, Lorentz-coordinate basis: Making use of method (2.23b) for computing the components of a vector or tensor, we see that in a specific, but arbitrary, Lorentz frame (where $\vec{\Sigma} = -\vec{e}_0$ is a volume vector representing a parallelepiped with unit volume $\Delta V = 1$, at rest in that frame, with its positive sense toward the future):

$$-T_{\alpha 0} = \boldsymbol{T}(\vec{e}_\alpha, -\vec{e}_0) = \vec{P}(\vec{e}_\alpha) = \begin{pmatrix} \alpha \text{ component of 4-momentum that} \\ \text{flows from past to future across a unit} \\ \text{volume } \Delta V = 1 \text{ in the 3-space } t = \text{const} \end{pmatrix}$$

$$= (\alpha \text{ component of density of 4-momentum}). \tag{2.67a}$$

Specializing α to be a time or space component and raising indices, we obtain the specialized versions of (2.67a):

T2

$$T^{00} = (\text{energy density as measured in the chosen Lorentz frame}),$$
$$T^{j0} = (\text{density of } j \text{ component of momentum in that frame}). \tag{2.67b}$$

Similarly, the αx component of the stress-energy tensor (also called the $\alpha 1$ component, since $x = x^1$ and $\vec{e}_x = \vec{e}_1$) has the meaning

$$T_{\alpha 1} \equiv T_{\alpha x} \equiv \mathbf{T}(\vec{e}_\alpha, \vec{e}_x) = \begin{pmatrix} \alpha \text{ component of 4-momentum that crosses} \\ \text{a unit area } \Delta y \Delta z = 1 \text{ lying in a surface of} \\ \text{constant } x, \text{ during unit time } \Delta t, \text{ crossing} \\ \text{from the } -x \text{ side toward the } +x \text{ side} \end{pmatrix}$$
$$= \begin{pmatrix} \alpha \text{ component of flux of 4-momentum} \\ \text{across a surface lying perpendicular to } \vec{e}_x \end{pmatrix}. \tag{2.67c}$$

The specific forms of this for temporal and spatial α are (after raising indices)

$$T^{0x} = \begin{pmatrix} \text{energy flux across a surface perpendicular to } \vec{e}_x, \\ \text{from the } -x \text{ side to the } +x \text{ side} \end{pmatrix}, \tag{2.67d}$$

$$T^{jx} = \begin{pmatrix} \text{flux of } j \text{ component of momentum across a surface} \\ \text{perpendicular to } \vec{e}_x, \text{ from the } -x \text{ side to the } +x \text{ side} \end{pmatrix}$$
$$= \begin{pmatrix} jx \text{ component} \\ \text{of stress} \end{pmatrix}. \tag{2.67e}$$

The αy and αz components have the obvious, analogous interpretations.

These interpretations, restated much more briefly, are:

$$\boxed{\begin{aligned} &T^{00} = (\text{energy density}), \quad T^{j0} = (\text{momentum density}), \\ &T^{0j} = (\text{energy flux}), \quad T^{jk} = (\text{stress}). \end{aligned}} \tag{2.67f}$$

components of stress-energy tensor

SYMMETRY

Although it might not be obvious at first sight, *the 4-dimensional stress-energy tensor is always symmetric:* in index notation (where indices can be thought of as representing the names of slots, or equally well, components on an arbitrary basis)

symmetry of stress-energy tensor

$$T^{\alpha\beta} = T^{\beta\alpha}. \tag{2.68}$$

This symmetry can be deduced by physical arguments in a specific, but arbitrary, Lorentz frame: Consider, first, the $x0$ and $0x$ components, that is, the x components of momentum density and energy flux. A little thought, symbolized by the following heuristic equation, reveals that they must be equal:

$$T^{x0} = \begin{pmatrix} \text{momentum} \\ \text{density} \end{pmatrix} = \frac{(\Delta\mathcal{E})dx/dt}{\Delta x \Delta y \Delta z} = \frac{\Delta\mathcal{E}}{\Delta y \Delta z \Delta t} = \begin{pmatrix} \text{energy} \\ \text{flux} \end{pmatrix}, \tag{2.69}$$

T2

and similarly for the other space-time and time-space components: $T^{j0} = T^{0j}$. [In the first expression of Eq. (2.69) $\Delta\mathcal{E}$ is the total energy (or equivalently mass) in the volume $\Delta x \Delta y \Delta z$, $(\Delta\mathcal{E})dx/dt$ is the total momentum, and when divided by the volume we get the momentum density. The third equality is just elementary algebra, and the resulting expression is obviously the energy flux.] The space-space components, being equal to the stress tensor, are also symmetric, $T^{jk} = T^{kj}$, by the argument embodied in Fig. 1.6. Since $T^{0j} = T^{j0}$ and $T^{jk} = T^{kj}$, all components in our chosen Lorentz frame are symmetric, $T^{\alpha\beta} = T^{\beta\alpha}$. Therefore, if we insert arbitrary vectors into the slots of $\boldsymbol{T}$ and evaluate the resulting number in our chosen Lorentz frame, we find

$$\boldsymbol{T}(\vec{A}, \vec{B}) = T^{\alpha\beta} A_\alpha B_\beta = T^{\beta\alpha} A_\alpha B_\beta = \boldsymbol{T}(\vec{B}, \vec{A}); \tag{2.70}$$

that is, $\boldsymbol{T}$ is symmetric under interchange of its slots.

Let us return to the physical meanings (2.67f) of the components of the stress-energy tensor. With the aid of $\boldsymbol{T}$'s symmetry, we can restate those meanings in the language of a 3+1 split of spacetime into space plus time: *When one chooses a specific reference frame, that choice splits the stress-energy tensor up into three parts. Its time-time part is the energy density T^{00}, its time-space part $T^{0j} = T^{j0}$ is the energy flux or equivalently the momentum density, and its space-space part T^{jk} is the stress tensor.*

2.13.2

2.13.2 4-Momentum Conservation

Our interpretation of $\vec{J}(\vec{\Sigma}) \equiv J^\alpha \Sigma_\alpha$ as the net charge that flows through a small 3-surface $\vec{\Sigma}$ from its negative side to its positive side gave rise to the global conservation law for charge, $\int_{\partial\mathcal{V}} J^\alpha d\Sigma_\alpha = 0$ [Eq. (2.58) and Fig. 2.11]. Similarly the role of $\boldsymbol{T}(\_\_, \vec{\Sigma})$ [$T^{\alpha\beta}\Sigma_\beta$ in slot-naming index notation] as the net 4-momentum that flows through $\vec{\Sigma}$ from its negative side to positive gives rise to the following equation for conservation of 4-momentum:

global law of 4-momentum conservation

$$\boxed{\int_{\partial\mathcal{V}} T^{\alpha\beta} d\Sigma_\beta = 0.} \tag{2.71}$$

(The time component of this equation is energy conservation; the spatial part is momentum conservation.) This equation says that all the 4-momentum that flows into the 4-volume $\mathcal{V}$ of Fig. 2.11 through its 3-surface $\partial\mathcal{V}$ must also leave $\mathcal{V}$ through $\partial\mathcal{V}$; it gets counted negatively when it enters (since it is traveling from the positive side of $\partial\mathcal{V}$ to the negative), and it gets counted positively when it leaves, so its net contribution to the integral (2.71) is zero.

This global law of 4-momentum conservation can be converted into a local law (analogous to $\vec{\nabla} \cdot \vec{J} = 0$ for charge) with the help of the 4-dimensional Gauss's theorem (2.57). Gauss's theorem, generalized in the obvious way from a vectorial integrand to a tensorial one, is:

$$\int_{\mathcal{V}} T^{\alpha\beta}{}_{;\beta}\, d\Sigma = \int_{\partial\mathcal{V}} T^{\alpha\beta} d\Sigma_\beta. \tag{2.72}$$

T2

Since the right-hand side vanishes, so must the left-hand side; and for this 4-volume integral to vanish for every choice of $\mathcal{V}$, the integrand must vanish everywhere in spacetime:

local law of 4-momentum conservation

$$\boxed{T^{\alpha\beta}{}_{;\beta} = 0; \quad \text{or} \quad \vec{\nabla} \cdot \boldsymbol{T} = 0.} \tag{2.73a}$$

In the second, index-free version of this local conservation law, the ambiguity about which slot the divergence is taken on is unimportant, since $\boldsymbol{T}$ is symmetric in its two slots: $T^{\alpha\beta}{}_{;\beta} = T^{\beta\alpha}{}_{;\beta}$.

In a specific but arbitrary Lorentz frame, the local conservation law (2.73a) for 4-momentum has as its temporal part

$$\frac{\partial T^{00}}{\partial t} + \frac{\partial T^{0k}}{\partial x^k} = 0, \tag{2.73b}$$

that is, the time derivative of the energy density plus the 3-divergence of the energy flux vanishes; and as its spatial part

$$\frac{\partial T^{j0}}{\partial t} + \frac{\partial T^{jk}}{\partial x^k} = 0, \tag{2.73c}$$

that is, the time derivative of the momentum density plus the 3-divergence of the stress (i.e., of momentum flux) vanishes. Thus, as one should expect, the geometric, frame-independent law of 4-momentum conservation includes as special cases both the conservation of energy and the conservation of momentum; and their differential conservation laws have the standard form that one expects both in Newtonian physics and in special relativity: time derivative of density plus divergence of flux vanishes; cf. Eq. (1.36) and associated discussion.

2.13.3 Stress-Energy Tensors for Perfect Fluids and Electromagnetic Fields

2.13.3

As an important example that illustrates the stress-energy tensor, consider a *perfect fluid*—a medium whose stress-energy tensor, evaluated in its *local rest frame* (a Lorentz frame where $T^{j0} = T^{0j} = 0$), has the form

perfect-fluid stress-energy tensor

$$T^{00} = \rho, \quad T^{jk} = P\delta^{jk} \tag{2.74a}$$

[Eq. (1.34) and associated discussion]. Here ρ is a short-hand notation for the energy density T^{00} (density of total mass-energy, including rest mass) as measured in the local rest frame, and the stress tensor T^{jk} in that frame is an isotropic pressure P. From this special form of $T^{\alpha\beta}$ in the local rest frame, one can derive the following geometric, frame-independent expression for the stress-energy tensor in terms of the 4-velocity $\vec{u}$ of the local rest frame (i.e., of the fluid itself), the metric tensor of spacetime $\boldsymbol{g}$, and the rest-frame energy density ρ and pressure P:

$$\boxed{T^{\alpha\beta} = (\rho + P)u^\alpha u^\beta + Pg^{\alpha\beta}; \quad \text{i.e., } \boldsymbol{T} = (\rho + P)\vec{u} \otimes \vec{u} + P\boldsymbol{g}.} \tag{2.74b}$$

See Ex. 2.26.

T2

In Sec. 13.8, we develop and explore the laws of relativistic fluid dynamics that follow from energy-momentum conservation $\vec{\nabla} \cdot \boldsymbol{T} = 0$ for this stress-energy tensor and from rest-mass conservation $\vec{\nabla} \cdot \vec{S}_{\rm rm} = 0$. By constructing the Newtonian limit of the relativistic laws, we shall deduce the nonrelativistic laws of fluid mechanics, which are the central theme of Part V. Notice, in particular, that the Newtonian limit ($P \ll \rho$, $u^0 \simeq 1$, $u^j \simeq v^j$) of the stress part of the stress-energy tensor (2.74b) is $T^{jk} = \rho v^j v^k + P\delta^{jk}$, which we met in Ex. 1.13.

Another example of a stress-energy tensor is that for the electromagnetic field, which takes the following form in Gaussian units (with $4\pi \to \mu_0 = 1/\epsilon_0$ in SI units):

electromagnetic stress-energy tensor

$$T^{\alpha\beta} = \frac{1}{4\pi}\left(F^{\alpha\mu}F^{\beta}{}_{\mu} - \frac{1}{4}g^{\alpha\beta}F^{\mu\nu}F_{\mu\nu}\right). \tag{2.75}$$

We explore this stress-energy tensor in Ex. 2.28.

EXERCISES

Exercise 2.25 *Example: Global Conservation of Energy in an Inertial Frame*

Consider the 4-dimensional parallelepiped $\mathcal{V}$ whose legs are $\Delta t\vec{e}_t$, $\Delta x\vec{e}_x$, $\Delta y\vec{e}_y$, $\Delta z\vec{e}_z$, where $(t, x, y, z) = (x^0, x^1, x^2, x^3)$ are the coordinates of some inertial frame. The boundary $\partial\mathcal{V}$ of this $\mathcal{V}$ has eight 3-dimensional "faces." Identify these faces, and write the integral $\int_{\partial\mathcal{V}} T^{0\beta} d\Sigma_\beta$ as the sum of contributions from each of them. According to the law of energy conservation, this sum must vanish. Explain the physical interpretation of each of the eight contributions to this energy conservation law. (See Ex. 2.23 for an analogous interpretation of charge conservation.)

Exercise 2.26 ***Derivation and Example: Stress-Energy Tensor and Energy-Momentum Conservation for a Perfect Fluid*

(a) Derive the frame-independent expression (2.74b) for the perfect fluid stress-energy tensor from its rest-frame components (2.74a).

(b) Explain why the projection of $\vec{\nabla} \cdot \boldsymbol{T} = 0$ along the fluid 4-velocity, $\vec{u} \cdot (\vec{\nabla} \cdot \boldsymbol{T}) = 0$, should represent energy conservation as viewed by the fluid itself. Show that this equation reduces to

$$\frac{d\rho}{d\tau} = -(\rho + P)\vec{\nabla} \cdot \vec{u}. \tag{2.76a}$$

With the aid of Eq. (2.65), bring this into the form

$$\frac{d(\rho V)}{d\tau} = -P\frac{dV}{d\tau}, \tag{2.76b}$$

where V is the 3-volume of some small fluid element as measured in the fluid's local rest frame. What are the physical interpretations of the left- and right-hand sides of this equation, and how is it related to the first law of thermodynamics?

(c) Read the discussion in Ex. 2.10 about the tensor $\boldsymbol{P} = \boldsymbol{g} + \vec{u} \otimes \vec{u}$ that projects into the 3-space of the fluid's rest frame. Explain why $P_{\mu\alpha}T^{\alpha\beta}{}_{;\beta} = 0$ should represent the law of force balance (momentum conservation) as seen by the fluid. Show that this equation reduces to

$$(\rho + P)\vec{a} = -\boldsymbol{P} \cdot \vec{\nabla} P, \tag{2.76c}$$

where $\vec{a} = d\vec{u}/d\tau$ is the fluid's 4-acceleration. This equation is a relativistic version of Newton's $\mathbf{F} = m\mathbf{a}$. Explain the physical meanings of the left- and right-hand sides. Infer that $\rho + P$ must be the fluid's inertial mass per unit volume. It is also the enthalpy per unit volume, including the contribution of rest mass; see Ex. 5.5 and Box 13.2.

T2

Exercise 2.27 ***Example: Inertial Mass per Unit Volume*

Suppose that some medium has a rest frame (unprimed frame) in which its energy flux and momentum density vanish, $T^{0j} = T^{j0} = 0$. Suppose that the medium moves in the x direction with speed very small compared to light, $v \ll 1$, as seen in a (primed) laboratory frame, and ignore factors of order v^2. The ratio of the medium's momentum density $G_{j'} = T^{j'0'}$ (as measured in the laboratory frame) to its velocity $v_i = v\delta_{ix}$ is called its total *inertial mass per unit volume* and is denoted ρ^{inert}_{ji}:

$$T^{j'0'} = \rho^{\text{inert}}_{ji} v_i. \tag{2.77}$$

In other words, ρ^{inert}_{ji} is the 3-dimensional tensor that gives the momentum density $G_{j'}$ when the medium's small velocity is put into its second slot.

(a) Using a Lorentz transformation from the medium's (unprimed) rest frame to the (primed) laboratory frame, show that

$$\rho^{\text{inert}}_{ji} = T^{00}\delta_{ji} + T_{ji}. \tag{2.78}$$

(b) Give a physical explanation of the contribution $T_{ji}v_i$ to the momentum density.

(c) Show that for a perfect fluid [Eq. (2.74b)] the inertial mass per unit volume is isotropic and has magnitude $\rho + P$, where ρ is the mass-energy density, and P is the pressure measured in the fluid's rest frame:

$$\boxed{\rho^{\text{inert}}_{ji} = (\rho + P)\delta_{ji}.} \tag{2.79}$$

See Ex. 2.26 for this inertial-mass role of $\rho + P$ in the law of force balance.

Exercise 2.28 ***Example: Stress-Energy Tensor, and Energy-Momentum Conservation for the Electromagnetic Field*

(a) From Eqs. (2.75) and (2.45) compute the components of the electromagnetic stress-energy tensor in an inertial reference frame (in Gaussian units). Your

T2

answer should be the expressions given in electrodynamics textbooks:

$$T^{00} = \frac{\mathbf{E}^2 + \mathbf{B}^2}{8\pi}, \qquad \mathbf{G} = T^{0j}\mathbf{e}_j = T^{j0}\mathbf{e}_j = \frac{\mathbf{E}\times\mathbf{B}}{4\pi},$$

$$T^{jk} = \frac{1}{8\pi}\left[(\mathbf{E}^2 + \mathbf{B}^2)\delta_{jk} - 2(E_j E_k + B_j B_k)\right]. \tag{2.80}$$

(In SI units, $4\pi \to \mu_0 = 1/\epsilon_0$.) See also Ex. 1.14 for an alternative derivation of the stress tensor T_{jk}.

(b) Show that the divergence of the stress-energy tensor (2.75) is given by

$$T^{\mu\nu}{}_{;\nu} = \frac{1}{4\pi}(F^{\mu\alpha}{}_{;\nu}F^{\nu}{}_{\alpha} + F^{\mu\alpha}F^{\nu}{}_{\alpha;\nu} - \frac{1}{2}F_{\alpha\beta}{}^{;\mu}F^{\alpha\beta}). \tag{2.81a}$$

(c) Combine this with Maxwell's equations (2.48) to show that

$$\nabla\cdot\boldsymbol{T} = -\boldsymbol{F}(\_, \vec{J}); \quad \text{i.e., } T^{\alpha\beta}{}_{;\beta} = -F^{\alpha\beta}J_\beta. \tag{2.81b}$$

(d) The matter that carries the electric charge and current can exchange energy and momentum with the electromagnetic field. Explain why Eq. (2.81b) is the rate per unit volume at which that matter feeds 4-momentum into the electromagnetic field, and conversely, $+F^{\alpha\mu}J_\mu$ is the rate per unit volume at which the electromagnetic field feeds 4-momentum into the matter. Show, further, that (as viewed in any reference frame) the time and space components of this quantity are

$$\frac{d\mathcal{E}_{\text{matter}}}{dtdV} = F^{0j}J_j = \mathbf{E}\cdot\mathbf{j}, \qquad \frac{d\mathbf{p}_{\text{matter}}}{dtdV} = \rho_e\mathbf{E} + \mathbf{j}\times\mathbf{B}, \tag{2.81c}$$

where ρ_e is charge density, and $\mathbf{j}$ is current density [Eq. (2.49)]. The first of these equations describes Ohmic heating of the matter by the electric field, and the second gives the Lorentz force per unit volume on the matter (cf. Ex. 1.14b).

Bibliographic Note

For an inspiring taste of the history of special relativity, see the original papers by Einstein, Lorentz, and Minkowski, translated into English and archived in Lorentz et al. (1923).

Early relativity textbooks [see the bibliography in Jackson (1999, pp. 566–567)] emphasized the transformation properties of physical quantities, in going from one inertial frame to another, rather than their roles as frame-invariant geometric objects. Minkowski (1908) introduced geometric thinking, but only in recent decades—in large measure due to the influence of John Wheeler—has the geometric viewpoint gained ascendancy.

In our opinion, the best elementary introduction to special relativity is the first edition of Taylor and Wheeler (1966); the more ponderous second edition (Taylor

T2

and Wheeler, 1992) is also good. At an intermediate level we strongly recommend the special relativity portions of Hartle (2003).

At a more advanced level, comparable to this chapter, we recommend Goldstein, Poole, and Safko (2002) and the special relativity sections of Misner, Thorne, and Wheeler (1973), Carroll (2004), and Schutz (2009).

These all adopt the geometric viewpoint that we espouse. In this chapter, so far as possible, we have minimized the proliferation of mathematical concepts (avoiding, e.g., differential forms and dual bases). By contrast, the other advanced treatments cited above embrace the richer mathematics.

Much less geometric than these references but still good, in our view, are the special relativity sections of popular electrodynamics texts: Griffiths (1999) at an intermediate level and Jackson (1999) at a more advanced level. We recommend avoiding special relativity treatments that use imaginary time and thereby obfuscate (e.g., earlier editions of Goldstein and of Jackson, and also the more modern and otherwise excellent Zangwill (2013)).

REFERENCES

Abbott, B. P., R. Abbott, T. D. Abbott, M. R. Abernathy, et al. (2016). Observation of gravitational waves from a binary black hole merger. *Physical Review Letters* **116**, 061102.

Abbott, B. P., R. Abbott, T.D. Abbott, F. Acernase, et al. (2017). GW170817: Observation of Gravitational Waves from a Binary Neutron Star Inspiral. *Physical Review Letters* **119**, 161101.

Aichelberg, P. C., and R. U. Sexl (1971). On the gravitational field of a massless particle. *General Relativity and Gravitation* **2**, 303–312.

Albrecht, A., and P. Steinhardt (1982). Cosmology for grand unified theories with radiatively induced symmetry breaking. *Physical Review Letters* **48**, 1220–1223.

Anderson, L., E. Aubourg, S. Bailey, F. Beutler, et al. (2014). The clustering of galaxies in the SDSS-III baryon oscillation spectroscopic survey: Baryon acoustic oscillations in the data releases 10 and 11 galaxy samples. *Monthly Notices of the Royal Astronomical Society* **441**, 24–62.

Andersson, N., and G. L. Comer (2007). Relativistic fluid dynamics: Physics for many different scales. *Living Reviews in Relativity* **10**, 1.

Archibald, A. M., N. V. Gusinskaia, J. W. T. Hessels, A. T. Deller, et al. (2018). Universality of free fall from the orbital motion of a pulsar in a stellar triple system. *Nature* **559**, 73–76.

Baumgarte, T. W., and S. Shapiro (2010). *Numerical Relativity: Solving Einstein's Equations on the Computer*. Cambridge: Cambridge University Press.

Beckwith, S. V., M. Stiavelli, A. M. Koekemoer, J. A. R. Caldwell, et al. (2006). The Hubble Ultra Deep Field. *Astronomical Journal* **132**, 1729–1755.

Bertotti, B. (1959). Uniform electromagnetic field in the theory of general relativity. *Physical Review* **116**, 1331–1333.

Bertotti, B., L. Iess, and P. Tortora (2003). A test of general relativity using radio links with the Cassini spacecraft. *Nature* **425**, 374–376.

Biedermann, G. W., X. Wu, L. Deslauriers, S. Roy, C. Mahadeswaraswamy, and M. A. Kasevich (2015). Testing gravity with cold-atom interferometers. *Physical Review A* **91**, 033629.

Birkhoff, G. (1923). *Relativity and Modern Physics*. Cambridge, Mass.: Harvard University Press.

Blanchet, L. (2014). Gravitational radiation from post-Newtonian sources and inspiraling compact binaries. *Living Reviews in Relativity* **17**, 2.

Blandford, R. D., and R. L. Znajek (1977). The electromagnetic extraction of energy from Kerr black holes. *Monthly Notices of the Royal Astronomical Society* **179**, 433–456.

Blandford, R. D., A. B. Saust, T. G. Brainerd, and J. Villumsen (1991). The distortion of distant galaxy images by large-scale structure. *Monthly Notices of the Royal Astronomical Society* **251**, 600–627.

Bondi, H. (1952). *Cosmology.* Cambridge: Cambridge University Press.

Brady, P. R., S. Droz, and S. M. Morsink (1998). The late-time singularity inside non-spherical black holes. *Physical Review D* **58**, 084034–084048.

Burke, W. L. (1971). Gravitational radiation damping of slowly moving systems calculated using matched asymptotic expansions. *Journal of Mathematical Physics* **12**, 402–418.

Cahn, R., and G. Goldhaber (2009). *The Experimental Foundations of Particle Physics.* Cambridge: Cambridge University Press.

Carroll, S. M. (2004). *Spacetime and Geometry. An Introduction to General Relativity*. New York: Addison-Wesley.

Carter, B. (1979). The general theory of the mechanical, electromagnetic and thermodynamic properties of black holes. In S. W. Hawking and W. Israel (eds.), *General Relativity, an Einstein Centenary Survey.* Cambridge: Cambridge University Press.

Chandrasekhar, S. (1983). *The Mathematical Theory of Black Holes*. Oxford: Oxford University Press.

Ciufolini, I., et al. (2019). An improved test of the general relativistic effect of frame-dragging using the LARES and LAGEOS satellites. *The European Physical Journal C* **79**, 872.

Cohen-Tannoudji, C., B. Diu, and F. Laloë (1977). *Quantum Mechanics*. New York: Wiley.

Cox, Arthur N., ed. (2000). *Allen's Astrophysical Quantities*. Cham, Switzerland: Springer.

Creighton, J. D. E., and W. G. Anderson (2011). *Listening to the Universe*. New York: Wiley.

Cutler, C., and K. S. Thorne (2002). An overview of gravitational wave sources. In N. Bishop and S. D. Maharaj (eds.), *Proc. GR16 Conference on General Relativity and Gravitation*, pp. 72–111. Singapore: World Scientific.

Cyburt, R. H., B. D. Fields, K. A. Olive, and T.-H. Yeh (2016). Big bang nucleosynthesis: Present status. *Reviews of Modern Physics* **88**, 015004.

Darwin, C. (1959). The gravity field of a particle. *Proceedings of the Royal Society A* **249**, 180–194.

Delva, P., N. Puchades, E. Schönemann, F. Dilssner, et al. (2018). A gravitational redshift test using eccentric Galileo satellites. *Physical Review Letters* **121**, 231101.

DeMorest, P. B., T. Pennucci, S. M. Ransom, M. S. E. Roberts, et al. (2010). A two solar mass neutron star measured using Shapiro delay. *Nature* **467**, 1081–1083.

Dicke, R. H. (1961). Dirac's cosmology and Mach's principle. *Nature* **192**, 440–441.

Dodelson, S. (2003). *Modern Cosmology*. New York: Academic Press.

Eckart, C. (1940). The thermodynamics of irreversible processes, III: Relativistic theory of the simple fluid. *Physical Review* **58**, 919–924.

Eddington, A. S. (1922). *The Mathematical Theory of Relativity*. Cambridge: Cambridge University Press.

Eddington, A. S. (1933). *The Expanding Universe*. Cambridge: Cambridge University Press.

Einstein, A. (1907). Über das Relativitätsprinzip und die ausdemselben gesogenen Folgerungen. *Jahrbuch der Radioaktivität und Elektronik* **4**, 411–462.

Einstein, A. (1915). Die Feldgleichungen der Gravitation. *Sitzungsberichte der Preussischen Akademie* **1915**, 844–847.

Einstein, A. (1916a). Die Grundlage der allgemeinen Relativitätstheorie. *Annalen der Physik* **49**, 769–822.

Einstein, A. (1916b). Näherungsweise Integration der Feldgleichungen der Gravitation. *Sitzungsberichte der Preussischen Akademie der Wissenschaften* **1916**, 688–696.

Einstein, A. (1917). Kosmologische Betrachtungen zur allgemeinen Relativitätstheorie. *Sitzungsberichte der Preussischen Akademie der Wissenschaften* **1917**, 142–152.

Einstein, A. (1918). Über Gravitationswellen. *Sitzungsberichte der Preussischen Akademie der Wissenschaften* **1918**, 154–167.

Einstein, A. (1931). Zum kosmologischen Problem der allgemeinen Relativitätstheorie. *Sitzungsberichte der Preussischen Akademie der Wissenschaften* **1931**, 235–237.

Einstein, A. (1934) On the method of theoretical physics. *Philosphy of Science* **1**, 163–169. Published version of Einstein's Herbert Spencer Lecture, delivered at Oxford, June 10, 1933. Chapter 25 epigraph reprinted with permission of the University of Chicago Press.

Einstein, A. (1989). *The Collected Papers of Albert Einstein*. Princeton, N.J.: Princeton University Press.

Everett, A., and T. Roman (2011). *Time Travel and Warp Drives: A Scientific Guide to Shortcuts through Time and Space*. Chicago: University of Chicago Press.

Everitt, C. W. F., D. B. DeBra, B. W. Parkinson, J. P. Turneaure, et al. (2011). Gravity Probe B: Final results of a space experiment to test general relativity. *Physical Review Letters* **106**, 221101.

Faraday, M. (1846). Thoughts on ray vibrations. *Philosophical Magazine* **140**, 147–161.

Feynman, R. P. (1966). *The Character of Physical Law*. Cambridge, Mass.: MIT Press.

Fierz, M., and W. Pauli (1939). On relativistic wave equations for particles of arbitrary spin in an electromagnetic field. *Proceedings of the Royal Society A* **173**, 211–232.

Finkelstein, D. (1958). Past-future asymmetry of the gravitational field of a point particle. *Physical Review* **110**, 965–967.

Flanders, H. (1989). *Differential Forms with Applications to the Physical Sciences*, corrected edition. Mineola, N.Y.: Courier Dover Publications.

Friedman, J., and A. Higuchi (2006). Topological censorship and chronology protection. *Annalen der Physik* **15**, 109–128.

Friedmann, A. A. (1922). Über die Krümmung des Raumes. *Zeitschrift für Physik* **10**, 377–386.

Frolov, V. P., and I. D. Novikov (1990). Physical effects in wormholes and time machines. *Physical Review D* **42**, 1057–1065.

Frolov, V. P., and I. D. Novikov (1998). *Black Hole Physics: Basic Concepts and New Developments*. Dordrecht: Kluwer.

Fuller, R. W., and J. A. Wheeler (1962). Causality and multiply connected spacetime. *Physical Review* **128**, 919–929.

Goldstein, H., C. Poole, and J. Safko (2002). *Classical Mechanics*. New York: Addison-Wesley.

Gourgoulhon, E. (2013). *Special Relativity in General Frames: From Particles to Astrophysics*. Berlin: Springer-Verlag.

Griffiths, D. J. (1999). *Introduction to Electrodynamics*. Upper Saddle River, N.J.: Prentice-Hall.

Guth, A. H. (1981). Inflationary universe: A possible solution to the horizon and flatness problems. *Physical Review D* **23**, 347–356.

Hafele, J. C., and R. E. Keating (1972a). Around-the-world atomic clocks: Predicted relativistic time gains. *Science* **177**, 166–168.

Hafele, J. C., and R. E. Keating (1972b). Around-the-world atomic clocks: Observed relativistic time gains. *Science* **177**, 168–170.

Harrison, E. R. (1970). Fluctuations at the threshold of classical cosmology. *Physical Review D* **1**, 2726–2730.

Hartle, J. B. (2003). *Gravity: An Introduction to Einstein's General Relativity*. San Francisco: Addison-Wesley.

Heymans, C., L. van Waerbeke, L. Miller, T. Erben, et al. (2012). CFHTLenS: The Canada-France-Hawaii telescope lensing survey. *Monthly Notices of the Royal Astronomical Society* **427**, 146–166.

Hilbert, D. (1915). Die Grundlagen der Physik. *Königliche Gesellschaft der Wissenschaften zu Göttingen. Mathematische-physikalische Klasse. Nachrichten* **1917**, 53–76.

Hobbs, G., A. Archibald, Z. Arzoumanian, D. Backer, et al. (2010). The International Pulsar Timing Array project: Using pulsars as a gravitational wave detector. *Classical and Quantum Gravity* **27**, 8, 084013.

Hobson, M. P., G. P. Efstathiou, and A. N. Lasenby (2006). *General Relativity: An Introduction for Physicists*. Cambridge: Cambridge University Press.

Illingworth, G. D., and the HUDF09 team (2013). The HST extreme deep field (XDF): Combining all ACS and WFC3/IR data on the HUDF region into the deepest field ever. *Astrophysical Journal Supplement Series* **209**, 6.

Iorio, L., M. L. Ruggiero, and C. Corda (2013). Novel considerations about the error budget of the LAGEOS-based tests of frame-dragging with GRACE geopotential models. *Acta Astronautica* **91**, 141–148.

Isaacson, R. A. (1968a). Gravitational radiation in the limit of high frequency. I. The linear approximation and geometrical optics. *Physical Review* **166**, 1263–1271.

Isaacson, R. A. (1968b). Gravitational radiation in the limit of high frequency. II. Nonlinear terms and the effective stress tensor. *Physical Review* **166**, 1272–1280.

Israel, W., and J. M. Stewart (1980). Progress in relativistic thermodynamics and electrodynamics of continuous media. In A. Held, (ed.), *General Relativity and Gravitation. Vol. 2. One Hundred Years after the Birth of Albert Einstein*, p. 491. New York: Plenum Press.

Jackson, J. D. (1999). *Classical Electrodynamics*. New York: Wiley.

James, O., E. von Tunzelmann, P. Franklin, and K. S. Thorne (2015a). Gravitational lensing by spinning black holes in astrophysics, and in the movie *Interstellar*. *Classical and Quantum Gravity* **32**, 065001.

James, O., E. von Tunzelmann, P. Franklin, and K. S. Thorne (2015b). Visualizing *Interstellar*'s Wormhole. *American Journal of Physics* **83**, 486–499.

Kachru, S., R. Kallosh, A. Linde, and S. Trivedi (2003). De Sitter vacua in string theory. *Physical Review D* **68**, 046005.

Kapitsa, P. L., and P. A. M. Dirac (1933). The reflection of electrons from standing light waves. *Proceedings of the Cambridge Philosophical Society* **29**, 297–300.

Kaspi, V., and M. Kramer (2016). Radio pulsars: The neutron star population and fundamental physics. In R. D. Blandford, D. Gross, and A. Sevrin (eds.), *Proceedings of the 26th Solvay Conference on Physics, Astrophysics and Cosmology*, pp. 21–62. Singapore: World Scientific.

Kay, B. S., M. J. Radzikowski, and R. M. Wald (1997). Quantum field theory on spacetimes with a compactly generated Cauchy horizon. *Communications in Mathematical Physics* **183**, 533–556.

Kazanas, D. (1980). Dynamics of the universe and spontaneous symmetry breaking. *Astrophysical Journal* **241**, L59–L63.

Kerr, R. P. (1963). Gravitational field of a spinning mass as an example of algebraically special metrics. *Physical Review Letters* **11**, 237–238.

Kim, S.-W., and K. S. Thorne (1991). Do vacuum fluctuations prevent the creation of closed timelike curves? *Physical Review D* **43**, 3929–3949.

Kolb, E. W., and M. S. Turner (1994). *The Early Universe*. New York: Addison-Wesley.

Komatsu, E., K. M. Smith, J. Dunkley, C. L. Bennett, et al. (2011). Seven-year Wilkinson Microwave Anisotropy Probe (WMAP) observations: Cosmological interpretation. *Astrophysical Journal Supplement* **192**, 18–35.

Kompaneets, A. (1957). The establishment of thermal equilibrium between quanta and electrons. *Journal of Experimental and Theoretical Physics* **4**, 730–737.

Kramer, M., I. H. Stairs, R. N. Manchester, M. A. McLaughlin, et al. (2006). Tests of general relativity from timing the double pulsar. *Science* **314**, 97–102.

Kruskal, M. D. (1960). The maximal extension of the Schwarzschild metric. *Physical Review* **119**, 1743–1745.

Landau, L. D., and E. M. Lifshitz (1941). Teoriya Polya. Moscow: Gosudarstvennoye Izdatel'stvo Tekhniko-Teoreticheskoi Literaturi. First Russian edition of Landau and Lifshitz (1951).

Landau, L. D., and E. M. Lifshitz (1951). *The Classical Theory of Fields*, first English edition. Cambridge, Mass.: Addison-Wesley.

Landau, L. D., and E. M. Lifshitz (1959). *Fluid Mechanics*. Oxford: Pergamon.

Lee, B. W., and S. Weinberg (1977). Cosmological lower bound on heavy-neutrino masses. *Physical Review Letters* **39**, 165–167.

Lemaître, G. (1927). Un univers homogène de masse constante et de rayon croissant rendant compte de la vitesse radiale des nébuleuses extra-galactiques. *Annales de la Société Scientifique Bruxelles A* **47**, 49–59.

Lemaître, G. (1933). La formation des nebuleuses dans l'univers en expansion. *Comptes Rendus* **196**, 903–904. Translated in *Cosmology and Controversy: The Historical Development of Two Theories of the Universe* by Helge Kragh. Copyright © 1996 by Princeton University Press. Chapter 28 epigraph reprinted with permission of Princeton University Press.

Lemaître, G. (1934). Evolution of the expanding universe. *Proceedings of the National Academy of Sciences* **20**, 12–17.

Levin, J., and G. Perez-Giz (2008). A periodic table for black hole orbits. *Physical Review D* **77**, 103005.

Liddle, A., and D. Lyth (2000). *Cosmological Inflation and Large Scale Structure*. Cambridge: Cambridge University Press.

Lightman, A. P., W. H. Press, R. H. Price, and S. A. Teukolsky (1975). *Problem Book in Relativity and Gravitation*. Princeton, N.J.: Princeton University Press.

Linde, A. (1982). A new inflationary universe scenario: A possible solution to the horizon, flatness, homogeneity, isotropy and primordial monopole problems. *Physics Letters B* **108**, 389–393.

López-Monsalvo, C. S. (2011). *Covariant Thermodynamics and Relativity*, PhD thesis, University of Southampton. Available at https://arxiv.org/pdf/1107.1005.pdf.

Lorentz, H. A. (1904). Electromagnetic phenomena in a system moving with any velocity smaller than that of light. *Proceedings of the Royal Netherlands Academy of Arts and Sciences (KNAW)* **6**, 809–831.

Lorentz, H. A., A. Einstein, H. Minkowski, and H. Weyl (1923). *The Principle of Relativity: A Collection of Original Memoirs on the Special and General Theory of Relativity*. Mineola, N.Y.: Courier Dover Publications.

Maggiore, M. (2007). *Gravitational Waves. Volume 1: Theory and Experiment*. Oxford: Oxford University Press.

Maggiore, M. (2018). *Gravitational Waves. Volume 2: Astrophysics and Cosmology*. Oxford: Oxford University Press.

Marolf, D., and A. Ori (2013). Outgoing gravitational shock-wave at the inner horizon: The late-time limit of black hole interiors. *Physical Review D* **86**, 124026.

McKinney, J. C., A. Tchekhovskoy, and R. D. Blandford (2012). General relativistic magnetohydrodynamical simulations of magnetically choked accretion flows around black holes. *Monthly Notices of the Royal Astronomical Society* **423**, 3083–3117.

Meier, D. L. (2012). *Black Hole Astrophysics: The Engine Paradigm*. Cham, Switzerland: Springer.

Merkowitz, S. M. (2010). Tests of gravity using lunar laser ranging. *Living Reviews in Relativity* **13**, 7.

Minkowski, H. (1908). Space and time. Address delivered at the 80th Assembly of German Natural Scientists and Physicians, at Cologne, Germany, September 21, 1908. First German publication: *Jahresbericht der Deutschen Mathematiker-Vereinigung* **1909**, 75–88. English translation in Lorentz et al. (1923).

Misner, C. W., K. S. Thorne, and J. A. Wheeler (1973). *Gravitation*. San Francisco: Freeman.

Morris, M., and K. S. Thorne (1988). Wormholes in spacetime and their use for interstellar travel: A tool for teaching general relativity. *American Journal of Physics* **56**, 395–416.

Morris, M. S., K. S. Thorne, and U. Yurtsever (1988). Wormholes, time machines, and the weak energy condition. *Physical Review Letters* **61**, 1446–1449.

Mroué, A. H., M. A. Scheel, B. Szilagyi, H. P. Pfeiffer, et al. (2013). Catalog of 174 black hole simulations for gravitational wave astronomy. *Physical Review Letters* **111**, 241104.

Mukhanov, V. (2005). *Physical Foundations of Modern Cosmology*. Cambridge: Cambridge University Press.

Ni, W.-T., and M. Zimmermann (1978). Inertial and gravitational effects in the proper reference frame of an accelerated, rotating observer. *Physical Review D* **17**, 1473–1476.

Nichols, D., R. Owen, F. Zhang, A. Zimmerman, et al. (2011). Visualizing spacetime curvature via frame-drag vortexes and tidal tendexes: General theory and weak-gravity applications. *Physical Review D* **84**, 124014.

Noether, E. (1918). Invariante Variationenprobleme. *Nachrichten von der Gesellschaft der Wissenschaften zu Göttingen* **1918**, 235–257.

Oort, J. H. (1932). The force exerted by the stellar system in the direction perpendicular to the galactic plane and some related problems. *Bulletin of the Astronomical Institute of the Netherlands* **238**, 249–287.

Oppenheimer, J. R., and H. Snyder (1939). On continued gravitational contraction. *Physical Review* **56**, 455–459.

Oppenheimer, J. R., and G. Volkoff (1939). On massive neutron cores. *Physical Review* **55**, 374–381.

Overduin, J., F. Everitt, P. Worden, and J. Mester (2012). STEP and fundamental physics. *Classical and Quantum Gravity* **29**, 184012.

Owen, R., J. Brink, Y. Chen, J. D. Kaplan, et al. (2011). Frame-dragging vortexes and tidal tendexes attached to colliding black holes: Visualizing the curvature of spacetime. *Physical Review Letters* **106**, 151101.

Padmanabhan, T. (1993). *Structure Formation in the Universe*. Cambridge: Cambridge University Press.

Pais, A. (1982). *Subtle Is the Lord. . . . The Science and Life of Albert Einstein*. Oxford: Oxford University Press.

Peacock, J. A. (1999). *Cosmological Physics*. Cambridge: Cambridge University Press.

Peccei, R. D., and H. R. Quinn (1977). CP conservation in the presence of pseudoparticles. *Physical Review Letters* **38**, 1440–1443.

Peebles, P. J. E. (1993). *Principles of Physical Cosmology*. Princeton, N.J.: Princeton University Press.

Peebles, P. J. E., and J. T. Yu (1970). Primeval adiabatic perturbation in an expanding universe. *Astrophysical Journal* **162**, 815–836.

Penzias, A. A., and R. W. Wilson (1965). A measurement of excess antenna temperature at 4080 Mc/s. *Astrophysical Journal* **142**, 419–421.

Perlmutter, S., M. Turner, and M. White (1999). Constraining dark energy with Type Ia supernovae and large-scale structure. *Physical Review Letters* **83**, 670–673.

Perlmutter, S., G. Aldering, G. Goldhaber, R. A. Knop, et al. (1999). Measurements of Ω and Λ from 42 high-redshift supernovae. *Astrophysical Journal* **517**, 565–586.

Pfister, H. (2007). On the history of the so-called Lense-Thirring effect. *General Relativity and Gravitation* **39**, 1735–1748.

Planck, M. (1949). *Scientific Autobiography and Other Papers*. New York: Philosophical Library. Chapter 24 epigraph reprinted with permission of the publisher.

Planck Collaboration (2016a). Planck 2015 results. I. Overview of products and scientific results. *Astronomy and Astrophysics* **594**, A1.

Planck Collaboration (2016b). Planck 2015 results. XIII. Cosmological parameters. *Astronomy and Astrophysics* **594**, A13.

Pontecorvo, B. (1968). Neutrino experiments and the problem of conservation of leptonic charge. *Soviet Physics JETP* **26**, 984–988.

Ransom, S. M., I. H. Stairs, A. M. Archibald, J.W.T. Hessels, et al. (2014). A millisecond pulsar in a stellar triple system. *Nature* **505**, 520–524.

Reitze, D., P. Saulson, and H. Grote (2019). *Advanced Interferometric Gravitational-Wave Detectors*. Singapore: World Scientific.

Riess, A. G., A. Filippenko, P. Challis, A. Clochiatti, et al. (1998). Observational evidence from supernovae for an accelerating universe and a cosmological constant. *Astronomical Journal* **116**, 1009–1038.

Roberts, M. S., and R. Whitehurst (1975). The rotation curve and geometry of M31 at large galactocentric distances. *Astrophysical Journal* **201**, 327–346.

Robertson, H. P. (1935). Kinematics and world structure I. *Astrophysical Journal* **82**, 248–301.

Robertson, H. P. (1936a). Kinematics and world structure II. *Astrophysical Journal* **83**, 187–201.

Robertson, H. P. (1936b). Kinematics and world structure III. *Astrophysical Journal* **83**, 257–271.

Robinson, I. (1959). A solution of the Maxwell-Einstein equations. *Bulletin of the Polish Academy of Sciences* **7**, 351–352.

Rubin, V. C., and W. K. Ford, Jr. (1970). Rotation of the Andromeda nebula from a spectroscopic survey of emission regions. *Astrophysical Journal* **159**, 379–403.

Ryan, M., and L. Shepley (1975). *Homogeneous, Relativistic Cosmology.* Princeton, N.J.: Princeton University Press.

Ryden, B. S. (2002). *Introduction to Cosmology*. New York: Addison-Wesley.

Sachs, R. K., and A. M. Wolfe (1967). Perturbations of a cosmological model and angular variations of the microwave background. *Astrophysical Journal* **147**, 73–90.

Sakharov, A. D. (1965). The initial stage of an expanding universe and the appearance of a nonuniform distribution of matter. *Journal of Experimental and Theoretical Physics* **49**, 345–358.

Sathyaprakash, B. S., and B. F. Schutz (2009). Physics, astrophysics and cosmology with gravitational waves. *Living Reviews in Relativity* **12**, 3.

Sato, K. (1981). Cosmological baryon number domain structure and the first order phase transition of the vacuum. *Physics Letters B* **33**, 66–70.

Saulson, P. (1994). *Fundamentals of Interferometric Gravitational Wave Detectors*. Singapore: World Scientific.

Schlamminger, S., K.-Y. Choi, T. A. Wagner, J. H. Gundlach, and E. G. Adelberger (2008). Test of the equivalence principle using a rotating torsion balance. *Physical Review Letters* **100**, 041101.

Schneider, P. (2015). *Extragalactic Astronomy and Cosmology.* Heidelberg: Springer.

Schneider, P., J. Ehlers, and E. Falco (1992). *Gravitational Lenses*. Berlin: Springer-Verlag.

Schutz, B. (2009). *A First Course in General Relativity*. Cambridge: Cambridge University Press.

Schwarzschild, K. (1903). Zur Elektrodynamik. 1. Zwei Formen des Princips der Action in der Elektrontheorie. *Nachrichten von der Gesellschaft der Wissenschaften zu Göttingen* **1903**, 126–131.

Schwarzschild, K. (1916a). Über das Gravitationsfeld eines Massenpunktes nach der Einsteinschen Theorie. *Sitzungsberichte der Preussischen Akademie der Wissenschaften* **1916**, 189–196.

Schwarzschild, K. (1916b). Über das Gravitationsfeld einer Kugel aus Inkompressibler Flüssigkeit nach der Einsteinschen Theorie. *Sitzungsberichte der Preussischen Akademie der Wissenschaften*
1916, 424–434.

Shapiro, S. L., and S. A. Teukolsky (1983). *Black Holes, White Dwarfs and Neutron Stars: The Physics of Compact Objects.* New York: Wiley.

Shibata, M. (2016). *Numerical Relativity.* Singapore: World Scientific.

Silk, J. (1968). Cosmic black-body radiation and galaxy formation. *Astrophysical Journal* **151**, 459–471.

Spivak, M. (1999). *A Comprehensive Introduction to Differential Geometry,* Volumes 1–5. Houston: Publish or Perish.

Starobinsky, A. (1980). A new type of isotropic cosmological model without singularity. *Physics Letters B* **91**, 99–102.

Straumann, N. (2013). *General Relativity*. Cham, Switzerland: Springer.

Sunyaev, R. A., and Ya. B. Zel'dovich (1970). Small-scale fluctuations of relic radiation. *Astrophysics and Space Science* **7**, 3–19.

Susskind, L. (2005). *The Cosmic Landscape: String Theory and the Illusion of Intelligent Design.* New York: Little, Brown.

Szekeres, G. (1960). On the singularities of a Riemann manifold. *Publicationes Mathematicae Debrecen* **7**, 285–301.

Taylor, E. F., and J. A. Wheeler (1966). *Spacetime Physics*, first edition. San Francisco: Freeman.

Taylor, E. F., and J. A. Wheeler (1992). *Spacetime Physics*, second edition. San Francisco: Freeman.

Thorne, K. S. (1980). Multipole expansions of gravitational radiation. *Reviews of Modern Physics* **52**, 299–340.

Thorne, K. S. (1983). The theory of gravitational radiation: An introductory review. In N. Dereulle and T. Piran (eds.), *Gravitational Radiation*, pp. 1–57. New York: North Holland.

Thorne, K. S. (2014). *The Science of Interstellar*. New York: W. W. Norton.

Thorne, K. S., and J. Hartle (1985). Laws of motion and precession for black holes and other bodies. *Physical Review D* **31**, 1815–1837.

Thorne, K. S., R. H. Price, and D. A. MacDonald (1986). *Black holes: the Membrane Paradigm*. New Haven, Conn.: Yale University Press.

Tolman, R. C. (1934). *Relativity, Thermodynamics and Cosmology.* Oxford: Oxford University Press.

Tolman, R. C. (1939). Static solutions of Einstein's field equations for spheres of fluid. *Physical Review* **55**, 364–373.

Touboul, P., G. Métris M., Rodrigues, A. Yves, et al. (2017) MICROSCOPE Mission: First Results of a Space Test of the Equivalence Principle. *Physical Review Letters* **119**, 231101.

Wald, R. M. (1984). *General Relativity*. Chicago: University of Chicago Press.

Walker, A. G. (1935). On Milne's theory of world structure. *Proceedings of the London Mathematical Society* **42**, 90–127.

Weinberg, S. (1972). *Gravitation and Cosmology. Principles and Applications of the General Theory of Relativity.* New York: Wiley.

Weinberg, S. (2008). *Cosmology*. Oxford: Oxford University Press.

Weissberg, J. M., D. J. Nice, and J. H. Taylor (2010). Timing measurements of the relativistic binary pulsar PSR B1913+16. *Astrophysical Journal* **722**, 1030–1034.

Will, C. M. (1993a). *Theory and Experiment in Gravitational Physics*. Cambridge: Cambridge University Press.

Will, C. M. (1993b). *Was Einstein Right?* New York: Basic Books.

Will, C. M. (2014). The confrontation between general relativity and experiment. *Living Reviews in Relativity* **17**, 4.

Zangwill, A. (2013). *Modern Electrodynamics*. Cambridge: Cambridge University Press.

Zee, A. (2013). *Einstein Gravity in a Nutshell*. Princeton, N.J.: Princeton University Press.

Zel'dovich, B. Ya., V. I. Popovichev, V. V. Ragul'skii, and F. S. Faizullov (1972). Connection between the wavefronts of the reflected and exciting light in stimulated Mandel'shtem-Brillouin scattering. *Journal of Experimental and Theoretical Physics Letters* **15**, 160–164.

Zel'dovich, Ya. B. (1968). The cosmological constant and the theory of elementary particles. *Soviet Physics Uspekhi* **11**, 381–393.

Zhang, F., A. Zimmerman, D. Nichols, Y. Chen, et al. (2012). Visualizing spacetime curvature via frame-drag vortexes and tidal tendexes II. Stationary black holes. *Physical Review D* **86**, 084049.

Zwicky, F. (1933). Die Rotverschiebung von extragalaktischen Nebeln. *Helvetica Physics Acta* **6**, 110–127.

NAME INDEX

Page numbers for entries in boxes are followed by "b," those for epigraphs at the beginning of a chapter by "e," those for figures by "f," and those for notes by "n."

SUBJECT INDEX

Second and third level entries are not ordered alphabetically. Instead, the most important or general entries come first, followed by less important or less general ones, with specific applications last.

Page numbers for entries in boxes are followed by "b," those for epigraphs at the beginning of a chapter by "e," those for figures by "f," for notes by "n," and for tables by "t."

CONTENTS OF THE UNIFIED WORK, *MODERN CLASSICAL PHYSICS*

T2 Track Two; see page xvii

PART V FLUID DYNAMICS 671

PREFACE TO *MODERN CLASSICAL PHYSICS*

The study of physics (including astronomy) is one of the oldest academic enterprises. Remarkable surges in inquiry occurred in equally remarkable societies—in Greece and Egypt, in Mesopotamia, India and China—and especially in Western Europe from the late sixteenth century onward. Independent, rational inquiry flourished at the expense of ignorance, superstition, and obeisance to authority.

Physics is a constructive and progressive discipline, so these surges left behind layers of understanding derived from careful observation and experiment, organized by fundamental principles and laws that provide the foundation of the discipline today. Meanwhile the detritus of bad data and wrong ideas has washed away. The laws themselves were so general and reliable that they provided foundations for investigation far beyond the traditional frontiers of physics, and for the growth of technology.

The start of the twentieth century marked a watershed in the history of physics, when attention turned to the small and the fast. Although rightly associated with the names of Planck and Einstein, this turning point was only reached through the curiosity and industry of their many forerunners. The resulting quantum mechanics and relativity occupied physicists for much of the succeeding century and today are viewed very differently from each other. Quantum mechanics is perceived as an abrupt departure from the tacit assumptions of the past, while relativity—though no less radical conceptually—is seen as a logical continuation of the physics of Galileo, Newton, and Maxwell. There is no better illustration of this than Einstein's growing special relativity into the general theory and his famous resistance to the quantum mechanics of the 1920s, which others were developing.

This is a book about classical physics—a name intended to capture the pre-quantum scientific ideas, augmented by general relativity. Operationally, it is physics in the limit that Planck's constant $h \to 0$. Classical physics is sometimes used, pejoratively, to suggest that "classical" ideas were discarded and replaced by new principles and laws. Nothing could be further from the truth. The majority of applications of physics today are still essentially classical. This does not imply that physicists or others working in these areas are ignorant or dismissive of quantum physics. It is simply that the issues with which they are confronted are mostly addressed classically. Furthermore, classical physics has not stood still while the quantum world was being explored. In scope and in practice, it has exploded on many fronts and would now be quite unrecognizable to a Helmholtz, a Rayleigh, or a Gibbs. In this book, we have tried to emphasize these contemporary developments and applications at the expense of historical choices, and this is the reason for our seemingly oxymoronic title, *Modern Classical Physics*.

This book is ambitious in scope, but to make it bindable and portable (and so the authors could spend some time with their families), we do not develop classical mechanics, electromagnetic theory, or elementary thermodynamics. We assume the reader has already learned these topics elsewhere, perhaps as part of an undergraduate curriculum. We also assume a normal undergraduate facility with applied mathematics. This allows us to focus on those topics that are less frequently taught in undergraduate and graduate courses.

Another important exclusion is numerical methods and simulation. High-performance computing has transformed modern research and enabled investigations that were formerly hamstrung by the limitations of special functions and artificially imposed symmetries. To do justice to the range of numerical techniques that have been developed—partial differential equation solvers, finite element methods, Monte Carlo approaches, graphics, and so on—would have more than doubled the scope and size of the book. Nonetheless, because numerical evaluations are crucial for physical insight, the book includes many applications and exercises in which user-friendly numerical packages (such as Maple, Mathematica, and Matlab) can be used to produce interesting numerical results without too much effort. We hope that, via this pathway from fundamental principle to computable outcome, our book will bring readers not only physical insight but also enthusiasm for computational physics.

Classical physics as we develop it emphasizes physical phenomena on macroscopic scales: scales where the particulate natures of matter and radiation are secondary to their behavior in bulk; scales where particles' statistical—as opposed to individual—properties are important, and where matter's inherent graininess can be smoothed over.

In this book, we take a journey through spacetime and phase space; through statistical and continuum mechanics (including solids, fluids, and plasmas); and through optics and relativity, both special and general. In our journey, we seek to comprehend the fundamental laws of classical physics in their own terms, and also in relation to quantum physics. And, using carefully chosen examples, we show how the classical laws are applied to important, contemporary, twenty-first-century problems and to everyday phenomena; and we also uncover some deep relationships among the various fundamental laws and connections among the practical techniques that are used in different subfields of physics.

Geometry is a deep theme throughout this book and a very important connector. We shall see how a few geometrical considerations dictate or strongly limit the basic principles of classical physics. Geometry illuminates the character of the classical principles and also helps relate them to the corresponding principles of quantum physics. Geometrical methods can also obviate lengthy analytical calculations. Despite this, long, routine algebraic manipulations are sometimes unavoidable; in such cases, we occasionally save space by invoking modern computational symbol manipulation programs, such as Maple, Mathematica, and Matlab.

This book is the outgrowth of courses that the authors have taught at Caltech and Stanford beginning 37 years ago. Our goal was then and remains now to fill what we saw as a large hole in the traditional physics curriculum, at least in the United States:

- We believe that every masters-level or PhD physicist should be familiar with the basic concepts of all the major branches of classical physics and should have had some experience in applying them to real-world phenomena; this book is designed to facilitate this goal.
- Many physics, astronomy, and engineering graduate students in the United States and around the world use classical physics extensively in their research, and even more of them go on to careers in which classical physics is an essential component; this book is designed to expedite their efforts.
- Many professional physicists and engineers discover, in mid-career, that they need an understanding of areas of classical physics that they had not previously mastered. This book is designed to help them fill in the gaps and see the relationship to already familiar topics.

In pursuit of this goal, we seek, in this book, to *give the reader a clear understanding of the basic concepts and principles of classical physics.* We present these principles in the language of modern physics (not nineteenth-century applied mathematics), and we present them primarily for physicists—though we have tried hard to make the content interesting, useful, and accessible to a much larger community including engineers, mathematicians, chemists, biologists, and so on. As far as possible, we emphasize theory that involves general principles which extend well beyond the particular topics we use to illustrate them.

In this book, we also seek to *teach the reader how to apply the ideas of classical physics.* We do so by presenting contemporary applications from a variety of fields, such as

- fundamental physics, experimental physics, and applied physics;
- astrophysics and cosmology;
- geophysics, oceanography, and meteorology;
- biophysics and chemical physics; and

- engineering, optical science and technology, radio science and technology, and information science and technology.

Why is the range of applications so wide? Because we believe that physicists should have enough understanding of general principles to attack problems that arise in unfamiliar environments. In the modern era, a large fraction of physics students will go on to careers outside the core of fundamental physics. For such students, a broad exposure to non-core applications can be of great value. For those who wind up in the core, such an exposure is of value culturally, and also because ideas from other fields often turn out to have impact back in the core of physics. Our examples illustrate how basic concepts and problem-solving techniques are freely interchanged across disciplines.

We strongly believe that classical physics should *not* be studied in isolation from quantum mechanics and its modern applications. Our reasons are simple:

- Quantum mechanics has primacy over classical physics. Classical physics is an approximation—often excellent, sometimes poor—to quantum mechanics.
- In recent decades, many concepts and mathematical techniques developed for quantum mechanics have been imported into classical physics and there used to enlarge our classical understanding and enhance our computational capability. An example that we shall study is nonlinearly interacting plasma waves, which are best treated as quanta ("plasmons"), despite their being solutions of classical field equations.
- Ideas developed initially for classical problems are frequently adapted for application to avowedly quantum mechanical subjects; examples (not discussed in this book) are found in supersymmetric string theory and in the liquid drop model of the atomic nucleus.

Because of these intimate connections between quantum and classical physics, quantum physics appears frequently in this book.

The amount and variety of material covered in this book may seem overwhelming. If so, keep in mind the key goals of the book: to teach the fundamental concepts, which are not so extensive that they should overwhelm, and to illustrate those concepts. Our goal is not to provide a mastery of the many illustrative applications contained in the book, but rather to convey the spirit of how to apply the basic concepts of classical physics. To help students and readers who feel overwhelmed, we have labeled as "Track Two" sections that can be skipped on a first reading, or skipped entirely—but are sufficiently interesting that many readers may choose to browse or study them. Track-Two sections are labeled by the symbol **T2**. To keep Track One manageable for a one-year course, the Track-One portion of each chapter is rarely longer than 40 pages (including many pages of exercises) and is often somewhat shorter. Track One is designed for a full-year course at the first-year graduate level; that is how we have mostly used it. (Many final-year undergraduates have taken our course successfully, but rarely easily.)

The book is divided into seven parts:

I. **Foundations**—which introduces our book's powerful *geometric* point of view on the laws of physics and brings readers up to speed on some concepts and mathematical tools that we shall need. Many readers will already have mastered most or all of the material in Part I and might find that they can understand most of the rest of the book without adopting our avowedly geometric viewpoint. Nevertheless, we encourage such readers to browse Part I, at least briefly, before moving on, so as to become familiar with this viewpoint. We believe the investment will be repaid. Part I is split into two chapters, Chap. 1 on Newtonian physics and Chap. 2 on special relativity. Since nearly all of Parts II–VI is Newtonian, readers may choose to skip Chap. 2 and the occasional special relativity sections of subsequent chapters, until they are ready to launch into Part VII, General Relativity. Accordingly, Chap. 2 is labeled Track Two, though it becomes Track One when readers embark on Part VII.

II. **Statistical Physics**—including kinetic theory, statistical mechanics, statistical thermodynamics, and the theory of random processes. These subjects underlie some portions of the rest of the book, especially plasma physics and fluid mechanics.

III. **Optics**—by which we mean classical waves of all sorts: light waves, radio waves, sound waves, water waves, waves in plasmas, and gravitational waves. The major concepts we develop for dealing with all these waves include geometric optics, diffraction, interference, and nonlinear wave-wave mixing.

IV. **Elasticity**—elastic deformations, both static and dynamic, of solids. Here we develop the use of tensors to describe continuum mechanics.

V. **Fluid Dynamics**—with flows ranging from the traditional ones of air and water to more modern cosmic and biological environments. We introduce vorticity, viscosity, turbulence, boundary layers, heat transport, sound waves, shock waves, magnetohydrodynamics, and more.

VI. **Plasma Physics**—including plasmas in Earth-bound laboratories and in technological (e.g., controlled-fusion) devices, Earth's ionosphere, and cosmic environments. In addition to magnetohydrodynamics (treated in Part V), we develop two-fluid and kinetic approaches, and techniques of nonlinear plasma physics.

VII. **General Relativity**—the physics of curved spacetime. Here we show how the physical laws that we have discussed in flat spacetime are modified to account for curvature. We also explain how energy and momentum generate this curvature. These ideas are developed for their principal classical applications to neutron stars, black holes, gravitational radiation, and cosmology.

It should be possible to read and teach these parts independently, provided one is prepared to use the cross-references to access some concepts, tools, and results developed in earlier parts.

Five of the seven parts (II, III, V, VI, and VII) conclude with chapters that focus on applications where there is much current research activity and, consequently, there are many opportunities for physicists.

Exercises are a major component of this book. There are five types of exercises:

1. *Practice.* Exercises that provide practice at mathematical manipulations (e.g., of tensors).
2. *Derivation.* Exercises that fill in details of arguments skipped over in the text.
3. *Example.* Exercises that lead the reader step by step through the details of some important extension or application of the material in the text.
4. *Problem.* Exercises with few, if any, hints, in which the task of figuring out how to set up the calculation and get started on it often is as difficult as doing the calculation itself.
5. *Challenge.* Especially difficult exercises whose solution may require reading other books or articles as a foundation for getting started.

We urge readers to try working many of the exercises, especially the examples, which should be regarded as continuations of the text and which contain many of the most illuminating applications. Exercises that we regard as especially important are designated by **.

A few words on units and conventions. In this book we deal with practical matters and frequently need to have a quantitative understanding of the magnitudes of various physical quantities. This requires us to adopt a particular unit system. Physicists use both Gaussian and SI units; units that lie outside both formal systems are also commonly used in many subdisciplines. Both Gaussian and SI units provide a complete and internally consistent set for all of physics, and it is an often-debated issue as to which system is more convenient or aesthetically appealing. We will not enter this debate! One's choice of units should not matter, and a mature physicist should be able to change from one system to another with little thought. However, when learning new concepts, having to figure out "where the 2πs and 4πs go" is a genuine impediment to progress. Our solution to this problem is as follows. For each physics subfield that we study, we consistently use the set of units that seem most natural or that, we judge, constitute the majority usage by researchers in that subfield. We do not pedantically convert cm to m or vice versa at every juncture; we trust that the reader can easily make whatever translation is necessary. However, where the equations are actually different—primarily in electromagnetic theory—we occasionally provide, in brackets or footnotes, the equivalent equations in the other unit system and enough information for the reader to proceed in his or her preferred scheme.

We encourage readers to consult this book's website, http://press.princeton.edu/titles/MCP.html, for information, errata, and various resources relevant to the book.

A large number of people have influenced this book and our viewpoint on the material in it. We list many of them and express our thanks in the Acknowledgments. Many misconceptions and errors have been caught and corrected. However, in a book of this size and scope, others will remain, and for these we take full responsibility. We would be delighted to learn of these from readers and will post corrections and explanations on this book's website when we judge them to be especially important and helpful.

Above all, we are grateful for the support of our wives, Carolee and Liz—and especially for their forbearance in epochs when our enterprise seemed like a mad and vain pursuit of an unreachable goal, a pursuit that we juggled with huge numbers of other obligations, while Liz and Carolee, in the midst of their own careers, gave us the love and encouragement that were crucial in keeping us going.

ACKNOWLEDGMENTS FOR *MODERN CLASSICAL PHYSICS*

This book evolved gradually from notes written in 1980–81, through improved notes, then sparse prose, and on into text that ultimately morphed into what you see today. Over these three decades and more, courses based on our evolving notes and text were taught by us and by many of our colleagues at Caltech, Stanford, and elsewhere. From those teachers and their students, and from readers who found our evolving text on the web and dove into it, we have received an extraordinary volume of feedback,[1] and also patient correction of errors and misconceptions as well as help with translating passages that were correct but impenetrable into more lucid and accessible treatments. For all this feedback and to all who gave it, we are extremely grateful. We wish that we had kept better records; the heartfelt thanks that we offer all these colleagues, students, and readers, named and unnamed, are deeply sincere.

Teachers who taught courses based on our evolving notes and text, and gave invaluable feedback, include Professors Richard Blade, Yanbei Chen, Michael Cross, Steven Frautschi, Peter Goldreich, Steve Koonin, Christian Ott, Sterl Phinney, David Politzer, John Preskill, John Schwarz, and David Stevenson at Caltech; Professors Tom Abel, Seb Doniach, Bob Wagoner, and the late Shoucheng Zhang at Stanford; and Professor Sandor Kovacs at Washington University in St. Louis.

Our teaching assistants, who gave us invaluable feedback on the text, improvements of exercises, and insights into the difficulty of the material for the students, include Jeffrey Atwell, Nate Bode, Yu Cao, Yi-Yuh Chen, Jane Dai, Alexei Dvoretsky, Fernando Echeverria, Jiyu Feng, Eanna Flanagan, Marc Goroff, Dan Grin, Arun Gupta, Alexandr Ikriannikov, Anton Kapustin, Kihong Kim, Hee-Won Lee, Geoffrey Lovelace, Miloje Makivic, Draza Markovic, Keith Matthews, Eric Morganson, Mike Morris, Chung-Yi Mou, Rob Owen, Yi Pan, Jaemo Park, Apoorva Patel, Alexander Putilin, Shuyan Qi, Soo Jong Rey, Fintan Ryan, Bonnie Shoemaker, Paul Simeon,

1. Specific applications that were originated by others, to the best of our memory, are acknowledged in the text.

Hidenori Sinoda, Matthew Stevenson, Wai Mo Suen, Marcus Teague, Guodang Wang, Xinkai Wu, Huan Yang, Jimmy Yee, Piljin Yi, Chen Zheng, and perhaps others of whom we have lost track!

Among the students and readers of our notes and text, who have corresponded with us, sending important suggestions and errata, are Bram Achterberg, Mustafa Amin, Richard Anantua, Alborz Bejnood, Edward Blandford, Jonathan Blandford, Dick Bond, Phil Bucksbaum, James Camparo, Conrado Cano, U Lei Chan, Vernon Chaplin, Mina Cho, Ann Marie Cody, Sandro Commandè, Kevin Fiedler, Krzysztof Findeisen, Jeff Graham, Casey Handmer, John Hannay, Ted Jacobson, Matt Kellner, Deepak Kumar, Andrew McClung, Yuki Moon, Evan O'Connor, Jeffrey Oishi, Keith Olive, Zhen Pan, Eric Peterson, Laurence Perreault Levasseur, Rob Phillips, Vahbod Pourahmad, Andreas Reisenegger, David Reis, Pavlin Savov, Janet Scheel, Yuki Takahashi, Clifford Will, Fun Lim Yee, Yajie Yuan, and Aaron Zimmerman.

For computational advice or assistance, we thank Edward Campbell, Mark Scheel, Chris Mach, and Elizabeth Wood.

Academic support staff who were crucial to our work on this book include Christine Aguilar, JoAnn Boyd, Jennifer Formicelli, and Shirley Hampton.

The editorial and production professionals at Princeton University Press (Peter Dougherty, Karen Fortgang, Ingrid Gnerlich, Eric Henney, and Arthur Werneck) and at Princeton Editorial Associates (Peter Strupp and his freelance associates Paul Anagnostopoulos, Laurel Muller, MaryEllen Oliver, Joe Snowden, and Cyd Westmoreland) have been magnificent, helping us plan and design this book, and transforming our raw prose and primitive figures into a visually appealing volume, with sustained attention to detail, courtesy, and patience as we missed deadline after deadline.

Of course, we the authors take full responsibility for all the errors of judgment, bad choices, and mistakes that remain.

Roger Blandford thanks his many supportive colleagues at Caltech, Stanford University, and the Kavli Institute for Particle Astrophysics and Cosmology. He also acknowledges the Humboldt Foundation, the Miller Institute, the National Science Foundation, and the Simons Foundation for generous support during the completion of this book. And he also thanks the Berkeley Astronomy Department; Caltech; the Institute of Astronomy, Cambridge; and the Max Planck Institute for Astrophysics, Garching, for hospitality.

Kip Thorne is grateful to Caltech—the administration, faculty, students, and staff—for the supportive environment that made possible his work on this book, work that occupied a significant portion of his academic career.